International Conference on Structural Nonlinear Dynamics and Diagnosis (CSNDD 2016)

MATEC Web of Conferences Volume 83 (2016)

Marrakech, Morocco
23 - 25 May 2016

Editors:

M. Belhaq

ISBN: 978-1-5108-3277-0

Printed from e-media with permission by:

Curran Associates, Inc.
57 Morehouse Lane
Red Hook, NY 12571

CURRAN ASSOCIATES INC. proceedings.com

Some format issues inherent in the e-media version may also appear in this print version.

Printed by Curran Associates, Inc. (2016)

For additional information, please contact EDP Sciences – Web of Conferences at the address below.

EDP Sciences – Web of Conferences
17, Avenue du Hoggar
Parc d'Activité de Courtabœuf
BP 112
F-91944 Les Ulis Cedex A
France

Phone: +33 (0) 1 69 18 75 75
Fax: +33 (0) 1 69 28 84 91

contact-edps@webofconferences.org

Additional copies of this publication are available from:

Curran Associates, Inc.
57 Morehouse Lane
Red Hook, NY 12571 USA
Phone: 845-758-0400
Fax: 845-758-2633
Email: curran@proceedings.com
Web: www.proceedings.com

TABLE OF CONTENTS

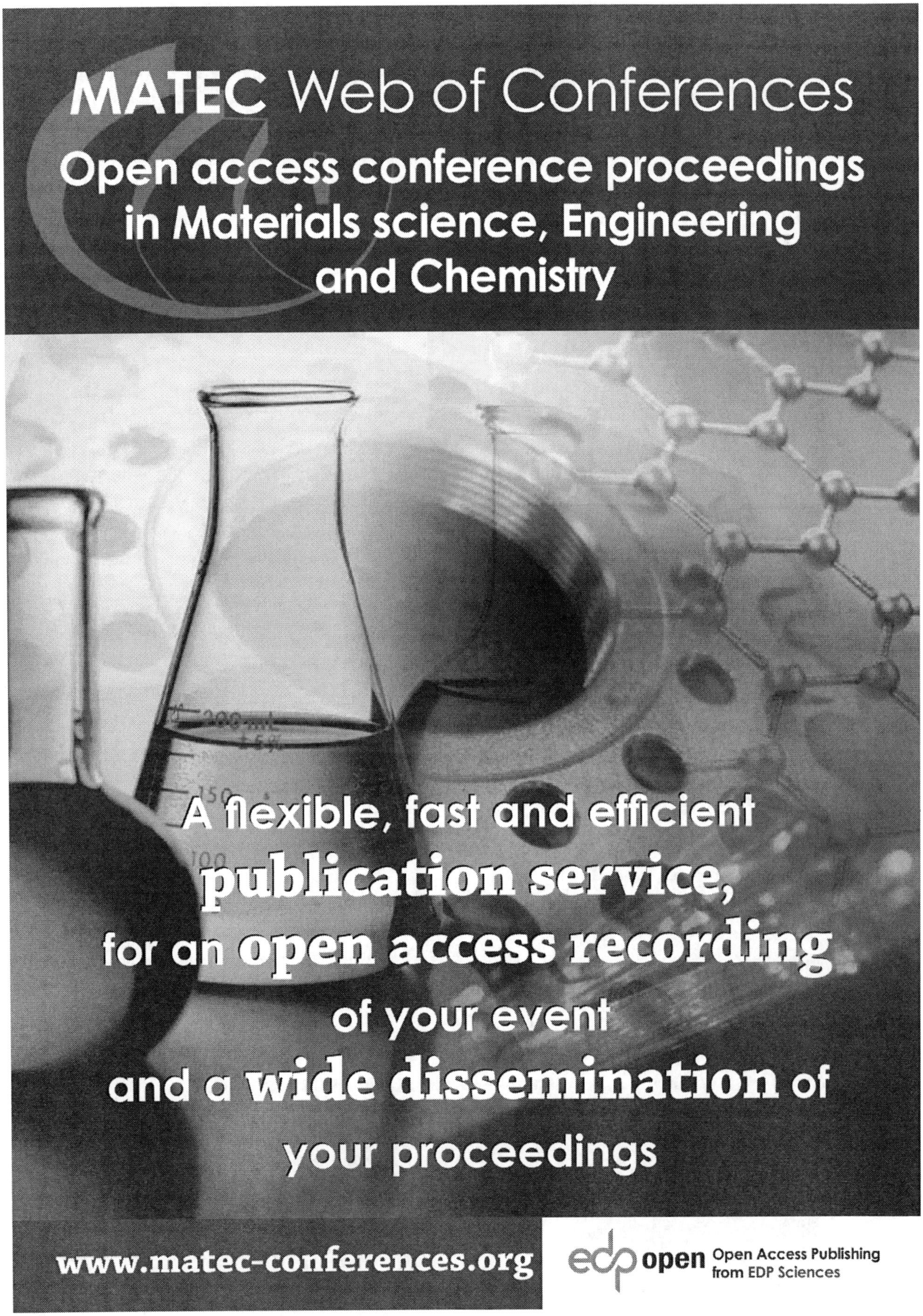

International Conference on Structural Nonlinear Dynamics and Diagnosis
Marrakech, May 23-25, 2016

Preface

The third International Conference on Structural Nonlinear Dynamics and Diagnosis CSNDD 2016 gathered scientists working in nonlinear dynamics and control in different branches of mechanics, applied mathematics and advanced physics to discuss and exchange ideas in theoretical, numerical and experimental techniques nonlinear dynamics related problems. Focuses were directed toward diverse topics, ranging from dynamical systems theory to different mechanical, physical and engineering applications. The CSNDD 2016 covered a large field of nonlinear dynamics, including:

Deterministic and stochastic dynamics
Energy harvesting
Mechanical, aeronautic, aerospace and naval structures
Stability of rotating machines
Model reduction methods dedicated to low/high frequency system dynamics
Nonlinear PDEs and their dynamics
Nonlinear dynamics in spatially extended systems
Structural health monitoring and damage detection
Multiple time scale dynamical systems
Optimization and reliability in structural vibrations
Nonlinear dynamics of flexible structures
Dynamics of non-smooth systems
Rogue waves, solitons, and supercontinuum in nonlinear fiber optics

The meeting has attracted more than 200 delegates from 35 countries. Ten keynote lectures were invited and 24 mini-symposia organized by leading experts in nonlinear dynamics. More than 120 oral presentations and 60 posters were scheduled. Sixty papers are published in MATEC Web of Conferences.

The organizing committee thanks the keynote lecturers, the invited speakers, the mini-symposia organizers as well as reviewers for their precious help in maintaining a high-quality scientific program of the conference.

Mohamed Belhaq
CSNDD 2016 Chair

September 30, 2016
University Hassan II - Casablanca

Mohamed Belhaq

Honorary Chairmen

Ali H. Nayfeh
Benjamin Franklin Medal
(University of Jordan and Virginia Tech)

Werner Schiehlen
(University of Stuttgart, Germany)

Chairman

Mohamed Belhaq
(Hassan II-Casablanca University, Morocco)

Co-Chairmen

Raouf Ibrahim
(Wayne State University, USA)

Muhammad R. Hajj
(Virginia Tech, USA)

Sami F. Masri
(University of Southern California, USA)

International Organizing Committee

A. Aziz-Alaoui	University LeHavre, France
M. Bendame	University of Waterloo, Canada
A. El Aroudi	Universitat Rovira i Virgili, Spain
M. N. Ichchou	Ecole Centrale Lyon, France
J. Mahfoud	INSA Lyon, France
S. M. Sah	KTH Royal Institute of Technology, Sweden

Local Organizing Committee

A. Azouani	ENSA, Mohamed I University, Al Houceima
M. Belhaq	FS, Hassan II University, Casablanca
M. El gadari	ENSAM, Moulay Ismail University, Meknes
A. Fahsi	FST, Hassan II University, Mohammedia
M. Houssni	RE Casablanca, Maroc
I. Kirrou	FS, Hassan II University, Casablanca
F. Lakrad	FS, Hassan II University, Casablanca
L. Mokni	FS, Hassan II University, Casablanca
S. Ouhimmou	FS, Hassan II University, Casablanca

Scientific Committee

Walter Arnold	Göttingen University, Germany
Jan Awrejcewicz	Technical University of Lodz, Poland
Ahmed Aziz-Alaoui	University of Le Havre, France
Vladimir Babitsky	Loughborough University, UK
José M. Balthazar	UNESP, São Paulo, Brasil
Olivier Bareille	Ecole Centrale Lyon, France
Noureddine Bouhaddi	Université de Franche-Comté, France
Rhali Benamar	University Mohamed V, Morocco
Kwok W. Chung	City University of Hong Kong, Hong Kong
Peter Eberhard	Stuttgart University, Germany
Abdelali El Aroudi	Universitat Rovira i Virgili, Spain
Abdelkhalak El Hami	INSA, Rouen, France
Bernold Fiedler	Free University of Berlin, Germany
Peter Hagedorn	TU Darmstadt, Germany
Muhammad R. Hajj	Virginia Tech, USA

edp sciences

MATEC Web of Conferences

Statement of Peer review

In submitting conference proceedings to MATEC Web of Conferences, I certify to the Publisher that I adhere to the **Policy on Publishing Integrity** of the journal in order to safeguard good scientific practice in publishing.

1. All articles have been subjected to peer review administered by the proceedings editors.
2. Reviews have been conducted by expert referees, who have been requested to provide unbiased and constructive comments aimed, whenever possible, at improving the work.
3. Proceedings editors have taken all reasonable steps to ensure the quality of the materials they publish and their decision to accept or reject a paper for publication has been based only on the merits of the work and the relevance to the journal.

Title, date and place of the conference

CSNDD 2016 May 23-25, 2016
Marrakech Maroc

Proceedings editor(s):

Mohamed Belhaq

Date and editor's signature

Sept. 26, 2016

17, avenue du Hoggar - PA de Courtabœuf - BP 112 - 91944 Les Ulis Cedex A (France)
Tél. : 33 (0)1 69 18 75 75 - Fax : 33 (0)1 69 07 45 17 - www.edpsciences.org

DOI: 10.1051/matecconf/20168301001

The role of parameters of smallness in deduction of approximated theories for deterministic dynamics of beams

Enrico Babilio[1,a] and Stefano Lenci[2,b]

[1] *Department of Structures for Engineering and Architecture (DiSt), University of Naples "Federico II", 80134 Naples, Italy*
[2] *Department of Civil and Building Engineering and Architecture (DICEA), Polytechnic University of Marche, 60131 Ancona, Italy*

Abstract. This paper considers the role played by the order of magnitude of terms involved in the equations of motion of a geometrically exact beam theory. Upon introducing a smallness parameter, a procedure for derivation of approximated models is sketched.

1 Introduction

In many real-world applications, technical rules and codes enforce structures to fulfill specific limitations. For instance, often structures of engineering interest are not allowed to exhibit displacements, rotations or deformations larger than assigned values. Thus, accordingly, their behavior can be investigated by using approximated models able to furnish information which is completely satisfactory from a practical point of view, at reduced computational cost in comparison with more refined models. The present short contribution is focused on the derivation of approximated beam models from a parent geometrically exact nonlinear Timoshenko theory, depending on axial and transversal displacements u, v and on a shear angle γ. The orders of magnitude of such functions, or their derivatives involved in the equations of motions, are relevant to understanding which term is negligible and which other should be retained in deducing approximated models. To this end, a parameter of smallness ϵ is introduced and u, v, γ, and their derivatives, are replaced by suitably rescaled functions of order of unity times powers of ϵ. By taking into account terms of same power of ϵ, a hierarchy of models can be derived.

2 The geometrically exact model

Let us consider an initially straight beam undergoing planar and twist-less deformed states and further assume that cross sections remains flat and undistorted, but not necessarily orthogonal to the beam axis during the deformation process. A small element of initial length dx, in reference and deformed configuration, is depicted in figure 1, to which we refer for notations: $u(x,t)$ and $v(x,t)$ stand for the axial and transversal displacements of the beam axis,

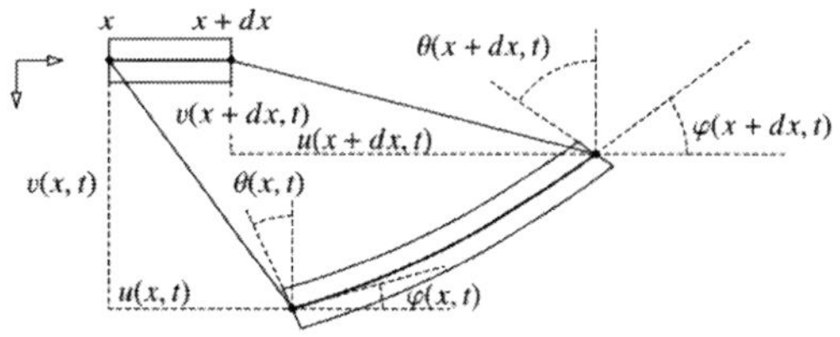

Figure 1. Element in reference and deformed configurations

$\varphi(x,t)$ and $\theta(x,t)$ are the beam axis rotation and the cross-sectional rotation. We assume that displacements (and forces) are positive if concordant with the reference (see vectors in figure 1) and rotations (and moments) are positive if counterclockwise oriented.

The length of the deformed axis of the beam element is given by

$$\mathrm{d}s = \sqrt{2\varepsilon + 1}\,\mathrm{d}x\,, \tag{1}$$

where ε is the axial strain written as

$$\varepsilon = u' + \frac{u'^2}{2} + \frac{v'^2}{2}\,, \tag{2}$$

the prime denoting derivative with respect to x.

The rotation φ is related to displacements u and v by

$$\varphi = -\arctan\left(\frac{v'}{1+u'}\right). \tag{3}$$

In order to measure the shear deformation, the angle $\gamma(x,t)$ is further introduced and its relationship with φ and θ is stated as

$$\gamma = \theta - \varphi\,. \tag{4}$$

[a] e-mail: enrico.babilio@unina.it
[b] e-mail: lenci@univpm.it

Let us finally introduce the function

$$\kappa = \gamma' + \frac{v'u'' - \left(1+u'\right)v''}{\left(1+u'\right)^2 + v'^2}\,, \tag{5}$$

which is nothing but θ', i.e. the derivative of θ with respect to x, and, on adopting nomenclatures introduced in [1], [2] or [3], it can be called the normalized, flexural or mechanical curvature, respectively.

On introducing the virtual variations of Eqs. (2), (4) and (5) and the properly conjugated generalized stresses N, T and M, the equations of motion can be derived from the Principle of Virtual Work as (see [4])

$$\rho\, A\, \ddot{u} + c\, \dot{u} = \left(N\frac{\mathrm{d}s}{\mathrm{d}x}(1+u') - Tv'\left(\frac{\mathrm{d}s}{\mathrm{d}x}\right)^{-1}\right)', \tag{6}$$

$$\rho\, A\, \ddot{v} + c\, \dot{v} = \left(N\frac{\mathrm{d}s}{\mathrm{d}x}v' + T(1+u')\left(\frac{\mathrm{d}s}{\mathrm{d}x}\right)^{-1}\right)', \tag{7}$$

$$M' + \rho\, I\, \ddot{\theta} = T\sqrt{2\varepsilon+1}\,, \tag{8}$$

where A, I, ρ and c are the cross-sectional area, the second moment of area, the mass density and the damping coefficient, respectively, and the overdot denotes derivative with respect to time t.

Notice that Eqs. (6)-(8) are geometrically exact.

Finally, as constitutive law we use the standard linear elastic relationship (see e.g. [5])

$$N = E\,A\,\varepsilon\,, \quad T = G\,A\,\gamma\,, \quad M = E\,I\,\kappa\,, \tag{9}$$

where E is the Young's modulus and G is the (second) Lamé's constant.

It is worth noting that although N, T and M by virtue Eqs. (9) are linearly dependent of the relevant kinematic variables ε, γ and κ, respectively, they however are nonlinear functions of u' and v'.

2.1 A remark on the moment-curvature relation

It is known that beam theories completely decoupled from the three-dimensional theory do not provide any structure for constitutive equations, as typically happens for models belonging to the class of Cosserat beam theories [6] (for a systematic treatment of beams as oriented bodies see also [7]). Thus, constitutive equations must be either introduced on an axiomatic way or experimentally established. For considerations on how to obtain constitutive equations, see [8].

For the sake of this communication, we axiomatically assume that Eqs. (9) are satisfactory relationship. While a deeper discussion on this aspect is beyond the scope of the present contribution (and is extensively treated in [4]), here we emphasize that, interestingly, in formulating a constitutive relationship for the bending moment, a key question treated in many papers and books is which, between the normalized curvature $\kappa = \theta'$ and the derivative of θ with respect to the arclength s, which is also a curvature, is the more suitable measure to adopt for beams undergoing large deflections and rotations.

Actually, both are used in the scientific literature. Arguments in favor of the former can be found in [1–3, 6, 9]. On the other hand, in [10–14] the latter is treated as the better choice.

3 A sketch on approximated models

The orders of magnitude of functions involved in Eqs. (2), (4) and (5) and in Eqs. (6)-(8) play relevant role in deriving approximated models of beams with different nonlinear dominant terms.

To compare orders of magnitude, let us introduce a real positive parameter ϵ much smaller than unity. Soon arises the question how small ϵ is. Indeed, the smallness of a certain function of engineering interest depends on the application under investigation and may differ from a technical problem to another.

In order to give a numerical estimation of ϵ in a practical application, let us consider a simply supported beam and assume that a single-wave deformed shape, such as

$$v(x,t) = \alpha(t)\sin\left(\frac{\pi x}{L}\right), \tag{10}$$

be in equilibrium with a given load. Shape (10) allows us to write

$$\sup_{x\in[0,L]} |v| = |\alpha|\,, \qquad \sup_{x\in[0,L]} |v'| = \frac{\pi}{L}|\alpha|\,. \tag{11}$$

If prescriptions about $|\alpha|$ are somehow known, we may agree to accept

$$\epsilon = \frac{\pi}{L}|\alpha|\,. \tag{12}$$

For instance, the Italian Code (NTC2008) enforces $|\alpha| \leq L/250$ for floors under exercise (dead plus live) loads in buildings. Hence, in such a specific case, ϵ can be taken as about 10^{-2}.

Now, let us assume that u, v and γ, and all their derivatives up to the requested order, may be approximated through their Taylor polynomials about a specific configuration, such as for instance the static equilibrium position or the undeformed configuration and introduce functions $\hat{u}, \hat{u}', \ldots, \hat{v}, \hat{v}', \ldots, \hat{\gamma}, \hat{\gamma}', \ldots,$ of order of unity (ϵ^0), which are rescaled versions of $u, u', \ldots, v, v', \ldots, \gamma, \gamma', \ldots,$ satisfying the same regularity requests enforced for the original counterparts.

Notice that, with these assumptions, since the function $\hat{u}'$ is of order of ϵ^0, it differs from the space derivative $\partial\hat{u}/\partial x$. Indeed, they have the same shape, but different order of magnitude.

3.1 Shear angle limitations and consequences

Recalling that the beam dynamics depends on functions u, v and γ and their derivatives, the order of magnitude of all of them must be known to derive approximated models. We discuss about the shear angle γ, first. Indeed, we assume that a suitable γ is in the range $(-\pi/18, \pi/18)$, which should be true for most practical applications. Through the Maclaurin expansion and retaining up to linear terms,

$$\sin\gamma \simeq \gamma\,, \qquad \cos\gamma \simeq 1\,, \tag{13}$$

where "$\simeq$" indicates approximated equalities. Notice that Eqs. (13) imply an absolute error of at most 0.5 and 1.5 percent, at $\pm\pi/18$, in estimating sine and cosine, respectively. Thus, we may accept that γ is of order $\sqrt{\epsilon}$, at most. Furthermore, we may assume that γ has a very small space variation (along the beam-axis) during the entire deformation process, implying that we retain γ' is of order ϵ.

In what follows, we prescribe the orders of magnitude for u' and v' and introduce approximations (ϵ^m, ϵ^n), being m and n the orders of ϵ chosen for u' and v', respectively.

3.2 Approximation (ϵ^1, ϵ^1)

On assuming that both u' and v' are of order ϵ, that is

$$u' = \epsilon\hat{u}' , \qquad v' = \epsilon\hat{v}' , \tag{14}$$

the axial strains ε in Eq. (2) turns out to be of order of ϵ, and, neglecting terms multiplied by powers of ϵ greater than one, is approximated as

$$\varepsilon \simeq \epsilon\hat{u}' = u' , \tag{15}$$

which is that usually adopted for small strain beam theory.

Beam axis rotations are of order of ϵ, too, and are approximated by

$$\varphi \simeq -\epsilon\hat{v}' , \tag{16}$$

as it can be seen developing Eq. (3) in Taylor series and retaining terms up to the first order of ϵ. If it is further assumed that also γ is of order ϵ, that is

$$\gamma = \epsilon\hat{\gamma} , \tag{17}$$

the cross-sectional rotation may be approximated as

$$\theta \simeq \epsilon(\hat{\gamma} - \hat{v}') . \tag{18}$$

The assumptions (14) inserted in Eq. (5), neglecting all terms whose order of magnitude is certainly smaller than ϵ, the bending curvature (5) takes the form

$$\kappa \simeq \epsilon\hat{\gamma}' - v'' + \epsilon\hat{v}'u'' . \tag{19}$$

Up to now, only hypotheses on smallness of u', v' and γ' have been assumed, but nothing has been said about u'' and v''. Hence, the order of magnitude of κ, which depends on the second derivatives of u and v, is still unknown.

Without specific information about u'' and v'', we may assume, at least in principle, that they are of any order. We restrict to five powers of ϵ, namely u'' and v'' are of order ϵ^1, $\epsilon^{\frac{1}{2}}$, ϵ^0 $\epsilon^{-\frac{1}{2}}$ or ϵ^{-1}, leading to the approximated expressions of curvature reported in Table 1, with the order of u'' on rows, that of v'' on columns. The order of magnitude of curvatures is also indicated in Table 1, recalling that all *hatted* functions are of order of magnitude of unity.

We observe that the curvature is of order ϵ if and only if v'' is ϵ and u'' is ϵ^0 at most. If u'' is of order of ϵ, or $\sqrt{\epsilon}$, we get back the curvature adopted in small strain theory, while if u'' is ϵ^0, curvature contains the nonlinear term $v'u''$. In all other cases reported, curvature has magnitude greater than ϵ, hence γ', which is of order of ϵ, is negligible. Table 1 shows also that curvature is nonlinear whenever u'' is greater of v'' of at least two order of ϵ.

3.3 Approximation $(\epsilon^1, \epsilon^{\frac{1}{2}})$

If u' is of order of v'^2, that is

$$u' = \epsilon\hat{u}' , \qquad v' = \sqrt{\epsilon}\hat{v}' , \tag{20}$$

the approximated ε, which is of order of ϵ, takes the form

$$\varepsilon \simeq \epsilon\left(\hat{u}' + \frac{\hat{v}'^2}{2}\right) , \tag{21}$$

known as von Kármán strain [15], usually adopted for beams undergoing moderately large deflections.

Rotations are of order of $\sqrt{\epsilon}$ and are written as

$$\theta \simeq -\sqrt{\epsilon}\hat{v}' , \quad \text{or} \quad \theta \simeq \sqrt{\epsilon}(\hat{\gamma} - \hat{v}') , \tag{22}$$

respectively depending on if it is assumed that γ is of order ϵ (as u') or $\sqrt{\epsilon}$ (as v'), that is

$$\gamma = \epsilon\hat{\gamma} , \quad \text{or} \quad \gamma = \sqrt{\epsilon}\hat{\gamma} . \tag{23}$$

The curvature takes the form

$$\kappa \simeq \epsilon\hat{\gamma}' - v'' + \sqrt{\epsilon}\hat{v}'u'' . \tag{24}$$

under the hypothesis that γ' is of order ϵ.

Table 2 reports terms which are dominant in the curvature expression. The order of magnitude is reported as well. We see that the curvature is of order ϵ if and only if v'' is ϵ and u'' is $\sqrt{\epsilon}$ at most. If u'' is of order of ϵ we get back the curvature adopted in both small strain and moderately large deflections theories, while if u'' is $\sqrt{\epsilon}$, curvature contains the nonlinear term $v'u''$. In all other cases reported, the derivative of the shear angle (recall that γ' is assumed of order ϵ) is negligible. Table 2 shows also that curvature is nonlinear whenever u'' is greater of v'' of at least one order of ϵ.

4 Summary and conclusions

The present short communication is focused on the derivation of approximated models from a geometrically exact beam theory, in which no assumptions on the order of magnitude of displacements, rotations and deformations are taken *a priori*.

Since the order of magnitude of each term allows understanding which one is negligible in approximated models, a parameter of smallness ϵ is introduced and relevant terms appearing in equations of motion are replaced by suitably rescaled functions of order of unity times powers of ϵ.

To explain the idea, axial strain, rotation and a hierarchy of approximated curvatures are derived under the hypothesis that v' and γ are of order ϵ or $\sqrt{\epsilon}$ and u' and γ' are of order ϵ.

It must be noted that to achieve the approximated versions of Eqs. (6)-(8), it is still necessary to infer order of magnitude of $\dot{u}$ and $\ddot{u}$ in Eq. (6), $\dot{v}$ and $\ddot{v}$ in Eq. (7), $\dot{u}', \ddot{u}', \dot{v}', \ddot{v}'$ in Eq. (8) (nested in $\ddot{\theta}$). The task can be performed as shown for other terms and then, by substituting

Table 1. The curvature in (ϵ^1, ϵ^1) approximation

<table>
<tr><td colspan="2"></td><td colspan="5">v''</td></tr>
<tr><td colspan="2"></td><td>ϵ</td><td>$\sqrt{\epsilon}$</td><td>1</td><td>$\frac{1}{\sqrt{\epsilon}}$</td><td>$\frac{1}{\epsilon}$</td></tr>
<tr><td rowspan="5">u''</td><td>ϵ</td><td rowspan="2">$\epsilon(\hat{\gamma}' - \hat{\vartheta}'')$</td><td rowspan="3">$-\sqrt{\epsilon}\hat{\vartheta}''$</td><td rowspan="4">$-\hat{\vartheta}''$</td><td rowspan="5">$-\frac{\hat{\vartheta}''}{\sqrt{\epsilon}}$</td><td rowspan="5">$-\frac{\hat{\vartheta}''}{\epsilon}$</td></tr>
<tr><td>$\sqrt{\epsilon}$</td></tr>
<tr><td>1</td><td>$\epsilon(\hat{\gamma}' - \hat{\vartheta}'' + \hat{\vartheta}'\hat{u}'')$</td></tr>
<tr><td>$\frac{1}{\sqrt{\epsilon}}$</td><td>$\sqrt{\epsilon}\hat{\vartheta}'\hat{u}''$</td><td>$\sqrt{\epsilon}(-\hat{\vartheta}'' + \hat{\vartheta}'\hat{u}'')$</td></tr>
<tr><td>$\frac{1}{\epsilon}$</td><td colspan="2">$\hat{\vartheta}'\hat{u}''$</td><td>$-\hat{\vartheta}'' + \hat{\vartheta}'\hat{u}''$</td></tr>
</table>

Table 2. The curvature in $(\epsilon^1, \epsilon^{\frac{1}{2}})$ approximation

<table>
<tr><td colspan="2"></td><td colspan="5">v''</td></tr>
<tr><td colspan="2"></td><td>ϵ</td><td>$\sqrt{\epsilon}$</td><td>1</td><td>$\frac{1}{\sqrt{\epsilon}}$</td><td>$\frac{1}{\epsilon}$</td></tr>
<tr><td rowspan="5">u''</td><td>ϵ</td><td>$\epsilon(\hat{\gamma}' - \hat{\vartheta}'')$</td><td rowspan="2">$-\sqrt{\epsilon}\hat{\vartheta}''$</td><td rowspan="3">$-\hat{\vartheta}''$</td><td rowspan="4">$-\frac{\hat{\vartheta}''}{\sqrt{\epsilon}}$</td><td rowspan="5">$\frac{-\hat{\vartheta}''}{\epsilon}$</td></tr>
<tr><td>$\sqrt{\epsilon}$</td><td>$\epsilon(\hat{\gamma}' - \hat{\vartheta}'' + \hat{\vartheta}'\hat{u}'')$</td></tr>
<tr><td>1</td><td>$\sqrt{\epsilon}\hat{\vartheta}'\hat{u}''$</td><td>$\sqrt{\epsilon}(-\hat{\vartheta}'' + \hat{\vartheta}'\hat{u}'')$</td></tr>
<tr><td>$\frac{1}{\sqrt{\epsilon}}$</td><td colspan="2">$\hat{\vartheta}'\hat{u}''$</td><td>$-\hat{\vartheta}'' + \hat{\vartheta}'\hat{u}''$</td></tr>
<tr><td>$\frac{1}{\epsilon}$</td><td colspan="3">$\frac{\hat{\vartheta}'\hat{u}''}{\sqrt{\epsilon}}$</td><td>$\frac{-\hat{\vartheta}'' + \hat{\vartheta}'\hat{u}''}{\sqrt{\epsilon}}$</td></tr>
</table>

approximated functions in Eqs. (6)-(8) and retaining terms of the same order, approximated equation are deduced.

Finally, we stress that the advantage of having a hierarchy of approximated models relies on the possibility of choosing the more suitable set of equations for a given problem, at the less computational cost for the desired level of accuracy.

References

[1] A.H. Nayfeh, P.F. Pai, *Linear and Nonlinear Structural Mechanics* (Wiley, New York, 2004)

[2] W. Lacarbonara, *Nonlinear Structural Mechanics. Theory, Dynamical Phenomena and Modeling* (Springer, New York, 2013)

[3] A. Luongo, D. Zulli, *Mathematical Models of Beams and Cables* (ISTE-Wiley, London UK, Hoboken NJ, 2013)

[4] E. Babilio, S. Lenci, submitted (2016)

[5] S. Timoshenko, 1995 *Strength of Materials* (Krieger Publishing Company, New York, 1995, third edition)

[6] S.S. Antman, *Problems of Nonlinear Elasticity* (Springer, New York, 2005)

[7] P. Villaggio, *Mathematical Models for Elastic Structures* (Cambridge, Cambridge University Press, 1997)

[8] E. Reissner, J. Appl. Math. Phys. ZAMP, **23**, 795-804 (1972)

[9] J. Gerstmayr, H. Irschik, J. Sound Vib., **318**, 461-487 (2008)

[10] D.H. Hodges, AIAA J. **22**(12), 1825-1827 (1984)

[11] M. Berzeri, A.A. Shabana, J. Sound Vib., **235**(4), 539-565 (2000)

[12] E. B. Magrab, *Vibrations of Elastic Systems With Applications to MEMS and NEMS* (Springer, New York, 2012)

[13] E. Babilio, Eur. Phys. J.-Spec. Top., **222**(7), 1519–1539 (2013)

[14] S. Lenci, F. Clementi, G. Rega, Meccanica, DOI 10.1007/s11012-016-0374-6 (2016)

[15] J.N. Reddy, P.Mahaffey, J. Coupled Syst. Multiscale Dyn., **1**(1), 120–134 (2013)

On nonlinear dynamic of a non-ideal Duffing system with fractional damping

J M Balthazar[1], A M Tusset[2], V Piccirillo[2], A Nabarrete[1], G Litak[3,4] and C Oliveira[1,5]
[1] ITA-Aeronautics Technological Institute, São José dos Campos, SP, Brazil,
[2] UTFPR – Federal University of Technology – Paraná, Ponta Grossa, PR, Brazil,
[3] Lublin University of Technology, Faculty of Mechanical Engineering, Lublin, Poland,
[4] University of Science and Technology, Kraków, Poland
[5] UFGD, Federal University of Grande Dourados, Dourados, MS, Brazil

Abstract: Fractional damping is appearing in different contexts in any systems with memory and hysteresis. Such damping is defined by a fractional derivative term, in contrary to classical viscous damping which takes into account the first order derivative. In this work, we characterize the nonlinear dynamics of a non-ideal Duffing system, with fractional damping using nonlinear dynamical tools. The non-ideal excitation originates from a DC electric motor with limited power supply driving an unbalanced rotating mass. The response of the system is investigated with the voltage as a control parameter. Numerical simulations show the occurrence of regular and non-regular motions, which are investigated via bifurcation diagrams and phase plane portraits.

1. Introduction

The study of problems that involve the coupling of several systems was explored widely in the last years in function of the change of constructive characteristics of the machines and structures. In this way, some phenomena are observed in dynamical systems composed by supporting structures and rotating machines, where the unbalancing of the rotating parts is the main cause of vibrations. In the study of these systems, for a more realistic formulation one has to consider the action of an energy source with limited power (non-ideal), that is, to consider the influence of the oscillatory system on the driving force and vice versa. Recently a number of non-ideal vibrating systems has been studied, for some examples, see [1-3], besides others. The above mentioned results consider a nonlinear stiffness, but in all of them the damping element of the structure exerts a linear viscous damping force. However, many materials show a behaviour that differs from a linear viscous damper. Different models for viscoelastic behaviour applying the concept of fractional derivatives were discussed [4]. Fractional calculus have been applied in several fields like as physics, chemistry, finances, and bioengineering [5-6]. In Ref. [7], it is shown that the fractional order duffing excited by a ideal system can produce a chaotic attractor. The idea in this work is study the nonlinear dynamic of a non-ideal Duffing system with fractional damping. The figure 1 shows the non-ideal fractional oscillator. The vibrating system under consideration consists of a main structure of mass m_1, constrained by a linearly viscous element with damping coefficient μ and a pseudoelastic SMD. The structure is also connected with a non-ideal DC motor, with

 DOI: 10.1051/matecconf/20168301002

driving rotor of moment of inertia J_M and an unbalanced mass m_0 with eccentricity r. The electrical scheme of the DC motor is shown in Fig. 1(b).

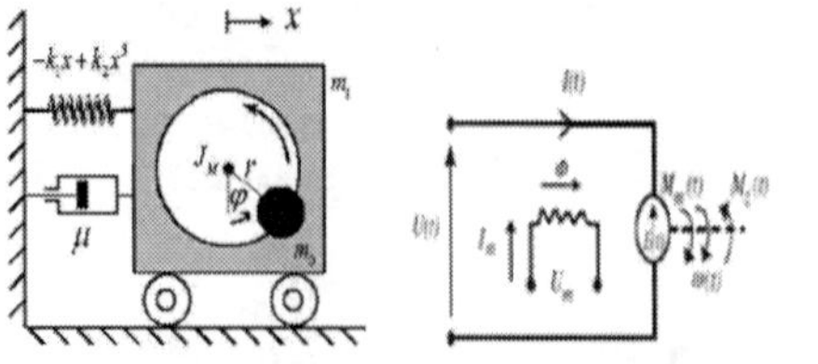

(a) (b)

Figure 1: (a) Non-ideal fractional oscillator and (b) the electrical schematic representation of the DC motor

2 System description and governing equations

The corresponding equations that govern the angular motion φ of the DC motor are typically written in the form [8]. The equations of motion are given by

$$\begin{cases} m_1 x'' + F_d - k_1 x + k_2 x^3 = \\ m_0 r\left((\varphi')^2 \sin\varphi - \varphi'' \cos\varphi\right) \\ \left(J + m_0 r^2\right)\varphi'' = C_M \Phi I(t) - m_0 r x'' \cos\varphi \\ I' = -\dfrac{R_t}{L_t} I(t) - \dfrac{C_E \Phi}{L_t}\varphi' + \dfrac{U(t)}{L_t} \end{cases} \tag{1}$$

The damping device exerts a force F_d given by

$$F_d = c\frac{d^v x}{dt^v} \tag{2}$$

and is thus proportional to the vth time derivative of the relative displacement. For $v = 1$ the force is a linear viscous damping force and for $v = 0$ the force is a linear restoring force. It is assumed that $0 \le v \le 1$ and in the interval $0 < v < 1$ the device contributes with both stiffness and damping. c is the coefficient and has the dimension mass·time $v-2$.

It is convenient to work with dimensionless position and time, namely $u = \frac{x}{x_0}$ and $\tau = \omega_0 t$, respectively, where x_0 is the static displacement, and introducing the following variable $\tilde{I} = \frac{I}{I_0}$, where I_0 is a rated current in the armature.
The non-dimensional fractional derivative is defined as: $\frac{d^v}{dt^v} \to \omega_0^v \frac{d^v}{d\tau^v}$. In this way is possible to rewrite the Eq. 1 in the dimensionless form:

$$\begin{cases} \ddot{u} + \beta(v) u^{(v)} - u + \delta u^3 = \\ w_1\left((\dot{\varphi})^2 \sin\varphi - \ddot{\varphi}\cos\varphi\right) \\ \ddot{\varphi} = p_3 \tilde{I}(\tau) - w_2 \ddot{u}\cos\varphi \\ \dot{\tilde{I}} = -p_1 \tilde{I} - p_2\dot{\varphi} + U_1(\tau) \end{cases} \tag{3}$$

Here the non-dimensional quantities are

$$\omega_0^2 = \frac{k_1}{m_1 + m_0}, \quad \beta(v) = \frac{c}{m_1 \omega_1^{2-v}},$$

$$\delta = \frac{k_2}{k_1} x_0^2, \; w_1 = \frac{m_0 r}{m_1 x_0}, \; w_2 = \frac{m_0 r x_0}{\left(J + m_0 r^2\right)},$$

$$p_1 = \frac{R_t}{L_t I_0 \omega_0}, \; U_1 = \frac{U}{L_t I_0 \omega_0}, \; p_2 = \frac{C_E \Phi}{L_t I_0}$$

$$p_3 = \frac{C_M \Phi I_0}{\left(J + m_0 r^2\right)\omega_0^2}.$$

$\beta(v)$ is a nondimensional coefficient for the damping force, and is possible to note that when $v = 1$ the coefficient $\beta(v)$ becomes $\beta(v) = \frac{c}{m_1 \omega_1}$ that is a classical nondimensional coefficient for the damping force.

3 Numerical Results

The numerical simulation of the vibrating system was carried out for parameters of the DC motor and mechanical parameters. Numerical dimensionless parameters are: $\delta = 0.1$, $w_1 = 0.2$, $w_2 = 0.3$, $p_1 = 0.3$, $p_2 = 3$, $p_3 = 0.15$, $h = \pi/200$ (integration step).

In order to characterize quantitatively the attractors involved in this study, we computed the 0-1 test proposed by [9]. In the last years, several researches have been applied this test to characterize the attractor of the dynamical system mainly in systems where exist difficulties to calculate the Lyapunov exponents. A complete overview about theoretical aspects and applications of the 0-1 test can be found in [10].

This test yields a binary type of result: 0 for regular (periodic/quasi periodic) and 1 for chaotic signals. Based on the time series u we define the dimensionless displacements in the (p,q) plane as follows [12]

$$p(n,\bar{c}) = \sum_{j=0}^{n} u(j)\cos(j\bar{c}) \qquad (4)$$

$$q(n,\bar{c}) = \sum_{j=0}^{n} u(j)\sin(j\bar{c}) \qquad (5)$$

where $\bar{c}$ is a constant, $\bar{c} \in (0,\pi)$. If the system dynamics is chaotic, the motion in the (p,q) plane would be unbounded, while it would be bounded for regular (periodic/quasiperiodic) motion. The mean square displacement is expressed as [10]

$$M(n,\bar{c}) = \lim_{N\to\infty} \frac{1}{N}\sum_{j=1}^{N}\left[(p(j+n,\bar{c}) - p(j,\bar{c}))^2\right] + \left[(q(j+n,\bar{c}) - q(j,\bar{c}))^2\right] \qquad (6)$$

For regular dynamics, it is quite likely that $M(n,\bar{c})$ will be a bounded function of n but not for chaotic dynamics. From this one can define the asymptotic growth rate of the mean square displacement using covariance and variance measures:

$$K_{\bar{c}} = \frac{\mathrm{cov}(Y, M(\bar{c}))}{\sqrt{\mathrm{var}(Y)\,\mathrm{var}(M(\bar{c}))}} \qquad (7)$$

where $Y = [1,2,\ldots,n_{\max}]$ and $M(\bar{c}) = [M(1,\bar{c}), M(2,\bar{c}),\ldots,M(n_{\max},\bar{c})]$ are vectors. Finally, the median is taken of K - values (Eq. 7) corresponding to 100 different values.

Figure 2 shows the bifurcation diagram for the non-ideal oscillator in terms of the control parameter U_1. For low values of it we have limit cycle (periodic attractors, see Fig. 3(a) and K = 0). This attractors suffer an abrupt change at $U_1 \approx 3.7$, where a bifurcation occurs, and after the bifurcation, there appears chaotic motion (see Fig. 3(b) and K = 0.99)

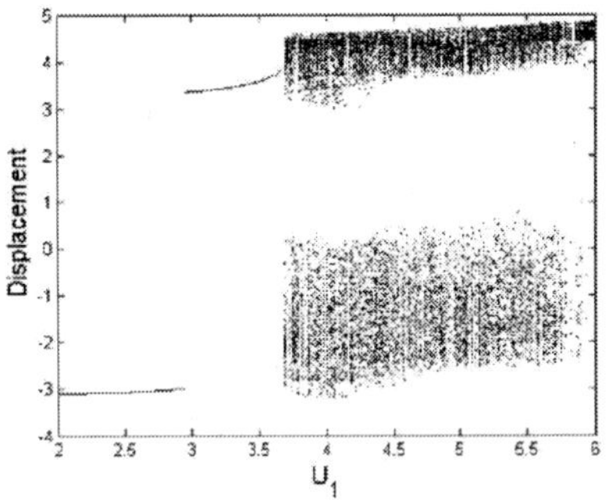

Figure 2: Bifurcation diagram showing displacement as a function of U_1 and a linear viscous damper $(\nu = 1)$.

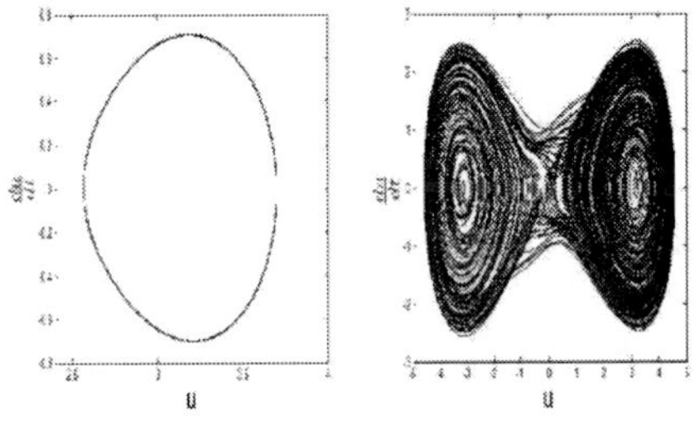

(a) (b)

Figure 3: (a) Phase portraits for $U_1 = 3.6$, and (b) Phase portraits for $U_1 = 3.8$.

Fig. 4 shows the results of the bifurcation diagram of the non-ideal oscillator in terms of the order of the derivative ν and $U_1 = 4$ (note chaotic behavior for $\nu = 1$). For low values of it we have limit cycle (chaotic attractors, see Fig. 5(a) and $K = 0.99$). This attractors suffer an abrupt change at $\nu \approx 1.35$, where a bifurcation occurs, and after the bifurcation, there appears periodic motion (see Fig. 5(b) and $K = 0$).

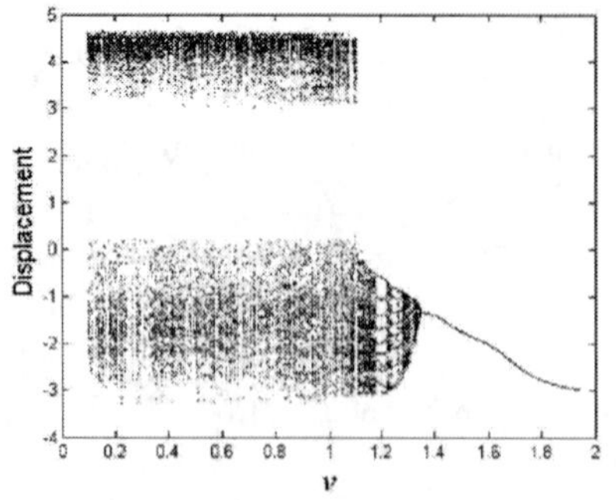

Figure 4: Bifurcation diagram showing displacement as a function of ν and $U_1 = 4$

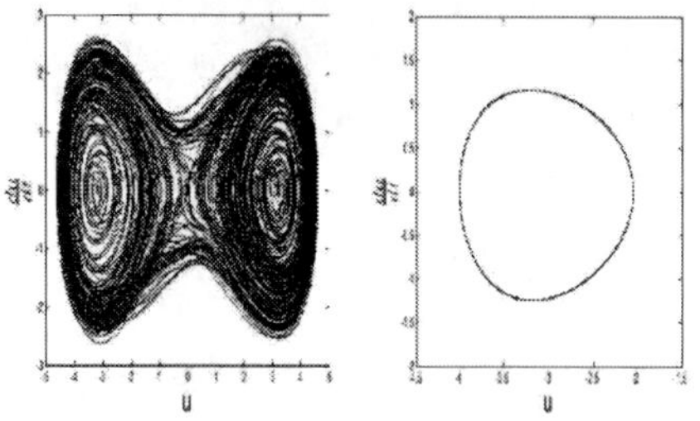

(a) (b)

Figure 5: Phase portraits for $U_1 = 4$: (a) $\nu = 0.8$, and (b) $\nu = 1.6$.

Figure 6 represents the resonance curve of non-ideal oscillator when U_1 is slowly increased, and the occurrence of the jump phenomenon during the passage through resonance $U_1 = 3$ is observed only $\nu = 1$.

4. Conclusions

In this paper we investigated the dynamics of the fractional order non-ideal Duffing oscillator by phase portraits, 0-1 test and bifurcation diagrams, and we made a comparison with a non-ideal Duffing oscillator with integer order. When the parameter $U_1 = 4$ the total orders of the system for the existence of chaos are $\nu < 1.35$.

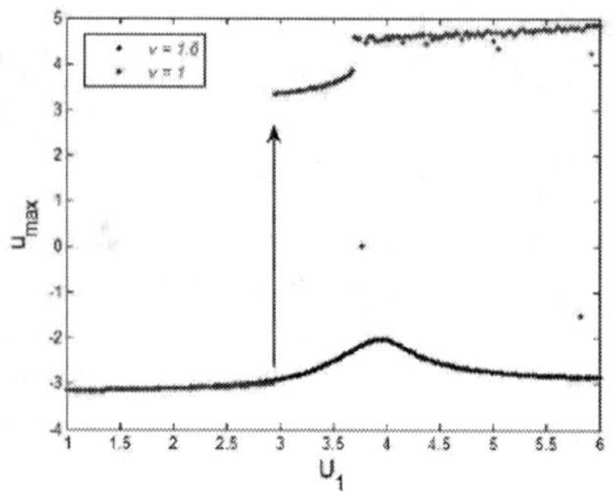

Figure 6: Frequency–response diagram in red $\nu = 1$, and $\nu = 1.6$.

Acknowledgements

GL is very grateful to prof. J.M. Balthazar and ITA- Aeronautics Technological Institute for hospitality, and also the Polish National Science Center for partial support under the grant agreement No. 2012/05/B/ST8/00080.

References

1. V. Piccirillo, J.M. Balthazar, A. Tusset, D. Bernardini, G. Rega, *Int J NonLinear Mech.* 77, 12-27 (2015).
2. V. Piccirillo, J. M. Balthazar, A. Tusset, D. Bernardini, G. Rega, *Proc Inst Mech Eng Part C J Mech Eng Sci* 230, 120-132 (2016).
3. S.L.T. de Souza, I.L. Caldas, R.L. Viana, J.M. Balthazar, R.M.L.R.F Brasil, *Chaos, Solitons & Fractals* 6, 1211–1220 (2005).
4. R.L. Bagley, P.J. Torvik, *Journal of Rheology* 27, 201 (1983).
5. M. Santhiah, P. Philominathan, M.I. Raja, K. Murali, *International Journal of Bifurcation and Chaos* 21, 161-175 (2011).
6. O. P. Agrawal, J. A. Tenreiro-Machado, and I. Sabatier (Eds.), *Nonlinear Dynamics* 38, Springer, Berlin 2004.
7. A. Syta, G. Litak, S. Lenci, M. Scheffler, *Chaos* 24, 013107 (2014).
8. J. Warminski, J.M. Balthazar, *Journal of the Brazilian Society of Mechanical Sciences and Engineering* 25, 413–419 (2003).
9. G. Gottwald, I. Melbourne, *Proc R Soc Lond* A 460, 603–611 (2004).
10. D. Bernardini, G. Litak, *Journal of the Brazilian Society of Mechanical Sciences and Engineering* doi: 10.1007/s40430-015-0453-y (2016).

Dynamics of structures of clusters of tethered satellites

R.-M.-L.-R.-F. Brasil[1], J.-M. Balthazar[2], R.-M.-O. Pauletti[3], V. Fallara[4]

[1]Federal University of ABC, Aerospace Department, 5001 Avenida dos Estados Unidos, Santo André, SP, Brazil
[2]Aeronautical Institute of Technology, Aeronautical Mechanics Department, S. Jose dos Campos, SP, Brazil
[3]Polytechnic School of University of São Paulo, Structural Engineering Department, São Paulo, SP, Brazil
[4]Federal University of ABC, Aerospace Department, 3 Arcturus, São Bernardo do Campo, SP, Brazil

Abstract. In this paper, we study the nonlinear dynamics of clusters of tethered satellites via numerical step-by-step time integration of the differential equations of motion. The basic algorithm is Central Difference explicit method. The masses of the space vehicles are considered lumped and the connection cables are considered massless and of linear elastic material. The considered nonlinearities are due to large displacements leading to changes in masses coordinates. The proposed algorithm does not involve assemblage of stiffness matrices. Instead, the elastic restoring forces are directly computed at each time step. A new time step optimization procedure is implemented. As an example, a cluster of four satellites in a tetrahedral disposition is considered. Free damped vibrations due to large initial conditions are computed.

1 Introduction

In this paper, we study the nonlinear dynamics of clusters of tethered satellites via numerical step-by-step time integration of the differential equations of motion. The basic algorithm is the Central Finite Difference explicit method, (Brasil [2][3]). The masses of the space vehicles are considered lumped and the connection cables are considered massless and of linear elastic material. The considered nonlinearities are due to large displacements leading to changes in masses coordinates. The proposed algorithm does not involve assemblage of stiffness matrices. Instead, the elastic restoring forces are directly computed at each time step. A new time step optimization procedure is implemented. As an example, a cluster of four satellites in a tetrahedral disposition is considered. Free damped vibrations due to large initial conditions are considered.

Cable structures have been widely used in Civil Engineering applications, as described by Pauletti [12] and Irvine [9]. Recent spacecraft applications, specifically for tethered satellites, are reported by Cosmo and Lorenzini, [5], Ellis and Hall [7], Kumar [10], Mankala and Agrawal [11], Stevens and Wiesel [18]. A problem to be overcome is the high strength and low weight necessary for cable spacecraft applications. Suggestions have been made to future use of carbon nanotubes, as commented by Baughman et al. [1].

2 Theoretical development

We write down the dynamic equilibrium equation of the masses of our tethered satellites system via Newton's Second Law, supposing that we have n generalized coordinates, the three orthogonal free displacements of each mass. The resulting system of second order Ordinary Differential Equations (Clough and Penzien, [4], Paz [17], Thomson [19]) is

$$\mathbf{f}_{inertia} = \mathbf{p}(t) - \mathbf{f}_{\text{dissipation}} - \mathbf{f}_{\text{elastic}} \qquad (1)$$

$\mathbf{f}_{inertia}$ is the n x 1 vector of inertia forces that we supposed to be linearly proportional to the accelerations vector as $\mathbf{f}_{inertia} = \mathbf{M\ddot{u}}$, $\mathbf{M}$ being the n x n mass matrix. This is a diagonal matrix in our formulation.

$\mathbf{p}(t)$ is the n x 1 vector of external applied forces, time dependent in general.

$\mathbf{f}_{\text{dissipation}}$ is the n x 1 vector of dissipation forces that we suppose to be linearly proportional to the velocities vector as $\mathbf{f}_{\text{dissipation}} = \mathbf{C\dot{u}}$, $\mathbf{C}$ being the n x n damping matrix. This is a diagonal matrix in our formulation.

$\mathbf{f}_{\text{elastic}}$ is the n x 1vector of elastic restoring forces, non linear, in general, due to the considered large displacements of the masses and the presence of previous tension of the cables and to the fact that the cables will not work on compression.

The Eq. (1) may be rewritten as

$$\mathbf{M\ddot{u}} + \mathbf{C\dot{u}} + \mathbf{f}_{\text{elastic}} = \mathbf{p}(t) \qquad (2)$$

DOI: 10.1051/matecconf/20168301003

This equation will be numerically integrated using a Central Finite Differences approach (Brasil [2][3]) as follows. Let us write down the expressions of Taylor expansions forward and backwards of the vector function $\mathbf{u} = \mathbf{u}(t)$ for the displacements of the masses of our tethered satellites system, involving time steps t_{i-1}, t_i, t_{i+1}, separated by a time interval h (in seconds).

$$\mathbf{u}_{i+1} = \mathbf{u}_i + h\dot{\mathbf{u}}_i + \frac{1}{2}h^2\ddot{\mathbf{u}}_i + \cdots \quad (3)$$

$$\mathbf{u}_{i-1} = \mathbf{u}_i - h\dot{\mathbf{u}}_i + \frac{1}{2}h^2\ddot{\mathbf{u}}_i - \cdots \quad (4)$$

Subtracting Eq. (4) of Eq. (3) we obtain a Central Finite Differences approximation of the velocity vector

$$\dot{\mathbf{u}}_i = \frac{\mathbf{u}_{i+1} - \mathbf{u}_{i-1}}{2h} \quad (5)$$

Adding Eq. (3) to Eq. (4) we obtain a Central Finite Differences approximation of the acceleration vector

$$\ddot{\mathbf{u}}_i = \frac{\mathbf{u}_{i+1} - 2\mathbf{u}_i + \mathbf{u}_{i-1}}{h^2} \quad (6)$$

Substituting approximations (5) and (6) in the Eq. (2) we get an algebraic linear system of equations

$$\hat{\mathbf{K}}\mathbf{u}_{i+1} = \hat{\mathbf{p}}_i \quad (7)$$

where

$$\hat{\mathbf{K}} = \frac{1}{h^2}\mathbf{M} + \frac{1}{2h}\mathbf{C} \quad (8)$$

and

$$\hat{\mathbf{p}}_i = \mathbf{p}_i + \frac{2}{h^2}\mathbf{M}\mathbf{u}_i - \left(\frac{1}{h^2}\mathbf{M} - \frac{1}{2h}\mathbf{C}\right)\mathbf{u}_{i-1} - \mathbf{f}_{\text{elastic},i} \quad (9)$$

is the pseudo static loading vector at this time step. No system solution procedure is actually needed in Eq. 7, as matrix of Eq. 8 is diagonal. The computation is carried out at vector components level.

The value of the elastic forces $\mathbf{f}_{\text{elastic},i}$ at any time step t_i is computed as follows. Let us consider one certain cable of the structure. For this particular cable, we get the 6 x 1 vector of the displacements of the two end nodes of Eq. (10), displayed in figure 1, considering the present coordinates of these nodes and the present inclination of the cable at this time step,

$$\mathbf{q} = [q_1\, q_2 \cdots q_6]^{\mathrm{T}} \quad (10)$$

the change of length of this cable will be

$$\Delta L = (q_4 - q_1)cx + (q_5 - q_2)cy + (q_6 - q_3)cz \quad (11)$$

where, *cx*, *cy* and *cz* are the direction cosines of the cable.

If N_0 is the initial traction of this cable, L_0 its initial length, E the modulus of elasticity of the material and A the section area, the axial force at this time step will be

$$N = N_0 + \frac{EA}{L_0}\Delta L \quad (12)$$

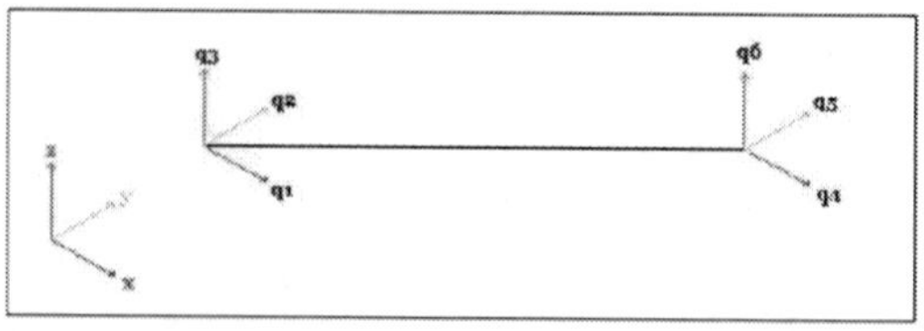

Fig. 1. Cable end nodes displacements

provided the value of Eq. (12) is positive. Otherwise it will be set to zero. Next, this axial force will be decomposed in the horizontal and vertical directions at each end node of the cable to be added to the elastic forces vector. The same routine will be repeated for all other cables.

One should note from Eq. (9) that $\mathbf{u}_{i-1}$ is needed at each time step. This poses a problem at the initial time t_0, when initial conditions u_0 and $\dot{u}_0$ are known, but not before. A simple scheme to solve it is to suppose a Uniformly Accelerated Motion in the first time interval, that is, $\ddot{u}_1 = \ddot{u}_0$, which allows for the computation of u_1 by writing the dynamical equilibrium Eq. (2) at time t_1.

3 Numerical simulations

We next analyse a sample model of four masses connected by cables initially forming a tetrahedral configuration, as displayed in figure 2.

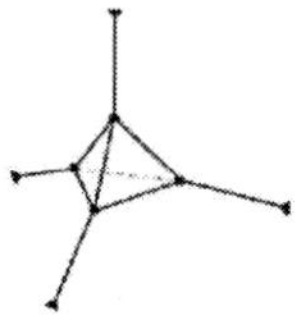

Fig. 2. The model

Damped free vibrations due to non-trivial large initial conditions lead to time histories of Fig. 3. Displacements in the three directions of one of the tetrahedron corner masses are shown. Severe nonlinearities may be observed due to large displacements and the possibility of cables becoming slack.

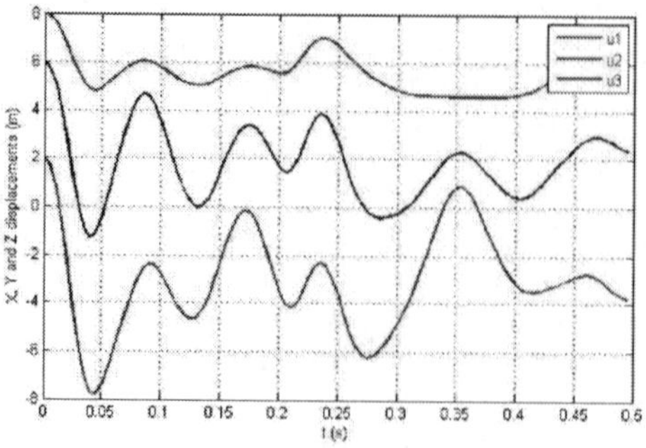

Fig. 3. Displacements time histories

4 Time step optimization

It can be shown that the central difference method is only conditionally stable, and, for linear multiple degrees of freedom systems, stability is guaranteed bounding the time-step by the Courant, Friedrichs and Lewys [6] condition

$$\Delta t \leq \frac{2}{\omega_{\max}^{0}}, \tag{13}$$

where $\omega_{\max}^{0}$ is the largest natural frequency of the system from the solution of the eigenvalue problem

$$\det(\mathbf{K}_0 - \omega^2\mathbf{M}) = 0, \tag{14}$$

where $\mathbf{K}_0$ and $\mathbf{M}$ are the system's stiffness and mass matrices. There is no guarantee that criterion (13) is on the safe side for non-linear systems.

The problem can be circumvented by repeatedly updating the stiffness matrix $\mathbf{K}_k$, evaluated at time t_k. Considering that the main attractiveness of the central difference method is the possibility of dealing exclusively with force vectors, this is in contradiction with the spirit of the method.

Irons and Treharne [8] have shown that an upper-limit estimative of the maximum frequency of a discrete system is given by the maximorum of its maximum elements' natural frequencies, i.e.,

$$\omega_{\max}^{k} = \underbrace{\overbrace{\max}^{n_e}}_{e=1}\{\omega_{e,\max}^{k}\} \tag{15}$$

where $\omega_{e,\max}^{k}$ is the maximum natural frequency of element e, evaluated at time t_k, by means of a series of eigenvalue problems

$$\det(\mathbf{K}^e - \omega^2\mathbf{M}^e) = 0, \text{ for all elements} \tag{16}$$

Taking into account that the central difference method only is fully vectorised when masses are lumped into the system's degrees of freed, Pauletti et al. [15][16] show that the maximum element frequency is given by

$$\omega_{e,\max}^{k} = \sqrt{\frac{n_n^e \lambda_{e,\max}}{m_e}}, \tag{17}$$

where m_e and n_n^e are respectively the total mass and the number of nodes of the element and $\lambda_{e,\max}$ is maximum eigenvalue of its stiffness matrix.

Let us consider a column-wise partition of the element stiffness matrix, evaluated at a given configuration $\mathbf{u}_k^e$ according to $\mathbf{K}_k^e = \mathbf{K}^e(\mathbf{u}_k^e) = [\mathbf{k}_1^k \ \mathbf{k}_2^k \ \dots \ \mathbf{k}_{n_{\mathrm{dof}}^e}^k]$, where n_{dof}^e is the number of degrees of freedom of the element. Then, an upper bound estimative of $\lambda_{e,\max}$ can be obtained according to

$$|\lambda_{\max}^e| \leq \underbrace{\overbrace{\max}^{n_{\mathrm{dof}}^e}}_{j=1}\left(\left\|\mathbf{k}_j^e\right\|_i\right) \tag{19}$$

where is the L1-norm of each column of the stiffness matrix, that is,

$$\left\|\mathbf{k}_j^e\right\|_i = \sum_{i=1}^{n_{\mathrm{dof}}^e}\left|\left(\mathbf{k}_j^e\right)_i\right|. \tag{20}$$

Besides, Pauletti and Almeida-Neto [13][14] have shown that columns can be conveniently evaluated by finite-difference approximations, according to

$$\mathbf{K}_j^k \cong \frac{1}{2h^e}\left[\mathbf{f}^e(\mathbf{u}_k^e + h^e\boldsymbol{\delta}_j) - \mathbf{f}^e(\mathbf{u}_k^e - h\boldsymbol{\delta}_j)\right],$$

$$j = 1, \ \dots \ , n_{\mathrm{dof}}^e \tag{21}$$

where $\mathbf{f}^e$ is the internal load vector for element e, h is a scalar parameter and $\pm h\boldsymbol{\delta}_j$ are backward and forward Kronecker-delta perturbations of the jth element degree of freedom, such that $\boldsymbol{\delta}_j^e = [\delta_{ij}]$, $i = 1, \dots, n_{\mathrm{dof}}^e$ and where δ_{ij} is the Kronecker delta. This finite-difference approximation provides excellent estimates for the stiffness matrices, as long as h is small enough.

In our present algorithm, no stiffness matrix is actually employed, but simply the internal force vectors are backward and forward perturbed, providing an upper-bound limit for $\lambda_{\max}^e$, according to Eqs. (21) and (19), leading to an upper-bound estimative of the system maximum frequency $\omega_{\max}^k$, according to Eq. (15), and thus a lower-bound estimative for the instantaneous critical time-step Δt_k given by Eq. (13).

References

1. R.H. Baughman, A.A. Zakhidov, W.A. de Heer. Science, *Carbon nanotubes – the route toward applications*, **297** (5582), 787-792 (2002)
2. R.M.L.R.F. Brasil. Bolteim Técnico do Departamento de Engenharia de estruturas e Fundações, *Programas para Análise Dinâmica de Estruturas: um grau de liberdade* (1992a)
3. R.M.L.R.F. Brasil. Bolteim Técnico do Departamento de Engenharia de estruturas e Fundações, *Programas para Análise Dinâmica de Estruturas: um grau de liberdade* (1992b)
4. R.W. Clough, J. Penzien. 2nd Ed. McGraw-Hill, *Dynamics of Structures* (1995)
5. M.L. Cosmo, E.C. Lorenzini. Third Edition NASA, *Tethers in Space Handbook* (1997)
6. R. Courant, K. Friedriehs, H. Lewy. Mathematische Annalen, *On the partial difference equations of mathematical physics*, **100** (1928), 32-74 (English translation in IBM, 1967)
7. J.R. Ellis, C.D. Hall. Celestial Mechanics and Dynamical Astronomy, *Out-of-plane librations of spinning tethered satellites systems*, **106** (1), 39-67 (2010)
8. B.M. Iron, A.C. Treharme. 2nd Conference on Matrix Methods in Structural Mechanics, *A bound theorem for eigenvalues and its practical applications*, 245-254 (Wright-Patterson Air Force Base, 1971)
9. M. Irvine. *Cable Structures* (1981)
10. K.D. Kumar. Journal of Spacecraft and Rockets, *Review of dynamics and control of*

nonelectrodynamic tethered satellite systems, **43** (4), 705-720 (2006)
11. K.K. Mankala, S.K. Agrawal. Journal of Vibrations and Acoustics, *Dynamic Modeling and Simulation of Satellite Tethered Systems*, **127**, 144-156 (2005)
12. R.M.O. Pauletti. Tese de livre Docência, Departamento de Engenharia de Estruturas e Fundações, *Estruturas Retesadas: história, projeto e contrução* (Escola Politécnica da USP, 1999)
13. R.M.O. Pauletti, A.E.S. Neto. XXXI Iberian Latin American Congress on Computational Methods in Engineering, II South American Congress on Computational Mechanics, *A Finite-Difference Approximation to Newton Method's Jacobian Matrices*, (2010)
14. R.M.O. Pauletti, A.E.S. Neto. IABSE-IASS Symposium, *A Finite-Difference Approximation to Newton's Method Jacobian Matrices* (2011)
15. R.M.O. Pauletti, D. M. Guirardi, S.M. Adriaenssens, L.G. Rodhe-Barbarigos. *An efficient mass tuning algorithm for the dynamic Relaxation Method* (to be published)
16. R.M.O. Pauletti, R.M.L.R.F. Brasil, D. M. Guirardi, J.M. Balthazar. *A practical time-step criterion for the central difference method for nonlinear dynamic analysis* (to be published)
17. M. Paz. 2nd Ed. Van Nostrand Reinhold, *Structural Dynamics* (1985)
18. R. Stevens, W. Wiesel. Journal of Guidance, Control, and Dynamics, *Large time scale optimal control of electrodynamic tether satellite*, **31** (6), 1716-1727 (2008)
19. W.T. Thomson. *Teoria da Vibração com aplicações* (1978)

In-Flight Parameter Estimation for Multirotor Aerial Vehicles

Davi Ferreira de Castro[1,a], Igor Afonso Acampora Prado[1,b], Mateus de Freitas Virgílio Pereira[1,c], Davi Antônio dos Santos[1,d] and José Manoel Balthazar[1,e]

Aeronautics Institute of Technology, ITA, Brazil.

Abstract. This paper proposes a method for in-flight parameter estimation for Multirotor Aerial Vehicles (MAV). This task is important because it provides parameters with better accuracy for the actual vehicle operation. In order to simulate a flight it is adopted a simulation environment Software-In-the-Loop (SIL).

1 Introduction

When dealing with MAVs, flight tests, regardless of being performed in outdoor or indoor environments, involve risks, both for equipment and for the people who are participating in the experiment. These risks can be explained by the MAVs high speed propellers [1] and, from the standpoint of control, they are unstable, due to nonlinearities, aerodynamic disturbances [2] and a complex and coupled dynamic [3].

To avoid hazards during flight tests of such vehicles, simulation is an alternative to be used. Nevertheless, the simulation does not eliminate the need for flight tests. Therefore, it is only a preliminary step.

In general, simulations save time during the project development, and they can be said to be more efficient than experimental tests time-wise. Even so, simulations may not truly represent reality. An usual approach is SIL simulations, in which all the components are simulated, such as sensors, actuators and vehicle model.

SIL simulations can validate a system before an experimental test, and it also allows the design of a compatible hardware with the software while avoiding damages of equipment. In this context, a simulation environment is useful for estimating in-flight parameters, since there are difficulties and risks to perform a real data acquisition.

This paper is organized as follows: in section 2 the MAV modeling is presented, in section 3 the simulation environment SIL is described, in section 4 the method for estimation is presented, in section 5 the simulation results are shown and finally in section 6 some conclusions and future work suggestions are proposed.

2 MAV Modeling

Consider an octocopter and three Cartesian coordinate systems (CCS) illustrated in Figure 1.

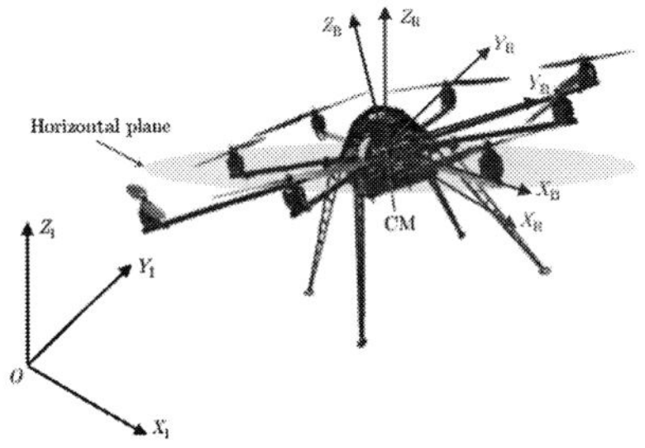

Fig. 1. Cartesian coordinate systems.

The CCS of the body $S_B \triangleq \{X_B, Y_B, Z_B\}$ is fixed to the vehicle structure and centered on the center of mass (CM). The inertial CCS $S_I \triangleq \{X_I, Y_I, Z_I\}$ is fixed to the earth at the point O. The CCS $S_R \triangleq \{X_R, Y_R, Z_R\}$ is parallel to S_I and centered in CM.

2.1 Mathematical Model

The differential equations that describe the dynamic behavior of the multirotor can be obtained by the Newton-Euler formalism. This work chooses to represent the attitude by the quaternion of rotation $\mathbf{q} \in \mathbb{R}^4$, however for visualization is adopted Euler angles (ϕ, θ and ψ).

The kinematic attitude model for the quaternion of rotation is given by the following differential equation [4]:

$$\dot{\mathbf{q}} = \boldsymbol{\Omega}\mathbf{q} \tag{1}$$

where

$$\boldsymbol{\Omega} = \frac{1}{2}\begin{bmatrix} 0 & -\omega^{\mathrm{T}} \\ \omega & -[\omega\times] \end{bmatrix}, \tag{2}$$

being $\omega \triangleq \begin{bmatrix}\omega_x \, \omega_y \, \omega_z\end{bmatrix}^{\mathrm{T}} \in \mathbb{R}^3$ the angular velocity of the vehicle represented in S_R and $[\omega\times]$ is given by:

$$[\omega\times] = \begin{bmatrix} 0 & -\omega_z & \omega_y \\ \omega_z & 0 & -\omega_x \\ -\omega_y & \omega_x & 0 \end{bmatrix}. \tag{3}$$

[a] e-mail: davifc@ita.br
[b] e-mail: igorap@ita.br
[c] e-mail: mateusfvp@hotmail.com
[d] e-mail: davists@ita.br
[e] e-mail: jmbaltha@ita.br

 DOI: 10.1051/matecconf/20168301004

Rewriting the equation (1) explicitly, one obtains:

$$\begin{bmatrix} \dot{q}_1 \\ \dot{q}_2 \\ \dot{q}_3 \\ \dot{q}_4 \end{bmatrix} = \frac{1}{2} \begin{bmatrix} 0 & -\omega_x & -\omega_y & -\omega_z \\ \omega_x & 0 & \omega_z & -\omega_y \\ \omega_y & -\omega_z & 0 & \omega_x \\ \omega_z & \omega_y & -\omega_x & 0 \end{bmatrix} \begin{bmatrix} q_1 \\ q_2 \\ q_3 \\ q_4 \end{bmatrix}. \quad (4)$$

Applying Newton's second law for rotational motion and neglecting disturbance torques, one can obtain:

$$\boldsymbol{\tau} = \dot{\mathbf{h}} + \boldsymbol{\omega} \times \mathbf{h}, \quad (5)$$

where $\boldsymbol{\tau} \triangleq [\tau_x\ \tau_y\ \tau_z]^{\mathrm{T}} \in \mathbb{R}^3$ is the resulting propulsion torque in the vehicle represented in S_{B} and $\mathbf{h} \triangleq \mathbf{I}\,\boldsymbol{\omega} \in \mathbb{R}^3$ is the angular momentum of the body represented in S_{B}, being $\mathbf{I} \in \mathbb{R}^{3x3}$ the matrix of inertia of the vehicle represented in S_{B}. It is assumed that the vehicle has symmetrical structure with respect to the coordinate axes. The matrix **I** resulting in a diagonal matrix defined by:

$$\mathbf{I} = \begin{bmatrix} I_x & 0 & 0 \\ 0 & I_y & 0 \\ 0 & 0 & I_z \end{bmatrix}. \quad (6)$$

Rewriting the equation (5) explicitly, one can obtain:

$$\dot{\omega}_x = \frac{(I_y - I_z)}{I_x}\omega_y\omega_z + \frac{\tau_x}{I_x}. \quad (7)$$

$$\dot{\omega}_y = \frac{(I_z - I_x)}{I_y}\omega_x\omega_z + \frac{\tau_y}{I_y}. \quad (8)$$

$$\dot{\omega}_z = \frac{(I_x - I_y)}{I_z}\omega_x\omega_y + \frac{\tau_z}{I_z}. \quad (9)$$

Applying the second Newton's law of motion and neglecting any perturbation, we obtain the following model of translational movement in S_{I}:

$$\ddot{\mathbf{r}} = \frac{1}{m} f \mathbf{n} + \begin{bmatrix} 0 \\ 0 \\ -g \end{bmatrix}, \quad (10)$$

where $\mathbf{r} \triangleq [r_x\ r_y\ r_z]^{\mathrm{T}} \in \mathbb{R}^3$ is the three-dimensional position of the CM represented in S_{I}, $\mathbf{n} \triangleq [n_x\ n_y\ n_z]^{\mathrm{T}} \in \mathbb{R}^3$ is the normal unit vector perpendicular to the rotor plane represented in S_{I}, f is the total thrust, g is the acceleration of gravity and m is the mass of the vehicle.

2.2 X-Plane Model

In this work it was considered the Gyro-200ED-X8 octocopter of the company Gyrofly Innovations [1], as illustrated in Figure 2(a). There was obtained a graphic model of Gyro-200ED-X8 octocopter for simulation in X-Plane. The tool used in the implementation of the model was the Plane-Maker. The Plane Maker is a software provided with X-Plane that allows the implementation of new aircraft models. The Figure 2(b) illustrates a graphical model of the Gyro-200ED-X8 octocopter implemented in the software Plane-Maker.

[1] http://www.gyrofly.com.br

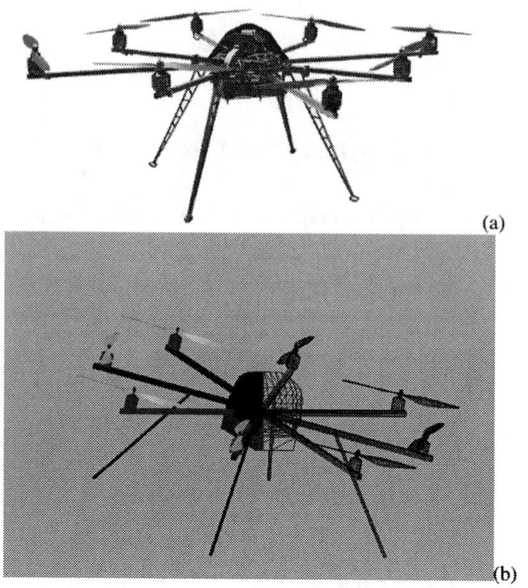

Fig. 2. (a)Illustration of Gyro-200ED-X8 UAV of Gyrofly Innovations and (b) Octocopter model Gyro-200ED-X8 implemented in Plane-Maker.

3 Simulation Environment SIL

The environment consists in a computer running two software: MATLAB/Simulink and X-Plane. The attitude control laws and position are implemented in Simulink, while the vehicle dynamics and the flight environment are simulated in X-Plane. The sequence of sending and receiving data by the software MATLAB/Simulink and XPlane, can be observed in Figure 3.

The standard method of X-Plane to communicate with external processes and machines is through User Datagram Protocol (UDP). The steps in this data flow are: the X-Plane sends the outputs of the octocopter model to the control system using UDP (in this work it was used a rate of 90 Hz); the MATLAB/Simulink receives the data and performs the calculation of the control input signals, which are sent to the X-Plane; the X-Plane receives control input signals via UDP with values for throttle command. The works of [5] and [6] show more details of the communication between MATLAB/Simulink and X-Plane.

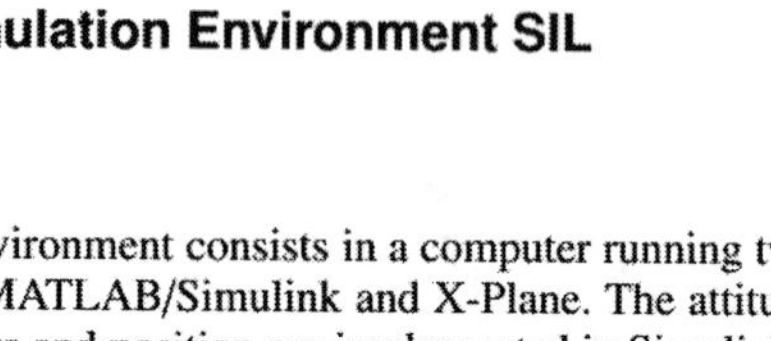

Fig. 3. Data flow between X-Plane and MATLAB/Simulink.

DOI: 10.1051/matecconf/20168301004

4 Parameter Estimation X-Plane Model

Define the set of identification indexes for the rotors of the octocopter as $\mathcal{J} \triangleq \{1, 2, ..., 8\}$. The analytical model of equations (7) - (10) has as inputs the total thrust magnitude f and the net propulsion torque $\boldsymbol{\tau}$. However, inputs of X-Plane model are throttles $T_j, \forall j \in \mathcal{J}$. The T_j has a magnitude which varies from 0 to 1. Since the control systems of both models are identical, it is necessary to convert f and $\boldsymbol{\tau}$ to T_j, in order to make the simulation with the X-Plane model equivalent to the analytical model, therefore allowing comparisons. The initial step of this conversion is to calculate the individual thrust f_j.

Consider the octocopter illustrated in Figure 4. In this figure, $\forall j \in \mathcal{J}$, f_j denotes the real thrust produced by the j-th rotor, τ_j denotes the torque reaction of the j-th rotor. The arm length of the octocopter is denoted by l.

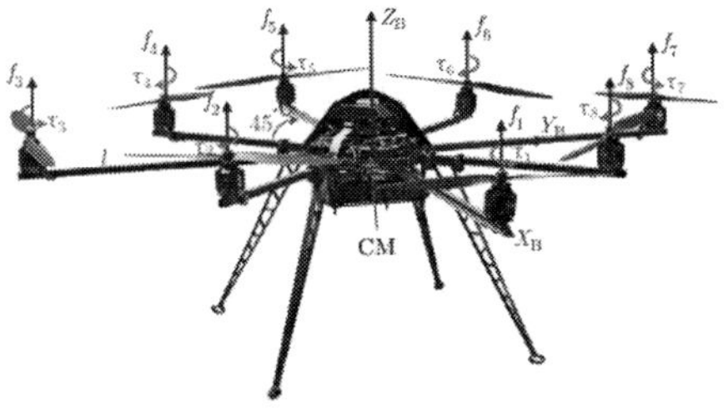

Fig. 4. Thrust and torque reaction of the rotors.

Through inspection of Figure 4, we can obtain the magnitude of the total thrust f and the components of the resultant propulsion torque $\boldsymbol{\tau} \triangleq [\tau_x\ \tau_y\ \tau_z]^{\mathrm{T}} \in \mathbb{R}^3$, as being

$$\begin{bmatrix} f \\ \boldsymbol{\tau} \end{bmatrix} = \boldsymbol{\Gamma}_O \mathbf{f}, \tag{11}$$

Based on equation (11), one can also affirm that the individual thrust commands $\overline{f}_j, \forall j \in \mathcal{J}$, are related to $\overline{f}$ and $\overline{\boldsymbol{\tau}}$, by:

$$\begin{bmatrix} \overline{f} \\ \overline{\boldsymbol{\tau}} \end{bmatrix} = \boldsymbol{\Gamma}_O \overline{\mathbf{f}}, \tag{12}$$

where

$$\overline{\mathbf{f}} \triangleq [\overline{f}_1\ \overline{f}_2\ \overline{f}_3\ \overline{f}_4\ \overline{f}_5\ \overline{f}_6\ \overline{f}_7\ \overline{f}_8]^{\mathrm{T}} \tag{13}$$

and

$$\boldsymbol{\Gamma}_O \triangleq \begin{bmatrix} 1 & 1 & 1 & 1 & 1 & 1 & 1 & 1 \\ 0 & -\frac{\sqrt{2}}{2}l & -l & -\frac{\sqrt{2}}{2}l & 0 & \frac{\sqrt{2}}{2}l & l & \frac{\sqrt{2}}{2}l \\ -l & -\frac{\sqrt{2}}{2}l & 0 & \frac{\sqrt{2}}{2}l & l & \frac{\sqrt{2}}{2}l & 0 & -\frac{\sqrt{2}}{2}l \\ k & -k & k & -k & k & -k & k & -k \end{bmatrix}, \tag{14}$$

where k is the coefficient of yaw torque, defined by $k \triangleq k_\tau / k_f$.

Using the Moore-Penrose pseudo-inverse matrix, the equation (12) is inverted, obtaining

$$\overline{\mathbf{f}} = \boldsymbol{\Gamma}_O^\dagger \begin{bmatrix} \overline{f} \\ \overline{\boldsymbol{\tau}} \end{bmatrix}. \tag{15}$$

The last step is to convert $\overline{f}_j$ to $\overline{T}_j$. It was made some tests in order to estimate how $\overline{T}_j$ relates to both $\overline{f}_j$ and $\overline{\tau}_j$.

The tests consisted in varying $\overline{T}_j$ and measuring both $\overline{f}_j$ and $\overline{\tau}_j$. It is expected that the measured values are equal to the command values. These tests have shown that $\overline{T}_j$ relates to both $\overline{f}_j$ and $\overline{\tau}_j$ by a linear equations below:

$$\overline{f}_j = k_f \overline{T}_j, \tag{16}$$

$$\overline{\tau}_j = k_\tau \overline{T}_j. \tag{17}$$

5 Simulation Tests

The dynamics of the mathematical model is integrated using the fourth order Runge-Kutta method with time step of $T = 0.002$ s, which is the same time step for the X-Plane model. The mass of the vehicle is $m = 2.132$ kg and the acceleration of gravity is $g = 9.796$ m/s^2.

In order to estimate the parameters k_f and k_τ, it was elaborated a test that makes use of a simulation environment that was presented Section 3.

The test is the following procedures: stabilize the vehicle attitude and height; add a random signal δ_1 to $\overline{T}_1$ to excite the models; measure f_1 and τ_1; estimate k_f and k_τ using the criterion of least squares and the measurements of f_1 and τ_1; and validate the estimated models. Only the data for rotor 1 are used, since all rotors are considered to be identical.

Let $\tilde{f}_1$ and $\tilde{\tau}_1$ denote measures of f_1 and τ_1, respectively. Modeling the measures by:

$$y(t) = x(t)\boldsymbol{\Theta} + \epsilon(t), \tag{18}$$

where $y(t) \triangleq [\tilde{f}_1(t)\ \tilde{\tau}_1(t)]^{\mathrm{T}} \in \mathbb{R}^2$, $x(t) \triangleq [\overline{T}_1(t)] \in \mathbb{R}$, $\boldsymbol{\Theta} \triangleq [k_f\ k_\tau]^{\mathrm{T}} \in \mathbb{R}^2$ and $\epsilon(t) \triangleq [\epsilon_1(t)\ \epsilon_2(t)]^{\mathrm{T}} \in \mathbb{R}^2$ represents eventual modeling errors or noise in the measurement.

Applying the data sets in the equation (18), for each sample at an instant of time $t = t_1, t_2, ..., t_n$ and writing in matrix form, one obtains:

$$\mathbf{Y} = \mathbf{X}\boldsymbol{\Theta} + \mathbf{E}, \tag{19}$$

where $\mathbf{Y} \triangleq [y(t_1)\ y(t_2)\ ...\ y(t_n)]^{\mathrm{T}}$, $\mathbf{X} \triangleq \begin{bmatrix} x(t_1) \\ x(t_2) \\ \vdots \\ x(t_n) \end{bmatrix}$ e $\mathbf{E} \triangleq [\epsilon(t_1)\ \epsilon(t_2)\ ...\ \epsilon(t_n)]^{\mathrm{T}}$.

Denoting the estimates of $\boldsymbol{\Theta}$ by $\widehat{\boldsymbol{\Theta}}$, one obtains $\widehat{\boldsymbol{\Theta}}$ according to the criterion of least squares (LS) [7], given by:

$$\widehat{\boldsymbol{\Theta}} = (\mathbf{X}^{\mathrm{T}}\mathbf{X})^{-1}\mathbf{X}^{\mathrm{T}}\mathbf{Y}. \tag{20}$$

For the models validation, a test was performed according to the following procedures: insertion of the estimated parameters in the models; stabilization of the vehicle attitude and height; random signal input μ_1 in $\overline{\tau}$; and comparison between the data for $\overline{f}_1$ and $\tilde{f}_1$.

In Figure 5 the random signals δ_1 and μ_1 are presented. In Figure 6 the comparison between $\overline{f}_1$ and $\tilde{f}_1$ is shown, in which the data for $\tilde{f}_1$ is expected to be equivalet to $\overline{f}_1$. In Figure 6(a) the test results before the models estimation are

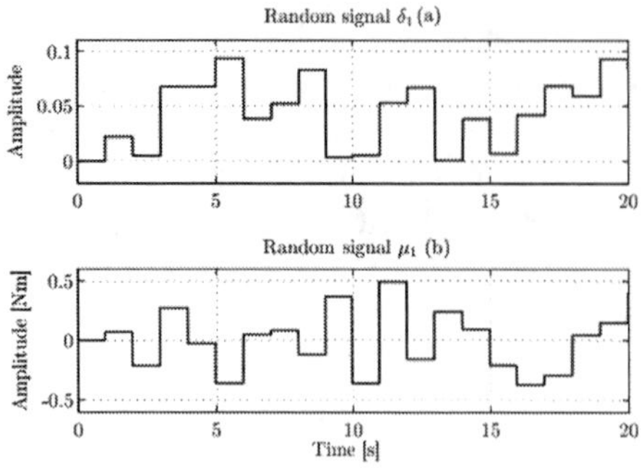

Fig. 5. Random signals δ_1 and μ_1.

presented and one can note that $\overline{f}_1$ and $\tilde{f}_1$ are not similar. For these tests the values for k_f and k_τ are arbitrary. In Figure 6(b) the results after the models estimation are shown and it can be seen that $\overline{f}_1$ and $\tilde{f}_1$ are equivalent, since the values used for k_f and k_τ were the estimated.

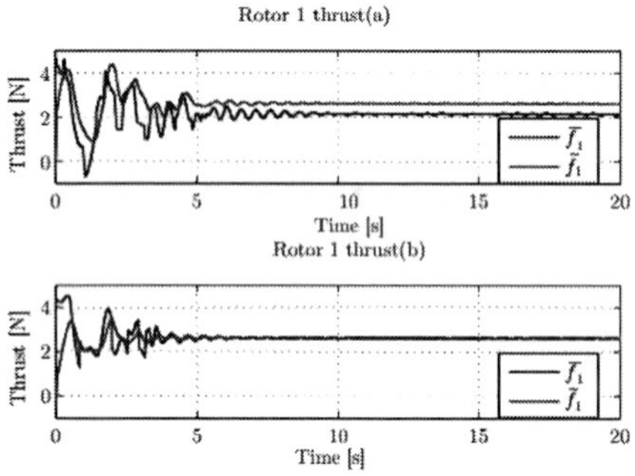

Fig. 6. (a) Tests before the parameters estimation with arbitrary values for k_f and k_τ. (b) Tests after the parameters estimation with the estimated values for k_f and k_τ.

For the estimation of $(I_x, I_y$ e $I_z)$, three tests are carried out. Each test consists in moving the vehicle along a single axis where the estimates have to be made, while the movement in the other axes are disregarded.

The Table 1 presents the estimated parameters.

Table 1. Parameter estimation k_f,k_τ,I_x, I_y and I_z

k_f [N]	k_τ [N m]	I_x [Kg m^2]	I_y [Kg m^2]	I_z [Kg m^2]
28,0269	0,3869	0,0628	0,0762	0,1014

6 Conclusion

The methodology presented proved to be effective, because it is possible to estimate the MAVs in-flight parameters through a simulation environment SIL. In this way we can minimize the risks in the operation of such vehicles. This methodology can be applied before flight tests. As future work, the presented procedures can be implemented in flight tests.

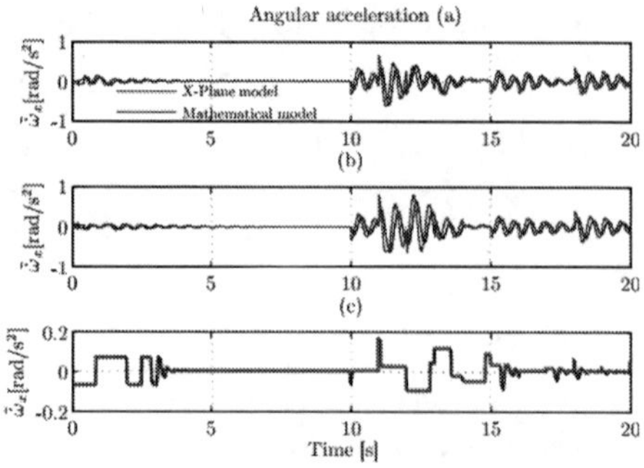

Fig. 7. (a) Data for $\tilde{\omega}_x$ for the X-Plane and the mathematical models obtained through tests carried out after the estimation of I_x in order to validate the estimated parameters. It should be noted that, after the stabilization of the models, μ_1 is added at 10 sec. (b) and (c) correspond, respectively, to the data for $\tilde{\omega}_y$ and $\tilde{\omega}_z$ for both models.

Acknowledgements

We would like to thanks Fundação de Amparo à Pesquisa do Amazonas (FAPEAM), Conselho Nacional de Desenvolvimento Científico e Tecnológico (CNPq) and Coordenação de Aperfeiçoamento de Pessoal de Nível Superior (CAPES) for the grants awarded.

References

[1] F. Mutter, S. Gareis, B. Schatz, A. Bayha, F. Gruneis, M. Kanis, and D. Koss, "Model-driven in-the-loop validation: Simulation-based testing of UAV software using virtual environments," in *Proceedings...*, (Piscataway), pp. 269–275, International Conference and Workshops on Engineering of Computer Based Systems, IEEE, Apr. 2011.

[2] G. V. Raffo, M. G. Ortega, and F. R. Rubio, "An integral predictive/nonlinear H_∞ control structure for a quadrotor helicopter," *Automatica*, vol. 46, pp. 29–39, 2010.

[3] Z. Fang, X. Y. Wang, and J. Sun, "Design and nonlinear control of an indoor quadrotor flying robot," in *Proceedings...*, (Piscataway), pp. 429–434, World Congress on Intelligent Control and Automation, IEEE, 2010.

[4] J. R. Wertz, *Spacecraft Attitude Determination and control*. The Netherlands: Kluwer Academic Publishers, 1978.

[5] L. R. Ribeiro and N. M. F. Oliveira, "UAV autopilot controllers test platform using Matlab/Simulink and X-Plane," pp. 1–6, ASEE/IEEE Frontiers in Education Conference, 2012.

[6] H. V. Figueiredo and O. Saotome, "Simulation platform for quadricopter: Using Matlab/Simulink and X-plane," pp. 51–55, Brazilian Robotics and Latin American Robotics Symposium, 2012.

[7] L. Ljung, *System Identification: Theory for the User*. New Jersey: Prentice Hall, Englewoord Clifs, 2 ed., 1987.

On the assessment of a tunable auxiliary mass damper with a friction damper in its suspension: numerical study

E.L.C. Guerineau[1], H.T. Coelho[1], F.P. Lepore Neto [1], M.B. Santos[1] and J. Mahfoud[2].

[1]Mechanical System Laboratory, Mechanical Engineering School, Federal University of Uberlandia, Brazil.
[2]LaMCoS, Université de Lyon, CNRS, INSA-Lyon, France.

Abstract: Auxiliary Mass Damper's (AMD) performance is susceptible to changes in the frequency or in the excitation force's nature. Therefore, to improve the robustness of the AMD it's necessary to design new systems which are tunable and that could be used over large frequency range. In this work a friction damper, which is an association in series of a spring and a scratcher, is used to tune the AMD at the same time it dissipates the mechanical energy of the principal mass by changing the normal force on the scratcher. Three normal force control strategies, and two combinations of them, are studied: i) The normal force is assumed constant; ii) The normal force is obtained from the solution of the equation of motion assuming null displacement for the principal mass; iii) The normal force is obtained based on the vibratory system's state variables condition, guarantying that the direction of the friction force promotes the movement of the principal mass toward its static equilibrium position. The effectiveness of the proposed tunable AMD, where the adaptability is obtained by controlling the normal force on the smart friction damper, is evaluated based on mass and frequency ratios variations for each strategy.

Keywords: Tunable Auxiliary Mass Damper, Variable Damped Absorber, Semi -Active Device Control Strategies, Vibration Attenuation and Friction Damper.

1 Introduction

Auxiliary masses are frequently attached to vibrating systems by springs and damping devices to reduce the amplitude of vibration of the system. Depending on the application, these auxiliary mass systems fall into one of two distinct classes. A Dynamic Vibration Absorber (DVA) is an auxiliary mass on a spring, which has a damping factor as lower as possible, once it is tuned to the frequency of the excitation force a system's antiresonance is introduced on this frequency reducing the primary system's vibration amplitudes. And, when it is necessary to provide damping, an auxiliary system is attached to the structure, so that, the auxiliary mass system works as a particular form of damper. This system is called Damped Absorber or Auxiliary Mass Damper (AMD) and it is an extension of the DVA concept [1].

To improve the AMD's performance some researchers use an active device on its suspension. These devices are active springs, done with memory shape alloys, piezo stacks and other actuators able to tune the AMD to a desired frequency [2]. Other solutions are the semi-active systems, which use friction dampers [3,4] or magnetorheological dampers [5] to tune the frequency of the AMD system and also to dissipate the mechanical energy.

AMD enable reducing vibration amplitudes without energy consumption, within a narrow frequency band for which the AMD has been tuned. Unfortunately, when changes in the excitation nature or in the system parameters occur, its performance will drop drastically. To improve AMD's robustness a suitable approach is a semi-active absorber (or adaptive, tunable), which changes its characteristics according to the necessities. Such a device has its physical parameters, as consequence also its impedance, adjustable. Associated to a suitable control law it is possible to adapt the system to different variety of excitations reducing the vibration amplitude. This way, the system becomes a Tunable Auxiliary Mass Damper (TAMD).

The energy necessary to tune the AMD is much less than the energy necessary to achieve the same attenuation using active actuators, once for the active systems the energy is expended to work against the excitation force.

The aim of this work is to develop a new TAMD where the adeptly is obtained by controlling the normal force of a smart friction damper. The numerical study demonstrates the effectiveness of the developed strategy.

2 Theoretical approach

A schema of the system studied is shown on Fig.1. It is a two DOFs that can be modeled as a one Degree of Freedom (DOF) linear vibratory system (m_1, in black) coupled to an AMD using a friction damper (m_2, in red).

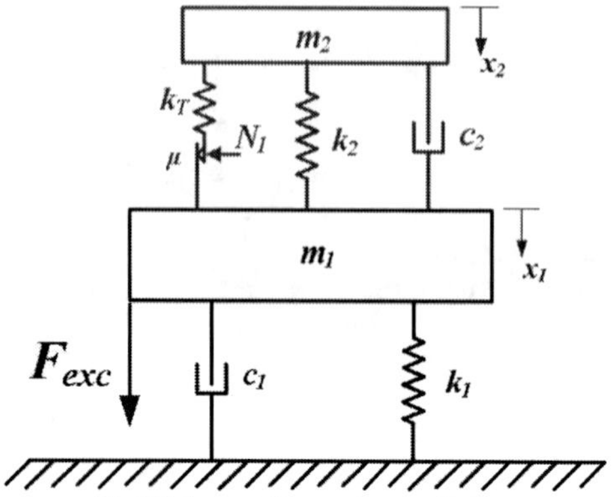

Fig.1 Schema of the system studied.

The adaptability is achieved by using a friction damper (Fig.2). It is an association in series of a spring and a scratcher which will tune the TAMD. This system enables to dissipate the mechanical energy of the principal mass, by changing the normal force on the scratcher, therefore are changed the apparent damping coefficient and stiffness of the TAMD's suspension. The force between nodes (1) and (2), indicated on Fig.2, is F_{12} and can be written as presented on Eq.(1).

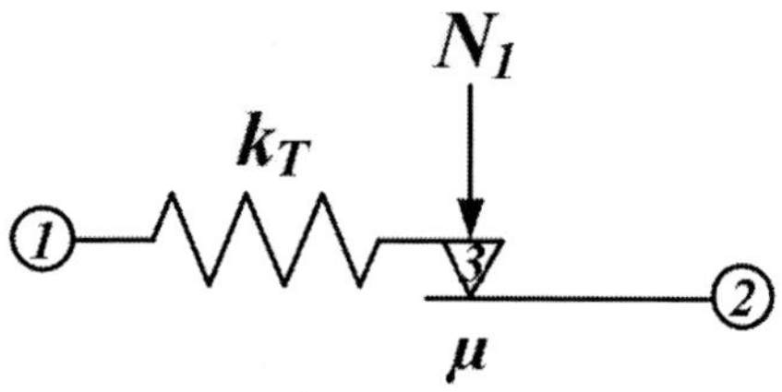

Fig.2 Friction damper model

$$F_{12} = \begin{cases} k_T(x_2 - x_1) & if \quad k_T(x_2 - x_1) \leq \mu N_1 \\ \mu N_1 & if \quad k_T(x_2 - x_1) > \mu N_1 \end{cases} \tag{1}$$

Points (1) and (2) from friction damper, as shown on Fig.2, are attached to mass m_1 and m_2 of the vibratory system respectively. The TAMD's suspension is composed by the stiffness k_2 and the damping c_2, as linear elements, and the nonlinear component characterized by the tangential stiffness k_T and the scratcher which has its force as defined in Eq.(1). The suspension between m_1 and the inertial frame is composed by the stiffness k_1 and the damping c_1. The motion equation for the entire system becomes:

$$\begin{bmatrix} m_1 & 0 \\ 0 & m_2 \end{bmatrix} \begin{Bmatrix} \ddot{x}_1 \\ \ddot{x}_2 \end{Bmatrix} + \begin{bmatrix} c_1 + c_2 & -c_2 \\ -c_2 & c_2 \end{bmatrix} \begin{Bmatrix} \dot{x}_1 \\ \dot{x}_2 \end{Bmatrix} + \begin{bmatrix} k_1 + k_2 & -k_2 \\ -k_2 & k_2 \end{bmatrix} \begin{Bmatrix} x_1 \\ x_2 \end{Bmatrix} = \begin{bmatrix} 1 \\ -1 \end{bmatrix} F_{12} + \begin{bmatrix} 1 \\ 0 \end{bmatrix} F_{exc} \tag{2}$$

The equation of motion has been integrated using the methodology proposed by Lu *et al* [3] which, uses the state space formulation from the linear system, and dispose the nonlinear force from friction damper as part of the excitation forces.
Knowing that mass (m_2/m_1) and frequency (ω_2/ω_1) ratios affect DVA's and AMD's behavior, the numerical assessment aimed to determine the best ratios to be used for the future experimental workbench, under design. As mentioned before TAMD works associated to a control law. Five control laws are used, that are three main normal force control strategies, and two combinations of them: i) The normal force is assumed constant [S1]; ii) The normal force is obtained, from Eq.(3), which is the solution of the equation of motion assuming null displacement for the principal mass, i.e. $x_1 = 0$ [S2]; iii) If F_{12} produces a movement of m_1 in the direction of its static equilibrium position, then the normal force is stemming from Eq.(4), otherwise it is null [S3]. The strategy [S4] uses the same logic described for [S3], but now the normal force is calculated using Eq.(3). And [S5] also uses the logic developed for [S3], however as for [S1] it uses a constant value for N_1.

$$N_1 = \frac{|m_2\ddot{x}_2 + c_2\dot{x}_2 + k_2x_2|}{\mu} \tag{3}$$

$$N_1 = \frac{k_T(x_2 - x_1)}{\mu} \tag{4}$$

To compare all methodologies it is necessary to establish a criteria that enable to show the reduction of the resonance amplitude peak simultaneously to the reduction of the amplitude of the receptance over the entire interested frequency band. It is clear from the literature that DVA split the original resonant peak in two resonant peaks, which, can be disastrous to the vibratory system if the excitation force contains harmonics with these new resonant frequencies. Normally, two parameters are used simultaneously to describe the performance of the control systems: Maximum Value $(L_\infty \; norm)$ and the $L_2 \; norm$. The former indicates the maximum amplitude expected for the system response and the second the overall mean value of the response. A similar statement is used in the control technique H_∞/H_2 [6]. Therefore, the best control law performance will provide the lowest values for both norms. In this work the performance parameter (P_p) is defined as:

$$P_p = \frac{maximum \; receptance \; amplitude}{receptance \; amplitude \; norm} \tag{5}$$

Assuming a column vector $A = [\varepsilon \, \varepsilon \, \varepsilon \ldots A_1 \ldots \varepsilon \, \varepsilon \, \varepsilon]_{1xN}$, where ε is a real constant closest to zero and A_1 a real positive constant, which represents the receptance with one peak only on the frequency spectrum $(A_1 = \delta(f_1)$,

 DOI: 10.1051/matecconf/20168301005

where $\delta(f_1)$ is the Dirac function on frequency f_1), the parameter P_p can be written as:

$$P_p = \lim_{\varepsilon\to 0} P_p = \lim_{\varepsilon\to 0}\frac{A_1}{\sqrt{(N-1)\varepsilon^2 + A_1^2}} = \frac{A_1}{A_1} = 1 \qquad (6)$$

The other extreme is a constant amplitude receptance, which are represented by the column vector $A = [\varepsilon\ \varepsilon\ \varepsilon \ldots \varepsilon\ \varepsilon\ \varepsilon]_{1xN}$ for this the parameter P_p is written as:

$$P_p = \lim_{\varepsilon\to 0} P_p = \lim_{\varepsilon\to 0}\frac{\varepsilon}{\sqrt{N\varepsilon^2}} = \frac{1}{\sqrt{N}} \qquad (7)$$

Equations (6) and (7) give the maximum and minimum of the performance parameters as defined on Eq. (5).

3 Numerical results

To obtain a reasonable comparison among the obtained results and those of the literature, mass and frequency ratios are the same described by Harris and Piersol (2002). In this work the ratios of mass used are $m_2/m_1 = [0.1, 0.2, 0.3, 0.4, 0.5]$ and for the frequency are $\omega_2/\omega_1 = [0.1, 0.5, 1]$. These ratios with the five control strategies will lead to 75 combinations of results.
The numerical results presented at this section has been obtained using the following physical parameters value, which represent the parameters from a designed modification of the experimental workbench used on previously works [7,8], m_1=3.0 kg, c_1=20 Ns/m and k_1=68k N/m for the vibratory system and the physical parameters for the secondary system (m_2, c_2 and k_2) are deduced from the mass and frequency ratios in use. The contact parameters are the tangential stiffness k_T=260k N/m and the friction coefficient μ=0.3. It is important to write that these parameters are theoretical ones, and should be estimated as soon as the test rig has been constructed.
To obtain the aforementioned receptances all strategies are applied for the control of the 2 DOF composed vibratory system excited with a $F_{exc} = 5N$ harmonic force. The excitation was harmonic with its frequency sweeping from 5 Hz up to 100 Hz, in steps of 1 Hz. The stead state response was obtained using 10 periods, and the steady state has been assumed to occur after 10 periods of oscillation. These parameters have been chosen after an observation of the system response at the resonance. For the strategies which uses constant normal force value the applied value was $N_1 = 10N$.
The values of the performance parameter P_p for each combination, for the 1 DOF free vibration and for the well-tuned DVA are shown on Figure 3. On this figure the color bar indicates the value of P_p and the arrows indicates the location of the best result for each strategy, also for the location of the DVA and the 1DOF free vibration.
Figure 4 presents the receptances, which give the lowest P_p values for each strategy. These are compared with the receptances for 1 DOF and for DVA.

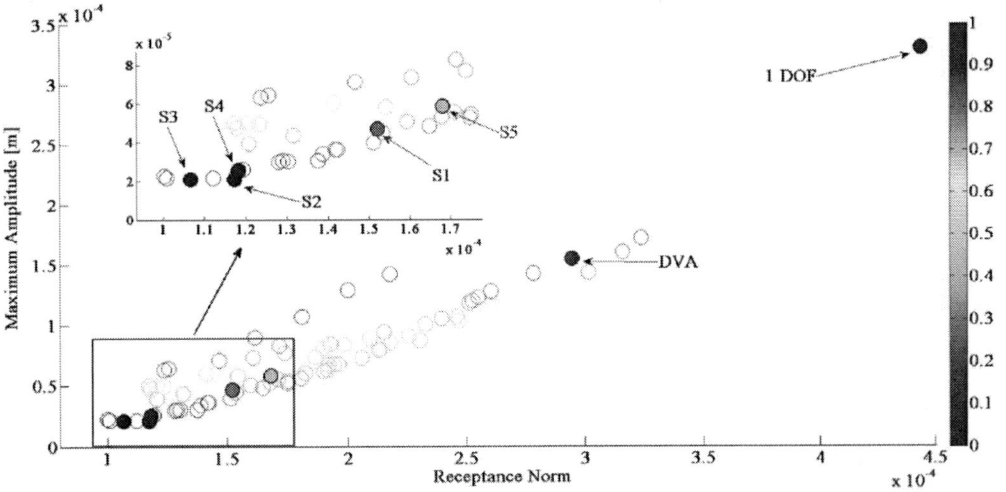

Fig.3 Maximum amplitude and receptance norm chart.

Table 1 summarizes the mass and frequency ratios combination which results in the lowest values for P_p for each strategy. Also presents the ratios used to tune the DVA and their respective performance parameter value. The 1 DOF free vibration have had its receptance also quantified by using the performance parameter P_p.
All strategies promote an improvement in the attenuation of the resonant peaks as well as in the L_2 $norm$ value over the analyzed frequency band. The best performance parameter is for Strategy S2, in fact, its receptance does not have any resonant peak and maintains the low value over the other frequency bands when compared to the 1 DOF curve. Strategies S3 and S4 give similar results when compared to S2, their highest values for P_p are due the maximum amplitude of the receptance which are

higher than S2 maximum amplitude. It should be noticed that all strategies give better results than DVA.

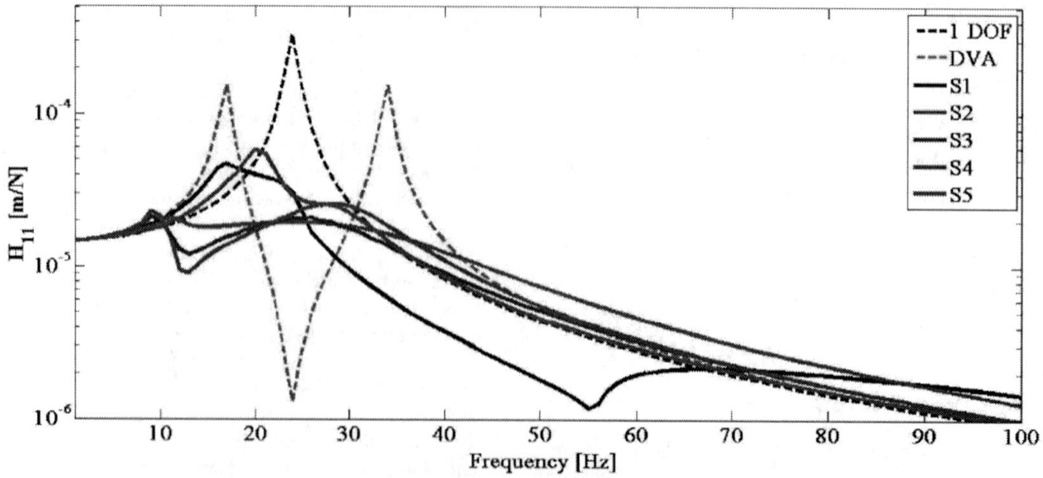

Fig.4 Receptance results.

Table 1 Ratios combination and P_p value for the bests of each strategy.

CONTROL STRATEGY	RATIOS COMBINATION		P_p
	ω_2/ω_1	m_2/m_1	
1 DOF	(-)	(-)	0.74
DVA	1.0	0.5	0.52
STRATEGY S1	0.5	0.5	0.30
STRATEGY S2	0.5	0.3	0.17
STRATEGY S3	0.5	0.4	0.19
STRATEGY S4	0.5	0.5	0.21
STRATEGY S5	1	0.1	0.34

4 Conclusions

A performance parameter has been defined and was shown to be effective to identify the receptance with the lowest peak and the lowest L_2 *norm* over the frequency band of analysis. Using the numerical simulations is possible to affirm that the proposed TAMD model presents better results than the traditional DVA. Future works, already undergoing, will confirm the numerical results experimentally.

Acknowledgements

The authors are grateful to the agencies and bureaus which had been supported this research project: Capes, Fapemig, CNPq and Brafitec.

References

1. HARRIS, C.M. and PIERSOL, A.G. Harris' Shock and Vibration Handbook. Fifth Edition. McGraw-Hill HANDBOOKS, 2002.
2. CHATTERJEE, S., Optimal active absorber with internal state feedback for controlling resonant and transient vibration, Journal of Sound and Vibration, 329, p. 5397–5414, 2010.
3. LU, L.-Y.; CHUNG, L.-L.; WU, L.Y. and LIN, G.-L. Dynamic Analysis of Structures with Friction Devices Using Discrete-Time State-Space Formulation. Computers and Structures. Taiwan, v. 84, p. 1049-1071, 2006.
4. LIN, C.-C, LIN, G.-L. and WANG, J.-F., Protection of seismic structures using semi-active friction TDM, Earthquake Engineering and Structural Dynamics, 39, p. 635–659, 2010.
5. WEBER, F., Semi-active vibration absorber based on real-time controlled MR damper, Mechanical Systems and Signal Processing, 46, p. 272–288, 2014.6. ROTEA, M.A. and KHARGONEKAR, P.P., Mixed H^2-optimal Control with an H^∞-constraint: The State Feedback Case, Automatica, 27(2): p. 307-316, 1991.
6. ROTEA, M.A. and KHARGONEKAR, P.P., Mixed H^2-optimal Control with an H^∞-constraint: The State Feedback Case, Automatica, 27(2): p. 307-316, 1991.
7. COELHO, H.T., SANTOS, M.B. and LEPORE NETO, F.P. Control Strategies for Friction Dampers: Numerical Assessment and Experimental Investigations. Second International Brazilian Conference on Tribology – TriboBR 2014, p.1-11, 2014.
8. COELHO, H.T., SANTOS, M.B., LEPORE NETO, F.P and J. MAHFOUD. Control strategies for friction dampers: numerical assessment and experimental investigations. MATEC Web of Conferences 16, 07007, p.1-4, 2014

Robust flatness-based switching reconfiguration control using state flow machines of electronic throttle valve

Hajer Gharsallaoui[1,a], Wafa Gritli[1], and Mohamed Benrejeb[1]

LA.R.A Automatique, Ecole Nationale d'Ingénieurs de Tunis, Université de Tunis El Manar, BP 37, Le Belvédère, 1002 Tunis, Tunisie.

Abstract. In this paper, a robust Fault Tolerant Control reconfiguration approach using State Flow Machines is proposed. Indeed, this reconfiguration strategy is based on robust flatness-based switching control using state machines and flow charts. This approach is developed in discrete time framework in order to track a reference trajectory starting from a flat output variable. For each model, a corresponding flatness-based controller is designed and consequently, a multi controller structure is obtained. The switching flatness- based control is based on switching between identified Operating Modes (OM) using state flow machines. The Luenberger observer's gains are determined using LMIs tools in order to identify the corresponding OM. The localization of the current OM is carried out by minimization of a performance test characterizing the distance between the system and the given OM. Study of the stability as well as the use of anti-windup devices related to switching between controllers have been considered in the proposed approach. The proposed approach is applied to the nonlinear system which is in our case of study an Electronic Throttle Valve (ETV) using state flow machines modeling.

1 Introduction

General multiple model-based control approaches have been developed in last decade. Thus, in the literature, there are many strategies that take in consideration the multi-models methods used for reconfiguration of control law, [1], [2] and [3]. Furthermore, Multiple Models (MM) method is proposed in [4]. In addition, controller switching approach represents a class of active fault tolerant control detailed in [5]. Moreover, Multiple Models Switching and Tuning (MMST) approach that concerns more particularly the reconfigurable control method is proposed in [3], [6] and [7]. In this paper, for a nonlinear process with multiple OMs, a robust control reconfiguration strategy, based on switching control, is used to ensure stability and desired performances. The LMI tools which are based on quadratic Lyapunov criteria are used in order to guaranty the stability of the global system. The performances obtained by switching in terms of tracking trajectory and fault rejection are discussed in this paper and the stability of the corresponding switched discrete-time linear systems is given. The case of ETV is studied.

2 Reconfiguration fault tolerant control approach

2.1 Description

A robust RST multicontroller based on flatness is proposed. The controller selection strategy is based on switching between identified Operating Modes (OMs). A stateflow is proposed in order to select the identified controller. The localization of the current OM is carried out by minimization of a performance test characterizing the distance between the system and the given OM.

2.2 Stability analysis

The stability analysis of a switching control strategy consists on two steps, [1]

- stability analysis of each subsystem, i.e., each controller must asymptotically stabilize each process operating mode,
- stability analysis of the overall system for arbitrary switching signals.

To guarantee the stability of the systems, Linear Matrix Inequalities, represented by 1, must be solved [8].

$$\begin{pmatrix} P & PA_j - Y_jC_j \\ (PA_j - Y_jC_j)^T & P \end{pmatrix} > 0 \tag{1}$$

with P a symmetric positive definite matrix given by $L_j = P^{-1}Y_j$ where $L_j \in \mathbb{R}^{n\times m}$ is the Lunberger observer gain for the jth model.

3 RST multicontroller design based on flatness

3.1 RST multicontroller structure

Let us denote the open-loop discrete-time process transfer function by

$$H_j(q^{-1}) = \frac{B_j(q^{-1})}{A_j(q^{-1})} \tag{2}$$

[a] e-mail: hajer.gharsallaoui@voila.fr

$A_j(q)$ and $B_j(q)$ are polynomials defined by (3) and (4) where the parameters $a_{j,i}$ and $b_{j,i}$ are constants, $j = 1, ..., m$, $i = 0, 1, ..., n-1$ and q^{-1} is the causal operator.

$$A_j(q^{-1}) = 1 + a_{j,n-1}q^{-1} + \ldots + a_{j,1}q^{-n+1} + a_{j,0}q^{-n} \quad (3)$$

$$B_j(q^{-1}) = b_{j,n-1}q^{-1} + \ldots + b_{j,1}q^{-n+1} + b_{j,0}q^{-n} \quad (4)$$

The realization of an RST controller based on flatness is carried out by considering the method of direct calculation of the state vector $Z_{j,k} = (z_{j,k}\ z_{j,k+1}\ \ldots\ z_{j,k+n-1})^T$, [9].
The state space representation of the system in its controllable form is such that

$$\begin{cases} Z_{j,k+1} = \mathbf{A}_j Z_{j,k} + \mathbf{B}_j u_{j,k} \\ y_{j,k} = \mathbf{C}_j Z_k \end{cases} \quad (5)$$

where the matrix $\mathbf{A}_j$, $\mathbf{B}_j$ and $\mathbf{C}_j$ are given by

$$\mathbf{A}_j = \begin{pmatrix} 0 & 1 & 0 & \cdots & 0 \\ 0 & 0 & 1 & \ddots & \vdots \\ \vdots & \ddots & \ddots & \ddots & 0 \\ 0 & \cdots & 0 & 0 & 1 \\ -a_{j,0} & -a_{j,1} & \cdots & -a_{j,n-2} & -a_{j,n-1} \end{pmatrix} \quad (6)$$

$$\mathbf{B}_j = \begin{pmatrix} 0 \cdots 0\ 1 \end{pmatrix}^T \quad (7)$$

$$\mathbf{C}_j = \begin{pmatrix} b_{j,0}\ b_{j,1} \cdots b_{j,n-1} \end{pmatrix} \quad (8)$$

The flatness control law is given by the following relation

$$u_{k,j} = K(q)z^d_{k,j} + (a-k)Z_{k,j} \quad (9)$$

where a and k are two constant vectors constituted by the $a_{j,i}$ and k_i coefficients of the $A_j(q)$ and $K(q)$ polynomials given by

$$a = \begin{pmatrix} a_{j,0}\ a_{j,1} \cdots a_{j,n-1} \end{pmatrix},\ k = \begin{pmatrix} k_0\ k_1 \cdots k_{n-1} \end{pmatrix} \quad (10)$$

The polynomial $K(q)$ is the denominator of the tracking dynamics which is a discrete-time n^{th} order model equivalent of a continuous-time one composed of second order with fixed damping factor ξ, fixed natural frequency ω_0, given by (11), and a first order model with a fixed time constant τ, given by (12), where s is the Laplace operator.

$$\frac{\omega_0^2}{s^2 + 2\omega_0\xi s + \omega_0^2} \quad (11)$$

$$\frac{1}{1+\tau s} \quad (12)$$

The structure of RST controller can be then obtained by

$$S_j(q^{-1})u_k = K(q)z^d_{j,k} - R_j(q^{-1})y_{j,k} \quad (13)$$

with

$$R_j(q^{-1}) = (k-a)\mathbf{A}_j^{n-1}O^{-1}_{(\mathbf{A}_j,\mathbf{C}_j)}Q \quad (14)$$

$$S_j(q^{-1}) = 1 + (a-k)(\mathbf{A}_j^{n-1}O^{-1}_{(\mathbf{A}_j,\mathbf{C}_j)}M_{(\mathbf{A}_j,\mathbf{B}_j,\mathbf{C}_j)} - (\mathbf{A}_j^{n-2}\mathbf{B}_j \cdots \mathbf{B}_j))Q^* \quad (15)$$

and

$$Q = \begin{pmatrix} q^{-(n-1)}\ q^{-(n-2)} \ldots q^{-1}\ 1 \end{pmatrix}^T \quad (16)$$

$$Q^* = \begin{pmatrix} q^{-(n-1)}\ q^{-(n-2)} \ldots q^{-1} \end{pmatrix}^T \quad (17)$$

$M_{(\mathbf{A}_j,\mathbf{B}_j,\mathbf{C}_j)}$ and $O_{(\mathbf{A}_j,\mathbf{C}_j)}$ are the controllability and the observability matrices.

3.2 Trajectories planning

The open loop control law can be determined by the following relations [10].

$$u^d_j(t) = f(z^d_j(t), ..., z_j^{d(r+1)}(t)) \quad (18)$$

$$y^d_j(t) = g(z^d_j(t), ..., z_j^{d(r)}(t)) \quad (19)$$

where f and g are vectorial functions. Then, it is sufficient to find a desired continuous flat trajectory $t \mapsto z^d_j(t)$ that must to be differentiable at the $(r+1)$ order.
In order to plan the desired flat trajectory $z^d_j(t)$, the polynomial interpolation technique is used. Let consider the state vector $Z^d_j(t) = (z^d_j(t)\ \dot{z}^d_j(t) \ldots z_j^{d(r+1)}(t))^T$ containing the desired continuous flat output and its successive derivatives. t_0 and t_f are the two moments known in advance. The expression of $Z^d_j(t)$ can be given as following [10]

$$Z^d_j(t) = M_{j,1}(t-t_0)c_{j,1}(t_0) + M_{j,2}(t-t_0)c_{j,2}(t_0,t_f) \quad (20)$$

where $M_{j,1}$ and $M_{j,2}$ are such as

$$M_{j,1} = \begin{pmatrix} 1 & t & \cdots & \frac{t^{n-1}}{(n-1)!} \\ 0 & 1 & \cdots & \frac{t^{(n-2)}}{(n-2)!} \\ \vdots & \ddots & \ddots & \vdots \\ 0 & \cdots & 0 & 1 \end{pmatrix} \quad (21)$$

$$M_{j,2} = \begin{pmatrix} \frac{t^n}{n!} & \frac{t^{n+1}}{(n+1)!} & \cdots & \frac{t^{2n-1}}{(2n-1)!} \\ \frac{t^{n-1}}{(n-1)!} & \frac{t^n}{n!} & \cdots & \frac{t^{(n-2)}}{(n-2)!} \\ \vdots & \ddots & \ddots & \vdots \\ t & \cdots & \frac{t^{n-1}}{(n-1)!} & \frac{t^n}{n!} \end{pmatrix} \quad (22)$$

and the vectors $c_{j,1}$ and $c_{j,2}$ defined by

$$c_{j,1} = Z^d_j(t_0) \quad (23)$$

$$c_{j,2} = M^{-1}_{j,2}(t_f - t_0)(Z_{j^d}(t_f) - M_{j,1}(t_f - t_0)Z^d_j(t_0)) \quad (24)$$

After planning a desired flat trajectory, the output desired trajectory $y^d_{j,k}$ is defined. In the discrete-time framework, the real output $y_{j,k}$ have asymptotically to track this such as (25), [10].

$$y^d_{j,k} = B_j(q)z^d_{j,k} \quad (25)$$

4 Application to the control of an electronic throttle valve

4.1 Electronic throttle valve modelling

The case of the Electronic Throttle Valve (ETV), of the Fig. 1, is studied to test the proposed controller.
The electrical part is modeled by (26)

$$u = L\frac{di}{dt} + Ri + E,\ E = k\frac{d\theta_m}{dt} = k\omega \quad (26)$$

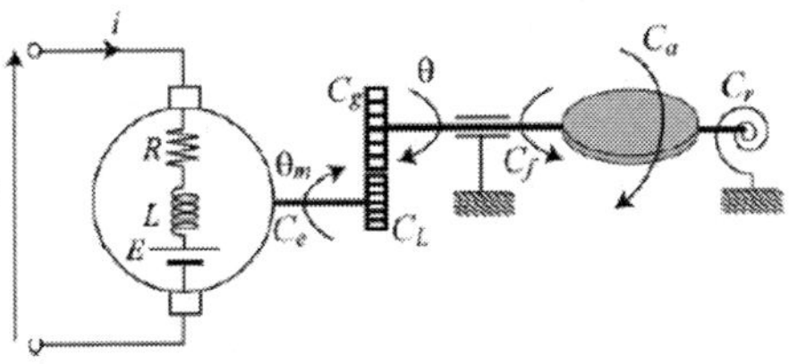

Fig. 1. Electronic throttle system [11]

where L is the inductance, R the resistance, E the electromotive force of its armature, u and i the voltage and the armature current respectively, k an electromotive force constant, θ_m the position of the motor shaft and ω the angular velocity of the motor rotor [12] and [13].
The mechanical part of the throttle is modeled by a gear reducer characterized by its reduction ratio η such as (27)

$$\eta = \frac{\theta_m}{\theta} = \frac{C_g}{C_L} \tag{27}$$

where C_L is the load torque and C_g the gear torque.
The mechanical part is modeled according to (28), such that, [12] and [13].

$$J\frac{d\Omega}{dt} = C_e - C_f - C_r - C_a \tag{28}$$

J is the overall moment of inertia, $\Omega = \frac{d\theta}{dt}$ the angular velocity of the throttle valve, $C_e = k_e i$ the electrical torque where k_e a constant, C_f the torque caused by mechanical friction, C_r the spring resistive torque and C_a the torque generated by the airflow.
By substituting in equation (28), the expressions C_e, C_f and C_r and by neglecting the torque generated by the airflow C_a, the transfer function of the model becomes such as (29)

$$H(s) = \frac{(180/\pi)k_e/l}{JLs^3 + JRs^2 + (kk_e + k_sL)s + Rk_s} \tag{29}$$

with $k_s = (180/\pi/l^2)k_r$, l a constant and s the Laplace operator.

The identified parameters of $H(s)$ are given in Table 1.

Table 1. Model's parameters

Parameters	Values
$R(\Omega)$	2.8
$L(H)$	0.0011
$k_e(N.m/A)$	0.0183
$k_r(v/rad/s)$	0.0183
$J(kg.m^2)$	4×10^{-6}
l	16.95

The default position θ_0 of the ETV creates a discontinuity in the dynamics of the system. The operating mode depends on the throttle position. Two linear models ($j = 1, 2$) have been chosen to carry out the study with the sampling period; $T_e = 0.002s$.

- A model representing the position of the plate above the position by default with $k_s = 1.877 \times 10^{-4} kg.m^2$. The corresponding discrete-time transfer function $H_1(q^{-1})$ is given by (30).

$$H_1(q^{-1}) = \frac{0.007833q^{-1} + 0.01396q^{-2} + 0.0007724q^{-3}}{1 - 1.948q^{-1} + 0.954q^{-2} - 0.006152q^{-3}} \tag{30}$$

- A model representing the position of the plate below the position by default with $k_s = 1.384 \times 10^{-3} kg.m^2$. The corresponding discrete-time transfer function $H_2(q^{-1})$ is given by (31).

$$H_2(q^{-1}) = \frac{0.007832q^{-1} + 0.01396q^{-2} + 0.0007724q^{-3}}{1 - 1.946q^{-1} + 0.954q^{-2} - 0.006152q^{-3}} \tag{31}$$

4.2 RST multicontroller based on flatness

The dynamics chosen are those of a fourth-order continuous-time system with a damping factor $\xi = 0.8$, a natural frequency $\omega_0 = 20\ rad/s$ and a time constant $\tau = 0.1$ s.
The polynomial $K(q)$ is then given by (32).

$$K(q) = q^4 - 3.897q^3 + 5.695q^2 - 3.699q + 0.9012 \tag{32}$$

The following polynomials were obtained for the RST multicontroller

$$R_1(q^{-1}) = -0.009138 + 0.06373q^{-1} - 0.09832q^{-2} + 0.04374q^{-3} \tag{33}$$

$$R_2(q^{-1}) = -0.03525 + 0.1149q^{-1} - 0.1236q^{-2} + 0.0439q^{-3} \tag{34}$$

$$S_1(q^{-1}) = 1 - 0.9495q^{-1} - 0.005626q^{-2} - 7.058 \times 10^{-6}q^{-3} \tag{35}$$

$$S_2(q^{-1}) = 1 - 0.9505q^{-1} - 0.006031q^{-2} - 2.722 \times 10^{-5}q^{-3} \tag{36}$$

4.3 Stability analysis

The Luenberger observers's gains are given by (37).

$$L_1 = \begin{pmatrix} -24.5159 \\ -59.3827 \\ -105.0460 \end{pmatrix}; \quad L_2 = \begin{pmatrix} -24.5175 \\ -59.3817 \\ -105.0352 \end{pmatrix} \tag{37}$$

The combination of the two controllers with the two process transfer functions, is found to be globally asymptotically stable with P definite positive, is given by (38).

$$P = \begin{pmatrix} 0.1839 & -0.1798 & 0.0587 \\ -0.1798 & 0.1759 & -0.0574 \\ 0.0587 & -0.0574 & 0.0188 \end{pmatrix} \tag{38}$$

4.4 Simulation results

The desired trajectories $z_1{}^d(t)$ and $z_2{}^d(t)$ are given by Fig. 2.
The simulation results are given in Fig. 3, 4 and 5. This figures show the effectiveness of the RST multicontroller based on flatness in terms of tracking a desired trajectory with high performance and disturbances rejection.

DOI: 10.1051/matecconf/20168301006

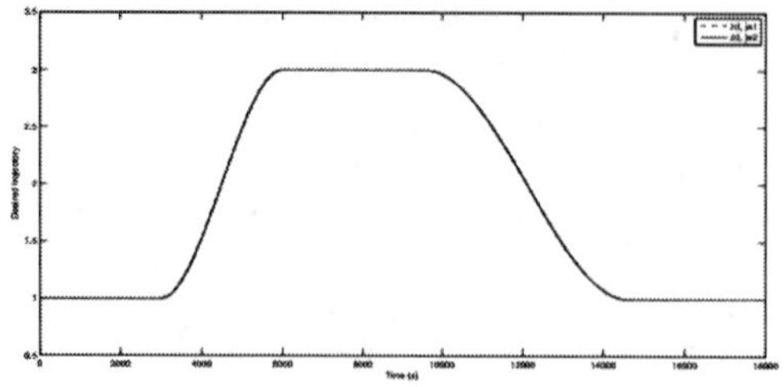

Fig. 2. Desired trajectory z_j^d

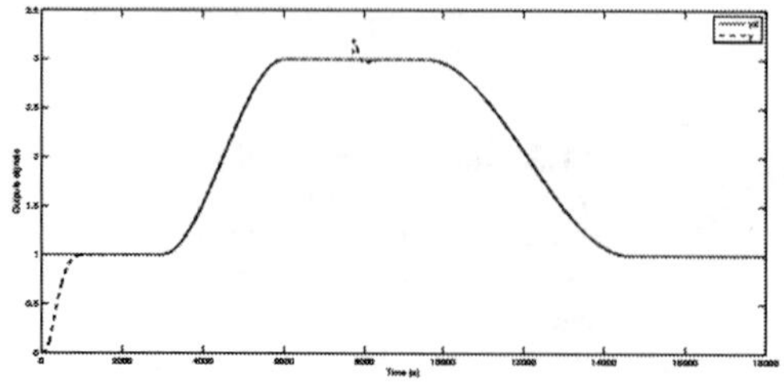

Fig. 3. System output y and desired output y^d

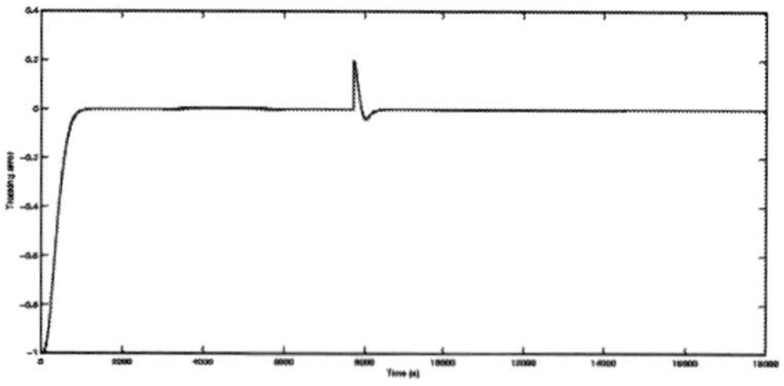

Fig. 4. Tracking error

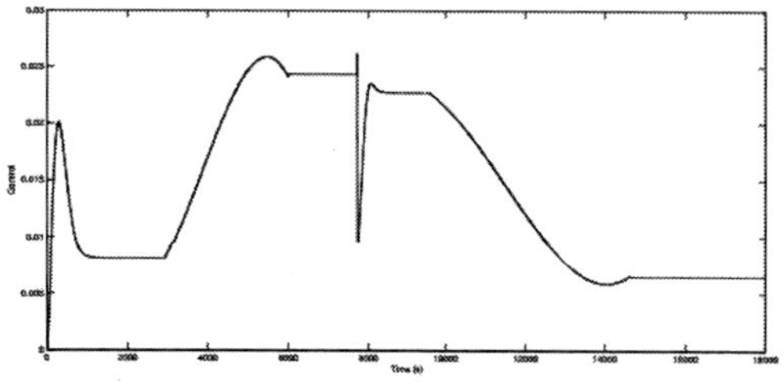

Fig. 5. Control

5 Conclusion

The proposed approach based on the flatness is applied to an Electronic Throttle Valve (ETV) with multiple Operating Modes (OMs). For each model, a corresponding RST controller based on flatness is designed and consequently, a reconfiguration strategy based on multi-controllers structure obtained. The switching between identified models of an ETV was based on the minimization of a performance test characterizing the distance between the system and the given OM. Simulation results proved the effectiveness of the proposed multicontroller.

References

1. P. Charbonnaud, J.F. Carrilo and S. Médar, *Robust control reconfiguration of a thermal process with multiple operating modes*, IEEE transactions on control systems technology, pp. 529-538, 2003.
2. S. Médar, P. Charbonnaud and F. Noureddine, *Active fault accommodation of a three tank system via switching control*, IFAC World Congress, Barcelona, 2002.
3. K. S. Narendra, O. A. Driollet, M. Feiler and K. George, *Adaptive control using multiple models, swtiching and tuning*, International Journal of Adaptative Control and Signal Processing, pp. 87-102, 2003.
4. J. Lemos, L. Ratos and J. Marques, *Switching reconfigurable control based on hidden markov models*, American Control Conference, San Diego, 1999.
5. M. Mahmoud, J. Jiang and Y. Zhang, *Stochastic stability analysis of fault tolerant control systems in the presence of noise*, IEEE Transactions on Automatic Control, pp. 1810-1815, 2001.
6. M. Blanke, M. Kinnaert,J. Lunze and M. Staroswiecki, *Diagnosis and Fault-Tolerant Control*, Springer, London, 2006.
7. Y. Zhang and J. Jiang, *Bibliographical review on reconfigurable fault-tolerant control systems*, Proceeding of the SAFEPROCESS: 5th IFAC Symposium on Detection and Safety for Technical Processes, Washington, 2003.
8. L. ElGhaoui and S. Niculescu, *Advances in Linear Matrix Inequality methods in control*, Philadelphia, 2000.
9. F. Rotella, F. J. Carrilo and M. Ayadi, *Digital flatness-based robust controller applied to a thermal process*, in Proceedings of the 2001 IEEE International Conference on Control Applications (CCA '01), pp. 936-941, Mexico City, 2001.
10. H. Gharsallaoui, M. Ayadi, M. Benrejeb and P. Borne, *Flatness-based Control and Conventional RST Polynomial Control of a Thermal Process*, in International Journal of Computers, Communications and Control (IJCCC), vol. 4, pp. 41-56, 2009.
11. I. Aidi, M. Ayadi, M. Benrejeb and P. Borne, *Flatness-based Control of Throttle Valve Using Neural Observer*, in International Journal of Research and Surveys, vol. 12, pp. 333-344, 2012.
12. M. Vasak, M. Baotic, I. Petrovic and N. Peric, *Hybrid Theory-Based Time-Optimal Control of an Electronic Throttle*, in IEEE Transactions on Industrial Electronics, vol. 54, pp. 1483-1494, 2007.
13. M. Lebbal, H. Chafouk, G. Hoblos and D. Lefebvre, *Modelling and Identification of Non-Linear Systems by a Multimodel Approach: Application to a Throttle Valve*, in International Journal Information and Systems Scienc, vol. 3, pp. 67-87, 2007.

A hyperbolic shear and normal deformation theory for deflection and stresses of FGM sandwich plate

H. Saidi, A. Tounsi, A. E.A. Adda.Bedia

Laboratoire des Matériaux et Hydrologie, Université de Sidi Bel Abbes, Département de Génie Civil, Faculté de Technologie, Algérie

Abstract. In the present paper, the static analysis of functionally graded sandwich plates subjected to thermo mechanical loads is studied. In this model, the displacements vary as a hyperbolic function through the thickness of the plate and satisfy stress boundary conditions on the top and the bottom of the plate. The material properties of the sandwich plate faces are assumed to be graded in the thickness direction according to a simple power-law distribution in terms of volume fraction of material constituents. The core layer is still homogeneous and made of an isotropic material. The governing equations of equilibrium for FG sandwich plates can be obtained using the virtual work principle, and the closed form solutions are obtained by using Navier technique. The accuracy of the present analysis is ascertained by comparing it with various results available in the literature. The influences played by side-to-thickness ratio, aspect ratio and volume fraction distributions are investigated.

1 Introduction

Functionally graded structures are those in which the volume fractions of two or more materials are varied continuously as a function of position along certain dimension(s) of the structure to achieve a required function. For example, thermal barrier plate structures for high-temperature applications may form from a mixture of ceramic and a metal. The composition is varied from a ceramic-rich surface to a metal-rich surface, with a desired variation of the volume fractions of the two materials in between the two surfaces. The ceramic constituent of the material provides the high-temperature resistance due to its low thermal conductivity [1]. The gradual change of material properties can be tailored to different applications and working environments. This makes functionally gradient materials preferable in many applications. The continuous change in the microstructure of Functionally Graded Materials (FGMs) distinguish them from the fibre-reinforced laminated composite materials, which have a mismatch of mechanical properties across an interface due to two discrete materials bonded together.

FGMs are widely used in many structural applications such as mechanics, civil engineering, optical, electronic, chemical, mechanical, biomedical, energy sources, nuclear, automotive fields and ship building industries to eliminate stress concentration and relax residual stresses and enhance bond strength. These materials also have great potential in applications where the operating conditions are severe, including spacecraft heat shields, heat exchanger tubes, fly wheels, plasma facings for fusion reactors, biomedical implants, engine components, and high power electrical contacts or even magnets.

In the past few decades, various shear deformation theories have been proposed and implemented to analysis of plates. Moreover, increased use of advanced materials in primary structures necessitates the development of precise theoretical model to accurately predict the behavior of the structures.

The researchers have paid much attention for modeling of the plates over the past few decades and a variety of plate theories have been introduced. The CPT can only provide reasonable results for thin plates since it disregards the effects of the transverse shear deformation and transverse normal stress. The FSDT [2, 3] considers the effect of transverse shear deformation, but this theory needs a shear correction factor in order to satisfy zero transverse shear stress boundary conditions on the top and bottom. The various higher order shear deformation theories (HSDTs) were proposed to analyze the plates by Ambartsumian [4], Levinson [5], Touratier [6], Soldatos [7], Karama et al. [8],

Aydogdu [9], Reddy [10], Xiang et al. [11] and Mantari et al. [12].

In the present paper, analysis of functionally graded sandwich plates under Thermo-mechanical loading is developed using a new hyperbolic shear strain shape function considering thickness stretching. This theory accounts for adequate distribution of the transverse shear strains in the thickness of the plate and satisfies the traction free boundary conditions on the top and bottom surface of the plates, thus a shear correction factor is not required. The accuracy of the present results is established by comparing those with well known trigonometric shear deformation theory.

2 Geometry and material properties of functionally graded sandwich plate

Let us consider the case of a sandwich plate composed of three microscopically heterogeneous layers as shown in Fig. 1. Rectangular Cartesian coordinates are used to describe infinitesimal deformations of a three-layer sandwich elastic plate occupying the region [0, a] · [0, b] · [_h/2, +h/2] in the unstressed reference configuration. The mid-plane of the composite sandwich plate is defined by z = 0 and its external bounding planes being defined by z=±h/2, the equations will be derived in tensorial notations and specialized afterwards for the problem under consideration. The face layers of the sandwich plate are made of an isotropic material with material properties varying smoothly in the z (thickness) direction only. The core layer is made of an isotropic homogeneous material.
The vertical positions of the bottom and top, and of the two interfaces between the layers are denoted by h0 = - h/2, h1, h2, h3 = h/2. The effective material properties for each layer, like Young's modulus and Poisson s ratio, can be expressed as:

$$P^{(n)}(z) = P_m + (P_c - P_m)V^{(n)} \quad (1)$$

where P_m and Pc denote the property of the bottom and top faces of layer 1, respectively, and vice versa for layer 3 depending on the volume fraction $V^{(n)}$ (n = 1,2,3). Note that Pm and Pc are, respectively, the corresponding properties of the metal and ceramic of the FGM sandwich plate. The volume fraction $V^{(n)}$.

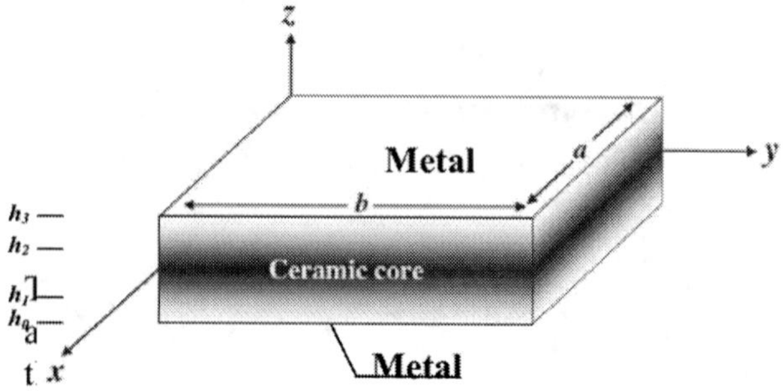

Figure1. Geometry of the FGM sandwich

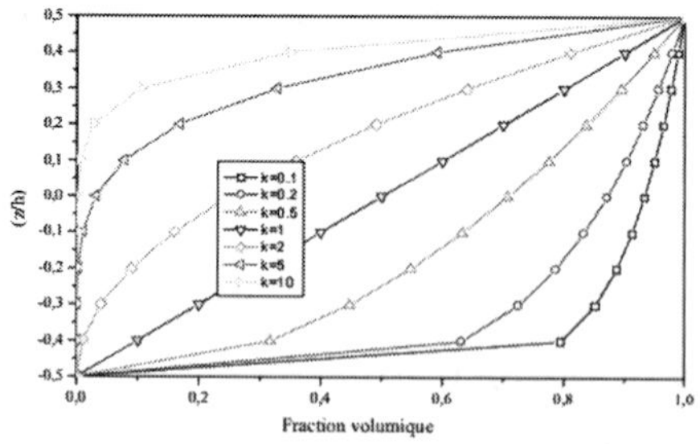

Figure 2: Variation in the volume fraction in a plate P-FGM.

3 Mathematical formulation and analytical solution

The displacement field for FG sandwich plates is described in the following equations:

$$u(x,y,z) = u_0(x,y) - zw_{,x} + f(z)\theta_x \quad (2a)$$
$$v(x,y,z) = v_0(x,y) - zw_{,y} + f(z)\theta_y \quad (2b)$$
$$w(x,y,z) = w_0(x,y) + f'(z)\theta_z \quad (2c)$$

where, u, v, w are displacements in the x, y, z directions, u_0, v_0 and w_0 are midplane displacements, θ_x, θ_y and θ_z rotations of the *yz*, *xz*, and *xy* planes due to bending, respectively. $f(z)$ represents shape function determining the distribution of the transverse shear strains and stresses along the thickness and $f'(z) = \partial f(z)/\partial z$.
The linear strain expressions derived from the displacement model are as follows:

$$\begin{Bmatrix} \varepsilon_x \\ \varepsilon_y \\ \gamma_{xy} \end{Bmatrix} = \begin{Bmatrix} \varepsilon_x^0 \\ \varepsilon_y^0 \\ \gamma_{xy}^0 \end{Bmatrix} + z \begin{Bmatrix} k_x \\ k_y \\ k_{xy} \end{Bmatrix} + f(z) \begin{Bmatrix} \eta_x \\ \eta_y \\ \eta_{xy} \end{Bmatrix} \quad (3a)$$

$$\begin{Bmatrix} \gamma_{yz} \\ \gamma_{xz} \end{Bmatrix} = f'(z) \begin{Bmatrix} \gamma_{yz}^0 \\ \gamma_{xz}^0 \end{Bmatrix} \quad (3b)$$

The linear constitutive relations are given as:

$$\begin{Bmatrix} \sigma_x \\ \sigma_y \\ \sigma_z \\ \tau_{yz} \\ \tau_{xz} \\ \tau_{xy} \end{Bmatrix} = \begin{bmatrix} Q_{11} & Q_{12} & Q_{13} & 0 & 0 & 0 \\ Q_{12} & Q_{22} & Q_{23} & 0 & 0 & 0 \\ Q_{13} & Q_{23} & Q_{33} & 0 & 0 & 0 \\ 0 & 0 & 0 & Q_{44} & 0 & 0 \\ 0 & 0 & 0 & 0 & Q_{55} & 0 \\ 0 & 0 & 0 & 0 & 0 & Q_{66} \end{bmatrix} \begin{Bmatrix} \varepsilon_x \\ \varepsilon_y \\ \varepsilon_z \\ \gamma_{yz} \\ \gamma_{xz} \\ \gamma_{xy} \end{Bmatrix} \quad (4)$$

where (σ_x, σ_y, σ_z, τ_{yz}, τ_{xz}, τ_{xy}) and (ε_x, ε_y, ε_z, γ_{yz}, γ_{xz}, γ_{xy}) are the stress and strain components, respectively, Q_{ij} are the stiffness coefficients.

The principle of virtual work, the following expressions can be obtained:

$$\int_{-h/2}^{h/2} \int_{\Omega} [\sigma_x \delta \varepsilon_x + \sigma_y \delta \varepsilon_y + \sigma_z \delta \varepsilon_z + \tau_{xy} \delta \gamma_{xy} + \tau_{yz} \delta \gamma_{yz} + \tau_{xz} \delta \gamma_{xz}] d\Omega \, dz - \int_{\Omega} q\delta \quad (5)$$

The governing equations of equilibrium can be derived easily from Equation (5)

$$[C]\{\Delta\} = \{F\} \quad (6)$$

where

$\{\Delta\} = \{U, V, W, X, Y, Z\}^t$ and $[C]$ is the stiffness matrix, $\{F\}$ is the force vector.

4 Numerical example and discussion

Numerous examples are solved to ensure the accuracy of the present theory for the prediction of static analysis under thermo mechanical loading , the closed form solution are obtained using the Navier solution of simply supported FGM sandwich plates.

The characteristic values for metal and ceramics used in the FG sandwich plate are listed in Table 1.

Table1. Material properties used in the FGM plate.

Properties	Metal (Aluminium)	Ceramic
E (GPa)	70	380
ν	0.3	0.3
ρ (Kg/m^3)	2707	3800

Figures 3 and 4 show the variation of the dimensionless deflection $\bar{w}$ with side-to-thickness ratio a/h and the aspect ratio a/b for one type of sandwich symmetric square plates (hFGM=1).

The deflection is maximum for the metallic plate and minimum for the ceramic plate. It decreases with the increase of a/h and a/b ratios.

Figures 5 and 6 show plots of the through-the-thickness distribution of the transverse shear stress $\bar{\tau}_{xz}$ and axial stress for $k = 0.0$, 1.5, and 3.5 of symmetric sandwich square plates.

The maximum value occurs at a point on the mid-plane of the plate and its magnitude for FGM sandwich plate is larger than that for homogeneous ceramic plate.

In Figures 7 and 8, we have plotted the through-the-thickness distributions of the dimensionless axial stress $\bar{\sigma}_x$ and the deflection $\bar{w}$ of the FG sandwich plate for $k = 3.5$ and $hFGM = 1$, respectively. These figures show the great influence played by the different thermal and bending loads on the axial stresses and deflection.

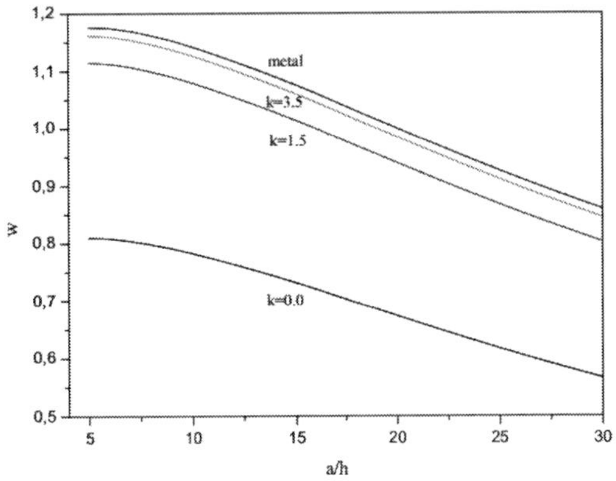

Figure3 Dimensionless center deflection w vs a function of side to thickness ratio a/h for FGM sandwich plate (hFGM/h=1)

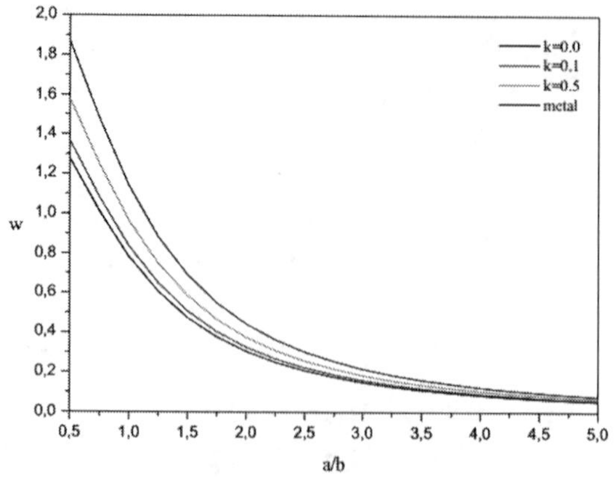

Figure4 Effect of the aspect ratio a/b on the dimensionless the center deflection w of FGM sandwich plates (hFGM=1)

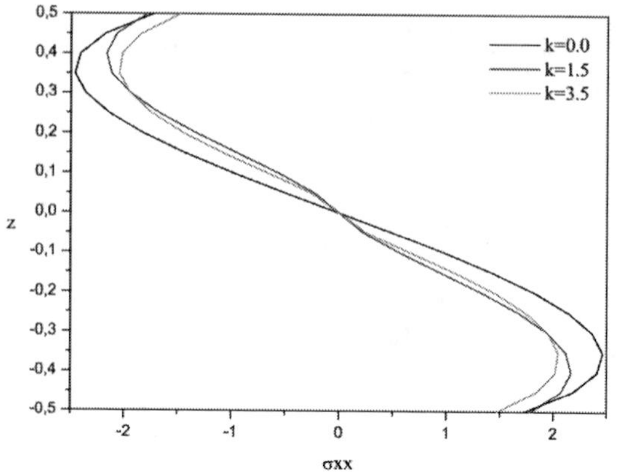

Figure6 Variation of axial stress sigmax through theplate thickness for FGM sandwich plate (hFGM=1)

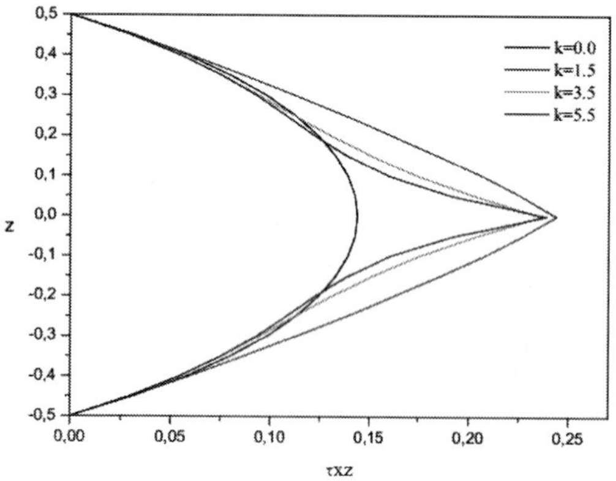

Figure5 Variation of transverse shear stress τxz through the plate thickness for FGM sandwich plate (hFGM=1)

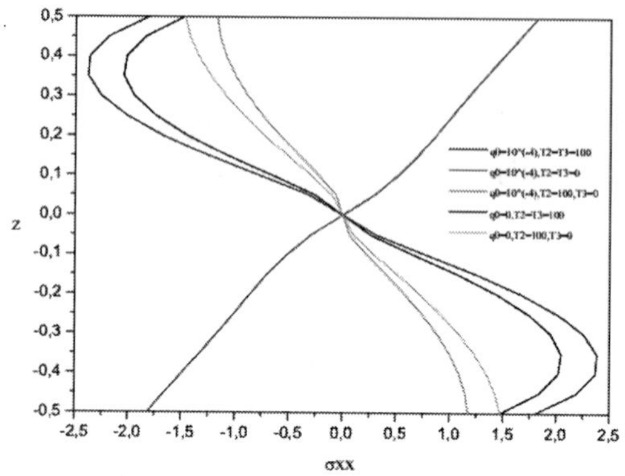

Figure7 Effect of the thermal load T_3 on axial stress σxx of the FGM sandwich plate (hFGM=1, k=3.5)

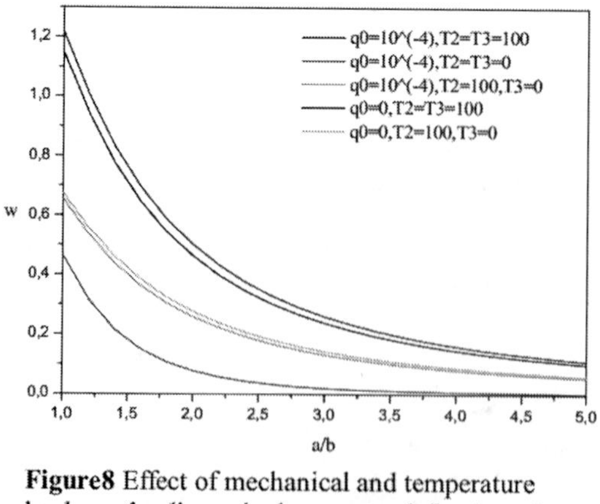

Figure8 Effect of mechanical and temperature loads on the dimensionless center deflection of an FGM sandwich plate (hFGM=1,k=3.5)

Conclusion

The static analysis of functionally graded sandwich plates subjected to thermo mechanical loads is studied. The displacement and stresses have been analyzed under thermo mechanical loadings. The gradation of the properties through the thickness is assumed to be of the power law distribution in terms of volume fractions of the constituents, the core layer is made of ceramic.

The governing equations are derived by employing the principle of virtual work. These equations are then solved via Navier –type solution. The results are presented for one type of simply supported rectangular sandwich plates under thermo-mechanical loading. The accuracy of the present analysis is ascertained by comparing it with various results available in the literature. The influences played by side-to-thickness ratio, aspect ratio and volume fraction distributions are investigated.

REFERENCES

[1] Hasselman DPH, Youngblood GE. Enhanced thermal stress resistance of structural ceramics with thermal conductivity gradient. Journal of the American Ceramic Society 1978; 61(1,2):49{53.

[2] Reissner E. Reflection on the theory of elastic plates. J Appl Mech 1945;38:1453–64.

[3] Mindlin RD. Influence of rotary inertia and shear on flexural motions of isotropic elastic plates. J Appl Mech 1951;18:31–8.

[4] Ambartsumian SA. On the theory of bending plates. Izv Otd Tech Nauk AN SSSR 1958;5:69–77.

[5] Levinson M. An accurate simple theory of static and dynamics of elastic plates. Mech Res Commun 1980;7:343–50.

[6] Touratier M. An efficient standard plate theory. Int J Eng Sci 1991;29:901–16.

[7] Soldatos KP. A transverse shear deformation theory for homogeneous monoclinic plates. Acta Mech 1992;94:195–200.

[8] Karama M, Afaq S, Mistou S. Mechanical behavior of laminated composite beam by new multi-layered laminated composite structures model with transverse shear stress continuity. Int J Solids Struct 2003;40:1525–46.

[9] Aydogdu M. A new shear deformation theory for laminated composite plates. Compos Struct 2003;89:94–101.

[10] Reddy JN. Mechanics of laminated composite plates and shells: theory and analysis. 2nd ed. Boca Raton: CRC Press; 2004.

[11] Xiang S, Jin YX, Bi ZY, Jiang SX, Yang MS. A n-order shear deformation theory for free vibration of functionally graded and composite sandwich plates.
Compos Struct 2011;93:2826–32.

[12] Mantari JL, Oktem AS, Soares CG. A new higher order shear deformation theory for sandwich and composite laminated plates. Composites Part B, 2012;43:1489–99.

Hysteretic damper based on the Ishlinsky-Prandtl model

Mikhail E. Semenov[1,2,3,a], Andrey M. Solovyov[2], Anton G. Rukavitsyn[2], Vladimir A. Gorlov[1], and Peter A. Meleshenko[1,2,b]

[1] Zhukovsky–Gagarin Air Force Academy, Starykh Bolshevikov st. 54 "A", 394064, Voronezh, Russia
[2] Digital Technologies Department, Voronezh State University, Universitetskaya sq. 1, 394006, Voronezh, Russia
[3] Voronezh State University of Architecture and Civil Engineering, XX-letiya Oktyabrya st. 84, 394006 Voronezh, Russia

Abstract. In this paper we consider the mathematical model of hysteretic damper based on the Ishlinsky-Prandtl model. The numerical results for the observable characteristics such as force transmission function and "force-displacement" transmission function are obtained and analyzed. The comparison of the efficiency of nonlinear viscous damper and hysteretic damper is also presented and discussed.

1 Introduction

The damper is a device used for damping the mechanical, electrical and other modes of vibration arising in the machines during its operation. Damping is an important task that has a wide range of applications. In general the *damping* is a process whereby energy is taken from the vibrating system and is being absorbed by the surroundings. The examples of damping include:

- internal forces of a spring;
- viscous force in a fluid;
- electromagnetic damping in galvanometers;
- shock absorber in a car;
- anti seismic damping devices in buildings etc.

Also the damping devices are widely used in modern avionics (damper of aero-elastic vibrations, which is the electronic system for automatic cancelation of short-aircraft vibrations that inevitably arise when the flight modes change).

Let us also list the types (relative to the physical nature of damping process) of damping devices:

- Structural;
- Coulomb Friction;
- Elastomer;
- Active Drivers;
- Passive Hydraulic;
- Semi-Active Hydraulic;
- Adaptive Hydraulic etc.

In the case of oscillations of mechanical systems we consider the linear viscous damping which is based on the energy dissipation due to viscous friction. However, this type of damping has a significant drawback, namely the low efficiency outside the region of resonance of the system. One way to solve this problem is to use a nonlinear viscous damper [1–5] or damper with hysteretic properties. It should also be pointed out the exciting and interesting (generally, from the fundamental point of view) model of viscous damping which is based on the technique of fractional derivatives and can be called as a fractional damping [6, 7].

Main purpose of this work is to study the dynamics of a mechanical system under external affection (forced oscillations) in the case of a damper with hysteretic nonlinearity. As a mathematical model of hysteretic damper we consider the Ishlinsky-Prandtl inverter which is a type of continual hysterons and can be presented as a system of parallel connected nonlinear links such as "stops" [8].

2 Hysteretic material

The materials that are used for hysteretic dampers in general are polymers (synthetic rubber). The composition of such materials is chosen appropriately to provide the high damping properties in a specific range of frequencies and temperatures. When the material is damped the energy dissipation occurs within the material itself. Such an effect is caused by friction between the inner layers that are "flow" when the damping occurs. In the case when such a structure is under damping vibrations the hysteresis loop appears in the "stress-strain". The area of the loop determines the energy loss (per unit volume of the body) in one cycle due to damping.

Let us briefly consider a mathematical model of hysteresis (based on the operator technique). The carrier of hysteretic nonlinearities is a converter W with scalar input $u(t)$ and outputs $x(t)$. The state of this converter is a pair $\{u, x\}$, i.e., an input-output pair. Let the set of possible states of the converter W is a strip $\Omega = \Omega(W)$ which is placed between the two horizontal lines Φ_l and Φ_r, as is shown in Fig. 1.

If the input $u(t)\,(t \geqslant t_0)$ is continuous and monotone then the output can be defined as:

$$x(t) = W[t_0, x_0]u(t)\,(t \geqslant t0) \quad (1)$$

so that the variable state $\{u(t), x(t)\}$ is a point in the broken line as is shown in Fig. 1 by thick line; this broken line (passing through the initial state $M_0 = \{u(t_0), x_0\}$ and the

[a] e-mail: mkl150@mail.ru
[b] e-mail: melechp@yandex.ru

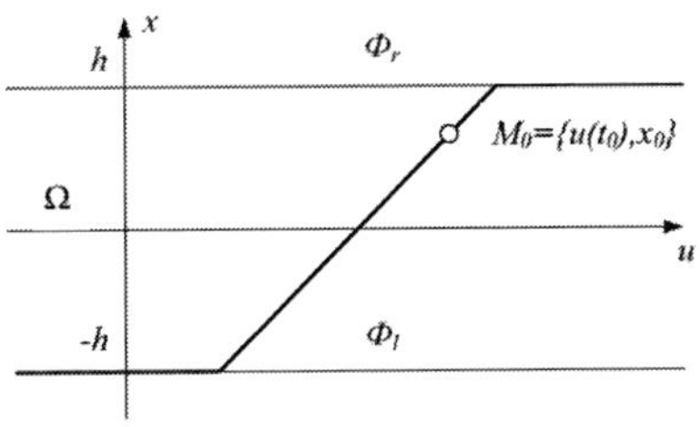

Fig. 1. Action of "stop"-nonlinearity

ends on the lines Φ_l and Φ_r) consists of segment with slope 1 and two horizontal half-lines. In other words, when the input is monotonic the output is determined by the equation:

$$x(t) = \begin{cases} \min\{h, u(t) - u(t_0) + x(t_0)\}, u(t) \text{ nondecreasing}, \\ \max\{-h, u(t) - u(t_0) + x(t_0)\}, u(t) \text{ nonincreasing}. \end{cases} \quad (2)$$

To determine the output (1) for piecewise monotonic continuous inputs we can use the semigroup identity

$$W[t_0, x_0]u(t) = W[t_1, W[t_0, x_0]u(t_1)]u(t) \; (t_0 \leqslant t_1 \leqslant t). \quad (3)$$

The described converter is called as a "stop" [8].

In the most common models of elastic-plastic fibers their states are completely determined by the variables of deformation u and stress x. The parameter h in this case is called as the yield strength of the material. Such fibers can be considered as converters with an input in the form of deformation and an output in the form of strain. In the Prandtl model the strain is determined by the deformation in the same manner as in "stop"-nonlinearity, but the trajectories of possible states between the boundary horizontal lines have the slope E which differs from 1 (for small deformations the fiber can be considered as an elastic material with the elastic modulus E).

Let us consider the inverter W which can be presented in the form of a simple flow diagram (without feedback) based on the finite number of hysterons $W^1, \ldots, W^N$ and elementary functional units (as is shown in Fig. 2). Usually

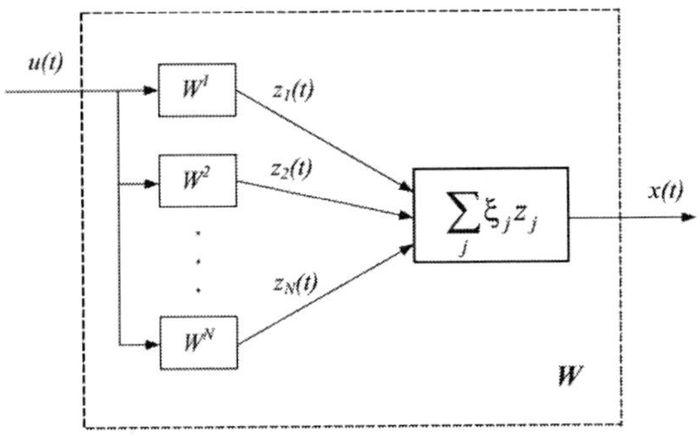

Fig. 2. Parallel connection of hysterons

such converters W are nondeterministic. Their condition can be described (instead of the input-output description) by the set $\{u, z_1, \ldots, z_N\} \in R^{N-1}$, where u is the converter's input and z_j is an output of hysteron W^j in the flow diagram.

Suppose that hysterons $W^1, \ldots, W^N$ with areas of permissible states $\Omega(W^1), \ldots, \Omega(W^N)$ are determined by the input-output relations

$$z_j(t) = W^j[t_0, z_j(t_0)]u(t) \, (j = 1, \ldots, N). \quad (4)$$

Let

$$\Omega(W) = \{\{u, z_1, \ldots, z_N\} : \{u, z_j\} \in \Omega(W^j), \, u \in R^1\}. \quad (5)$$

Parallel connection of hysterons W^j with weights ξ_j we call the converter W with an area of permissible states (5) such that for every initial state

$$q(t_0) = \{u_0, z_0\} = \{u(t_0), z_1(t_0), \ldots, z_N(t_0)\} \in \Omega(W) \subset R^{N+1} \quad (6)$$

all the continuous scalar inputs $u(t)\,(t \geqslant t_0)$ that satisfy the condition $u(t_0) = u_0$ are allowed. The output is determined by the relation

$$x(t) = W[t_0, z_0]u(t) = \sum_{j=1}^{N} \xi_j W^j[t_0, z_j(t_0)]u(t), \; (t \geqslant t_0). \quad (7)$$

The considered converter W is called as Ishlinsky-Prandtl material [6]. Of course the "classical" Ishlinsky-Prandtl converter is determined as a continuos system (sum should be replaced by an integral). However in our numerical simulations it is more appropriate to use the discrete analog of the Ishlinsky-Prandtl converter (7). In this way in the following consideration we will call the converter (7) as an Ishlinsky-Prandtl converter.

As an example we consider the reaction of the Ishlinsky-Prandtl material on sinusoidal impact. We use the converter W with the parameters $E = 1$, $W^j[t_0, z_j(t_0)] = 0$, $\xi_j = 1$, $W^j : h = \{-j, j\}$, $j = 1, \ldots, 10$ and the input in the form $u(t) = 12\sin(t)$. "Stress-strain" diagram of such a converter is shown in Fig. 3.

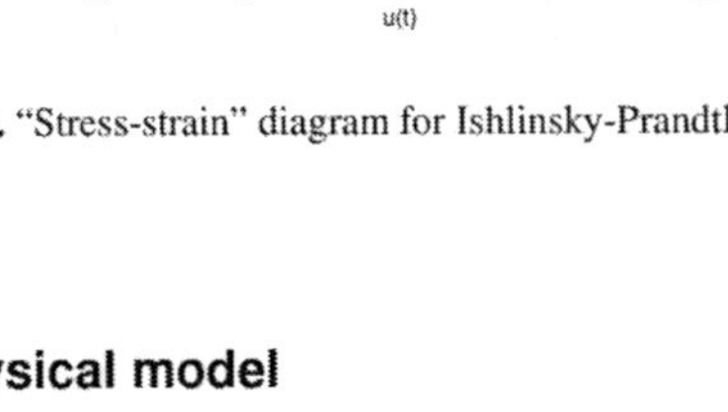

Fig. 3. "Stress-strain" diagram for Ishlinsky-Prandtl material

3 Physical model

Let us consider a mechanical system under external affection and in the presence of the damping part $f_d(t)$ as is shown in Fig. 4. The system is presented as a cylinder with mass M under external affection $f(t)$ of harmonic nature.

In the cylinder there is a car of mass m (moving without friction in the horizontal plane) connected to the border by a spring with stiffness k. For simplicity we assume that the system is one-dimensional.

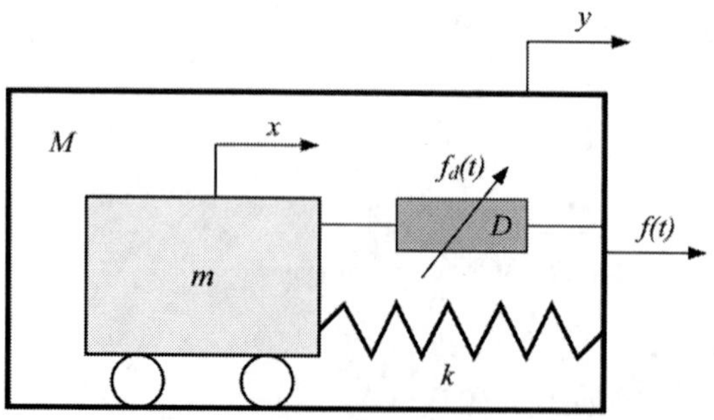

Fig. 4. The considered physical system

Suppose that the affection force $f(t)$ is described by the following relation:

$$f(t) = Y\omega^2 \sin(\omega t), \tag{8}$$

where $Y\omega^2$ is an amplitude and ω is a frequency of the affection force.

The equation of motion for considered system is

$$\begin{cases} M\ddot{y} + kz + f_d(t) = f(t), \\ m\ddot{x} - kz - f_d(t) = 0, \\ f(t) = Y\omega^2 \sin(\omega t),\ z(t) = y(t) - x(t). \end{cases} \tag{9}$$

Here overdot determines the time derivative and $z(t)$ is a relative displacement.

3.1 Viscous damping

Let us consider the case of viscous damping. In general case the viscous friction can be described as follows:

$$f_d(t) = c(1 + z(t))^n \dot{z},\ n \geqslant 0, \tag{10}$$

where c is a damping coefficient. In the case $n = 0$ we have a linear viscous friction. For $n > 0$ the nonlinear viscous fricition takes place.

The equation of motion for relative displacement $z(t)$ becomes

$$\ddot{z}(t) + \frac{M+m}{Mm}\left[c(1 + z(t))^n \dot{z}(t) + kz(t)\right] = \frac{Y}{M}\omega^2 \sin(\omega t). \tag{11}$$

It is more suitable to introduce the dimensionless variables

$$\Omega = \frac{\omega}{\omega_0},\ \mu = \frac{mM}{m+M},\ A = \frac{Y}{M},\ \tau = \omega_0 t,$$

$$\omega_0 = \sqrt{\frac{k}{\mu}},\ \zeta = \frac{c}{2\omega_0\mu},\ u = z.$$

After such notations the final equation of motion has the form:

$$\frac{d^2u}{d\tau^2} + 2\zeta(1+u)^n \frac{du}{d\tau} + u = A\Omega^2 \sin(\Omega\tau). \tag{12}$$

3.2 Hysteretic damping

Now we consider the hysteretic damper. In this case (using the notations introduced above) the damping force can be presented as:

$$f_d(\tau) = W[\tau, z_j(\tau)]u,\ j = \overline{1, N}, \tag{13}$$

where W is an Ishlinsky-Prandtl operator which is determined by the relation (7).

In this case the equation of motion for the considered system becomes:

$$\frac{d^2u}{d\tau^2} + \alpha W[\tau, z_j(\tau)]u + u = A\Omega^2 \sin(\Omega\tau),\ j = \overline{1, N}, \tag{14}$$

where $\alpha = S/k$ and S is an area of damping material's section, k is a spring stiffness.

3.3 Main characteristics

Let us consider the main characteristics reflecting the efficiency of the damper in the resonance system and beyond. Such characteristics are force transmission function and "force-displacement" transmission function.

The *force transmission function* is determined by the ratio of the force applied to the cylinder M and the force applied to the car m (Fig. 4). This function reflects the efficiency of suppression of the external affection by the force transmission from an external source to the load. This characteristic is expressed as follows:

$$T_{ff} = \frac{1}{Y\omega^2} \max\left|m\omega_0^2 \frac{d^2x}{d\tau^2}\right|. \tag{15}$$

The *"force-displacement" transmission function* is determined by the relation of the motion of car m relative to the cylinder M and the force applied to the cylinder. This quantity reflects the efficiency of vibration absorption by the ability of the damper to reduce the relative motion of the car under influence of external forces. This characteristic is expressed as

$$T_{ff} = \frac{\max|x(\tau)|}{Y\omega^2}. \tag{16}$$

During the following simulations we use these quantities for comparison of the linear viscous, nonlinear viscous and hysteretic dampers.

4 Numerical simulation

4.1 Difference scheme

The dynamics of the considered system can be simulated using the explicit difference scheme applied to equations (12) and (14). Finite-difference equations for viscous damper is:

$$\frac{u_{i+2} - 2u_{i+1} + u_i}{h_\tau^2} + \\ +2\zeta(1 + u_{i+1})^n \tfrac{u_{i+1} - u_i}{h_\tau} + u_{i+1} = A\Omega^2 \sin(\Omega i h_\tau), \tag{17}$$

and for hysteretic damper is:

$$\frac{u_{i+2} - 2u_{i+1} + u_i}{h_\tau^2} + \qquad (18)$$

$$+\alpha W[ih_\tau, z_j(ih_\tau)]u_{i+1} + u_{i+1} = A\Omega^2 \sin(\Omega i h_\tau).$$

Here $i = \overline{0, T}$, $j = \overline{1, N}$, h_τ is a step of a grid by τ axis, N is a number of the Ishlinsky-Prandtl hysterons, T is a time period.

The explicit solutions for the viscous damper and for hysteretic damper are:

$$u_{i+2} = A\Omega^2 h_\tau^2 \sin(\Omega i h_\tau) - \qquad (19)$$

$$-2\zeta h_\tau (1 + u_{i+1})^n (u_{i+1} - u_i) - h_\tau^2 u_{i+1} + 2u_{i+1} - u_i,$$

$$u_{i+2} = A\Omega^2 h_\tau^2 \sin(\Omega i h_\tau) - \qquad (20)$$

$$-\alpha h_\tau^2 W[ih_\tau, z_j(ih_\tau)]u_{i+1} - h_\tau^2 u_{i+1} + 2u_{i+1} - u_i,$$

respectively.

4.2 Results

We make the numerical simulations using the explicit results (19) and (20). In order to compare the viscous damper and hysteretic damper we present the numerical results for two characteristics of the system the force transmission function and "force-displacement" transmission function. For the nonlinear viscous damper we use the following set $n = \{0, 2, 4\}$.

For the hysteretic damper we use the Ishlinsky-Prandtl material (which corresponds to rubber) with the parameters $E = 10000$, $W^j[t_0, z_j(t_0)] = 0$, $\xi_j = 1$, W^j: $h = \{-j, j\}$, $j = 1, \ldots, 50$ (with a step 0.1), $\alpha = 0,0001$.

The characteristics of the mechanical system (per dimensionless units): $M = 1$, $m = 1$, $\zeta = 0.8$, $\omega_0 = 10$; the external affection with parameters $A = 1$, $\omega = 0, \ldots, 30$ (with a step 0.2); parameters of the difference scheme: $h_\tau = 0.0167$, $T = 10000$.

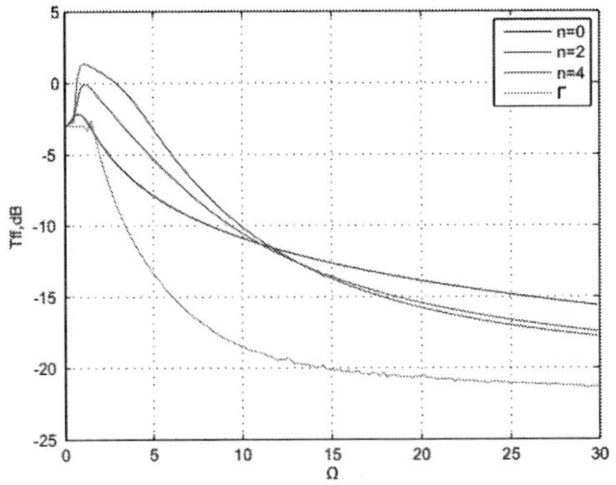

Fig. 5. Force transmission function

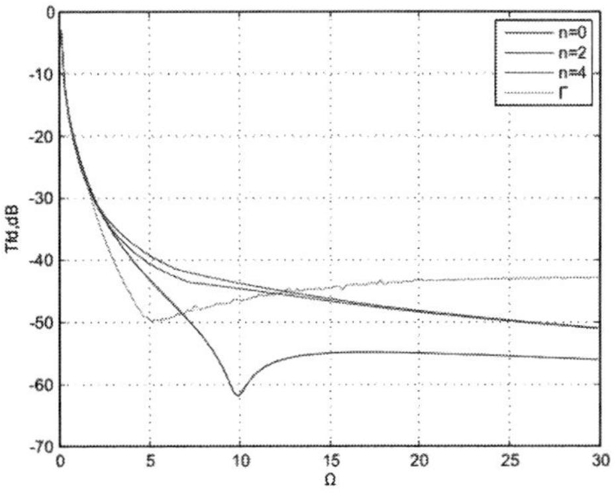

Fig. 6. "Force-displacement" transmission function

The simulation results are shown in Fig. 5 and Fig. 6. As it can be seen from these figures the linear viscous damper (dark blue curve $n = 0$) has a high efficiency in the resonance region of the system, however outside the resonance region the damping properties sharply decrease. At the same time the nonlinear viscous damper (green $n = 2$ and red $n = 4$ curves) has a wide range of effective use, but loses in efficiency to linear damper in the resonance region of the system.

The hysteretic damper (light blue curve) based on Ishlinsky-Prandtl material has a high efficiency both in the resonance region and beyond. The disadvantage of the hysteretic damper is in decreasing of the ability to reduce the relative movement of car under external forces outside the resonance region of the system.

5 Conclusions

In the presented paper we have considered the hysteretic damper based on the Ishlinsky-Prandtl model. We presented the numerical results for the observable characteristics of the system under consideration such as force transmission function and "force-displacement" transmission function. These results allow to compare the various types of viscous dampers (linear and nonlinear) and hysteretic damper. Based on the obtained numerical results we can formulate the following concluding notes:

- **Linear viscous damper** has a high efficiency in the resonance region of the system, however outside the resonance region damping properties sharply decrease.
- **Nonlinear viscous** damping has a wide range of effective use, but loses in efficiency to linear damper in the resonance region of the system.
- **Hysteretic damper** based on Ishlinsky-Prandtl material **has a high efficiency both in the resonance region and beyond.** The disadvantage of the hysteretic damper is in decreasing of the ability to reduce the relative movement of car under external forces outside the resonance region of the system.

This work is supported by the RFBR grant 16-08-00312-a.

References

1. Lv Qibao, Yao Zhiyuan, Nonlinear Dynam. **79**, (2015) 2325
2. Z. Milovanovic, I. Kovacic, M.J. Brennan, J. Vib. Acoust. **131**, (2009) 054502
3. Z.Q. Lang, S.A. Billings, R. Yue, J. Li, Automatica **43**, (2007) 805
4. Z.K. Peng, G. Meng, Z.Q Lang, W.M. Zhang, F.L. Chu, Int. J. Non-Linear Mech. **47**, (2012) 1073
5. E. Rigaud, J. Perret-Liaudet, J. Sound Vib. **265**, (2003) 289
6. F. Rüdinger, Eng. Struct. **28**, (2006) 1774
7. A.Y.T. Leung, Zhongjin Guo, H.X. Yang, Commun. Nonlinear Sci. Num. Sim. **18**, (2013) 2900
8. M.A. Krasnosel'skii, A.V. Pokrovskii, *Systems with Hysteresis* (Springer Verlag, 1989) 410 pp.

Stochastic collocation-based finite element of structural nonlinear dynamics with application in composite structures

K. Sepahvand[a]

Department of Mechnaical Engineering, Technical University of Munich, 85748 Garching, Germany

Abstract Stochastic analysis of structures having nonlinearity by means of sampling methods leads to expensive cost in term of computational time. In contrast, non-sampling methods based on the spectral representation of uncertainty are very efficient with comparable accurate results. In this paper, the application of spectral methods to nonlinear dynamics of structures with random parameters is investigated. The impact of the parameter randomness on structural responses has been considered. To this end, uncertain parameters and the structure responses are represented using the gPC expansions with unknown deterministic coefficients and random orthogonal polynomial basis. The deterministic finite element model of the structure is used as black-box and it is executed on a set of random collocation points. As the sample structure responses are estimated, a nonlinear optimization process is employed to calculate the unknown coefficients. The method has this main advantage that can be used for complicated nonlinear structural dynamic problems for which the deterministic FEM model has been already developed. Furthermore, it is very time efficient in comparison with sampling methods, as MC simulations. The application of the method is applied to the nonlinear transient analysis of composite beam structures including uncertain quadratic random damping. The results show that the proposed method can capture the large range of uncertainty in input parameters as well as in structural dynamic responses while it is too time-efficient.

1 Introduction

The modeling of real structural systems includes some level of nonlinearity and, as well as , some degrees of uncertainty. The main sources of structural nonlinearity may pre-loaded bilinear springs e.g. bolted flanges with gap, free-play joints, elastomer springs, damping from the polymer materials or due to the aerodynamics of moving objects, large deformation and etc. The predictive capabilities becoming always faster to consider these nonlinearity in simulation but limited by the parameter uncertainties. The uncertainty account for two major sources, i.e. parametric and modeling uncertainties. Former is due to the uncertainty in material and geometry parameters and later due to the lack of information on model, e.g. boundary conditions, mathematical approach. The major challenge is, however, how does these nonlinearity and uncertainty can be quantified and modeled in analytical and numerical simulation.

Even the nonlinear stochastic analysis of such structures with random excitation has been well developed in past decades, predicting nonlinear response under parameter randomness is still a challenging issue [1]. The linear structural behavior including parameter uncertainty has been well developed, cf. [2–5]. On the other side, the most of published works on of stochastic nonlinear dynamics are limited to the application of the sampling-based probabilistic methods, e.g. Monte Carlo (MC), or perturbation methods [6–10]. The methods are straightforward in application, but limited to the small range of uncertainty and very expensive in the term of computational costs. The non-sampling spectral methods, in contrast, are very efficient and exhibit the same level of accuracy sampling methods.The application of the methods is well known in linear structural dynamics [11–16] where random parameters are approximated by spectral expansions to give a complete characterization of the uncertain parameters.

In this paper, the spectral stochastic finite element method (sSFEM) analysis of the nonlinear structural dynamics with parametric uncertainty is developed. This extension will rely in particular type of structural nonlinearity which can be formulated as separate uncertain term. To this end, the theory of spectral representation of uncertain parameters using generalized polynomial chaos (gPC) expansion is presented. The sSFEM formulation of the structural nonlinear dynamics involving uncertain nonlinear terms is developed. The uncertain nonlinear term is approximated by means of a gPC expansion having deterministic nonlinear functions and random orthogonal basis. Similar expansion is used to represent the FE nodal responses with unknown deterministic coefficients. A collocation based FE analysis, then, provides this facility to use any deterministic numerical simulation procedure to estimate these coefficients. The application of the proposed method is practiced on composite beam with polymer core. The primarily experimental modal analysis of the beam sample shows nonlinear behavior due to increasing the force amplitude.

[a] e-mail: k.sepahvand@tum.de

2 Spectral representation of uncertain parameters

Let X be an uncertain parameter with limited variance $\sigma^2 < \infty$ in random space Ω, so that

$$X : \Omega \xrightarrow{f_X(x)} \mathbb{R}$$

The spectral method allows us to represent the parameter by means of truncated expansion such as the generalized polynomial chaos (gPc)as

$$X(\boldsymbol{\xi}) = a_0 + \sum_{i=1}^{N} a_i \Psi_i(\boldsymbol{\xi}) = \sum_{i=o}^{n} a_i \Psi_i(\boldsymbol{\xi}) \tag{1}$$

i which a_i are unknown deterministic coefficients, $f_X(x)$ is the probability density function (PDF), $\boldsymbol{\xi}$ denotes vector of random variables $= \{\xi_1, \xi_3, \ldots, \xi_n\}$, e.g. $\xi_i \in N(0,1)$ and $\Psi_i(\boldsymbol{\xi})$ are random orthogonal polynomials with $\Psi_0 = 1$. The coefficients are estimated based on minimization of the approximation error on entire random space, known as Galerkin projection, or on sample collocation points in the random space. The details of the both procedure has been discussed in [17]. Multiplying the both side of the Eq. (1) by test function $\Psi_k(\boldsymbol{\xi})$ and integrating both side on random space Ω leads to

$$a_i = \frac{\langle X(\boldsymbol{\xi}) \ , \ \Psi_k(\boldsymbol{\xi})\rangle}{\langle \Psi_i(\boldsymbol{\xi})^2\rangle}, \qquad k = 0, 1, \ldots, n \tag{2}$$

3 Nonlinear stochastic finite element analysis

It is assumed nonlinear term of the structure can be separately represented as $\boldsymbol{G}(\dot{\boldsymbol{u}}, \boldsymbol{u})$ in the FE model, i.e.

$$\boldsymbol{M}\ddot{\boldsymbol{u}} + (\boldsymbol{C} + \boldsymbol{H})\,\dot{\boldsymbol{u}} + \boldsymbol{K}\boldsymbol{u} + \boldsymbol{G}(\boldsymbol{u}, \dot{\boldsymbol{u}}) = \boldsymbol{F} \tag{3}$$

in which $\boldsymbol{M}$ and $\boldsymbol{K}$ are linear mass and stiffness matrices, $\boldsymbol{C}$ and $.\boldsymbol{H}$ are viscous and structural damping matrices and $\boldsymbol{F}$ is the external force vector. The nodal displacement vector is denoted as $\boldsymbol{u}$. In the case of any uncertainty in the nonlinear term, the stochastic FE model can be given as

$$\boldsymbol{M}\ddot{\boldsymbol{u}} + (\boldsymbol{C} + \boldsymbol{H})\,\dot{\boldsymbol{u}} + \boldsymbol{K}\boldsymbol{u} + \boldsymbol{G}(\boldsymbol{u}, \dot{\boldsymbol{u}}, \boldsymbol{\xi}) = \boldsymbol{F} \tag{4}$$

where random vector $\boldsymbol{\xi}$ is used to represent the structure uncertainty. Accordingly, the nodal displacements are function of the random vector, i.e. $\boldsymbol{u} = \boldsymbol{u}(t, \boldsymbol{\xi})$. The random nonlinear matrix is represented by the spectral expansion as

$$G(u, \dot{u}, \boldsymbol{\xi}) = g_0(u, \dot{u}) + \sum_{i=1}^{n_1} g_i(u, \dot{u})\Psi_i(\boldsymbol{\xi}) = \boldsymbol{g}\boldsymbol{\Psi}^T \tag{5}$$

The zeroth order term $g_0(u, \dot{u})$ denotes the nominal nonlinear term and the coefficient $g_i(u, \dot{u})$ are deterministic functions. The same orthogonal basis is employed to represent the propagation of the uncertainty in the structural responses. Hence, the spectral representation of the random nonlinear term and the nodal displacements are written as

$$\boldsymbol{u}(t, \boldsymbol{\xi}) = \boldsymbol{U}_0(t) + \sum_{i=1}^{n_2} \boldsymbol{U}_i(t)\Psi_i(\boldsymbol{\xi}) = \boldsymbol{U}\boldsymbol{\Psi}^T \tag{6}$$

The expansions in Eqs. (5) and (6) are substituted in stochastic FE model in Eq. (4), this leads to

$$\boldsymbol{M}\ddot{\boldsymbol{U}}\boldsymbol{\Psi}^T + (\boldsymbol{C} + \boldsymbol{H})\,\dot{\boldsymbol{U}}\boldsymbol{\Psi}^T + \boldsymbol{K}\boldsymbol{U}\boldsymbol{\Psi}^T + \boldsymbol{g}\boldsymbol{\Psi}^T = \boldsymbol{F} \tag{7}$$

This discretized equation is the spectral stochastic FE model (sSFEM) of the structure having random nonlinear term and accordingly yields to an additional error due to the random space discretization among the FE discretization error. This error is given by

$$\begin{aligned}\varepsilon(t, \boldsymbol{\xi}) \ &= \boldsymbol{M}\ddot{\boldsymbol{U}}\boldsymbol{\Psi}^T + (\boldsymbol{C} + \boldsymbol{H})\,\dot{\boldsymbol{U}}\boldsymbol{\Psi}^T + \boldsymbol{K}\boldsymbol{U}\boldsymbol{\Psi}^T \\ &+ \boldsymbol{g}\boldsymbol{\Psi}^T - \boldsymbol{F}\end{aligned} \tag{8}$$

Generally, there are two broad classes of methods that can be used to minimize the above mentioned error: intrusive and non–intrusive methods. The implementation of intrusive methods requires access to the data structures of the FE model or analytical governing equations, whereas in the non–intrusive method the FE model or the governing equations are employed as third party solver or black–box. The choice between these methods depends on the problem under investigation and the information available on the system model. For instance, if the governing equations of the system are linear and available, the intrusive method can be used to generate the deterministic equivalent model. However, the application of the non–intrusive method to the nonlinear equations is easier.

3.1 Non–intrusive sSFEM

The so–called non–intrusive SFEM method looks for a set of deterministic model responses corresponding to some specific points of $\boldsymbol{\xi}$ realizations from which the system response can be approximated. In the concept of stochastic variational, the non–intrusive SFEM method is the same as the collocation method which forces the residual error to be deterministically zero at specific points, i.e. in Eq. (8)

$$\int_{\{\xi_i\}} \varepsilon(t, \boldsymbol{\xi})\ \delta(\xi_i - p_i) f(\xi_i) d\xi_i = 0 \tag{9}$$

In which δ is the delta function and p_i denotes the set of specific collocation points. The method allows the use of third–party commercial or free license deterministic FE codes as a black–box to get the system responses associated with each realization of random vector. These characteristics make the method very attractive for parametric quantification in complex models and industrial applications where the deterministic FE model has been already developed. The key idea is to use deterministic FE codes to perform

DOI: 10.1051/matecconf/20168301009

spatial discretization to arrive at a system of random algebraic equations to employ in the SFEM. Normally, the roots of orthogonal polynomials are used as realizations of the random variables. An optimization process, usually based on least squares techniques, is employed if the number of realizations are more than the number of unknown coefficients in the gPC expansion of the system response.

4 Case study

In this section the application of the proposed method is implemented to perform the uncertainty analysis in stochastic structural nonlinear vibration of composite beam having random quadratic damping, cf. Fig. 1. The damping core has a thickness of 30 μm

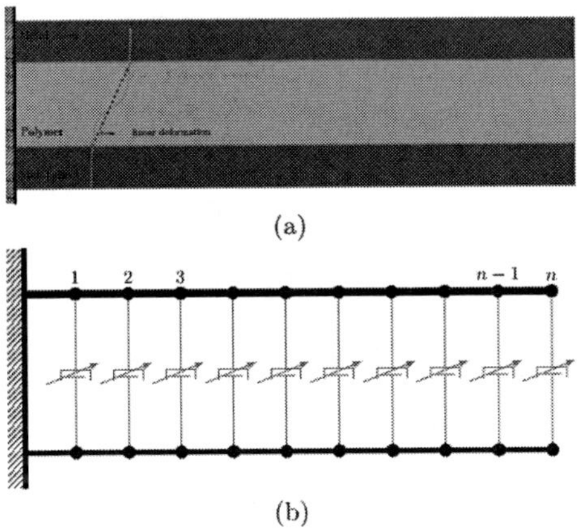

Figure 1. (a) Double elastic composite beam with polymer core, (b) FE model with nonlinear quadratic damping

and the elastic sheet metals have a thickness of 1 mm. The beam length is set to 180 mm. The experimental modal analysis shows that the composite beam possesses a high damping level by increasing the excitation frequency compared to the simple elastic beam without polymer core. as shown in Fig. 2. The initial experimental modal analysis of the sample beam demonstrates also some level of nonlinearity where the frequency response shows amplitude dependency, see Fig. 3. It is shown that increasing the force amplitude, and accordingly the structural responses, will change the frequency responses and the damping level. The general model of such nonlinear damping is given by

$$G(\dot{u}, \xi) = c_s(\xi)\dot{u}|\dot{u}|^{n-1} \qquad (10)$$

in which $n = 0$ denotes the Coulomb friction, $n = 1$ represents the classical linear viscous damping and $n = 2$ gives nonlinear quadratic damping. It is assumed that the damping coefficient c_s has some level of uncertainty which is approximated by the 3rd order gPC expansion having random orthogonal Her-

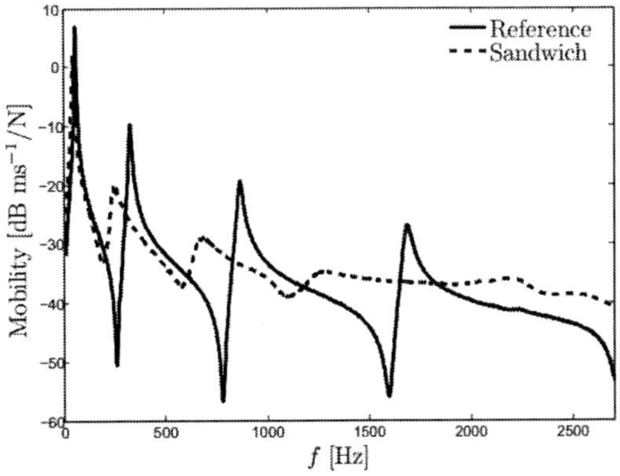

Figure 2. The composite beam with polymer core shows high damping level compared to the elastic beam without damping layer having the same thickness.

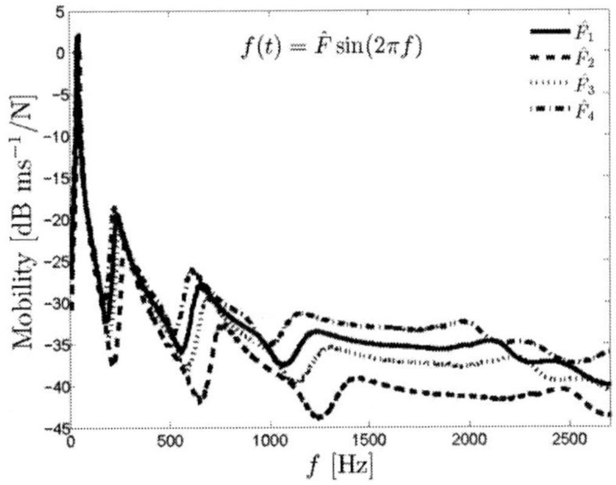

Figure 3. Increasing the force amplitude will change the frequency response and the damping level of the composite beam with polymer core.

mite polynomials, i.e.

$$c_s(\xi) = \sum_{i=0}^{3} a_i H_i(\xi) \qquad (11)$$

It is also assumed that the uncertainty in the damping coefficient can be represented by the lognormal PDF as $LN(\mu, \sigma)$ with $\mu = 1$ and $\sigma = 0.25$. The sSFEM uses a deterministic FE model of the double beam as a black–box where the interlayer nodes are connected to each other by discrete nonlinear damping elements. For each realization of the structural response, the collocation points are produced from the combination of roots of one order higher than the order of the gPC expansion used to approximate the responses. The combinations are selected in such a manner to give high probability. The sample points are passed through the deterministic FE code to realize the tip displacement of the beam. Here a set of 17 collocation points is employed. An optimization process based on the least squares is utilized since the number of collocation points is more than the number of unknown gPC coefficients. The deterministic ($\xi = 0$) responses of the beam tip for the linear and

DOI: 10.1051/matecconf/20168301009

nonlinear damping are shown in Fig. 4. The difference

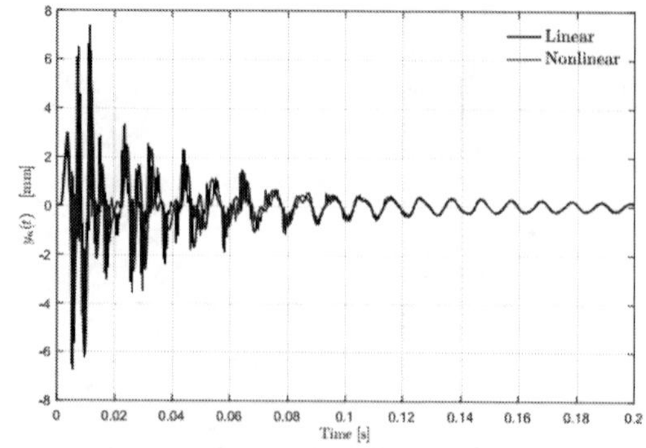

Figure 4. Linear and nonlinear deterministic responses, $\xi = 0$.

between the responses in deterministic case is not considerable, particularly, when the time increases. This is however different when damping uncertainties are considered, as shown in Fig. 5. The bold line denotes

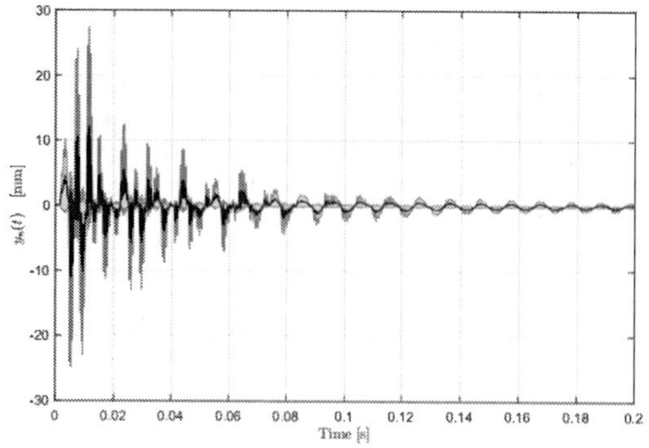

Figure 5. Linear and nonlinear stochastic responses.

the expected value of the response. It is seen that the parameter uncertainty will propagate in stochastic nonlinear responses, shown as gray area in Fig. 5. To investigate the impact of parameter uncertainty, the PDF of the maximum tip displacement is shown in Fig. 6 As shown, the PDF is asymmetry with varying degrees of skewness w.r.t. the mean value shown by the dashed lines. It can be seen that the longer tile ERP is on the right side of the deterministic value, meaning that uncertainties in the damping parameters will increase the level of the displacement.

5 Conclusions

The paper presented the theory and application of spectral stochastic FE modeling nonlinear structural dynamics problems with uncertain parameters. Attention was focused on the application of non–sampling probability method in combination with the stochastic

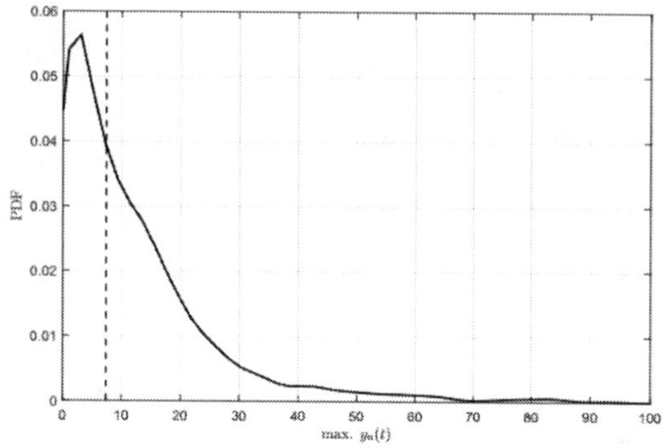

Figure 6. Probability density function (PDF) of the maximum displacement of the beam tip. The dashed line show the position of the deterministic solution.

collocation simulation. The paper aimed to provide a brief review of the uncertainty quantification methods in nonlinear structures using the generalized polynomial chaos expansion. The spectral stochastic FEM formulation for nonlinear structural vibration based on the non–intrusive techniques were developed. The non–intrusive method is useful due to the fact that the available deterministic FE model of the nonlinear structure can be used to investigate the stochastic analysis as black-box. The use of these methods remarkably enhances the computational efficiency. This issue has been pointed out as a significant advantage for numerical simulation of the practical engineering problem. The potential of each technique for numerical simulation of stochastic structural dynamic problems has been discussed on composite beam with polymer core in which the experimental results show some degree of nonlinearity in damping parameters.

References

1. T. L. Paez and J. Red-Horse. Structural dynamics challenge problem: Summary. *Computer Methods in Applied Mechanics and Engineering*, 197(29-32):2660–2665, 2008.
2. R. A. Ibrahim. Structural dynamics with parameter uncertainties. *Applied Mechanics Reviews*, 40(3):309–328, 1987.
3. C. Soize. A nonparametric model of random uncertainties for reduced matrix models in structural dynamics. *Probabilistic Engineering Mechanics*, 15(3):277–294, 2000.
4. M. Hanss. *Applied Fuzzy arithmetic, an introduction with engineering applications*. Springer–Verlag, Heidelberg, 2005.
5. K. Sepahvand, S. Marburg, and H.-J. Hardtke. Stochastic free vibration of orthotropic plates using generalized polynomial chaos expansion. *Journal of Sound and Vibration*, 331:167–179, 2012.
6. W. K. Liu, T. Belytschko, and A. Mani. Probabilistic finite elements for nonlinear structural dy-

namics. *Computer Methods in Applied Mechanics and Engineering*, 56(1):61–81, 1986.

7. A. Der Kiureghian and J.-B. Ke. The stochastic finite element method in structural reliability. *Probabilistic Engineering Mechanics*, 3(2):83–91, 1988.
8. Z. Lei and C. Qiu. A stochastic variational formulation for nonlinear dynamic analysis of structure. *Computer Methods in Applied Mechanics and Engineering*, 190(5-7):597–608, 2000.
9. Y.-G. Zhao and T. Ono. Moment methods for structural reliability. *Structural Safety*, 23(1):47–75, 2001.
10. J. M. Balthazar, P. B. Goncalves, and R. F. Brasil. Uncertainties in nonlinear structural dynamics. *Mathematical Problems in Engineering*, 2008:538725–9, 2008.
11. R. G. Ghanem and P. D. Spanos. *Stochastic finite elements: A spectral approach.* Dover, Mineola, 1991.
12. C. Soize. A comprehensive overview of a non–parametric probabilistic approach of model uncertainties for predictive models in structural dynamics. *Journal of Sound and Vibration*, 288(3):623–652, 2005.
13. K. Sepahvand, S. Marburg, and H.-J. Hardtke. Stochastic structural modal analysis involving uncertain parameters using generalized polynomial chaos expansion. *International Journal of Applied Mechanics*, 3:587–606, 2011.
14. K. Sepahvand, S. Marburg, and H.-J. Hardtke. Stochastic free vibration of orthotropic plates using generalized polynomial chaos expansion. *Journal of Sound and Vibration*, 331(1):167–179, 2012.
15. K. Sepahvand, K. Nabih, and S. Marburg. Dynamical analysis of multibody systems with design uncertainties collocation-based stochastic modeling of uncertain geometric mistuning in bladed rotor. *Procedia IUTAM*, 13:53–62, 2015.
16. K. Sepahvand. Spectral stochastic finite element vibration analysis of fiber-reinforced composites with random fiber orientation. *Composite Structures*, 145:119–128, 2016.
17. K. Sepahvand, S. Marburg, and H.-J. Hardtke. Uncertainty quantification in stochastic systems using polynomial chaos expansion. *International Journal of Applied Mechanics*, 2(2):305–353, 2010.

Non-planar motions of a string vibrating against a smooth unilateral obstacle

Harkirat Singh[1,a], Ashok K. Mandal[1], and Pankaj Wahi[1]

Indian Institute of Technology, Kanpur-208016, India

Abstract. Earlier theoretical studies on string vibrations against smooth unilateral obstacle assumed planar motions. The present paper analyzes the non planar motions of a string vibrating against a smooth unilateral curved obstacle. In particular, we study the effect of obstacle in determining different types of non planar motions of a string. The tension is assumed to be variable along the length due to the stretching of the string, which introduces nonlinear coupling between the perpendicular modes. The system of equations has been discretized by assuming functional form of the modes which satisfies all the geometrical boundary conditions. The mathematical model has been numerically investigated for different initial conditions and chosen value of obstacle parameters. The trajectory of a typical point on the string is oscillating ellipse and it transforms into rotating ellipses with time varying major and minor axis for larger amplitudes.

1 Introduction

Strings are the simplest continuous systems which exhibit vibrations. The applications of strings are quite diverse, for example, in cranes to lift loads and transmit power, in ropeways to suspend heavy weights, in elevators as a support system, in a classical Atwood machine as a classroom demonstration to verify laws of motion, etc. At high school level, the physics behind all these exercises is conceived by treating the string to be inextensible and massless entity. However, in reality the string deforms as well as possess certain mass density which immediately usher in the idea of vibrating strings. The mathematical model for small vibrations of a tightly stretched elastic string proposed by D'Alembert is given by

$$\frac{\partial^2 y}{\partial t^2} = c^2 \frac{\partial^2 y}{\partial x^2}, \qquad c^2 = \frac{T}{\rho}, \tag{1}$$

where "tightly stretched" implies that the tension T is a constant, and ρ denotes the constant mass density per unit length. Thus, dynamics of a string is ingested in a linear partial differential equation. Given initial and boundary conditions, this hyperbolic PDE can be solved exactly by D'Alembert solution, using Fourier transform, or via. separation of variables. Thus, the problem of string vibrations boils down to solving wave equation under appropriate boundary conditions which has been dealt very elegantly by mathematicians over the years. The real complication arises in situations where either the transverse motions are large enough that the assumption of small vibrations no longer holds or the supports are not ideal giving rise to non-trivial boundary conditions. Under these situations, we can not get exact solutions to the string vibrations problem. Interestingly, real practical applications of string vibrations often take us into this realm.

[a] e-mail: `harkirat@iitk.ac.in`

String vibrations are inevitable when we discuss about musical instruments. Indian stringed musical instruments have been subjected to investigation in the works of C.V. Raman[1] and B.C. Deva[2]. The peculiar feature of the Indian stringed musical instruments is the presence of a finite-sized curved bridge. The string wraps and unwraps around the curved bridge, therefore changing the effective length of the vibrating string. The effect of curved bridge is prominent in determining the tonal quality of the instrument. Theoretically, the problem of vibration of string against a rigid wall was mathematically formulated by Amerio[3]. Besides this, sundry theoretical studies have been reported in the literature[4–6]. However, all these studies have been restricted to planar motions of vibrating string. In the present paper, we have studied the non-planar motions of string vibrating against a smooth unilateral obstacle. We have further explored the effect of the obstacle on the dynamics of the string. We have developed a mathematical model taking into consideration the holonomic constraint and derive the governing equation using Hamilton's principle. In addition to this, we analyze the dynamics of a string, explore the possibility of different types of motion and numerically investigate the results.

2 Mathematical Model

A schematic representation of the physical system under consideration is shown in figure 1. It comprises one ideal(no bending stiffness) string which vibrates against a smooth obstacle at one of the ends. The string starts from O which is the origin of our system and it is tied to the support at other end which is at distance L. The obstacle is assumed to be rigid and fixed in our model. The mathematical form of the

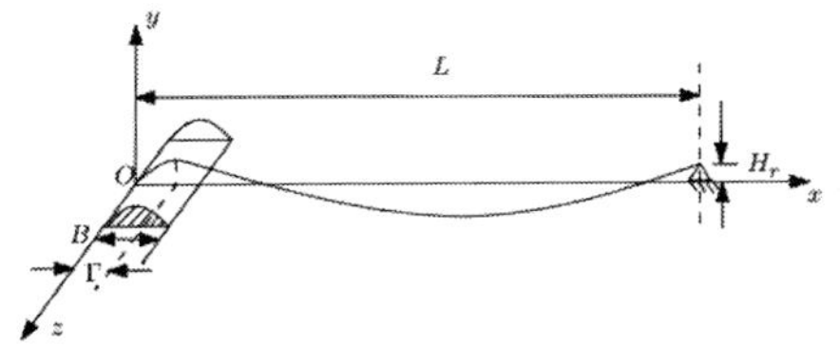

Fig. 1. Schematic representation of the system under consideration

bridge is given by

$$Y_B(x) = A_P x(B - x), \tag{2}$$

where A_P is a constant (related to the curvature of the bridge), and B is the length of the bridge. The shape of the obstacle is assumed such that it is curved only along one direction and the cross section in the plane parallel to the string is parabolic. The string has uniform density ρ per unit length, however, the tension varies along the length due to stretching of the string. We further assume that the string wraps perfectly around the obstacle, and thus, remains tangent to the obstacle at the point of separation. The Lagrangian of the system can be written as

$$\int_0^L \left(\frac{1}{2}\rho(y_t^2 + z_t^2) - \frac{1}{2}T_0(y_x^2 + z_x^2) + \frac{EA}{8}(y_x^2 + z_x^2)^2\right)dx. \tag{3}$$

The wrapped portion of the string is constrained to move on the surface of the obstacle. Now, introducing the non-dimensional parameters as

$$\bar{x} = \frac{x}{L},\ \bar{y}(\bar{x},\bar{\tau}) = \frac{y(x,t)}{h},\ \bar{z}(\bar{x},\bar{\tau}) = \frac{z(x,t)}{h},\ \gamma = \frac{\Gamma(t)}{L},$$
$$b = \frac{B}{L},\ a = A_p\frac{L^2}{h},\ \tau = t\sqrt{\frac{T}{\rho L^2}},\ \sigma = \frac{EAh^2L^2}{2L^2T_0}, \tag{4}$$

where h is the height of the bridge $(A_p B^2/4)$ and σ is the non-dimensional stretching nonlinearity parameter of the system. Using the Hamilton's principle, the governing equations for the system can be written as

$$\bar{y}(\bar{x},\tau) = a\bar{x}(b - \bar{x}) \qquad 0 \le x \le \gamma^-, \tag{5}$$

and for $\gamma^+ \le x \le L$,

$$-\frac{\partial^2\bar{y}(\bar{x},\tau)}{\partial\tau^2} + \frac{\partial^2\bar{y}(\bar{x},\tau)}{\partial\bar{x}^2} + \sigma\frac{\partial}{\partial\bar{x}}\ \bar{y}_{\bar{x}}(\bar{y}_{\bar{x}}^2 + \bar{z}_{\bar{x}}^2) = 0. \tag{6}$$

Since, there is no constraint in the z-direction, we can obtain a single PDE for the entire length, i.e, $0 \le x \le L$, which is given by

$$-\frac{\partial^2\bar{z}(\bar{x},\tau)}{\partial\tau^2} + \frac{\partial^2\bar{z}(\bar{x},\tau)}{\partial\bar{x}^2} + \sigma\frac{\partial}{\partial\bar{x}}\ \bar{z}_{\bar{x}}(\bar{y}_{\bar{x}}^2 + \bar{z}_{\bar{x}}^2) = 0. \tag{7}$$

The boundary conditions in terms of non-dimensional variables are given as

$$\bar{y}(\gamma,\tau) = a\gamma(b-\gamma),\ \bar{y}(1,\tau) = 0,\ \bar{z}(0,\tau) = 0,\ \bar{z}(1,\tau) = 0, \tag{8}$$

and

$$\frac{\partial\bar{y}(\bar{x},\tau)}{\partial\bar{x}}|_{\bar{x}=\gamma} = a(b - 2\gamma). \tag{9}$$

3 Model order reduction

The physical system has been reduced to non-linear coupled partial differential equations. The dynamics of the string can be fully understood by finding solutions of these PDEs. However, there is no standard approach to find the closed form exact solutions. In this section, we use Galerkin projection approach to discretize the system by assuming functional form of the approximating functions which satisfies all the geometrical boundary conditions. Moreover, in order to write the expressions succinctly, we will write x instead of $\bar{x}$, t in place of τ and drop the explicit dependence of y on the variables x and τ. Restricting our analysis to one mode, the assumed functional form for y-direction motion is given by

$$y = \tfrac{a\gamma(b-\gamma)}{1-\gamma}(1 - x) + \beta(t)\sin\left(\tfrac{\pi(x-\gamma)}{1-\gamma}\right)$$
$$for\ \gamma_1^+ \le x \le 1, \tag{10}$$

and for z-direction motion, we assume

$$z = \alpha(t)\sin(\pi x) \quad for\ 0 \le x \le 1. \tag{11}$$

Substituting (10)-(11) in (6), we obtain residual term denotes by R_y. Again, we substitute (5)-(11) followed by (10)-(11) in (7) to get residual forms R_{z1} and R_{z2}, respectively. To minimize the error, we use Galerkin projection in the following manner

$$\int_0^{\gamma_1} R_{z1}\sin(\pi x) + \int_{\gamma_1}^1 R_{z2}\sin(\pi x) = 0, \tag{12}$$

$$\int_{\gamma_1}^1 R_y \sin\left(\frac{\pi(x-\gamma_1)}{1-\gamma_1}\right) = 0. \tag{13}$$

Using the assumed functional forms, slope continuity conditions at the point of contact given by

$$\frac{\partial y}{\partial x}|_{x=\gamma_-} = a(b - 2\gamma), \tag{14}$$

$$\tag{15}$$

which leads to

$$\frac{\partial y}{\partial x}|_{x=\gamma_+} = -\frac{a\gamma(b-\gamma)}{1-\gamma} + \frac{\beta(t)\pi}{1-\gamma}. \tag{16}$$

Equating the above two equations, we get

$$-a\gamma(b-\gamma) + \beta(t)\pi = a(1-\gamma)(b-2\gamma), \tag{17}$$

which can be simplifies as

$$a\gamma^2 - 2a\gamma + ab - \pi\beta(t) = 0. \tag{18}$$

The above quadratic equation in terms of γ suggests that there are two roots possible. However the only relevant root is

$$\gamma = 1 - \sqrt{1 - \left(b - \frac{\pi\beta(t)}{a}\right)}. \tag{19}$$

The above equation holds for all time instants, differentiating leads to

$$\dot{\gamma} = -\frac{\pi\dot{\beta}(t)}{2a(1-\gamma)} \tag{20}$$

DOI: 10.1051/matecconf/20168301010

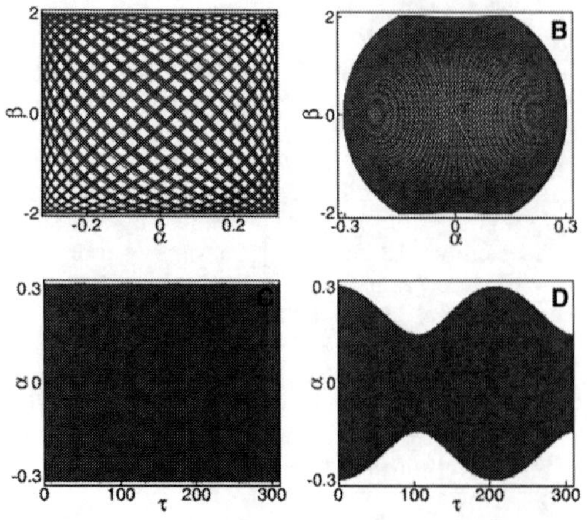

Fig. 2. Representation of stretching nonlinearity. Fig(A) shows the trajectory of a typical point in the absence of stretching nonlinearity whereas fig(B) shows the trajectory of the same point in the presence of stretching nonlinearity under same set of initial conditions($\alpha(0) = 0$, $\dot{\alpha}(0) = 1$, $\beta(0) = 2$ and $\dot{\beta}(0) = 0$). Fig(C) and fig(D) shows the α variation corresponding to fig(A) and fig(B), respectively.

and

$$\ddot{\gamma} = \frac{\dot{\gamma}^2}{1-\gamma} - \frac{\pi\ddot{\beta}(t)}{2a(1-\gamma)}. \tag{21}$$

Thus, we have obtained the point of contact in terms of modal functions. These expressions can be used to obtain the ODEs, however, ODEs being lengthy are omitted here.

4 Numerical Simulations

In this section, we numerically investigate the ODEs, obtained in section 3, under different set of initial conditions. The obstacle parameters are pertinent to the bridge of Indian classical stringed instruments. In particular, we choose $b = 0.06$, $\sigma = 0.001$ and $a = 1111$, such that $\frac{ab^2}{4}$ is 1, for our numerical simulations. Any non-zero initial condition in both α and β simultaneously results in non-planar motions. The effect of stretching non-linearity on the dynamics of string have also been observed numerically in figure 2. It is clear from figure 2 that the modal interaction between the perpendicular modes occur only in the presence of stretching nonlinearity. In addition to this, we observe that nature of trajectory of a typical point on the string slowly transforms as we increase the amplitude of oscillations. We explore the different types of non-planar motions in the subsequent discussion.

4.1 Oscillating trajectories

It has been numerically found that for small values of β, the trajectories are oscillating ellipses with time varying major and minor axis. The exchange of energy between the perpendicular modes is quite significant and its strength increases as we increase the amplitude of vibrations. The trajectory of a point and the variation of α and β is shown in figure 3A. These types are also observed when the initial disturbances are given in terms of velocity and displacement in z and y direction, respectively. However, these types of non-planar motions are possible only below certain value of β.

4.2 Rotating trajectories

Numerical simulations also divulge that the larger values of β engender rotating elliptical trajectories with varying major and minor axis. Thus, any non-zero initial condition in terms of α and β give rise to rotating trajectories above certain value of β. In this case, the extent of energy exchange is much higher as compared to the previous case. We further observe that the maximum value attains by α variation remains same irrespective of the magnitude of the perturbation. The trajectory of a point and the variation of α and β is shown in figure 3B. This transformation of

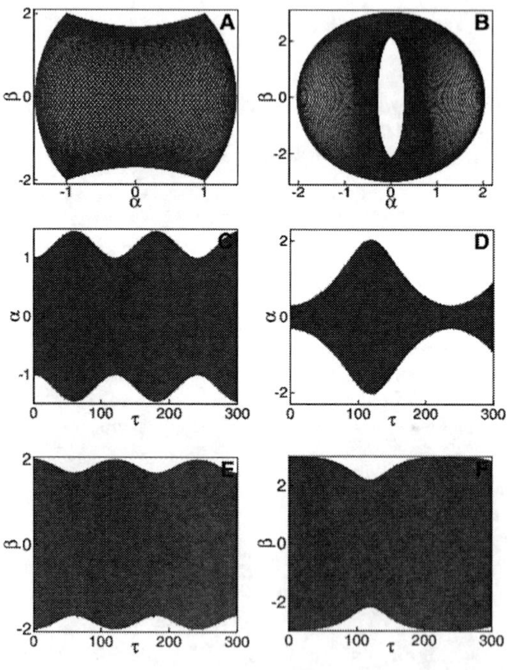

Fig. 3. Some representations of non-planar motions for different set of initial conditions. Fig(A) shows the oscillating elliptical trajectory corresponds to $\alpha(0) = 1$ and $\beta(0) = 2$, fig(B) shows the rotating elliptical trajectory corresponds to $\dot{\alpha}(0) = 1$ and $\beta(0) = 3$. Fig(C,E) and fig(D,F) shows the α and β variation with respect to non-dimensional time corresponding to fig(A) and fig(B), respectively.

trajectories denotes the onset of unstable regime for planar motions.

5 Conclusion

It is evident from our analysis that assumption of planar motion does not capture the holistic dynamics of the string. Moreover, in almost all practical scenarios, the motion of the string is undoubtedly non planar. For example, in stringed musical instrument, the string vibrates erratically in space and amplitude of motion is large enough to include the nonlinear effects due to variable tension. Thus, our model is very relevant to such cases and it aptly captures the underlying physics behind 3-dimensional motions of a string vibrating against a unilateral obstacle. It has been found that the oscillating elliptical trajectories transforms into rotating elliptical trajectories for larger amplitudes. The qualitative change in dynamics of string marks the onset of unstable regime which will be analyzed in detail in our future study. Moreover, the obstacle is assumed to be curved only along one direction, taking complex geometry and considering the effect of friction between the string and obstacle will indeed capture the more realistic dynamics.

References

1. C.V. Raman, On some indian stringed instruments. Proc.Indian Assoc.Cultiv.Sci., **7**(1921), 29-33.
2. B.C. Deva, Musical instruments of india: Their history and development. 2nd ed. (Munshiram Manoharlal), New Delhi, India, (1987).
3. L. Amerio. Continuous solutions of the problem of a string vibrating against an obstacle. Rend.Sem.Mat.Univ.Padova, **59**:(1978) 67-96.
4. R. Burridge, J. Kappraff, and C. Morshedi. The sitar string, a vibrating string with a one-sided inelastic constraint. SIAM Journal of Applied Mathematics, **42(6)**:(1982) 1231- 1251.
5. C. P. Vyasarayani, S. Birkett, and J. McPhee. Modeling the dynamics of a vibrating string with a finite distributed unilateral constraint: Application to the sitar. Journal of the Acoustical Society of America, **125(6)**:(2009) 3673-3682.
6. A.K. Mandal and P. Wahi. Natural frequencies, modeshapes and modal interactions for strings vibrating against an obstacle: Relevance to sitar and veena. Journal of Sound and Vibration, **338**:(2015) 42-59.

MATEC Web of Conferences **83**, 02001 (2016) DOI: 10.1051/matecconf/20168302001
CSNDD 2016

Energy harvesting in a delayed and excited Duffing harvester device

Z. Ghouli[1], M. Hamdi[2], F. Lakrad[1], M. Belhaq[1]

[1]Laboratory of Renewable Energy and Structural Dynamics, University Hassan II-Casablanca, Morocco
[2]University Mohammed I Oujda, FST-Al Hoceima, Morocco

Abstract. We explore quasi-periodic (QP) vibration-based energy harvesting (EH) in a forced nonlinear oscillator under time delay. The energy harvesting system consists in a delayed Duffing oscillator subject to harmonic excitation coupled to an electric circuit through an electromechanical coupling mechanism. We consider the case of primary resonance for which the frequency of the excitation is near the natural frequency of the oscillator. Application of the double-step perturbation method enables the approximation of the amplitude of the QP vibrations used to extract the average powers from the harvester device. Results shown that for a small value of delay amplitude, the periodic vibration-based EH can be extracted in a narrow range near the resonance, while QP vibration-based EH can be extracted over broad ranges of excitation. For relatively increasing value of the delay amplitude, only QP vibration-based EH can be extracted.

1 Introduction

To improve the EH capability, the limitation of the linear attachment has been overcome by using nonlinearities, as for example, a cubic nonlinear component. This essential nonlinearity extends the bandwidth of the harvester over broad range of excitation frequencies, either in the case where the harvesting system is designed as monostable with hardening characteristic [1] or in the case of bistablity; see for instance [2] and references therein. However, the nonlinear attachment is no longer ensured, especially when the response suffers jump phenomena near the resonance[3].

To overcome this problem, the idea of using QP vibrations is proposed. Indeed, it was reported recently that introducing a modulated delay amplitude in a pure self-excited oscillator coupled to an electromagnetic energy harvesting system produces large amplitude QP vibrations performing in broader range of parameters [4,5].

The purpose of this work is to enhance the energy harvesting performance in a harvester system with Duffing-type nonlinear attachment and subject to harmonic excitation using time delay. The energy harvester system consists in a delayed lumped-parameter Duffing oscillator coupled to an electrical circuit through an electromagnetic coupling mechanism. Special attention is paid to the effect of different delay and coupling parameters on the energy harvesting performance taking advantage of large-amplitude QP vibrations induced by time delay.

in Section 2 we present the electromagnetic VEH system and derive the periodic response as well as the harvested power near the primary resonance using the multiple scales method. Section 3 is devoted to the investigation of the QP response and the QP vibration-based EH. The conclusion is given in the last Section.

2 Model description and periodic energy harvesting

Consider an energy harvester device under time delay in the position such that the governing system of equations can be written in the dimensionless form as

$$\ddot{x}(t) + \delta\dot{x}(t) + \omega_0^2 x(t) + \gamma x(t)^3 - \chi v(t) = \alpha x(t-\tau) + f\cos(\lambda t) \quad (1)$$

$$\dot{v}(t) + \beta v(t) + \kappa\dot{x}(t) = 0 \quad (2)$$

where $x(t)$ is the relative displacement of the rigid mass m, $v(t)$ is the voltage across the load resistance, δ is the mechanical damping ratio, γ is the stiffness parameter, χ is the piezoelectric coupling term in the mechanical equation, κ is the piezoelectric coupling term in the electrical circuit equation, β is the reciprocal of the time constant of the electrical circuit, α and τ are, respectively, the gain and time delay, while f and λ are, respectively, the amplitude and the frequency of the excitation.

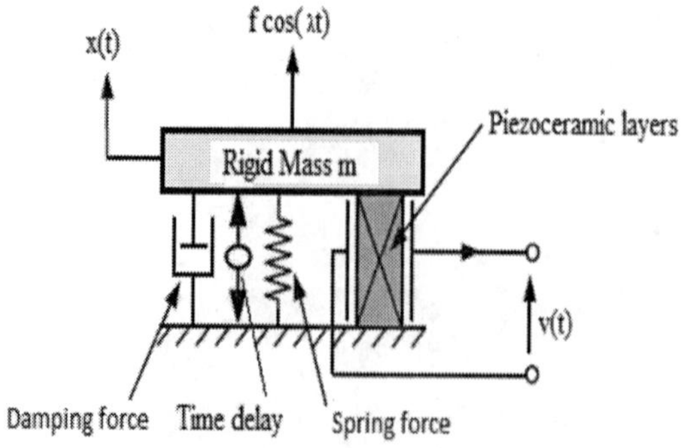

Fig. 1. Schematic of the energy harvesting system

To study the response of the system near the primary resonance, we introduce the resonance condition $\lambda = \omega_0 + \sigma$ where σ is a detuning parameter and we use the multiple scales technique [6] by introducing a bookkeeping parameter ϵ such that parameters can be scaled as $\delta = \epsilon\tilde{\delta}, \gamma = \epsilon\tilde{\gamma}, \chi = \epsilon\tilde{\chi}, \alpha = \epsilon\tilde{\alpha}, f = \epsilon\tilde{f}, \sigma = \epsilon^2\tilde{\sigma}$. Thus, Eqs. (1) and (2) take the form

$$\ddot{x}(t)+\omega_0^2 x = \epsilon[-\tilde{\delta}\dot{x}(t)-\tilde{\gamma}x(t)^3+\tilde{\chi}v(t)+\tilde{\alpha}x(t-\tau)+\tilde{f}\cos(\lambda t)] \tag{3}$$

$$\dot{v}(t) + \beta v(t) + \kappa\dot{x}(t) = 0 \tag{4}$$

Applying the multiple scale technique, a solution to Eqs. (3) and (4) can be sought in the forme

$$x(t) = x_0(T_0, T_1, T_2)+\epsilon x_1(T_0, T_1, T_2)+\epsilon^2 x_2(T_0, T_1, T_2)+O(\epsilon^3) \tag{5}$$

$$v(t) = v_0(T_0, T_1, T_2)+\epsilon v_1(T_0, T_1, T_2)+\epsilon^2 v_2(T_0, T_1, T_2)+O(\epsilon^3) \tag{6}$$

where $T_0 = t, T_1 = \epsilon t$ and $T_2 = \epsilon^2 t$. Substituting (5) and (6) into (3) and (4), extracting systems of equation at different order of ϵ and eliminating secular terms as usual, we obtain up to the first order the modulation equations

$$\begin{cases} \dfrac{da}{dt} = \dfrac{C_1}{2\omega_0}a + \dfrac{f}{2\omega_0}\sin(\varphi) \\ a\dfrac{d\varphi}{dt} = \dfrac{C_2}{2\omega_0}a - \dfrac{C_3}{2\omega_0}a^3 + \dfrac{f}{2\omega_0}\cos(\varphi) \end{cases} \tag{7}$$

where a and θ are the amplitude and the phase, and $C_1 = -\delta\omega_0 - \alpha\sin(\omega_0\tau) - \frac{\kappa\beta\chi\omega_0}{(\beta^2+\omega_0^2)}$, $C_2 = 2\omega_0\sigma + \alpha\cos(\omega_0\tau) - \frac{\kappa\chi\omega_0^2}{(\beta^2+\omega_0^2)}$, $C_3 = \frac{3\gamma}{4}$ and $\varphi = \tilde{\sigma}T_2 - \theta$.
The solution to the first order can be written as $x_0(T_0, T_1, T_2) = a\cos(\omega_0 t+\theta)$ and $v_0(T_0, T_1, T_2) = V\cos(\omega_0 t+\theta+\arctan\frac{\beta}{\omega_0})$ and the relation between the voltage amplitude V and the vibration amplitude a is given by

$$V = \frac{\kappa\omega_0}{\sqrt{\beta^2 + \omega_0^2}}a \tag{8}$$

The steady state response of system (7) are determined by setting $\frac{da}{dt} = \frac{d\varphi}{dt} = 0$. We obtain the following sixth-order algebraic equation in a

$$C_3^2a^6 - 2C_2C_3a^4 + (C_1^2 + C_2^2)a^2 - f^2 = 0 \tag{9}$$

An expression for the average power is obtained by integrating the dimensionless form of the instantaneous power $P(t) = \beta v(t)^2$ over the period of the excitation T. This is given by

$$P_{av} = \frac{1}{T}\int_0^T \beta v^2 dt \tag{10}$$

where $T = \frac{2\pi}{\lambda}$. Then, the average power reads

$$P_{av} = \frac{1}{2}\left(\frac{\beta\kappa^2\omega_0^2}{\beta^2 + \omega_0^2}\right)a^2 \tag{11}$$

where a is obtained from Eq. (9).

3 Quasi-periodic energy harvesting

In this Section, we investigate the QP response as well as the QP vibration-based energy harvesting using the second-step perturbation method [7]. Up to the second order the modulation equations are given by

$$\begin{cases} \dfrac{da}{dt} = S_1a - 3S_3a^2sin(\varphi) + S_4a^3 + \beta_1 sin(\varphi) + \beta_2 cos(\varphi) \\ a\dfrac{d\varphi}{dt} = S_2a - S_3a^2cos(\varphi) + S_5a^3 + S_6a^5 + \beta_1 cos(\varphi) - \\ \beta_2 sin(\varphi) \end{cases} \tag{12}$$

where S_1, S_2, S_3, S_4, S_5, S_6, β_1, and β_2 are given in Appendix. We transform the polar form (12) to the following Cartesian system using the variable change $u = a\cos\varphi$ and $w = -a\sin\varphi$

$$\begin{cases} \dfrac{du}{dt} = S_2w + \beta_2 + \eta\{S_1u + 2S_3uw + (S_4u + S_5w)(u^2 + w^2) + \\ S_6(u^2 + w^2)^2w\} \\ \dfrac{dw}{dt} = -S_2u - \beta_1 + \eta\{S_1w + S_3w^2 + S_3(u^2 + w^2) + \\ (S_4u - S_5w)(u^2 + w^2) - S_6(u^2 + w^2)^2w\} \end{cases} \tag{13}$$

Applying the method of multiple scales on the slow flow (13), one obtains the following autonomous slow-slow flow system

$$\begin{cases} \dfrac{dR}{dt} = S_4R^3 + (S_1 - \alpha_2S_3 + 2\alpha_1^2S_4 + 2\alpha_2^2S_4)R \\ R\dfrac{d\psi}{dt} = S_6R^5 + (S_5 + 6(\alpha_1^2 + \alpha_2^2)S_6)R^3 + \\ ((2\alpha_1^2 + 2\alpha_2^2)S_5 + (3\alpha_1^4 + 3\alpha_2^4 + 6\alpha_1^2\alpha_2^2)S_6)R \end{cases} \tag{14}$$

where $\alpha_1 = \frac{\beta_1}{S_2}, \alpha_2 = \frac{\beta_2}{S_2}$ and R and ψ are, respectively, the amplitude and the phase of the slow flow limit cycle.

Equilibria of this slow-slow flow system (14) determine the QP solutions of the original system (1),(2). The nontrivial equilibrium is obtained by setting $\frac{dR}{dt} = 0$ and given by

$$R = \sqrt{\frac{-S_1 + \alpha_2S_3 - 2\alpha_1^2S_4 - 2\alpha_2^2S_4}{S_4}} \tag{15}$$

Thus, the approximate periodic solution of the slow flow (13) is given by

$$u(t) = Rcos(\varphi t) - \alpha_1 \tag{16}$$

$$w(t) = -Rsin(\varphi t) - \alpha_2 \tag{17}$$

The amplitude $a(t)$ of the QP oscillations reads

$$a(t) = \sqrt{[R^2 + \alpha_1^2 + \alpha_2^2] - [2\alpha_1Rcos(\varphi t) - 2\alpha_2Rsin(\varphi t)]} \tag{18}$$

and the envelope of the QP modulation is delimited by a_{min} and a_{max} given by

$$a_{min} = min\{\sqrt{[R^2 + \alpha_1^2 + \alpha_2^2] \pm [2\alpha_1R \pm 2\alpha_2R]}\} \tag{19}$$

$$a_{max} = max\{\sqrt{[R^2 + \alpha_1^2 + \alpha_2^2] \pm [2\alpha_1 R \pm 2\alpha_2 R]}\} \quad (20)$$

The average QP power can be written as

$$P_{avQP} = \frac{1}{2}\left(\frac{\beta\kappa^2\nu^2}{\beta^2 + \nu^2}\right)a^2 \quad (21)$$

where now a is derived from Eqs. (19) and (20) and ν represents the frequency of slow flow limit cycle corresponding to the frequency of the QP modulation.

Figures 2 and 3 show, respectively, the variation of the amplitude of the periodic and QP responses and the output average power amplitude (P_{av}, P_{avQP}) versus the external frequency λ for fixed values of parameters and for $\alpha = 0.2$ (Fig. (2)) and $\alpha = 0.4$ (Fig. (3)). The periodic response is given by (9) and the QP modulation envelope is obtained from Eqs. (19) and (20). Also, the average power for the periodic response is given by Eq. (11), while that for the QP response is deduced from (21). The solid lines represent analytical approximations, dashed lines correspond to unstable solution, while circles indicate numerical simulations. Inset in the figures are shown time histories of the amplitudes and the average power responses. Figure 2 shows that for a small value of delay amplitude $\alpha = 0.2$, the periodic vibration-based energy harvesting can be extracted in a narrow range near the resonance (solid line in the power response), thereby limiting the bandwidth of the harvester. Instead, QP vibration-based energy harvesting can be extracted over broad range of excitation frequencies far from the resonance, as depicted inset in Fig. 2b.

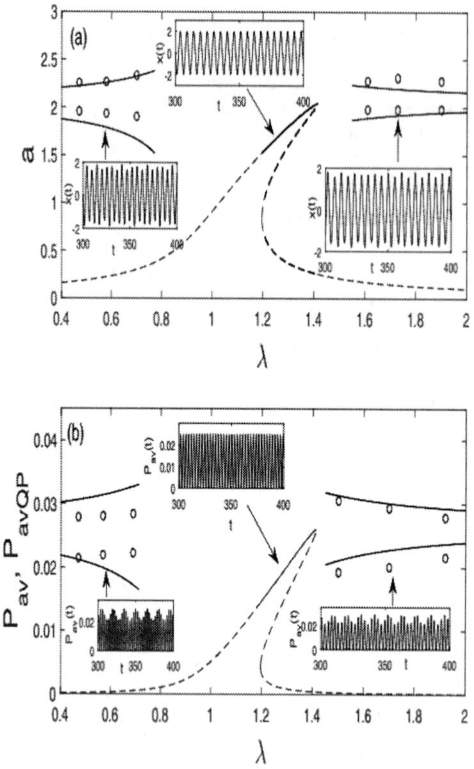

Fig. 2. Vibration and powers amplitudes vs λ for $f = 0.2, \chi = 0.05, \kappa = 0.5$ and $\alpha = 0.2$. Analytical prediction (solid lines for stable and dashed line for unstable) and numerical simulation (circles).

Figure 3 indicates that for relatively increased value of the delay amplitude, $\alpha = 0.4$, periodic vibrations turn to unstable, whereas the QP-based energy harvesting can be extracted away from the resonance with improved performance, as shown inset in Fig. 3b by time histories of the average power responses.

4 Conclusion

The EH performance in a delayed forced Duffing oscillator coupled to an piezoelectric harvesting device is examined near the primary resonance in the case of a mono-stable oscillator. The method of multiple scales is performed twice to obtain the slow-slow flows of the EH system. The periodic and QP solutions as well as the corresponding power amplitudes are obtained and the influence of different system parameters on the EH performance is reported. It was shown that for a small value of delay amplitude, the periodic vibration-based EH can be extracted in a narrow range near the resonance, while QP vibration-based EH can be extracted over broad ranges of excitation away from the resonance. For relatively increasing value of the delay amplitude, EH can be extracted only from QP vibrations over broad bands of excitation frequencies.

5 Appendix

$$S_1 = -\frac{\delta}{2} - \frac{1}{2}\alpha sin(\tau) - \frac{1}{4}\alpha^2 cos(\tau) sin(\tau) - \frac{\chi\kappa\beta}{2(\beta^2 + \omega_0^2)}$$
$$-\frac{\chi\kappa\alpha(\beta cos(\tau) - \omega_0 sin(\tau))}{8(\beta^2 + \omega_0^2)} + \frac{\chi^2\kappa^2\beta\omega_0}{4[(\beta^2 - \omega_0^2)^2 + 4\beta^2\omega_0^2]}$$

$$S_2 = \lambda - \omega_0 + \frac{\delta^2}{8} + \frac{1}{2}\alpha cos(\tau) + \frac{1}{8}\alpha^2 cos^2(\tau) - \frac{1}{8}\alpha^2 sin^2(\tau)$$
$$-\frac{\chi\kappa\omega_0}{2(\beta^2 + \omega_0^2)} - \frac{\chi\kappa\alpha(\beta sin(\tau) + \omega_0 cos(\tau))}{8(\beta^2 + \omega_0^2)} - \frac{\chi^2\kappa^2(\beta^2 - \omega_0^2)}{4[(\beta^2 - \omega_0^2)^2 + 4\beta^2\omega_0^2]}$$

$$S_3 = \frac{3}{8}\gamma f$$

$$S_4 = \frac{3}{16}\gamma\delta + \frac{3}{8}\alpha\gamma sin(\tau) + \frac{3\gamma\kappa\beta}{8(\beta^2 + \omega_0^2)} + \frac{3\gamma\kappa\beta\chi}{16(\beta^2 + \omega_0^2)}$$

$$S_5 = -\frac{3}{8}\gamma - \frac{3}{16}\alpha\gamma cos(\tau) + \frac{3\gamma\kappa\omega_0}{8(\beta^2 + \omega_0^2)} - \frac{3\gamma\kappa\omega_0}{16(\beta^2 + \omega_0^2)}$$

$$S_6 = \frac{15}{256}\gamma^2$$

$$\beta_1 = \frac{1}{2}f + \frac{1}{8}\alpha f cos(\tau) - \frac{f\chi\kappa\omega_0}{8(\beta^2 + \omega_0^2)}$$

$$\beta_2 = \frac{1}{8}\delta f - \frac{1}{8}\alpha f sin(\tau) - \frac{f\chi\kappa\beta}{8(\beta^2 + \omega_0^2)}$$

 DOI: 10.1051/matecconf/20168302001

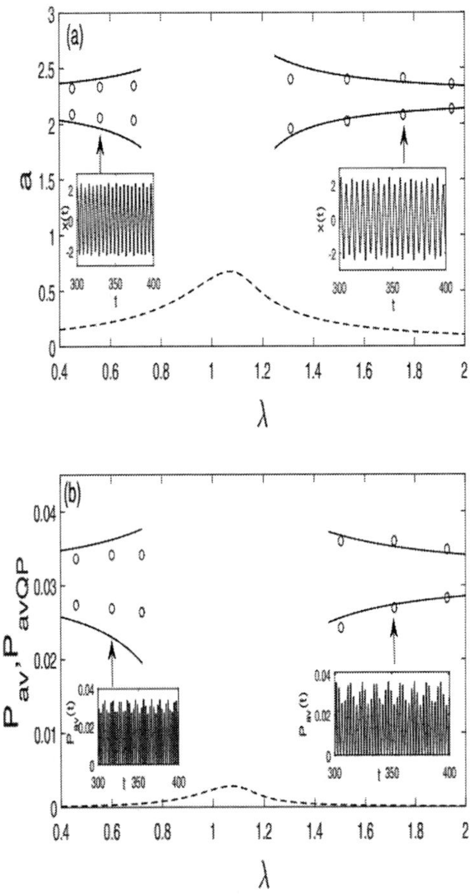

Fig. 3. Vibration and power amplitudes vs λ for $f = 0.2, \chi = 0.05, \kappa = 0.5$ and $\alpha = 0.4$. Analytical prediction (solid lines for stable and dashed line for unstable) and numerical simulation (circles).

References

1. Mann, B.P. Sims, N.D.: Energy harvesting from the nonlinear oscillations of magnetic levitation. J. Sound Vib. **319**, 515-530 (2009)
2. Daqaq, M.: Transduction of a bistable inductive generator driven by white and exponentially correlated gaussian noise. J. Sound Vib. **330**, 2554-2564 (2011)
3. Quinn, D.D., Triplett, A.L., Vakakis, A.F., Bergman, L.A.: Energy harvesting from impulsive loads using intestinal essential nonlinearities. J. Vib. Acoust. **133**, 011004 (2011)
4. Hamdi, M., Belhaq, M.: Quasi-periodic vibrations in a delayed van der Pol oscillator with time-periodic delay amplitude. J. Vib. Control (2015). doi: 10.1177/1077546315597821
5. Belhaq, M., Hamdi, M.: Energy harversting from quasi-periodic vibrations. Nonlinear Dyn. DOI 10.1007/s11071-016-2668-6 (2016)
6. Nayfeh, A.H., Mook, D.T.: Nonlinear Oscillations. Wiley, New York (1979)
7. Belhaq, M., Houssni, M.: Quasi-periodic oscillations, chaos and suppression of chaos in a nonlinear oscillator driven by parametric and external excitations. Nonlinear Dyn. **18**, 1-24 (1999)

Energy harvesting of nonlinear damping system under time delayed feedback gain

A. Bichri[1], I. Kirrou[2,3,a], M. Belhaq[2]

[1] Faculty of Science and Technology-Errachidia, University Moulay Ismaïl, Morocco
[2] Laboratory of Renewable Energy and Structural Dynamics, University Hassan II-Casablanca, Morocco
[3] Laboratory of Engineering Science, ESTEM Research Center, Casablanca, Morocco

Abstract. This paper presents the application of delayed feedback velocity for optimizing the harvested power in cubic nonlinear damper system. We consider a harvester consisting of a nonlinear single degree of freedom system (spring-masse-damper) subjected to a base excitation near the primary resonance. Analytical investigation using the multiple scales method is performed to obtain approximation of the amplitude response. This amplitude can be used to extract the average power. Results show that for appropriate values of the feedback gain, energy harvesting is more efficient at resonance compared to the cubic nonlinear damper system without time delay.

1 Introduction

This paper focuses on energy harvesting (EH) performance extracted from vibrations provided by ambient energy as, for instance, wind induced vibrations and sailing ships oscillations. To increase the frequency range where EH operates, various models have been suggested [2–4].

In the case where the harvester operates within the linear regime, the maximum power is generated when the system is excited at resonance [5,6]. To overcome this limitation, different alternative approaches have been explored by introducing nonlinear components into the system. In some systems [7,8] the nonlinear stiffness is accompanied by nonlinear damping.

On the other hand, to enhance the energy harvesting performance a new concept based on delayed feedback control has been performed. A recent work reported on this issue considered a delayed van der Pol oscillator with modulated delay amplitude [4]. It was shown that modulating the delay amplitude may produce large amplitude quasi-periodic vibrations in self-excited oscillators.

In the present work, we study the EH in a cubic nonlinear damper system using a delayed feedback velocity. The harvester consists of a nonlinear single degree of freedom system (spring-masse-damper) subjected to a base excitation near the primary resonance. This study extends the results obtained in the case without the time delay reported in [1].

The rest of the manuscript is organized as follows: In section 2, we present the mathematical model of the energy harvesting system. In section 3 we perform the method of multiple scales to obtain the amplitude-phase response near the primary resonance. The effect of delay amplitude on the frequency response and the average power is reported and discussed. Section 4 concludes the work.

[a] e-mail: ilhamkirrou@gmail.com

2 The dynamic model and perturbation analysis

We consider a harvester device consisting of a nonlinear single degree of freedom system (spring-masse-damper) subjected to a base excitation near the primary resonance. The quantities F_K, F_{dp} and F_d are, respectively, the nonlinear restoring force, the nonlinear damping force and the delay feedback in the velocity. The schematic model of the system is shown in Fig.1 and the equation of motion can be written in the form

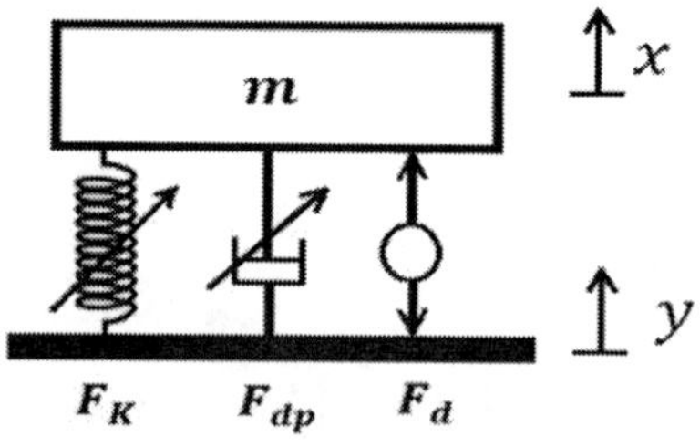

Fig. 1. Single degree of freedom, base excited, energy harvesting system.

$$m\ddot{\bar{z}}(t) + c_1\dot{\bar{z}}(t) + c_3\dot{\bar{z}}^3(t) + K_1\bar{z}(t) + K_3\bar{z}^3(t) = -m\ddot{y}(t) -\alpha_d\dot{\bar{z}}(t-\bar{T}) \quad (1)$$

where the overdot represents a derivative with respect to time t. The variable $\bar{z}$ represents the relative displacement of the mass m, c_1 and c_3 are damping coefficients, k_1 and k_3 are, respectively, the linear and nonlinear stiffness of the restoring force, α_d is the delay amplitude, $\bar{T}$ is the time delay and $y(t) = Y\cos\Omega t$ is the harmonic base motion. The relative normal displacement is given by

$$\bar{z}(t) = x(t) - y(t) \quad (2)$$

DOI: 10.1051/matecconf/20168302002

To obtain a dimensionless form of Eq. (1), we introduce the following nondimensional parameters: $\xi_1 = \frac{c_1}{2m\omega_0}$, $\xi_3 = \frac{c_3\omega_0 Y^2}{m}$, $\gamma = \frac{K_3 Y^2}{m\omega_0^2}$, $\lambda_d = \frac{\alpha_d}{m\omega_0}$, $\omega_0 = \sqrt{\frac{K_1}{m}}$, $T = \omega_0 \overline{T}$, $\tau = \omega_0 t$, $\omega = \frac{\Omega}{\omega_0}$, $z = \frac{\tilde{z}}{Y}$, $\alpha_1 = 2\xi_1$, $\alpha_2 = \xi_2$. The dimensionless equation of motion takes the form

$$\ddot{z} + \alpha_1\dot{z} + \alpha_2\dot{z}^3 + z + \gamma z^3 = \omega^2\cos(\omega\tau) - \lambda_d\dot{z}(\tau - T) \quad (3)$$

To analyze the effect of the time-delayed feedback on the frequency response near the primary resonance, we use the method of multiples scales [10, 11]. To this end, we express the resonance condition by introducing a detuning parameter σ according to $\Omega_0^2 = \omega^2 + \varepsilon\sigma$ and $\Omega_0^2 = 1$. Introducing a small parameter ε, the equation of motion reads

$$\ddot{z} + \omega^2 z = \varepsilon\Big(- \sigma z - \alpha_1\dot{z} - \alpha_2\dot{z}^3 + \omega^2\cos(\omega\tau) - \lambda_d\dot{z}(\tau - T)\Big) - \varepsilon^2\gamma z^3 \quad (4)$$

To ultimately solve Eq. (4), steady-state solution are expanded as

$$z(\tau) = z_0(\tau_0, \tau_1, \tau_2) + \varepsilon z_1(\tau_0, \tau_1, \tau_2) + \varepsilon^2 z_2(\tau_0, \tau_1, \tau_2) + O(\varepsilon^3) \quad (5)$$

where $T_0 = \tau$, $T_1 = \epsilon\tau$ and $T_2 = \epsilon^2\tau$. In terms of the variables $T_i (i = 0, 1, 2)$, the time derivatives become $\frac{d}{d\tau} = D_0 + \varepsilon D_1 + \varepsilon^2 D_2 + O(\varepsilon^3)$ and $\frac{d^2}{d\tau^2} = D_0^2 + 2\varepsilon D_0 D_1 + \varepsilon^2 D_1^2 + 2\varepsilon^2 D_0 D_2 + O(\varepsilon^3)$, where $D_i = \frac{\partial}{\partial T_i}$. Substituting (5) into (4), equating terms of different order of ε, resolving and eliminating the secular terms, we obtain the following slow flow modulation equation of the amplitude and the phase:

$$\begin{cases} \dfrac{dr}{d\tau} = h\sin\theta + Ar + Cr^3 \\ r\dfrac{d\theta}{d\tau} = h\cos\theta + Br + Dr^3 \end{cases} \quad (6)$$

Equilibria of this slow flow, corresponding to periodic solutions of Eq. (4), are determined by setting $\frac{dr}{dt} = \frac{d\theta}{dt} = 0$. This leads to the amplitude-frequency response equation

$$(D^2 + C^2)r^6 + 2(AC + BD)r^4 + (A^2 + B^2)r^2 - h^2 = 0 \quad (7)$$

where $A = \frac{\alpha_1}{2} - \frac{\lambda_d}{2}\cos\phi$, $B = \frac{\sigma}{2\omega} + \frac{\lambda_d}{2}\sin\phi$, $C = \frac{3\alpha_2\omega^2}{8}$, $D = \frac{3\gamma}{8\omega}$, $h = -\frac{\omega}{2}$, $\phi = \omega T$.

The average power is obtained by averaging over one period of the excitation. This leads to

$$P_{gen} = \frac{\omega}{2\pi}\int_0^{\frac{2\pi}{\omega}} (\alpha_1\dot{z} + \alpha_3\dot{z}^3)\dot{z}d\tau \quad (8)$$

Using Eq. (8), we obtain the generated average power

$$P_{gen}^{av} = \frac{\alpha_1\omega^2 r^2}{2} + \frac{3\alpha_3\omega^4 r^4}{8} \quad (9)$$

The power required for the active control action is given by

$$P_{act} = \frac{\omega}{2\pi}\int_0^{\frac{2\pi}{\omega}} \lambda_d\dot{z}(t - \tau)\dot{z}d\tau \quad (10)$$

Then, the consumed average power is given by

$$P_{act}^{av} = \frac{\lambda_d\omega^2 r^2}{2}\cos\phi \quad (11)$$

where $\psi = \omega\tau + \theta$ and $\phi = \omega T$.
The harvested (or net) average power defined as the difference of generated and consumed average powers reads

$$P_{harv}^{av} = P_{gen}^{av} - P_{act}^{av} \quad (12)$$

3 Main results

The influence of different system parameters on the dynamic response and the maximum average output powers is examined. We fix the damping parameters $\alpha_1 = 0.25$, $\alpha_2 = 0.025$ and $\gamma = 0.45$.
The frequency-response curve, as expressed by Eq. (7), is depicted in Fig. 2 for increasing delay amplitude λ_d. The solid lines correspond to stable branches while the dashed line corresponds to the unstable one. The plots indicate that as the delay amplitude is increased the maximum deflection displays an increasing in the hardening nonlinear response and shift toward higher frequencies.

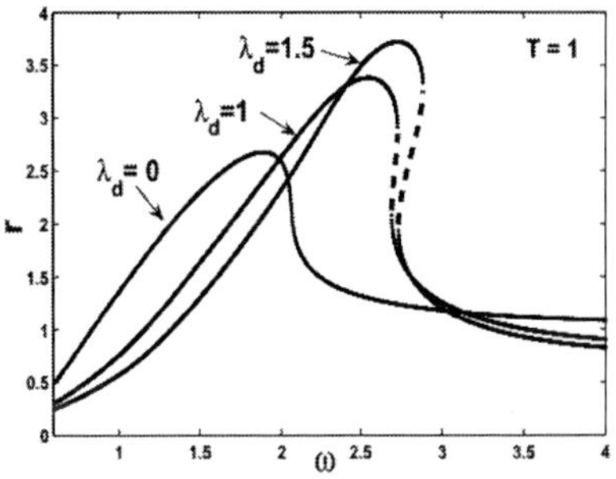

Fig. 2. Amplitude-frequency response versus ω for different values of the delay amplitude λ_d.

Figure. 3 illustrates the variation of the amplitude response versus the delay amplitude λ_d for a fixed values of the time delay T and the excitation frequency ω. Analytical approximations (solid lines) are compared to numerical simulations (circles). It can be seen that increasing the amplitude λ_d causes an increase in the amplitude response. Figure. 4 represents the generated average power response versus the delay amplitude λ_d for a fixed value of T and ω. The figure shows an increase of the output power by increasing λ_d.

Figure. 5 depicts the variation of the average powers quantities with respect to the delay amplitude λ_d for a given values of ω and T. It is observed that the harvested average power increases in a certain range of λ_d.

4 Conclusions

The application of delayed feedback velocity for optimizing the harvested power in cubic nonlinear damper system

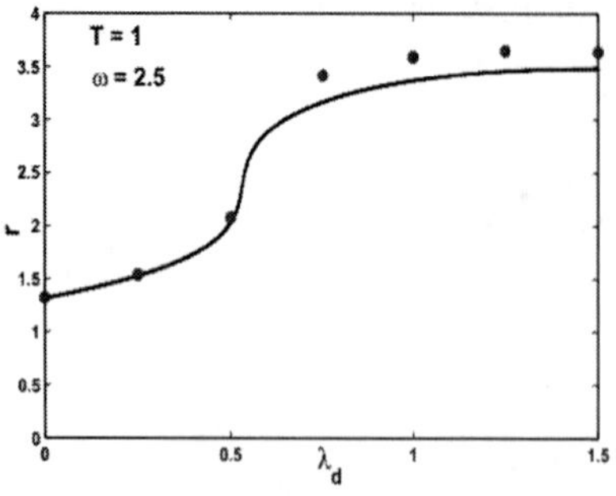

Fig. 3. Amplitude response versus λ_d for $T = 1$ and $\omega = 2.5$; Solid lines: analytical approximation; circle: numerical simulation.

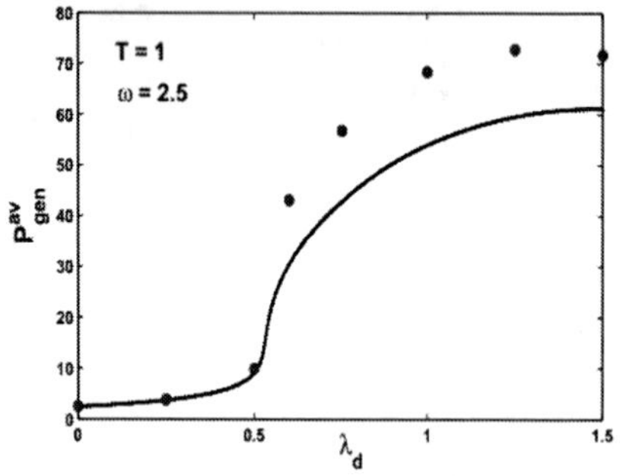

Fig. 4. Generated average power versus λ_d for $T = 1$ and $\omega = 2.5$; Solid lines: analytical approximation; circle: numerical simulation.

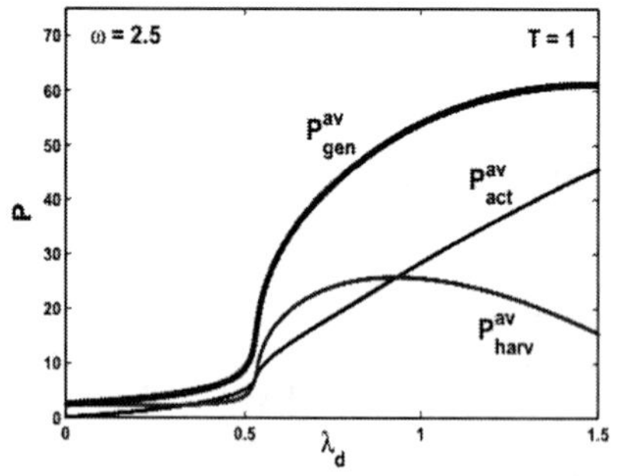

Fig. 5. Average powers quantities versus the delay amplitude λ_d for $T = 1$ and $\omega = 2.5$.

is presented. The harvester device consists of a nonlinear single degree of freedom system (spring-masse-damper) subjected to a base excitation is considered near the primary resonance. Analytical investigation using the multiple scales method was performed to obtain approximation of the amplitude response. This amplitude is used to obtain the evolution of the extract average power as the delayed feedback gain is increased. Results shown that in the cubic nonlinear damper system, time delay introduced in the velocity enhance significantly the performance of the net energy harvesting.

References

[1] M.G., Tehrani, S.J., Elliott, Exetending the dynamic range of an energy harvester using nonlinear damping, Journal of Sound and Vibration **333**: 23-629, (2014).
[2] T.S., Osorio, M.F., Daqaq, Energy harvesting under excitations of time varying frequency, Journal of Sound and Vibration, **329**: 2497-2515, (2010).
[3] A., Erturk, D.J., Inman, Broadband piezoelectric power generation on high-energy orbits of the bistable Duffing oscillator with electromechanical coupling, Journal of Sound and Vibration, **330**: 2339-2353, (2011).
[4] M., Hamdi, M., Belhaq, Energy from harvesting from quasi-periodic vibrations, Nonlinear dynamics, 1-13, (2016). DOI 10.1007/s11071-016-2668-6.
[5] N.G., Stephen, On energy harvesting from ambient vibration, Journal of Sound and Vibration, **293**: 409-425 (2006).
[6] H.A., Sodano, D.J., Inman, G., Park, Generation and storage of electricity from power harvesting devices, Journal of Intelligent Material Systems and Structures **16**: 67-75, (2005).
[7] B.P., Mann, N., Sims, Using nonlinearity to improve the performance of vibration-based energy harvesting devices, proceedings of the 7th European Conference on Structural Dynamics, Southampton July 2008.
[8] B.P., Mann, N.D., Sims, Energy harvesting from the nonlinear oscillations of magnetic levitation, Journal of Sound and Vibration **319**: 515-530, (2009).
[10] A.H., Nayfeh, D.T., Mook, Nonlinear Oscillations. Wiley, New York (1979).
[11] A.H., Nayfeh, Introduction to Perturbation Techniques. New York: Wiley (1981).

Homotopy analysis approach for nonlinear piezoelectric vibration energy harvesting

Shahram Shahlaei-Far[1,*], Airton Nabarrete[1], and José Manoel Balthazar[1]

[1]Instituto Tecnológico de Aeronáutica - ITA, Enghenaria Aeronáutica e Mecânica, 12208-900, São José dos Campos, Brazil

Abstract. Piezoelectric energy harvesting from a vertical geometrically nonlinear cantilever beam with a tip mass subject to transverse harmonic base excitations is analyzed. One piezoelectric patch is placed on the slender beam to convert the tension and compression into electrical voltage. Applying the homotopy analysis method to the coupled electromechanical governing equations, we derive analytical solutions for the horizontal displacement of the tip mass and consequently the output voltage from the piezoelectric patch. Analytical approximation for the frequency response and phase of the geometrically forced nonlinear vibration system are also obtained. The research aims at a rigorous analytical perspective on a nonlinear problem which has previously been solely investigated by numerical and experimental methods.

1 Introduction

Piezoelectric transduction for vibration-based energy harvesting plays an important role for remote/wireless sensor systems. This vibration-to-electric energy conversion mechanism uses the ambient vibration energy to power and/or to recharge such electronic components. Usually, as in the case of resonance-based energy harvesting, the piezoelectric effect is achieved by fixing a cantilever beam with one or two piezoceramic layers on a host structure and using single excitation frequency. The piezoelectric effect converts mechanical strain induced in the layers into an electric current generating alternating voltage [1,2].

Requiring an energy harvester to harvest a reasonable amount of energy when the excitation frequency is low can be a difficult undertaking as the physical dimensions of the devices are small. Maximizing the harvested energy over a wide range of excitation frequencies is achieved by implementing nonlinear structural systems and using a double potential well function [3].

In this study, an inverted cantilever beam with piezoelastic patch and a tip mass is mounted vertically on an energy harvesting device and is subjected to transverse base excitations. [4]

Applying the homotopy analysis method (HAM), introduced by Shijun Liao [5], to the coupled electromechanical governing equations of motion, novel analytical solutions of the horizontal displacement coordinate, of its amplitude and phase as well as the output voltage are derived. The aim is to address the subject of piezoelectric energy harvesting from a purely analytical perspective emphasizing the capabilities of a first-order approximation of HAM to present highly accurate closed-form solutions.

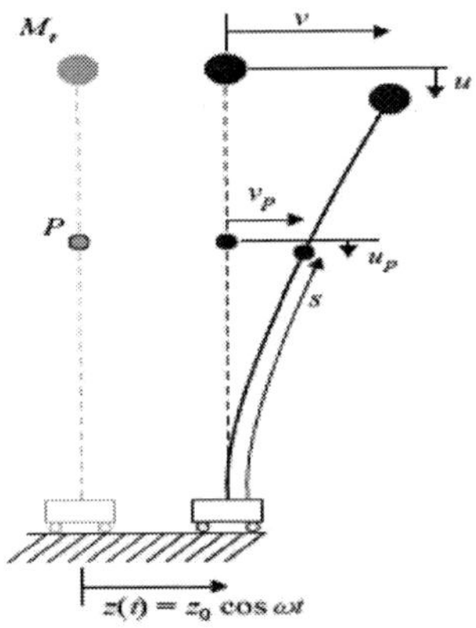

Fig. 1. Inverted cantilever beam with tip mass on host structure subject to transverse excitation [4]

2 Governing equations

The coupled electromechanical governing equations for the beam-mass system are [4]

$$[N_5^2 I_t + M_t + \rho \bar{A} N_1 + (\rho \bar{A} N_3 + M_t N_4^2 + N_5^4 I_t) v^2] + (\rho \bar{A} N_3 + M_t N_4^2 + N_5^4 I_t) v \dot{v}^2 + \theta_1 V - \theta_2 v^2 V = -(\rho \bar{A} N_2 + M_t) \ddot{z} \quad (1)$$

and

$$C_p \dot{V} + \frac{V}{R_l} + \theta_1 \dot{v} + \theta_2 v^2 \dot{v} = 0 \quad (2)$$

*Corresponding author: shazzz_85@yahoo.de

subject to

$$v(0) = A,\ \dot{v}(0) = 0 \text{ and } V(0) = 0 \quad (3)$$

where v is the horizontal elastic displacement at the tip mass, V is the voltage output of the piezoelectric patch, L is the length of the beam, $\bar{A}$ is its cross-sectional area, ρ is the mass density, E is the equivalent Young's modulus, I is the second moment of area, M_t is the tip mass of the beam, I_t is its moment of inertia, θ_1 and θ_2 are parameters for work done by the piezoelectric patch, C_p is its capacitance, R_l is the load resistor and g is the gravitational constant. The values of the parameters N_i, $i = 1,2,\ldots,6$, θ_1 and θ_2 are given in Table 1.

For our analysis, we consider the time domain between $t = 1300s$ and $t = 1500s$ to demonstrate nonlinear dynamical characteristics of the steady state region. First, we solve Eq. (2) to obtain a solution for V as a function of v and $\dot{v}$ considering when the tip mass oscillates close to one of the equilibrium positions. The amplitude of oscillation of the horizontal displacement is very small (the velocity being even considerably smaller), thus the expression containing v and $\dot{v}$ in (2) can be approximated by a constant value. Substituting the obtained solution expression of V into Eq. (1) we gain a single differential equation dependent on v. Finally, we substitute the solution of v into the expression for V. Solving Eq. (2) we obtain

$$V(\dot{v}, v, t) = KC_pR_l\left(1 - e^{-\frac{1}{C_pR_l}t}\right) \quad (4)$$

where

$$K = -\frac{\theta_1\dot{v} + \theta_2 v^2\dot{v}}{C_p} \quad (5)$$

For large values of t as considered here, the term $e^{-\frac{1}{C_pR_l}t}$ becomes negligible (see values for C_p and R_l), thus, in our case, Eq. (4) is simplified and reduced to

$$V(\dot{v}, v, t) = -R_l\dot{v}(\theta_1 + \theta_2 v^2) \quad (6)$$

Inserting Eq. (6) into Eq. (1) we have

$$C_1\ddot{v} + C_2v^2\ddot{v} + C_3v\dot{v}^2 + C_4v + C_5v^3 + \theta_1^2R_l\dot{v} + 2\theta_1\theta_2R_l\dot{v}v^2 + \theta_2^2R_l\dot{v}v^4 = C_6z_0\Omega^2\cos(\Omega t - \varphi) \quad (7)$$

where the constants C_i, $i = 1,2,\ldots,6$ are given in Table 1. Note that, without loss of generality, we can introduce the phase angle φ in the expression of the harmonic load as a quantity to be determined. The solution derivation uses the approach described in [6] for nonlinear problems of forced vibrations.

Defining the variables

$$\tau = \Omega t,\ v(t) = Au(\tau) \quad (8)$$

and inserting them into Eq. (7), we obtain

$$C_1A\Omega^2\frac{\partial^2 u(\tau)}{\partial\tau^2} + C_2A^3\Omega^2\frac{\partial^2 u(\tau)}{\partial\tau^2}u^2 + C_3A^3\Omega^2u\left(\frac{\partial u(\tau)}{\partial\tau}\right)^2 + C_4Au + C_5A^3u^3 + \theta_1^2R_lA\Omega\frac{\partial u(\tau)}{\partial\tau} + +2\theta_1\theta_2R_lA^3\Omega\frac{\partial u(\tau)}{\partial\tau}u^2 + \theta_2^2R_lA^5\Omega\frac{\partial u(\tau)}{\partial\tau}u^4 = \mathrm{F}_1\cos(\tau) + \mathrm{F}_2\sin(\tau) \quad (9)$$

subject to the initial conditions

$$u(0) = 1,\ \frac{\partial u(0)}{\partial\tau} = 0 \quad (10)$$

with the constants F_1 and F_2 satisfying

$$(F_1^2 + F_2^2)^{\frac{1}{2}} = C_6z_0\Omega^2,\ \varphi = \tan^{-1}\left(\frac{F_2}{F_1}\right) \quad (11)$$

3 Homotopy analysis method

The periodic solution to Eq. (9) can be expressed by a set of base functions

$$\{\sin(m\tau), \cos(m\tau)\,|\,m = 1,2,3,\ldots\} \quad (12)$$

such that

$$u(\tau) = \sum_{k=1}^{\infty}(\alpha_k\sin(k\tau) + \beta_k\cos(k\tau)) \quad (13)$$

where α_k and β_k are coefficients to be determined. We choose the initial guess as

$$u_0(\tau) = \cos(\tau) \quad (14)$$

which satisfies the initial conditions of Eq. (10). We choose the linear operator to be

$$\mathcal{L}[\phi(\tau, q)] = \Omega^2\left(\frac{\partial^2\phi(\tau,q)}{\partial\tau^2} + \phi(\tau, q)\right) \quad (15)$$

with the property

$$\mathcal{L}[D_1\sin(\tau) + D_2\cos(\tau)] = 0 \quad (16)$$

where D_1 and D_2 are constants of integration. According to Eq. (9), we define the nonlinear operator

$$\mathcal{N}[\phi(\tau, q), \Lambda(q)] = C_1\Lambda(q)\Omega^2\frac{\partial^2\phi(\tau,q)}{\partial\tau^2} + C_2\Lambda(q)^3\Omega^2\frac{\partial^2\phi(\tau,q)}{\partial\tau^2}\phi(\tau, q)^2 + C_3\Lambda(q)^3\Omega^2\phi(\tau, q)\left(\frac{\partial\phi(\tau,q)}{\partial\tau}\right)^2 + C_4\Lambda(q)\phi(\tau, q) + C_5\Lambda(q)^3\phi(\tau, q)^3 + \theta_1^2R_l\Lambda(q)\Omega\frac{\partial\phi(\tau,q)}{\partial\tau} + 2\theta_1\theta_2R_l\Lambda(q)^3\Omega\frac{\partial\phi(\tau,q)}{\partial\tau}\phi(\tau, q)q)^2 + \theta_1^2R_l\Lambda(q)^5\Omega\frac{\partial\phi(\tau,q)}{\partial\tau}\phi(\tau, q)^4 - \mathrm{F}_1\cos(\tau) - \mathrm{F}_2\sin(\tau) \quad (17)$$

where $q \in [0,1]$ is an embedding parameter, $\phi(\tau, q)$ is a function of τ and q, $\Lambda(q)$ is a function of q. The zero-order deformation equation is given by

$$(1 - q)\mathcal{L}[\phi(\tau, q) - u_0(\tau)] = qc_0H(\tau)\mathcal{N}[\phi(\tau, q), \Lambda(q)] \quad (18)$$

where $c_0 \neq 0$ is the convergence-control parameter and $H(\tau)$ a nonzero auxiliary function. For $q = 0$ and $q = 1$ we have

$$\phi(\tau, 0) = u_0(\tau),\ \phi(\tau, 1) = u(\tau) \tag{19}$$

$$\Lambda(0) = A_0,\ \Lambda(1) = A \tag{20}$$

Thus, the function $\phi(\tau, q)$ varies from the initial guess $u_0(\tau)$ to the desired solution as q varies from 0 to 1. The Taylor expansions of $\phi(\tau, q)$ and $\Lambda(q)$ with respect to q are

$$\phi(\tau, q) = u_0(\tau) + \sum_{m=1}^{\infty} u_m(\tau)\, q^m \tag{21}$$

$$\Lambda(q) = A_0 + \sum_{m=1}^{\infty} A_m\, q^m \tag{22}$$

where

$$u_m(\tau) = \frac{1}{m!} \frac{\partial^m \phi(\tau, q)}{\partial q^m}\Big|_{q=0} \tag{23}$$

$$A_m = \frac{1}{m!} \frac{\partial^m \Lambda(q)}{\partial q^m}\Big|_{q=0} \tag{24}$$

Choosing properly the auxiliary function $H(\tau)$ and the convergence-control parameter c_0, the series in Eqs. (23) and (24) converge when $q = 1$, such that

$$u(\tau) = u_0(\tau) + \sum_{m=1}^{\infty} u_m(\tau) \tag{25}$$

$$A = A_0 + \sum_{m=1}^{\infty} A_m \tag{26}$$

Differentiating the zero-order equation (18) m times with respect to q, dividing it by $m!$ and setting $q = 0$, the mth-order deformation equation is obtained as

$$\mathcal{L}[u_m(\tau) - \chi_m u_{m-1}(\tau)] = c_0 H(\tau) R_m(\vec{u}_{m-1}, \vec{A}_{m-1}) \tag{27}$$

subject to the initial conditions

$$u_m(0) = \frac{\partial u_m(0)}{\partial \tau} = 0 \tag{28}$$

where

$$\chi_m = \begin{cases} 0 & m \leq 1 \\ 1 & m > 1 \end{cases} \tag{29}$$

$$\vec{u}_{m-1} = \{u_0(\tau), u_1(\tau), \ldots, u_{m-1}(\tau)\},\quad \vec{A}_{m-1} = \{A_0, A_1, \ldots, A_{m-1}\} \tag{30}$$

$$R_m(\vec{u}_{m-1}, \vec{A}_{m-1}) = \frac{1}{(m-1)!} \frac{d^{m-1} \mathcal{N}[\phi(\tau, q), \Lambda(q)]}{d q^{m-1}}\Big|_{q=0} \tag{31}$$

For the nonzero auxiliary function to obey the *rule of solution expression* and the *rule of coefficient ergodicity*, we choose it to be

$$H(\tau) = \cos(2\kappa\tau) \tag{32}$$

where κ is an integer. It can be determined uniquely as $H(\tau) = 1$.

The general solution to Eq. (26) for $m \geq 1$ is derived to be

$$u_m(\tau) = D_1 \sin(\tau) + D_2 \cos(\tau) + \chi_m u_{m-1}(\tau) + \frac{c_0}{\Omega^2} \sum_{k=1}^{l(m)} \frac{(c_{m,k} \cos((2k+1)\tau) + d_{m,k} \sin((2k+1)\tau))}{1 - (2k+1)^2} \tag{33}$$

where D_1 and D_2 need to be determined by the initial conditions in Eq. (28).

For a first-order homotopy approximation ($m = 1$), we obtain from Eq. (31) the right-hand side of the first-order deformation equation (27) as

$$\begin{aligned} R_1(\vec{u}_0, \vec{A}_0) = \Big(&-C_1 A_0 \Omega^2 - \tfrac{3}{4} C_2 A_0^3 \Omega^2 + \tfrac{1}{4} C_3 A_0^3 \Omega^2 + C_4 A_0 + \tfrac{3}{4} C_5 A_0^3 - F_1\Big) \cos(\tau) + \\ &\Big(-\theta_1^2 R_l A_0 \Omega - \tfrac{1}{2} \theta_1 \theta_2 R_l A_0^3 \Omega - \tfrac{1}{8} \theta_2^2 R_l A_0^5 \Omega - F_2\Big) \sin(\tau) + \Big(-\tfrac{1}{4} C_2 A_0^3 \Omega^2 - \tfrac{1}{4} C_3 A_0^3 \Omega^2 + \\ &\tfrac{1}{4} C_5 A_0^3\Big) \cos(3\tau) + \Big(-\tfrac{1}{2} \theta_1 \theta_2 R_l A_0^3 \Omega - \tfrac{3}{16} \theta_2^2 R_l A_0^5\Big) \sin(3\tau) - \Big(\tfrac{1}{16} \theta_2^2 R_l A_0^5 \Omega\Big) \sin(5\tau) \end{aligned} \tag{34}$$

Enforcing the coefficients of $\sin(\tau)$ and $\cos(\tau)$ to be equal to zero in order to avoid secular terms in the solution expression, and using the conditions in Eq. (11), we obtain the steady state solutions of the amplitude A_0 and the phase φ, respectively, as

$$\left[\left(-C_1 A_0 \Omega^2 - \tfrac{3}{4} C_2 A_0^3 \Omega^2 + \tfrac{1}{4} C_3 A_0^3 \Omega^2 + C_4 A_0 + \tfrac{3}{4} C_5 A_0^3\right)^2 + \left(-\theta_1^2 R_l A_0 \Omega - \tfrac{1}{2} \theta_1 \theta_2 R_l A_0^3 \Omega - \tfrac{1}{8} \theta_2^2 R_l A_0^5 \Omega\right)^2\right]^{\frac{1}{2}} = C_6 z_0 \Omega^2 \tag{35}$$

$$\varphi = \tan^{-1}\left(\frac{-\theta_1^2 R_l A_0 \Omega - \frac{1}{2} \theta_1 \theta_2 R_l A_0^3 \Omega - \frac{1}{8} \theta_2^2 R_l A_0^5 \Omega}{-C_1 A_0 \Omega^2 - \frac{3}{4} C_2 A_0^3 \Omega^2 + \frac{1}{4} C_3 A_0^3 \Omega^2 + C_4 A_0 + \frac{3}{4} C_5 A_0^3}\right) \tag{36}$$

Finally, solving the first-order deformation equation and using Eqs. (24) and (25), the first-order homotopy approximation of $v(t)$ becomes

$$\begin{aligned} v(t) = A u(\tau) \approx A_0(u_0(\tau) + u_1(\tau)) = \frac{c_0}{\Omega^2}\Big[&\Big(\frac{\Omega^2}{c_0} A_0 - \tfrac{1}{32} C_2 A_0^3 \Omega^2 - \tfrac{1}{32} C_3 A_0^3 \Omega^2 + \tfrac{1}{32} C_5 A_0^3\Big) \cos(\Omega t) + \\ &\Big(-\tfrac{3}{16} \theta_1 \theta_2 R_l A_0^3 \Omega - \tfrac{9}{128} \theta_2^2 R_l A_0^5 - \tfrac{5}{384} \theta_2^2 R_l A_0^5 \Omega\Big) \sin(\Omega t) + \tfrac{1}{32}(C_2 A_0^3 \Omega^2 + C_3 A_0^3 \Omega^2 - \\ &C_5 A_0^3) \cos(3\Omega t) + \Big(-\tfrac{1}{16} \theta_1 \theta_2 R_l A_0^3 \Omega - \tfrac{3}{128} \theta_2^2 R_l A_0^5\Big) \sin(3\Omega t) + \Big(\tfrac{1}{384} \theta_2^2 R_l A_0^5 \Omega\Big) \sin(5\Omega t)\Big] \end{aligned} \tag{37}$$

To obtain the solution for the voltage V, insert Eq. (37) into Eq. (6) considering the steady-state solution for large values of $t$as specified.

4 Discussion of results

The parameters for the beam, the tip mass and the energy harvester are given in Table 1. The model parameters were obtained from [4]. Due to the use of a double potential well, the post-buckled response has two equilibrium positions. The variation of the tip mass allows for three cases corresponding to pre-buckled, buckled and post-buckled responses. For our analysis we have chosen a value of 10.5g for the tip mass which lies in the post-buckling region.

Table 1. Parameter values for the beam and energy harvester

Beam and tip mass		Energy harvester	
ρ	7850 kg/m³	L_c	0.028 m
E	210 GN/m²	b_c	0.014 m
b	0.016 m	h_c	0.0003 m
h	0.000254 m	γ_c	-4× 10^{-5}Nm/V
$L = L_t$	0.3 m	C_p	51.4 nF
$\frac{I_t}{M_t}$	0.00004087 m²	R_l	10^5 Ω
Model parameters			
θ_1	-9.7703× 10^{-6}	N_8	1.0592× 10^4
θ_2	-2.9146× 10^{-7}	N_9	0.3669
N_1	0.0680	C_1	2.5738
N_2	0.1090	C_2	34.8460
N_3	0.9201	C_3	34.8460
N_4	4.1123	C_4	2.4244× 10^6
N_5	5.2360	C_5	3.3233× 10^7
N_6	112.7420	C_6	4.1182
N_7	772.7215	z_0	0.016

Figure 2 and 3 show the nonlinear time response and voltage output of the energy harvester for $t = 1400$ to $t = 1460$, respectively. They demonstrate that the output is considerably higher when the tip mass moves between the two equilibrium positions rather than when it is close to one of them. A value of $c_0 = 1$ was chosen.

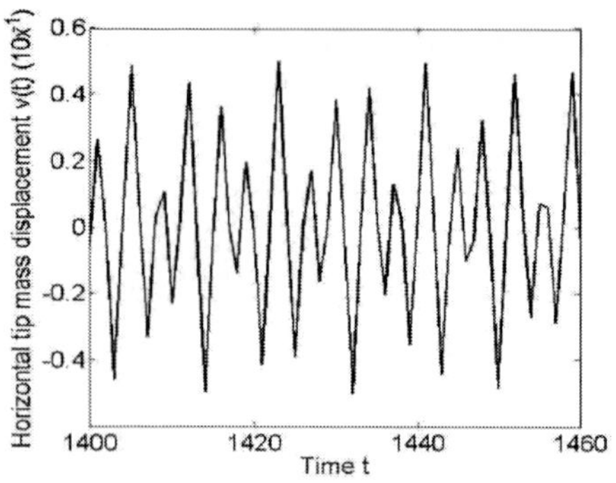

Fig. 2. Nonlinear time response of the energy harvester

In Fig. 4 the normalized amplitude-frequency curve is presented demonstrating the classical jump phenomenon for hardening-type nonlinearity. With respect to results obtained in [4], the outcomes are in close agreement the main difference being that an experimental result does not exhibit two coexisting solutions simultaneously and therefore jump phenomena in the amplitude-frequency curve may only be hinted at.

For the problem at hand, the standard simultaneous application of HAM for two coupled differential equations does not achieve unique solutions for the amplitude A and phase φ of the horizontal displacement $v(t)$ since the number of zero terms required to avoid secular terms in the final solution is greater than the number of unknown parameters. By this approach the amplitude of the horizontal displacement is derived to be a function of the external excitation frequency Ω and at the same time a fixed value independent from the harmonic base excitation.

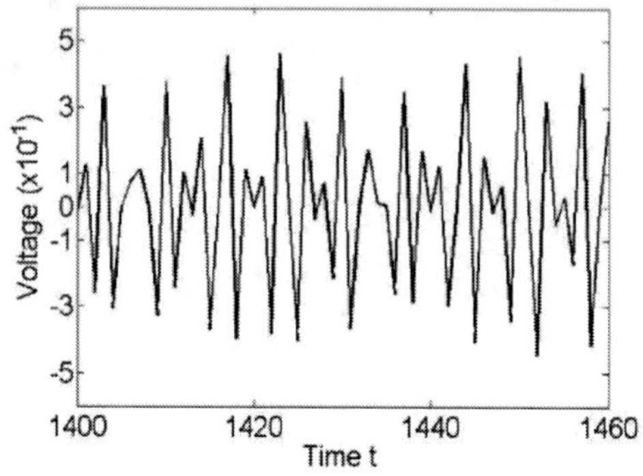

Fig. 3. Voltage output of the energy harvester

Moreover, zero initial conditions for the displacement and velocity of the tip mass used for the numerical analysis in [4] are not feasible when applying HAM, thus at $t = 0$, the vertical beam is displaced by an amplitude A to be determined. This is compatible with the beam's natural behavior.

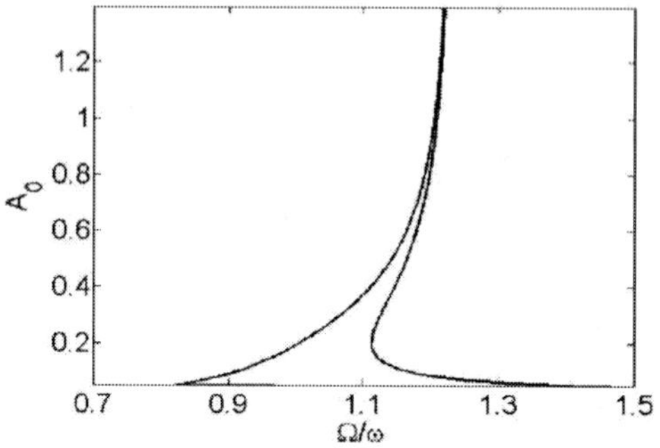

Fig. 4. Amplitude-frequency curve for values in Table 1

References

[1] A. Erturk, J. Hoffmann, D.J. Inman, Appl. Phys. Letters, **94**, 25, 1-3 (2009)
[2] A. Rysak, M. Müller, m. Borowiec, J. zubrzycki, G. Litak, Defence Sci Tech Agency, 508-512, (2013)
[3] I. Gammaitoni, I. Neri, H. Vocca, Appl. Phys. Lett. **94**, 16, 1-3 (2009).
[4] M.I. Friswell, S.F. Ali, O. Bilgen, S. Adhikari, A.W. Lees, G. Litak, J. Intel. Mat. Sys. Struct. **23**, 13, 1505-1521 (2012)
[5] S. Liao, *Beyond perturbation: Introduction to the homotopy analysis method* (CRC Press/Chapman Hall, Boca Rotan, 2003)
[6] S. Shahlaei-Far, A. Nabarrete, J.M. Balthazar, J. Theor. Appl. Mech. (to be published)

DOI: 10.1051/matecconf/20168303001

Modeling and position control of tethered octocopters

Davi Ferreira de Castro[a], Igor Afonso Acampora Prado[b], and Mateus de Freitas Virgílio Pereira [c], Davi Antônio dos Santos[d] and José Manoel Balthazar[e]
* Aeronautics Institute of Technology, ITA, Brazil.

Abstract. This work presents the modeling and control of a multirotor aerial vehicle with tethered configuration. It is considered an octocopter with a saturated proportional-plus-derivative position control. A viscoelastic model is considered for the tether, which has a tension control. Numerical simulations are carried out to compare the performance of the tethred configuration with the vehicle in free flight.

1 Introduction

Multirotor Aerial Vehicles (MAVs) have been the subject of many academic studies and have attracted a lot of attention from industry in recent years. MAVs have flight capabilities such as hovering, Vertical Take-Off and Landing (VTOL) and agile maneuvering capability, which cannot be achieved by conventional fixed wing aircraft. Some drawbacks of multicopter vehicles are the reduced endurance, which hardly exceeds some minutes and limited payload. Tethered MAVs have been introduced to the aerospace sector as a possible alternative. The tether, even though limiting the vehicle flight range, can be used for data transmission and power supply during operation, which improves the vehicle endurance.

In academia, research works on tethered multicopters are still rare by the present day and there are few companies and institutions that have invested in the development of such vehicles. The first references to consider the tethered configuration are found in [7] and [4]. Some applications of this kind configuration are found in aerostats [1] and robot satellites [2]. The work in [3] presents a navigation strategy in partially mapped environments. This strategy is applied to the collaboration between an UAV [1] and an UGV [2], in which power is provided to the former by a cable connecting the two vehicles. The work [5] presents a strategy to control a tethered helicopter with a combination of classical PID control laws.

This paper presents a modeling and position control strategy for tethered MAV. A viscoelastic model is considered for the cable, in order to reproduce its dynamic behavior. The controller is based on a saturated state feedback control, thus simplifying the controller. This paper is organized as follows: in Section 2 we present the tethered configuration model, in Section 3 a control system is proposed, in Section 4 we present the simulation results and Section 5 presents the conclusions.

[a] e-mail: davifc@ita.br
[b] e-mail: igorap@ita.br
[c] e-mail: mateusfvp@hotmail.com
[d] e-mail: davists@ita.br
[e] e-mail: jmbaltha@ita.br
[1] Unmanned Aerial Vehicle (UAV)
[2] Unmanned Ground Vehicle (UGV)

2 Tethered configuration model

This Section presents the multicopter model and tether model, which together describe the dynamics of the tethered multicopter configuration.

2.1 Multicopter model

Consider a octocopter and three Cartesian coordinate systems (CCS) illustrated in Figure 1.

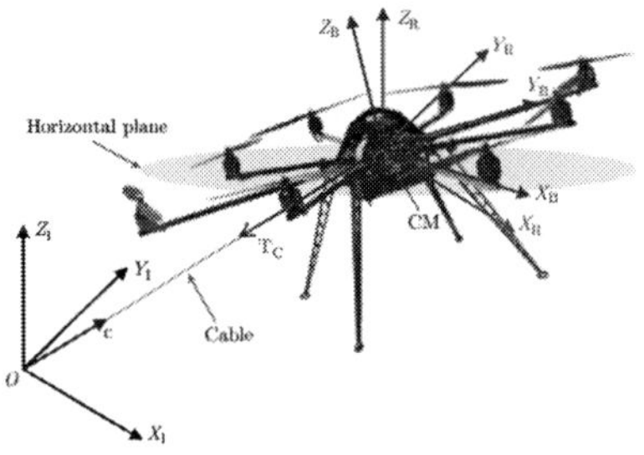

Fig. 1. Cartesian coordinate systems.

The CCS of the body $S_B \triangleq \{X_B, Y_B, Z_B\}$ is fixed to the vehicle structure and centered on the center of mass (CM). The inertial CCS $S_I \triangleq \{X_I, Y_I, Z_I\}$ is fixed to the earth at the point O. The CCS $S_R \triangleq \{X_R, Y_R, Z_R\}$ is parallel to S_I and centered in CM. T_C is tether tension and **c** defined tether line vector.

The differential equations that describe the dynamic behavior of the multirotor can be obtained by the Newton-Euler formalism. This work chooses to represent the attitude by the quaternion of rotation $\mathbf{q} \in \mathbb{R}^4$, however for visualization is adopted Euler angles (ϕ, θ and ψ).

The kinematic attitude model for the quaternion of rotation is given by the following differential equation [8]:

$$\dot{\mathbf{q}} = \boldsymbol{\Omega}\mathbf{q} \tag{1}$$

where

$$\boldsymbol{\Omega} = \frac{1}{2}\begin{bmatrix} 0 & -\omega^T \\ \omega & -[\omega\times] \end{bmatrix}, \tag{2}$$

being $\omega \triangleq \left[\omega_x\ \omega_y\ \omega_z\right]^{\mathrm{T}} \in \mathbb{R}^3$ the angular velocity of the vehicle represented in S_{R} and $[\omega\times]$ is given by:

$$[\omega\times] = \begin{bmatrix} 0 & -\omega_z & \omega_y \\ \omega_z & 0 & -\omega_x \\ -\omega_y & \omega_x & 0 \end{bmatrix}. \tag{3}$$

Rewriting the equation (1) explicitly, one obtains:

$$\begin{bmatrix} \dot{q}_1 \\ \dot{q}_2 \\ \dot{q}_3 \\ \dot{q}_4 \end{bmatrix} = \frac{1}{2}\begin{bmatrix} 0 & -\omega_x & -\omega_y & -\omega_z \\ \omega_x & 0 & \omega_z & -\omega_y \\ \omega_y & -\omega_z & 0 & \omega_x \\ \omega_z & \omega_y & -\omega_x & 0 \end{bmatrix}\begin{bmatrix} q_1 \\ q_2 \\ q_3 \\ q_4 \end{bmatrix}. \tag{4}$$

Applying Newton's second law for rotational motion and neglecting disturbance torques, one can obtain:

$$\boldsymbol{\tau} = \dot{\mathbf{h}} + \omega \times \mathbf{h}, \tag{5}$$

where $\boldsymbol{\tau} \triangleq [\tau_x\ \tau_y\ \tau_z]^{\mathrm{T}} \in \mathbb{R}^3$ is the resulting propulsion torque in the vehicle represented in S_{B} and $\mathbf{h} \triangleq \mathbf{I}\ \omega \in \mathbb{R}^3$ is the angular momentum of the body represented in S_{B}, being $\mathbf{I} \in \mathbb{R}^{3x3}$ the matrix of inertia of the vehicle represented in S_{B}. It is assumed that the vehicle has symmetrical structure with respect to the coordinate axes. The matrix **I** resulting in a diagonal matrix defined by:

$$\mathbf{I} = \begin{bmatrix} I_x & 0 & 0 \\ 0 & I_y & 0 \\ 0 & 0 & I_z \end{bmatrix}. \tag{6}$$

Rewriting the equation (5) explicitly, one can obtain:

$$\dot{\omega}_x = \frac{(I_y - I_z)}{I_x}\omega_y\omega_z + \frac{\tau_x}{I_x}. \tag{7}$$

$$\dot{\omega}_y = \frac{(I_z - I_x)}{I_y}\omega_x\omega_z + \frac{\tau_y}{I_y}. \tag{8}$$

$$\dot{\omega}_z = \frac{(I_x - I_y)}{I_z}\omega_x\omega_y + \frac{\tau_z}{I_z}. \tag{9}$$

Applying the second Newton's law of motion and neglecting any perturbation, we obtain the following model of translational movement in S_{I}:

$$\ddot{\mathbf{r}} = \frac{1}{m}f\mathbf{n} + \begin{bmatrix} 0 \\ 0 \\ -g \end{bmatrix}, \tag{10}$$

where $\mathbf{r} \triangleq [r_x\ r_y\ r_z]^{\mathrm{T}} \in \mathbb{R}^3$ is the three-dimensional position of the CM represented in S_{I}, $\mathbf{n} \triangleq [n_x\ n_y\ n_z]^{\mathrm{T}} \in \mathbb{R}^3$ is the normal unit vector perpendicular to the rotor plane represented in S_{I}, f is the total thrust, g is the acceleration of gravity and m is the mass of the vehicle.

2.2 Tether model

A viscoelastic model for the cable is considered in order to reproduce its behavior, as illustrated in Figure 2. More precisely, the model consists of a damper with coefficient d_a and a spring with stiffness k_a in parallel, both connected in series with a spring of different stiffness k_b.

Here x_1 is an unobservable variable which describes the inner friction, x_2 represents the elongation of that segment, L denotes the unstressed tether segment lenght that is reeled out by the ground device and q_9 is deployed tether segment. It is important to note that in the model adopted the mass of the cable is considered negligible in relation to the vehicle mass. The equations and parameters are proposed by Sandino *et al* [5]. The variables that describe the tether state are shown in Figure 2. The transfer function associated tether model is proposed by Sandino *et al* [5] is given by:

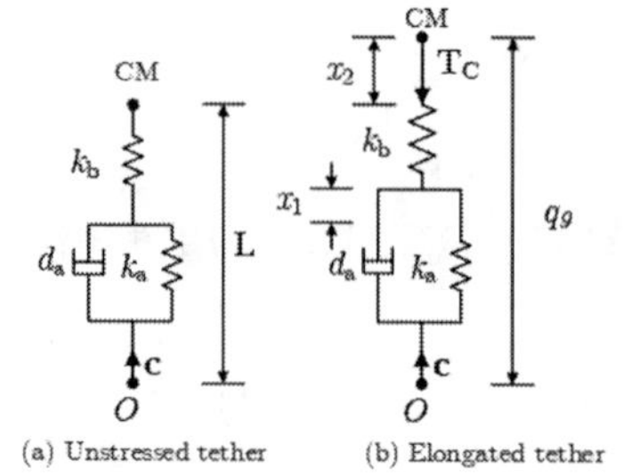

Fig. 2. Viscoelastic model of the tether, adapted from [5].

$$G(s) = \frac{T_C(s)}{U_C(s)} = \frac{d_a k_b\left(\frac{k_a}{d_a} + s\right)}{s(k_a + k_b + d_a s)} \tag{11}$$

where $U_C(s)$ represents the tether velocity of frequency domain. Regarding model parameters d_a, k_a and k_b, they are usually expressed as:

$$k_a = \frac{q.E_1}{L},\ k_b = \frac{q.E_2}{L},\ d_a = \frac{q.\eta}{L} \tag{12}$$

Parameter values that have been used in this study for simulation purposes are listed in Table 1.

Table 1. Parameters of the tether model [5].

	Parameter	Value	Units
Diameter	d_C	$10e^{-3}$	m
Cross section	q	$\pi\left(\frac{d_C}{2}\right)^2$	m^2
Modulus of elasticity	E_1	0.194	GPa
	E_1	0.543	GPa
Viscosity	η	0.096	GPa.s
Initial tether length	L	10	m

3 Control system

3.1 Multicopter control

The multicopter control strategy presented in this section is described by Santos *et al* [6]. Let us define the total thrust vector $\in \mathbb{R}^3$ of the octocopter to be:

$$\mathbf{f} \triangleq f\mathbf{n}. \tag{13}$$

Let us define the inclination angle $\varphi \in \mathbb{R}$ of the octocopter to be:

$$\varphi \triangleq \cos^{-1} n_z. \tag{14}$$

Note that φ consists of the angle between $\mathbf{n}$ and Z_R (see Figure 3).

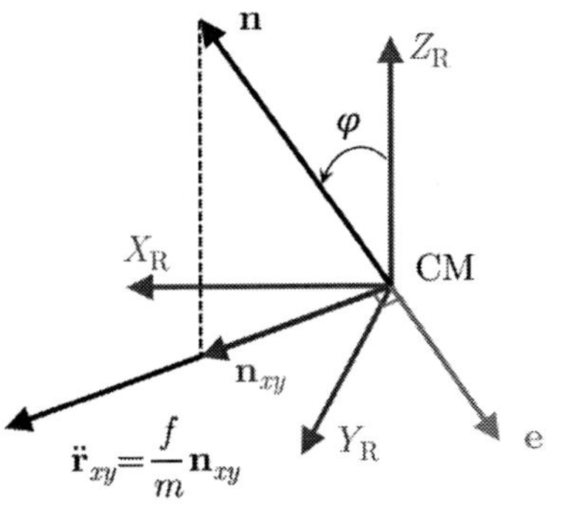

Fig. 3. Relationship between the normal unit vector $\mathbf{n}$ and the horizontal acceleration $\ddot{\mathbf{r}}_{xy}$.

Let $\bar{\mathbf{r}} \triangleq [\bar{r}_x\ \bar{r}_y\ \bar{r}_z]^T \in \mathbb{R}^3$ denote the position desired for the octocopter and define the corresponding tracking error $\tilde{\mathbf{r}}^{(i)} \in \mathbb{R}^3$ by

$$\tilde{\mathbf{r}} \triangleq \mathbf{r} - \bar{\mathbf{r}}. \tag{15}$$

Let $\varphi_{max} \in \mathbb{R}$ be the maximum allowable values of the inclination angle φ, $f_{min} \in \mathbb{R}$ and $f_{max} \in \mathbb{R}$ be, respectively, the minimum and the maximum allowable values of the total thrust magnitude f. The solution to the problem of position control is to find a feedback control law for $\mathbf{f}$, which makes $\mathbf{r} \to \bar{\mathbf{r}}$ or $\tilde{\mathbf{r}} \to \mathbf{0}$, while respecting the constraints $f_{min} \le f \le f_{max}$ and $\varphi \le \varphi_{max}$. The control position is partitioned into control height and horizontal position control.

3.1.1 Height control

From Equation (10), the dynamic model of the height of the vehicle r_z is given by

$$\ddot{r}_z = \frac{n_z}{m} f - g. \tag{16}$$

Consider the system modeled by Equation (16), and the minimum $f_{min} > 0$ and maximum $f_{max} > f_{min}$ allowable values of the command total thrust magnitude f. Assume that m, g and $n_z \neq 0$ are exactly known. The feedback control law is given by:

$$f = \begin{cases} f_{min}, & \alpha_z < f_{min} \\ \alpha_z, & \alpha_z \in [f_{min}, f_{max}] \\ f_{max}, & \alpha_z > f_{max} \end{cases} \tag{17}$$

with

$$\alpha_z \triangleq \frac{m}{n_z}\left(g - k_1(r_z - \bar{r}_z) - k_2\dot{r}_z\right), \tag{18}$$

respecting the total thrust magnitude constraint $f_{min} \le f \le f_{max}$ and is such that if $f_{min} \le \alpha_z \le f_{max}$, then the vehicle height r_z responds to the height desired $\bar{r}_z$, as if it were governed by the following second order linear time-invariant (LTI) system:

$$\ddot{r}_z + k_2\dot{r}_z + k_1 r_z = k_1\bar{r}_z, \tag{19}$$

where $k_1 > 0 \in \mathbb{R}$ e $k_2 > 0 \in \mathbb{R}$ are controller coefficients.

3.1.2 Horizontal position control

Let $\mathbf{r}_{xy} \triangleq [r_x\ r_y]^T \in \mathbb{R}^2$ and $\mathbf{n}_{xy} \triangleq [n_x\ n_y]^T \in \mathbb{R}^2$ denote the horizontal projections of $\mathbf{r}$ and $\mathbf{n}$, respectively. From equation (10), one can write the horizon position dynamic model as

$$\ddot{\mathbf{r}}_{xy} = \frac{f}{m}\mathbf{n}_{xy}. \tag{20}$$

Consider the system modeled by equation (20) and the maximum allowable value $\varphi_{max} > 0$ of the inclination angle φ. Assume that m and $f \neq 0$ are exactly known. The feedback control law is given by:

$$\mathbf{n}_{xy} = \begin{cases} \alpha_{xy}, & \|\alpha_{xy}\| \le \sin\varphi_{max} \\ \sin\varphi_{max}\frac{\alpha_{xy}}{\|\alpha_{xy}\|}, & \|\alpha_{xy}\| > \sin\varphi_{max} \end{cases}, \tag{21}$$

with

$$\alpha_{xy} \triangleq -\frac{m}{f}\left[k_z\left(\mathbf{r}_{xy} - \bar{\mathbf{r}}_{xy}\right) + k_4\dot{\mathbf{r}}_{xy}\right], \tag{22}$$

respecting the inclination angle constraint $\varphi \le \varphi_{max}$ and is such that if $\|\alpha_{xy}\| \le \sin\varphi_{max}$, then $\mathbf{r}_{xy}$ responds to the horizontal position desired $\bar{\mathbf{r}}_{xy} \triangleq [\bar{r}_x\ \bar{r}_y]^T$ as if it were governed by the following pair of second order LTI systems:

$$\ddot{\mathbf{r}}_{xy} + k_4\dot{\mathbf{r}}_{xy} + k_3\mathbf{r}_{xy} = k_3\bar{\mathbf{r}}_{xy}, \tag{23}$$

where $k_3 > 0 \in \mathbb{R}$ e $k_4 > 0 \in \mathbb{R}$ are controller coefficients.

In practice, the attitude desired for the internal attitude control loops need to be computed from the normal unit vectors $\mathbf{n}$. In order to specify a unique attitude, it is necessary to select a heading. For example, one can choose a zero heading just by taking into consideration the attitude represented by the principal Euler angle/axis $(\varphi, \mathbf{e})$, where φ is the angle of inclination own and $\mathbf{e} \in \mathbb{R}^3$ is a unit vector (see Figure 3) is given by:

$$\mathbf{e} = \frac{\mathbf{n} \times \mathbf{n}_{xy}}{\|\mathbf{n} \times \mathbf{n}_{xy}\|} \tag{24}$$

From $(\varphi, \mathbf{e})$ one can thus represent the attitude of S_B with respect to S_R, using , for example, Euler angles parameterization that permits obtains $\bar{\phi}$, $\bar{\theta}$ and $\bar{\psi}$.

3.2 Tether tension control

The tether tension is controlled by a proportional control configuration. Equation 25 and Figure 4 show the architecture of the tether control, which is described by Sandino *et al*[5].

$$U_C^{\star} = k_p(T_C^{\star} - T_C) \tag{25}$$

where $U_C^{\star}$ is the desired velocity tether, $T_C^{\star}$ is the desired tether tension and k_p is proportional gain.

DOI: 10.1051/matecconf/20168303001

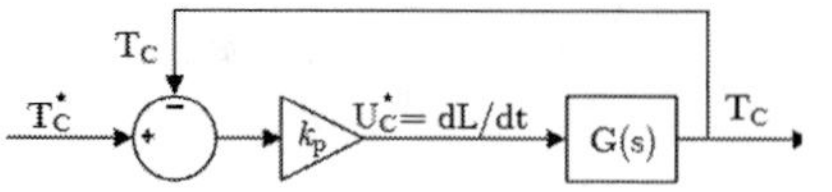

Fig. 4. Control system for simplified tether model given by G(s) as proposed by Sandino *et al* [5].

3.3 Tethered multicopter control

In order to control the height, a term T_C was added in Equation 18. Therefore, Equation 26 represents the proposed controller.

$$\alpha_z \triangleq \frac{m}{n_z}(g - k_1(r_z - r_z) - k_2\dot{r}_z) + T_C \quad (26)$$

4 Simulation results

The multicopter dynamics is integrated by using the fourth order Runge-Kutta method with time step of $T = 0.002$ s. The Reference trajectory follows the way-points : $\mathbf{w(1)} = [0\ 0\ 0]^T$m, $\mathbf{w(2)} = [4.24\ 4.24\ 8]^T$m, $\mathbf{w(3)} = [4.24\ -4.24\ 8]^T$m, $\mathbf{w(4)} = [-4.24\ -4.24\ 8]^T$m, $\mathbf{w(5)} = [-4.24\ 4.24\ 8]^T$m and $\mathbf{w(6)} = [4.24\ 4.24\ 8]^T$m. The mass of vehicle is $m = 2.132$ kg and the gravity acceleration is $g = 9.796$ m/s^2. The control parameters of vehicle, k_1, k_2, k_3 and k_4, are chosen based on equations (19) and (23). The maximum overshoot of $M_p^{ref} = 0.02$ m, and a peak time of $t_p^{ref} = 2$ s, were selected, which resulted in $k_1 = 6.29$ and $k_2 = 3.91$. Similarly, the maximum overshoot of $M_p^{ref} = 0.02$ m and a peak time of $t_p^{ref} = 3$ s were selected and the $k_3 = 2.78$ and $k_4 = 2.61$ gains were obtained. The proportional gain tether control selected was $k_p = 0.01$. The simulation was ran for 70 seconds, and the following bounds were selected: $\varphi_{max} = 30°$, $f_{max} = 40$ N e $f_{min} = 2$ N. A sinusoidal disturbance input was added in axes X and Z at $t = 15$s with amplitude of 5N . The simulations provide comparisons between free and tethered flight. The control error was quantified by the root mean square error, given by:

$$E^{RMS} = \sqrt{\frac{1}{3}[(\text{error } x)^2 + (\text{error } y)^2 + (\text{error } z)^2]} \quad (27)$$

Figure 5 presents the position tracking error of the way-points. We can note that the disturbance results in a lower amplitude in the position error for tethered vehicle when compared with the MAV in free flight.

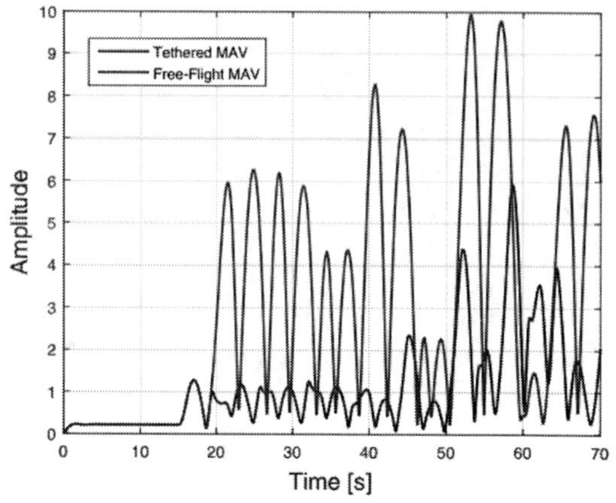

Fig. 5. Position error

5 Conclusion

This work has presented the modeling and control of a tethered MAV configuration. Numerical simulations showed that, when compared with the traditional vehicle in free flight, the tethered configuration has a better performance under external disturbances. The tether reduces the position error amplitude, improving the robustness of the vehicle. For future work, we consider a lumped mass model for the tether, as well as a pertubation model that corresponds to actual wind profiles in outdoor environments.

Acknowledgements

We would like to thanks Fundação de Amparo à Pesquisa do Amazonas (FAPEAM), Conselho Nacional de Desenvolvimento Científico e Tecnológico (CNPq) and Coordenação de Aperfeiçoamento de Pessoal de Nível Superior (CAPES) for the grants awarded.

References

1. S. Khaleefa, S. Alsamhi, and N. Rajput. Tethered balloon technology for telecommunication, coverage and path loss. IEEE Students Conference on Electrical, Electronics and Computer Science, 2014.
2. M. Nohmi. Mission design of a tethered robot satellite stars for orbital experiment. 18th IEEE International Conference on Control Applications, 2009.
3. C. Papachristos and A. Tzes. The power tethered uav ugv team: A collaborative strategy for navigation in partially mapped environments. 22nd Mediterranean Conference on Control and Automation (MED), 2014.
4. D. Rye. Longitudinal stability of a hovering tethered rotorcraft. *Journal of Guindance, Control and Dynamics*, 8:743–752, 1985.
5. L. A. Sandino, M. Bejar, K. Kondak, and A. Ollero. Advances in Modeling and Control of Tethered Unmanned Helicopters to Enhance Hovering Performance. *Journal of Intelligent and Robotic Systems*, 73:3–18, 2014.
6. D. A. Santos, O. Saotome, and A. Cela. Position Formation Control of Multirotors using Thrust Vector Constraints. *IEEE Transactions on Aerospace and Electronic Systems*, 2013.
7. G. Schmidt and R. Swik. Automatic hover control of an unmanned tethered rotoplatform. *Automatica*, 10:393–403, 1974.
8. J. R. Wertz. *Spacecraft Attitude Determination and control*. Kluwer Academic Publishers, The Netherlands, 1978.

MATEC Web of Conferences **83**, 03002 (2016) DOI: 10.1051/matecconf/20168303002
CSNDD 2016

Asymptotically stable control for a nonlinear-based multirotor aerial vehicle model

Igor Afonso Acampora Prado[a], Davi Ferreira de Castro[b], and Mateus de Freitas Virgílio Pereira[c], Davi Antônio dos Santos[d] and José Manoel Balthazar[e]
[1]Instituto Tecnológico de Aeronáutica, ITA, Brazil.

Abstract. The interest for multirotor aerial vehicles (MAVs) is currently growing due to their low cost, high manoeuvrability, simplified mechanics, capability to perform vertical take-off and landing as well as hovering flight. These characteristics make them a promising technology suitable for applications such as surveillance of indoor and urban environments. The present work faces the problem of controlling the attitude of a MAV by means of a linear feedback control which guarantees asymptotic stability when controlling nonlinear dynamics. The simulations show the effectiveness of the method.

1 Introduction

Multirotor Aerial Vehicles (MAVs) have motivated many researches in different fields of knowledge such as sensors fusion [3], computer vision [7] and control strategies [6]. Although there is a massive amount of concluded and ongoing research works on MAVs, the design of control laws for such vehicles still has challenges to be overcome. Most of those challenges are related to design control laws that guarantee asymptotic stability, mostly because this system is nonlinear.

Reference [2] presents two nonlinear control techniques, backstepping and sliding mode, in order to control the attitude of the MAV. On the other hand, reference [4] utilizes a nonlinear H_∞ controller to stabilize the rotational movements. Both strategies have a high degree of complexity. Therefore, it would be interesting to design a linear feedback control which also guarantees stability of system.

The Linear Quadratic Regulator (LQR) is an usual control strategy applied to MAVs [1], which requires a linear design model. Such linear model does not take into account the nonlinear phenomena that are present in the real operation of these vehicles. However, for a more effective control law, nonlinearities have to be considered in order to avoid unstable regions of operation. One of the nonlinearities included in the motion equations is the gyroscopic effects which come from the rotation of the rigid body and the four propellers.

This paper uses the [5] theorem to stabilize the attitude of a MAV. This theorem showed that is possible to construct a linear feedback control law for a nonlinear dynamic that is globally asymptotically stable.

a e-mail: igorap@ita.br
b e-mail: davifc@ita.br
c e-mail: mateusfvp@hotmail.com
d e-mail: davists@ita.br
e e-mail: jmbaltha@ita.br

2 MAV modeling

This section presents the equations of motion of a multirotor aerial vehicle (MAV), considering all the relevant effects necessary to be accounted for in ground-truth model for simulation-based evaluation of control laws. We start with preliminary definitions in Subsection 2.1, then we present the rotor dynamics in 2.2 and the MAV rotational dynamics in 2.3.

2.1 Preliminary definitions

We define two Cartesian coordinate systems (CCS) as illustrated in Figure 1. The body CCS, $S_B \triangleq \{X_B, Y_B, Z_B\}$ is attached to the vehicle's body with the origin at the vehicle's center of mass (CM), the X_B axis pointing forward, the Z_B axis pointing upward, perpendicular to the plane of the rotors, and the Y_B axis completing a dextrogeous frame. The CCS, $S_G = \{X_G, Y_G, Z_G\}$ is fixed to the ground at a known point O, with the Z_G axis pointing upward, aligned with the local vertical.

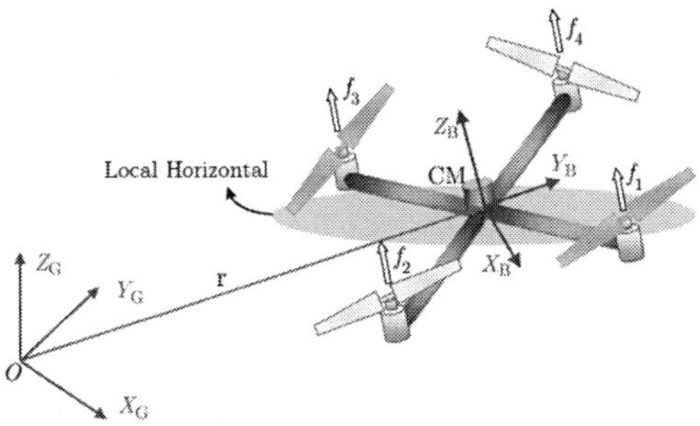

Fig. 1. The Cartesian coordinate systems (CCS). $S_B = \{X_B, Y_B, Z_B\}$ is the body CCS and $S_G = \{X_G, Y_G, Z_G\}$ is the ground CCS.

2.2 Rotor dynamics

The thrust force f_i and reaction torque τ_i produced by each individual rotor are modeled, respectively, by the aerodynamic models

$$f_i = k_f \omega_i^2, \quad (1)$$

$$\tau_i = k_\tau \omega_i^2, \quad (2)$$

for $i = 1, ..., 4$, where k_f is the force coefficient, k_τ is the torque coefficient, and ω_i is the rotation speed of the ith rotor (positive in counterclockwise direction), whose dynamics can be modeled by the following first-order linear model:

$$\dot{\omega}_i = -\frac{1}{\tau_\omega}\omega_i + \frac{k_\omega}{\tau_\omega}\bar{\omega}_i \quad (3)$$

where $\bar{\omega}_i \in [0, \bar{\omega}_{\max}]$ is the rotation speed command for the ith rotor, k_ω is the rotation coefficient, and τ_ω is rotor time constant.

2.3 Resultant control thrust and torque

Although the formulation given here could be generalized to any multirotor vehicle configuration, for illustration purpose, we consider a quadrotor configuration with the longitudinal axis X_B pointing between rotor 1 and rotor 2, and with 45 degree of separation angle between adjacent arms; we call it the "x" configuration. Moreover, we consider that rotor 1 and rotor 3 rotate in clockwise direction, while rotor 2 and rotor 4 rotate in counterclockwise direction. For the aforementioned configuration, one can show that the relationship between the individual rotor thrusts f_i and the resultant thrust magnitude and torque vector is given by

$$\begin{bmatrix} F^c \\ \mathbf{T}^c \end{bmatrix} = \boldsymbol{\Gamma}\mathbf{f}, \quad (4)$$

with $\mathbf{f} \triangleq \begin{bmatrix} f_1 \; f_2 \; f_3 \; f_4 \end{bmatrix}^{\mathrm{T}}$ and

$$\boldsymbol{\Gamma} \triangleq \begin{bmatrix} 1 & 1 & 1 & 1 \\ l/\sqrt{2} & -l/\sqrt{2} & -l/\sqrt{2} & l/\sqrt{2} \\ -l/\sqrt{2} & -l/\sqrt{2} & l/\sqrt{2} & l/\sqrt{2} \\ k_\tau/k_f & -k_\tau/k_f & k_\tau/k_f & -k_\tau/k_f \end{bmatrix}, \quad (5)$$

where l is the length of each vehicle's arm. The superscript c in Eq. (5) stands for the control command defined by the operator and is used to distinguish from the other sources of force and torque that will be presented soon.

By inverting Eq. (5), one can compute the command $\bar{\mathbf{f}} \triangleq \begin{bmatrix} \bar{f}_1 \; \bar{f}_2 \; \bar{f}_3 \; \bar{f}_4 \end{bmatrix}^{\mathrm{T}}$ as

$$\bar{\mathbf{f}} = \Xi \begin{bmatrix} \bar{F}^c \\ \bar{\mathbf{T}}^c \end{bmatrix}, \quad (6)$$

where $\Xi \triangleq \Gamma^{-1}$, $\bar{F}^c$ is the thrust magnitude command, and $\bar{\mathbf{T}}^c$ is the control torque command.

2.4 Rotational motion

Representing the attitude using the Euler angles $\alpha \triangleq \begin{bmatrix} \phi \; \theta \; \psi \end{bmatrix}^{\mathrm{T}}$ in the rotation sequence 1-2-3, we have the following rotational kinematics equations:

$$\dot{\alpha} = \mathcal{A}\boldsymbol{\Omega} \quad (7)$$

where $\boldsymbol{\Omega} \triangleq \begin{bmatrix} \Omega_x \; \Omega_y \; \Omega_z \end{bmatrix}^{\mathrm{T}}$ is the S_{B} representation of the vehicle's angular velocity w.r.t. S_{G} and

$$\mathcal{A} \triangleq \begin{bmatrix} \cos\psi/\cos\theta & -\sin\phi/\cos\theta & 0 \\ \sin\psi & \cos\psi & 0 \\ -\cos\psi\sin\theta/\cos\theta & \sin\psi\sin\theta/\cos\theta & 1 \end{bmatrix} \quad (8)$$

Assume that the vehicle has a rigid structure and S_{G} is an inertial frame. Considering the existence of gyroscopic effects, due to the rotors, and disturbance torques, and using the Newton-Euler formulation, one can model the rotational dynamics of the MAV by

$$\dot{\boldsymbol{\Omega}} = \mathbf{J}^{-1}(\mathbf{J}\boldsymbol{\Omega})\times\boldsymbol{\Omega} + I_r\mathbf{J}^{-1}\boldsymbol{\Omega}\times\mathbf{e}_3\sum_{i=1}^{4}(-1)^i\omega_i + \mathbf{J}^{-1}\mathbf{T}^c + \mathbf{J}^{-1}\mathbf{T}^d \quad (9)$$

where $\mathbf{T}^c$ and $\mathbf{T}^d$ are the S_{B} representations of the resultant control torque and disturbance torque, respectively, $\mathbf{e}_3 \triangleq [0 \;\; 0 \;\; 1]^{\mathrm{T}}$, I_r is the moment of inertia of the rotors w.r.t. the rotation axis, and $\mathbf{J}$ is the body inertia matrix. Consider that the vehicle has a symmetric structure with known mass m and inertia matrix in S_{B}

$$\mathbf{J} = \begin{bmatrix} J_x & 0 & 0 \\ 0 & J_y & 0 \\ 0 & 0 & J_z \end{bmatrix}. \quad (10)$$

This paper makes use of the AR Drone 2 parameters in order to emulate a real platform. The values can be seen in Table 1.

3 State-space models

This section presents the nonlinear state-space model for rotation (Section 2). This model will be adopted for designing control laws by using the Rafikov and Balthazar's theorem. Firstly, note that the dynamics in Eq. (7) and (9) can be formulated as

$$\dot{\mathbf{x}}(t) = \mathbf{A}\mathbf{x}(t) + \mathbf{h}(\mathbf{x}) + \mathbf{B}\mathbf{u}(t) \quad (11)$$

where $\mathbf{x} \in \mathbb{R}^n$ is the state vector, $\mathbf{u} \in \mathbb{R}^m$ is the control inputs, $\mathbf{A} \in \mathbb{R}^{n\times n}$ and $\mathbf{B} \in \mathbb{R}^{n\times m}$ are constant real matrices, and $\mathbf{h}(\mathbf{x}) \in \mathbb{R}^n$ is a vector whose elements are nonlinear functions of $\mathbf{x}$.

3.1 Rotational state-space model

The rotational MAV state-space model is given by

$$\mathbf{x} \triangleq \begin{bmatrix} \phi \; \theta \; \psi \; \Omega_x \; \Omega_y \; \Omega_z \end{bmatrix}^{\mathrm{T}} \in \mathbb{R}^6, \quad (12)$$

$$\mathbf{u} \triangleq \mathbf{T}^c \in \mathbb{R}^3, \quad (13)$$

and

$$\mathbf{A} = \begin{bmatrix} 0&0&0&1&0&0 \\ 0&0&0&0&1&0 \\ 0&0&0&0&0&1 \\ 0&0&0&0&0&0 \\ 0&0&0&0&0&0 \\ 0&0&0&0&0&0 \end{bmatrix} \quad (14)$$

Table 1. Parameters of AR Drone 2.

Variables	Values	Unit
Mass of AR Drone 2, m	0.429	kg
Acceleration of gravity, g	9.80665	m/s^2
Arm length, l	0.1785	m
Moment of inertia for x-axis, J_x	2.238×10^{-3}	$kg.m^2$
Moment of inertia for y-axis, J_y	2.986×10^{-3}	$kg.m^2$
Moment of inertia for z-axis, J_z	4.804×10^{-3}	$kg.m^2$
Moment of inertia of each rotor, I_r	2.030×10^{-5}	$kg.m^2$
Force coefficient k_f	8.050×10^{-6}	$N/(rad/s)^2$
Torque coefficient, k_τ	2.423×10^{-7}	$N.m/(rad/s)^2$
Time constant, τ	4.718×10^{-3}	s
Maximum speed of the motor, $\omega_{\max}$	1047	rad/s

$$\mathbf{B} = \begin{bmatrix} 0 & 0 & 0 \\ 0 & 0 & 0 \\ 0 & 0 & 0 \\ \frac{1}{J_x} & 0 & 0 \\ 0 & \frac{1}{J_y} & 0 \\ 0 & 0 & \frac{1}{J_z} \end{bmatrix} \tag{15}$$

$$\mathbf{h}(\mathbf{x}) = \begin{bmatrix} \frac{1}{2}(-\psi^2\Omega_x + \theta^2\Omega_x) - \phi\Omega_y \\ \psi\Omega_x - \frac{\psi^2\Omega_y}{2} \\ -\theta\Omega_x + \psi\theta\Omega_y \\ \frac{(J_y-J_z)}{J_x}\Omega_y\Omega_z - \frac{I_r\Omega_y}{J_x}(\omega_1 - \omega_2 + \omega_3 - \omega_4) \\ \frac{(J_z-J_x)}{J_y}\Omega_x\Omega_z + \frac{I_r\Omega_x}{J_y}(\omega_1 - \omega_2 + \omega_3 - \omega_4) \\ \frac{(J_x-J_y)}{J_z}\Omega_x\Omega_y \end{bmatrix} \tag{16}$$

4 Control design

In this section, we present a control strategy to drive the MAV's nonlinear dynamics from a certain initial condition to a desired state. Following the control method proposed by [5], we formulate an optimal control problem that, given the nonlinear dynamics in Eq. (11), aims at minimizing the following LQR-like cost function:

$$J = \int_0^\infty (l(\mathbf{x}) + \mathbf{u}^T\mathbf{R}\mathbf{u})dt \tag{17}$$

where $\mathbf{R} \in \mathbb{R}^{m \times m}$ is a positive definite matrix and $l(\mathbf{x}) \in \mathbb{R}$ is a function related to the weights of the dynamical states over time. Function $l(\mathbf{x})$ will be selected as follows to account for the nonlinear terms in the dynamical model.

Propostion: Let $l(\mathbf{x})$ be

$$l(\mathbf{x}) = \mathbf{x}^T\mathbf{Q}\mathbf{x} - \mathbf{h}(\mathbf{x})^T\mathbf{P}\mathbf{x} - \mathbf{x}^T\mathbf{P}\mathbf{h}(\mathbf{x}) \tag{18}$$

where $\mathbf{Q} \in \mathbb{R}^{n \times n}$ and $\mathbf{P} \in \mathbb{R}^{n \times n}$ are real symmetric positive definite matrices and $\mathbf{P}$ is the solution of the matrix algebraic Riccati equation

$$\mathbf{PA} + \mathbf{A}^T\mathbf{P} - \mathbf{PBR}^{-1}\mathbf{B}^T\mathbf{P} + \mathbf{Q} = 0. \tag{19}$$

Then, as long as $l(\mathbf{x})$ is positive definite, the linear feedback control law

$$\mathbf{u} = -\mathbf{R}^{-1}\mathbf{B}^T\mathbf{P}\mathbf{x} \tag{20}$$

is optimal in order to transfer the nonlinear system in Eq. (2) from an initial condition $\mathbf{x}(0)$ to the origin while minimizing the cost function in Eq (3). Additionally, the controlled system is locally asymptotically stable in the neighborhood $\Gamma_0 \subset \Gamma, \Gamma \subset \mathbb{R}^n$, of the origin if $\mathbf{x}(0) \in \Gamma_0$. If $\Gamma = \mathbb{R}^n$, then the controlled system is globally asymptotically stable.

Proof: See Rafikov and Balthazar [5].

5 Simulation tests

The simulation uses three-degree-of-freedom nonlinear equations of the attitude of the MAV and are implemented in Matlab/Simulink. In this section, numerical simulations are presented to verify the effectiveness of the proposed method. In these numerical simulations, the initial condition $\mathbf{x}(0)$ is given by

$$\mathbf{x}(0) = [30\ 20\ 10\ 0\ 0\ 0]^T. \tag{21}$$

Choose

$$\mathbf{Q} = \begin{bmatrix} 17 & 0 & 0 & 0 & 0 & 0 \\ 0 & 17 & 0 & 0 & 0 & 0 \\ 0 & 0 & 17 & 0 & 0 & 0 \\ 0 & 0 & 0 & 1 & 0 & 0 \\ 0 & 0 & 0 & 0 & 1 & 0 \\ 0 & 0 & 0 & 0 & 0 & 1 \end{bmatrix} \tag{22}$$

and

$$\mathbf{R} = \begin{bmatrix} 1 & 0 & 0 \\ 0 & 1 & 0 \\ 0 & 0 & 1 \end{bmatrix} \tag{23}$$

we obtain

$$\mathbf{P} = \begin{bmatrix} 4.16 & 0 & 0 & 0.01 & 0 & 0 \\ 0 & 4.17 & 0 & 0 & 0.01 & 0 \\ 0 & 0 & 4.20 & 0 & 0 & 0.02 \\ 0.01 & 0 & 0 & 0.002 & 0 & 0 \\ 0 & 0.01 & 0 & 0 & 0.003 & 0 \\ 0 & 0 & 0.02 & 0 & 0 & 0.01 \end{bmatrix} \tag{24}$$

by solving the Riccati Eq. (19). Then, from Eq. (20) one obtains

 DOI: 10.1051/matecconf/20168303002

$$\mathbf{u} = \begin{bmatrix} 4.123 & 0 & 0 & 1.01 & 0 & 0 \\ 0 & 4.123 & 0 & 0 & 1.012 & 0 \\ 0 & 0 & 4.12 & 0 & 0 & 1.02 \end{bmatrix} \mathbf{x} \qquad (25)$$

The behavior of the attitude angles can be seen in Figure 2. The simulations show that the function $l(\mathbf{x})$ calcu-

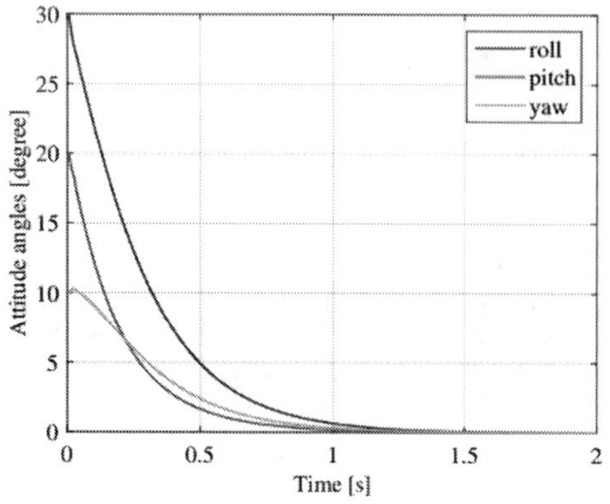

Fig. 2. Euler angles over time.

lated for the attitude dynamics of the MAV is positive, as illustrated in Figure 3, which guarantees asymptotic stability according to Rafikov and Balthazar's theorem.

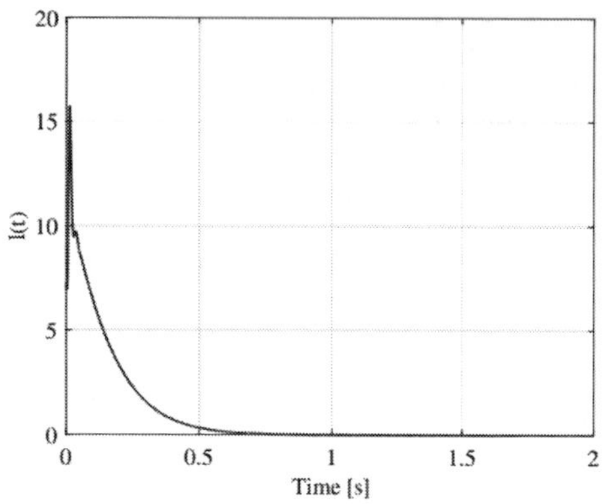

Fig. 3. Positive function $l(t)$, calculated for the attitude dynamics of the MAV.

The control signals, presented in Figure 4, lie within the acceptable range for a real MAV, i.e. ±2.28 N.m.

6 Conclusion

This paper has presented the attitude control of a MAV by designing the linear feedback controller. The linear feedback control problem was formulated under the optimal control theory viewpoint. Asymptotic stability of the closed-loop nonlinear system is guaranteed by means of a Lyapunov function as presented in [5].

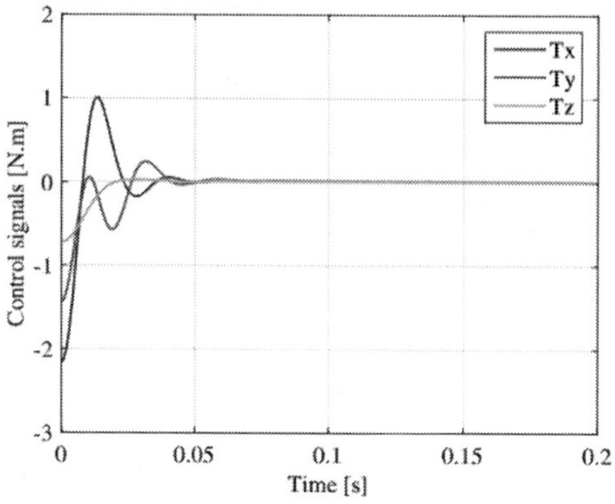

Fig. 4. Control signals over time.

Acknowledgments

The authors thank Coordenação de Aperfeiçoamento de Pessoal de Nível Superior (CAPES) and Conselho Nacional de Desenvolvimento Científico e Tecnológico (CNPq) for financial supports.

References

1. S. Bouabdallah, A. Noth, and R. Siegwart. Pid vs lq control techniques applied to an indoor micro quadrotor. In *Proceedings of the IEEE/RSJ International Conference on Intelligent Robots and Systems*, pages 2451–2456, New Orleans, September 2004.
2. S. Bouabdallah and R. Siegwart. Backstepping and sliding-mode techniques applied to an indoor micro quadrotor. In *Proceedings of the IEEE International Conference on Robotics and Automation*, pages 2247–2252, Barcelona, April 2005.
3. T. Cheviron, T. Hamel, R. Mahony, and G. Baldwin. Robust nonlinear fusion of inertial and visual data for position, velocity and attitude estimation of uav. In *Proceedings of the IEEE International Conference on Robotics and Automation*, pages 2010–2016, Roma, April 2007.
4. G. V. Raffo, M. G. Ortega, and F. R. Rubio. An integral predictive/nonlinear $\mathcal{H}_\infty$ control structure for a quadrotor helicopter. *Automatica*, 46(1):29–39, 2010.
5. M. Rafikov and J. M. Balthazar. On control and synchronization in chaotic and hyperchaotic systems via linear feedback control. *Communications in Nonlinear Science and Numerical Simulation*, 13(4):1246–1255, 2008.
6. D. A. Santos, O. Saotome, and A. Cela. Trajectory control of multirotor helicopters with thrust vector constraints. In *21st Mediterranean Conference on Control and Automation*, pages 375–379, Chania, 2013.
7. S. Saripalli, J. F. Montgomery, and G. S. Sukhatme. Visually-guided landing of an unmanned aerial vehicle. *IEEE Transactions on Robotics and Automation*, 10(3):371–381, 2003.

Paris law parameter identification based on the Extended Kalman Filter

M. Melgar[1], C. Gomez-Jimenez[2], L.D. Cot[1,a], S. Dejean[3], C. Mabru[1], and J. Martinez-Vega[4]

[1] Institut Clment Ader ; Universit de Toulouse ; INSA ; 3, rue Caroline Aigle, 31400 Toulouse, France
[2] INSA ; Universit de Toulouse ; 135, avenue de rangueil, 31077 Toulouse, France
[3] Institut de Mathmatiques de Toulouse ; Universit de Toulouse ; UPS ; 118, route de Narbonne, 31062 Toulouse, France
[4] LAPLACE INP ; Universit de Toulouse ; UPS ; 118, route de Narbonne, 31062 Toulouse, France

Abstract. Aircraft structures are commonly subjected to repeated loading cycles leading to fatigue damage. Fatigue data can be extrapolated by fatigue models which are adopted to describe the fatigue damage behaviour. Such models depend on their parameters for accurate prediction of the fatigue life. Therefore, several methods have been developed for estimating the model parameters for both linear and nonlinear systems. It is useful for a broad class of parameter identification problems when the dynamic model is not known. In this paper, the Paris law is used as fatigue-crack-length growth model on a metallic component under loading cycles. The Extended Kalman Filter (EKF) is proposed as estimation method. Simulated crack length data is used to validate the estimation method. Based on experimental data obtained from fatigue experiment, the crack length and model parameters are estimated. Accurate model parameters allow a more realistic prediction of the fatigue life, consequently, the remaining useful life (RUL) of component can be accurately computed. In this sense, maintenance performance could be improved.

1 Introduction

Fatigue damage is one of the most probable modes of failure in aeronautical structures. During ground-air-ground cycles, in-service aircraft undergoes repeated pressurization and depressurization leading to fatigue loading cycles. Prediction of fatigue damage can prevent catastrophic failures. Currently, structural maintenance consists in time-based inspections integrated to overall aircraft timetable. The issue to improve operational reliability and reduce maintenance costs led to more cost-effective condition-based inspection approaches. Many investigations on structural health monitoring have shown that SHM is an appropriate tool for condition-based maintenance. On-board sensors and actuators assess structure health in real-time and allow to detect structural damages. Our purpose addresses accurate fatigue damage prognostic based on state-parameter estimation and experimental data in a metal specimen. The appropriate fatigue model for metal is the nonlinear Paris law depending upon uncertain parameters that describes the crack growth as a function of loading cycles [1] [2]. Usually, parameters are unknown because they are not measurable. Parameters associated to material properties are determined from experimental data. Parameter uncertainty can be observed if an experiment is repeated several times.

Many contributions based on parameter estimation have been presented. Among them, we can mention Coppe et al. [3] who have determined one of the two Paris law parameter using Bayesian inference and least-squares method with simulated noisy measurements; then, the RUL is computed. Wu and Ni [4] evaluated a stochastic fatigue damage parametric model whose parameters are deduced from experimental data and linear regression. Khelil et al.[5] have determined Paris law parameters using a linear regression method and experimental data from two aluminum alloy specimens. On the other hand, measurement instruments are tainted by random white noise which is combined to measurements and thus, affects the parameter estimation process. Such situation has led researchers to develop filtering methods such as the extended Kalman filter (EKF) for nonlinear systems [6][7][8]. The EKF progressively reduces the white noise in measurements and estimates both system state and associated parameters. Szabat et al. [7] have developed an algorithm to estimate various mechanical parameters of a mechanical control system. Lei et al.[8] proposed an estimating algorithm for detecting structural damage with limited measurements for civil structures. Singleton et al.[9] used the EKF to learn model parameters and to predict the RUL of bearing faults. In-service structures are subjected to fatigue loading under real-time and in-situ variable conditions. Probabilistic-statistical methods have been introduced to estimate parameters from experimental measurements [10]. Estimated parameters must reach true parameter values through real measurements. Combining physics-based damage models and available real-time measurements, parameter estimation might be performed. The estimation of both state and parameters is based on a joint system state-parameter vector also called the joint extended Kalman filter (JEKF) [11]. Corigliano et al. [12][13] showed that the JEKF provides the state variable and parameters estimation of a damage model for a structural system subject to dynamic loadings. Thus, the JEKF can be used to estimate both crack length and Paris law unknown parameters. Contributions of EKF estimation methods using fatigue crack models based on Paris law can be seen in [14] [15]. Based on accurate estimated parameters, the Paris law is able to provide a high-quality fatigue crack growth prediction. Knowing the behaviour of the crack growth, enables to predict the time to failure and

[a] e-mail: lea.cot@insa-toulouse.fr

therefore, the remaining useful life (RUL) of the component [10][16][3]. The identification of accurate parameters allows an accurate RUL prediction. As a result, the structural maintenance might be improved. In this paper, the JEKF is used to estimate both state and Paris law parameters using experimental data provided by fatigue loading experiment on steel specimen. The JEKF provides accurate parameter values. A sensitivity analysis is performed by first studying influence of process and measurement noises. The impact of initial estimation on parameters and initial variances of estimation error on parameters have been also studied. It is shown that the proposed modelling under initial conditions lead to accurate estimation results consistent with experimental data. The paper is organized as follows. In section 2, a single edge crack propagation experiment of a steel sample and measurement collection is explained. Section 3 details the JEKF estimation problem modelling. In section 4, results obtained are discussed followed by conclusions.

2 Fatigue damage experiment

Crack size real data have been obtained from a crack propagation experiment of a X38CrMoV5 steel sample under fatigue loading [17]. A single edge crack propagation experiment is considered. All specimens were machined and polished to obtain the final thickness (0.6 mm). More details of the specimen geometry and the experimental procedure can be consulted in [17]. Fatigue measurements obtained from this experiment (Fig. 1) have been used in the estimation process. Fatigue model is nonlinear dynamical

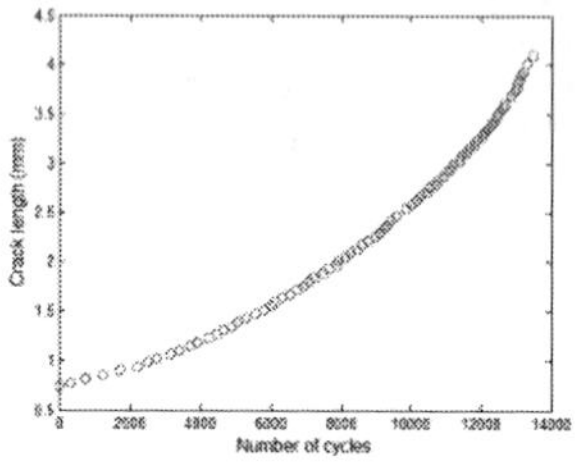

Fig. 1. Experimental fatigue crack length growth.

systems depending on model parameters. In the case of homogenous materials such as metal, the fatigue damage increases with applied cycles in a cumulative way. Currently, the Paris law is considered as fatigue crack growth model based on an ordinary differential equation expressing the crack size growth l with the number of loading cycles t applied to the structure

$$\dot{l}(n) = C(\Delta K(l))^m. \qquad (1)$$

C and m are uncertain parameters also called Paris law parameters. Because of manufacturing process of structure, parameters C and m are variables and uncertainties arise. The stress intensity factor can be expressed as $\Delta K(l) = g(l)l^{\frac{1}{2}}$ where function g depends on the geometry of the specimen. The crack growth length is therefore expressed as

$$\dot{l}(n) = C(g(l))^m l^{\frac{m}{2}}. \qquad (2)$$

In the case of the fatigue experiment explained in section 2, function g is defined as

$$g(l) = \Delta\Sigma\,\sqrt{\pi}F(\frac{l}{w})$$

where w is the specimen width, $\Delta\Sigma$ the loading applied and $F(\frac{l}{w})$ the correction factor defined as

$$F(\frac{l}{w}) = 1.0869 + 0.2383\frac{l}{w} + 1.9830(\frac{l}{w})^2 - 2.8373(\frac{l}{w})^3$$
$$+2.5771(\frac{l}{w})^4. \qquad (3)$$

In this case

$$K(l) = \Delta\Sigma\,\sqrt{\pi}F(\frac{l}{w})\,\sqrt{l}. \qquad (4)$$

3 Estimation problem modelling

3.1 Joint extended Kalman filter in brief

Let us set the general parametric model represented by the differentiable nonlinear function f defined on $\mathbb{R}^p \times \mathbb{R}^q$ and values in $\mathbb{R}^p$ as

$$\begin{cases} \dot{y} = f(y, \Lambda) \\ y_{(0)} = y_0 \end{cases} \qquad (5)$$

where $\Lambda = (\lambda^{(1)}, ..., \lambda^{(q)})$ is the unknown parameter vector.

3.2 Extended Kalman Filter formulation

Let us consider the discrete-time process governed by the nonlinear stochastic equations at time k

$$\begin{cases} x_k = f(x_{k-1}, \Lambda_{k-1}) + w_{k-1} \\ z_k = h(x_k) + v_k \end{cases} \qquad (6)$$

where f and h are nonlinear differentiable functions, $x \in \mathbb{R}^p$ the state variable, $\Lambda \in \mathbb{R}^q$ the unknown parameter vector and $z \in \mathbb{R}^p$ the measurement vector. w and v are process and measurement noise assumed to be additive and zero mean Gaussian white. The estimation of both state and parameters requires the use of the joint extended Kalman filter by constructing the augmented state vector $X \in \mathbb{R}^{p+q}$ such that $X_k = (x_k\ \Lambda_k)^T$ at time k. Assuming that parameters are constant during the process, the JEKF is now based on equations

$$\begin{cases} x_k = f(x_{k-1}) + w_{k-1} \\ \Lambda_k = id(\Lambda_{k-1}) \\ z_k = h(X_k) + v_k \end{cases} \qquad (7)$$

The JEKF requires the computation of the Jacobian matrix $J_k(\hat{X}_{k-1})$ at time k relative to the augmented state X_k evaluated at the last estimate $\hat{X}_{k-1} = (\hat{x}_{k-1}, \hat{\Lambda}_{k-1})^T$, i.e. at time $k-1$. Because $J_{id,k}(\hat{x}_{k-1}) = 0_{q,n}$ and $J_{id,k}(\hat{\Lambda}_{k-1}) = I_q$, the Jacobian matrix is expressed as

$$J_k(\hat{X}_{k-1}) = \begin{pmatrix} J_{f,k}(\hat{x}_{k-1}) & J_{f,k}(\hat{\Lambda}_{k-1}) \\ 0_{q,n} & I_q \end{pmatrix} \qquad (8)$$

where $J_{f,k}(\hat{x}_{k-1}) = \frac{\partial f^{(i)}(x,\Lambda)}{\partial x^{(j)}}|_{(x=\hat{x}_{k-1},\Lambda=\hat{\Lambda}_{k-1})}$, $\forall 1 \le i, j \le n$,
and $J_{f,k}(\hat{\Lambda}_{k-1}) = \frac{\partial f^{(i)}(x,\Lambda)}{\partial \lambda^{(j)}}|_{(x=\hat{x}_{k-1},\Lambda=\hat{\Lambda}_{k-1})}$, $\forall 1 \le i \le q,\ 1 \le j \le n$.
Only the crack length increase is observed. So, in 7, the third equation become

$$z_k = HX_k + v_k$$

where the sensitivity matrix $H \in M_{p,p+q}(\mathbb{R})$ is $H = (1\,0\,0)^T$. The EKF standard algorithm is then applied to solve system (7).

3.3 Application to the Paris law

In this case, the joint vector is $X = (l, C, m)^T$ and the vector to be estimated is $\Lambda = (C, m)^T$. The first recurrent equation in (7) is obtained by approximating the rate of damage growth in Eq.1 by a finite difference. We therefore obtain $l(n_k) = f(l(n_{k-1}), \Lambda_{k-1})$ where function f is defined as

$$f(l(n_{k-1}), \Lambda_{k-1}) \approx l(n_{k-1}) + Cg((n_{k-1}))^m l(n_{k-1})^{\frac{m}{2}}(n_k - n_{k-1}),$$
$$f(l_{k-1}, \Lambda_{k-1}) \approx l_{k-1} + Cg(l_{k-1})^m l_{k-1}^{\frac{m}{2}}(n_k - n_{k-1}), \quad (9)$$

by setting $l(n_k) = l_k$, $\forall k \in \mathbb{N}$. The Jacobian matrix is expressed as

$$J_k(\hat{X}_{k-1}) = \begin{pmatrix} J_{f,k}(\hat{l}_{k-1}) & J_{f,k}(\hat{C}_{k-1}) & J_{f,k}(\hat{m}_{k-1}) \\ 0 & 1 & 0 \\ 0 & 0 & 1 \end{pmatrix} \quad (10)$$

$J_{f,k}(\hat{l}_{k-1}) = \frac{\partial f(l,\Lambda)}{\partial l}\big|_{(l=\hat{l}_{k-1}, \Lambda=\hat{\Lambda}_{k-1})}$, $J_{f,k}(\hat{C}_{k-1}) = \frac{\partial f(l,\Lambda)}{\partial C}\big|_{(l=\hat{l}_{k-1}, \Lambda=\hat{\Lambda}_{k-1})}$ and $J_{f,k}(\hat{m}_{k-1}) = \frac{\partial f(l,\Lambda)}{\partial m}\big|_{(l=\hat{l}_{k-1}, \Lambda=\hat{\Lambda}_{k-1})}$.

As shown in section (4.1), this modelling do not lead to satisfying results due to exponent of parameter m in Eq. 9. Another modelling is therefore addressed by considering the joint vector $X = (\log l, \log C, m)^T$ and the vector of parameters to be estimated $\Lambda = (\log C, m)^T$. In this case, the new state function is defined as

$$f(\log l_{k-1}, \Lambda_{k-1}) \approx \log l_{k-1} + \exp(\log C + m \log g(l) + (\frac{m}{2} - 1)\log l_{k-1})\Delta n_k \quad (11)$$

where $g(l) = \Delta\Sigma\sqrt{\pi}F(\frac{l}{w})$ and $\Delta n_k = n_k - n_{k-1}$.
By taken into account expressions (3) of function F and its first derivative $\dot{F}$, and setting $\beta = \Delta\Sigma\sqrt{\pi}$, the Jacobian partial derivatives are expressed as follow

$$J_{f,k}(\hat{l}_{k-1}) = 1 + [\hat{m}_{k-1}\frac{\dot{F}(\frac{\hat{l}_{k-1}}{w})}{F(\frac{\hat{l}_{k-1}}{w})} + \frac{\hat{m}_{k-1}}{2} - 1] \quad (12)$$
$$.\exp(\hat{C}_{k-1} + \hat{m}_{k-1}\log(\beta F(\frac{\hat{l}_{k-1}}{w})) + (\frac{\hat{m}_{k-1}}{2} - 1)\hat{l}_{k-1})\Delta n_k,$$
$$J_{f,k}(\hat{C}_{k-1}) = \exp(\hat{C}_{k-1} + \hat{m}_{k-1}\log(\beta F(\frac{\hat{l}_{k-1}}{w})) + (\frac{\hat{m}_{k-1}}{2} - 1)\hat{l}_{k-1})\Delta n_k,$$
$$J_{f,k}(\hat{m}_{k-1}) = [\log(\beta F(\frac{\hat{l}_{k-1}}{w})) + \frac{1}{2}\hat{l}_{k-1}]$$
$$.\exp(\hat{C}_{k-1} + \hat{m}_{k-1}\log(\beta F(\frac{\hat{l}_{k-1}}{w})) + (\frac{\hat{m}_{k-1}}{2} - 1)\hat{l}_{k-1})\Delta n_k.$$

Based on equations (7), state function of Eq. 11 and Jacobian coefficients of Eq. 12, the EKF algorithm is applied to estimate the joint vector X that includes the Paris parameters. Results are provided in section 4.2.

4 Results and discussion

4.1 Results

Using real data allows us to recreate a real situation where measurements can be obtained by an SHM system. Our main interest is identified the better results in terms of the estimation error on real measurements. The estimations of crack length and parameters are carry out using around 78 real measurements of crack length. In general terms, the crack length estimation can presents a slight convergence to measurement over time, it is taking into account that there is a few number of measurements below 4000 cycles. However, parameter estimations don't converge to expected values satisfactorily. The JEKF algorithm was initiated with the process noise covariance matrix $Q = 10^{-3} * I_3$, measurement noise variance $R = 10^{14}$ and the diagonal elements of the estimation error variance matrix $P_0 = \{1, 2.5E3, 2E4\}$. High values for R and P_0 were considered with the propose to reach an accuracy on the estimation of the parameters C and m. However, the estimation error on C is unstable without reaching a notable trend. The estimation error on m increases with very small variations without achieving satisfactory results. In term of the computational charge, the estimation process takes more time by finding the appropriate initial conditions. Furthermore, the estimators of both parameters don't converge to expected values. We can say that both the initial conditions and nonlinearity of the system affect the estimation process. It is known that the logarithms are one of the most important mathematical tools for the linearisation of exponential functions. In this paper, we propose another modelling based on the logarithmic transformation of the state function, see Eq.11, to improve the accuracy of the parameter estimations.

4.2 Logarithmic transformation based new results

Here results are presented using JEKF algorithm based on logarithmic transformation. The JEKF algorithm is adjusted to start up with initial conditions considered, around 50 tests were developed. The estimations can be unstable and difficult convergence to expected values ($C = 2\times10^{-10}$ and $m = 1.63$). However, the estimators of the parameters can achieve satisfactory results. The simulations were organised by increasing the initial conditions for the parameters C and m, and the initial estimation error variance of both parameters. Values of Q and R are fixed to $Q/R = 0.01$ ratio to learn and to analyse the impact of the initial conditions of the parameters in the estimation process. We show that the precision on both parameter estimation errors depends on the process initialization which also impact the cycle for which a required precision is reached. The estimation process was observed by the last estimations, estimation error variances and absolute errors of both parameters. In Fig. 2-a depicts the last estimations made of parameter C when the initial estimation error variance on the parameter C ($P_{0_{22}}$) is increased. When $P_{0_{22}}$ grows, the estimation of C moves away from to expected value over time. For $P_{0_{22}} < 1$ approximately, the estimator of C is closed to expected value. In another way, if estimation error variance on parameter m ($P_{0_{33}}$) is increased, the estimator of m converge to expected value over time, see Fig. 2-b. That is to say, the estimator of m closes to expected value when the $P_{0_{33}}$ increases over time. We observe that the initial variances for both parameters can affect the speed of the estimation process. In other words, $P_{0_{22}}$ and $P_{0_{33}}$ can influence that the JEKF algorithm achieves large or small values of the last estimation error variance. For instances, Fig. 3-a

 DOI: 10.1051/matecconf/20168303003

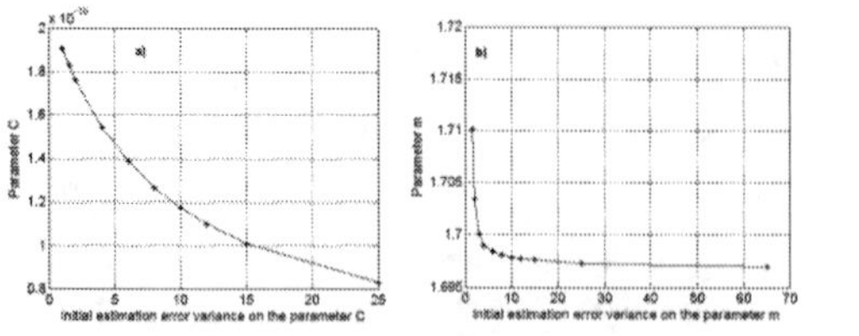

Fig. 2. a) Last estimation of parameter C and b) Last estimation of parameter m.

depicts that estimation error variance on C reaches large values when the $P_{0_{22}}$ increases. In contrast with the last estimation error variance on m, see Fig. 3-b, this reaches small values with low variations. Therefore, the variation of the initial estimation error variance on C considerably can affects the parameter estimation process with increasing the last estimation error variances on C and m. It means that appropriate initialisation of variance on the parameter C can has more influence in the estimation process due to that it is a coefficient of the model. In this paper, we have

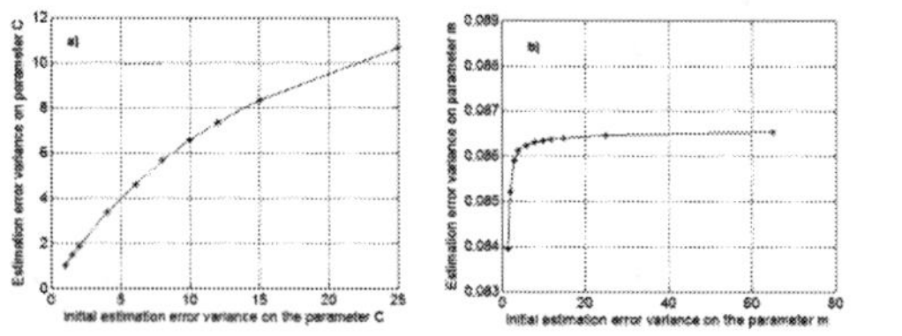

Fig. 3. a) Estimation error variance on C and b) Estimation error variance on m.

used the finite difference approximations for modelling the crack length evolution. Fig. 4 depicts the crack length evolution by exact solution and finite difference of Paris law model. There is that evolution trends are closed over time. Therefore, estimation of l should converges to the expected trend. In Fig. 5-a shows a summary of the estimation errors on l. We can observed that the estimation error of T3, T4 and T5 quickly converge to zero and has closed values. It means that they have the same Q/R ratio. The trend of estimation error on l can be minimized for small values of R. After 4000 cycles, the estimation error on l convergences and after 10000 cycles, it is better. However, estimations of the parameters C and m present o=another trends to expected values but in some instant they approximate to best values. In Fig. 5-b, we can observed that best estimations of parameter C are found for T6 and T7 with values $Q = 0.0001 * I_3$ and $R = 0.1$ before to 4000 cycles, and $Q = 0.0001 * I_3$ and $R = 1$ before 7000 cycles, respectively. Another trends show not a decreasing until zero, the estimation errors remain above 2. On the other hand, the best estimations of parameter m are found for T6 before 4000 cycles and T2, T4, T5 and T7 after 7000 cycles. Therefore, we can select the estimation of T6 and T7 because both the estimation errors of parameter C and m converge to zero before at 4000 cycles and between 6500 and 7500 cycles, respectively. However, both estimation errors of parameter C and m by the T7, the trends show that the estimators are more stable than the estimators by test T6 which increase quickly. Estimation error on C remains stable and relatively small after 6000 cycles compared with the other tests. If we stop the updating of the estimation of m when its estimation error reaches the value more near to zero, best estimation is identified between 6500 and 7500 cycles.

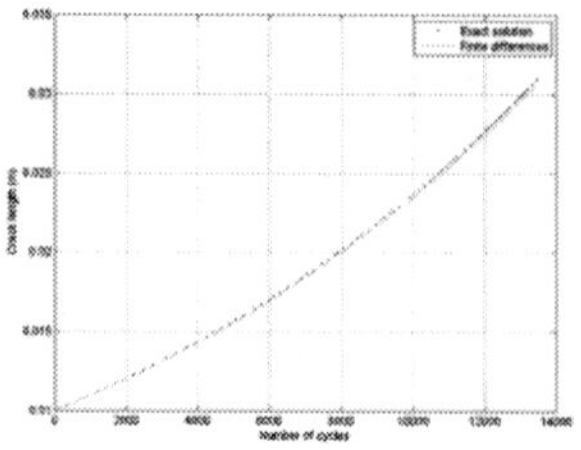

Fig. 4. Crack length evolution by exact solution and finite-difference method.

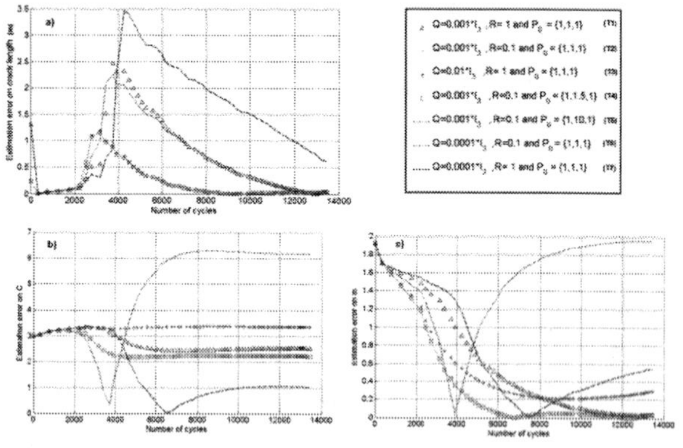

Fig. 5. a) Summary of estimation error on l. b) Summary of estimation errors on C. c) Summary of estimation errors on m.

5 Conclusion

In this paper real data was used to estimate Paris law parameters. Finite-difference method used to discretised the Paris law model and approximating the crack length evolution. An adaptive estimation algorithm was applied to the problem of fatigue crack propagation, in order to identify the best model parameters for improving of the prediction of the crack length. Using real data, results have been presented and discussed. The Extended Kalman Filter (EKF) by the joint formulation and logarithmic transformation was developed with promising estimations. Paris law parameters were estimated in order to analyse the impact of the initial conditions in the estimation process. The results show that estimators eventually converge to the expected values in specific instants. Here pretty accuracies can be obtained by appropriate initialisation of the parameters C and m and their estimation error variances. The initial estimation error variance plays an important roll to estimate the parameters with small variation in the the estimation error variances. Results show that the estimators of C and m can be lightly confidence at a specific instant. Important contributions have been made for the parameter identification problems. Considering real data allows simulating real situations where the important measurements given by an SHM system are used to estimate unknown model parameters for predicting the future condition of a monitored structural component and estimating its RUL.

References

1. SM Beden, S Abdullah, and AK Ariffin. *European Journal of Scientific Research*, 28:364–397, 2009.
2. JC Newman, EP Phillips, and MH Swain. *International Journal of fatigue*, 21:109–119, 1999.
3. A Coppe, MJ Pais, RT Haftka, and NH Kim. *Journal of Aircraft*, 49:1965–1973, 2012.
4. WF Wu and CC Ni. *Probabilistic Engineering Mechanics*, 18:107–118, 2003.
5. F Khelil, B Aour, M Belhouari, and N Benseddiq. *Engineering, Technology & Applied Science Research*, 3:pp–488, 2013.
6. J Hu, Z Wang, H Gao, and LK Stergioulas. *Automatica*, 48:2007–2015, 2012.
7. K Szabat and T Orlowska-Kowalska. *Industrial Electronics, IEEE Transactions on*, 55:1075–1084, 2008.
8. Y Lei, Y Jiang, and Z Xu. *Mechanical Systems and Signal Processing*, 28:229–243, 2012.
9. RK Singleton, EG Strangas, and S Aviyente. *Industrial Electronics, IEEE Transactions on*, 62:1781–1790, 2015.
10. E Zio and F Di Maio. *Expert Systems with Applications*, 39:10681–10692, 2012.
11. LD Cot and R Lozi. *International Journal of Chaotic Computing*, 2:3–13, 2013.
12. A Corigliano and S Mariani. *Computer Methods in Applied Mechanics and Engineering*, 193:3807–3835, 2004.
13. S Mariani and A Corigliano. *Computer methods in applied mechanics and engineering*, 194:5242–5272, 2005.
14. A Ray and S Tangirala. *Control Systems Technology, IEEE Transactions on*, 4:443–451, 1996.
15. AC Cobb, JE Michaels, TE Michaels, Donald O Thompson, and Dale E Chimenti. In *Aip Conference Proceedings*, volume 975, page 1252, 2008.
16. A Coppe, RT Haftka, and NH Kim. In *12th AIAA Non-Deterministic Approaches Conference*, pages 12–15, 2010.
17. M Shah, C Mabru, and F Rezaï-Aria. *Fatigue & Fracture of Engineering Materials & Structures*, 38:742–754, 2015.

BAS MULTI-ROBOT FORMATION COORDINATION CONTROL FOR COLLECTIVE MISSIONS

Elżbieta M. Jarzębowska

Warsaw University of Technology, Institute of Aeronautics and Applied Mechanics,
00-665 Warsaw, Nowowiejska 24 St., Poland, e-mail:elajarz@meil.pw.edu.pl

Abstract. The paper presents a study of a mobile multi-robot formation (MRF) control problem. The robots are nonholonomic car-like vehicles and perform their mission while coordinating the formation shape when needed. The MRF control problem consists then of tracking reshaping a formation. Control design is based upon the centralized strategy assuming that information from a leader robot can be distributed to formation members. The main motivations for the work are that control strategies designed for single robots when applied to formations become complex and grow in terms of computational time, many of new control algorithms are dedicated to different formation coordination schemes and many of them were never verified experimentally. Two control levels are considered, i.e. kinematics and dynamics. Kinematics controllers applied are the input-output linearization and the Samson algorithm. They are simple in implementation and proved good performance in controlling single car-like robots. We examine them with regard to their potential applications for MRF. The dynamics controllers are based upon the generalized programmed motion equations method used successfully for a leader-follower control strategy for a fixed shape formation. Results of simulation studies, comparisons and discussions of implemented control algorithms are presented.

1 Introduction

The paper presents a study of a mobile multi-robot formation (MRF) control problem. The robots are nonholonomic car-like vehicles and perform their mission while coordinating the formation shape when needed. The MRF control problem consists then of performing a mission and being able to reshape the formation. Control design is based upon the centralized strategy assuming that information from robot sensors enables determination of instantaneous relative position of each robot in the workspace, for example, by means of cameras that watch and monitor the entire environment. Also, a decentralized formation control scenario is studied when each member of the formation has information of its neighbors and not of the other robots in the formation. Altogether, three mission scenarios are presented using kinematics and dynamics based controllers. Problems of controlling and coordinating an autonomous MRF gain popularity due to number of benefits of coordinated collective missions in comparison with a single robot mission or a human activity, especially in a hazardous environment or when a task is distributed. Popularity of the MRF also comes from their numerous applications, e.g. transportation of payloads, reconnaissance, surveillance and rescue missions, cleaning and inspections in hazardous environment and many more [7,8,9]. In some situations, formation maintaining is not enough and a MRF may need to change its shape to conform a mission.

Different approaches to MRF modeling, coordination and control are presented in the literature. Usually, members of a MRF are modeled as kinematic or dynamic car-like mobile vehicles or as particle agents that form a swarm equipped with various sensing possibilities [4,8]. There are three major approaches to formation control, i.e. behavior-based, leader-follower and a virtual structure formation control. In the behavior-based formation control [6], a desired set of behaviors is implemented to individual robots. By grading the behaviors, the overall behavior of a robot is specified. In the leader-follower control strategy [2,5], followers follow the selected leaders and they position themselves relative to them. In the virtual structure control [1], a formation is considered as a single structure and desired motions are assigned to it. The formation in the virtual structure approach is rigid such that geometric relationships between robots have to be fixed during motion. Thus, this formation cannot reshape. Formation reshaping control strategies were implemented in some studies together with the leader-follower approach. For example, in [2] a framework for cooperative control of a group of mobile robots that allows building a complex system from simple controllers is described. The switching between simple decentralized controllers allows formation reshaping while following a leader or performing a task. In [3] nonlinear control and the graph theory approach are used to study the formation control of mobile robots. The formation reshaping can be achieved by adding or deleting edges in the formation graph.

One of motivations for the presented work is that the problems with the control strategies in [2], [3] and also in other works are that their level of computational complexity grows with the increase of the number of robots in the formation. Also, new control algorithms are generated for different formation coordination schemes. They often suffer from complexity and often cannot be reused for other MRF when different tasks are specified. Also, many of them were never verified experimentally. In this work, a MRF coordination control as well as control of performing a desired task is designed using existing controllers at the kinematics and dynamics levels. The kinematics controllers selected to the study are the input-output linearization and the Samson algorithm. They are simple in implementation and proved good performance in controlling single car-like robots. We examine them with regard to their potential applications for MRF. The dynamics controllers are based upon the generalized programmed motion equations method used successfully for a leader-follower control strategy for a fixed shape formation [5]. The other motivation is to study formation reshaping scenarios when a change of a leader is needed. It may happen though that a leader robot fails during a mission and another robot is to take its role. Three mission scenarios are presented for reshaping the MRF and changing a leader.

Results of simulation studies as well as comparisons and discussions of implemented control algorithms, at the kinematics and dynamics levels, are presented in this work.

2 Leader – follower formation control algorithms applied at kinematics and dynamics levels

There are quite efficient and easy implementable control algorithms designed for a single robot and two of them were verified if they are applicable to MRF coordination as well as for performing desired tasks. The formation consists of Pioneer 3DX two wheeled robots, whose kinematics is equivalent to this of a unicycle, see Fig. 1.

Figure 1. Pioneer 3DX robot for experiments

The first examined algorithm, at the kinematics level, is the input-output dynamics linearization, which provides full linearization, i.e. the resulting kinematic control model has the form

$$\ddot{z}_1 = u_1, \quad \ddot{z}_2 = u_2, \tag{1}$$

where z_i are new states and u_i are new inputs, e.g. PD inputs of the form

$$\begin{aligned} u_1 &= \ddot{x}_d(t) + k_d(\dot{x}_d(t) - \dot{x}) + k_p(x_d(t) - x) \\ u_2 &= \ddot{y}_d(t) + k_d(\dot{y}_d(t) - \dot{y}) + k_p(y_d(t) - y) \end{aligned} \tag{1a}$$

Advantages of the input-output dynamics linearization are the exponential convergence, easy numerical and real robot implementation. A disadvantage is a structural singularity of the nonlinear state vector transformation, which requires a non-zero vehicle velocity.

The second examined algorithm is the Samson tracking control algorithm, which is based upon an

 DOI: 10.1051/matecconf/20168303004

error dynamics, which is obtained from the robot kinematics $A(q)\dot{q}=0$ by differentiating tracking errors determined in the robot frame, i.e.

$$\begin{bmatrix} x_e \\ y_e \\ \varphi_e \end{bmatrix} = \begin{bmatrix} \cos\varphi & \sin\varphi & 0 \\ -\sin\varphi & \cos\varphi & 0 \\ 0 & 0 & 1 \end{bmatrix} \begin{bmatrix} x_d - x \\ y_d - y \\ \varphi_d - \varphi \end{bmatrix} \tag{2}$$

Inserting $\dot{x}_d = V_d \cos\varphi_d,\ \dot{y}_d = V_d \sin\varphi_d,\ \dot{\varphi}_d = \omega_d$ to (2) along a desired trajectory one gets

$$\begin{bmatrix} \dot{x}_e \\ \dot{y}_e \\ \dot{\varphi}_e \end{bmatrix} = \begin{bmatrix} 0 & \omega & 0 \\ -\omega & 0 & 0 \\ 0 & 0 & 0 \end{bmatrix} \begin{bmatrix} x_e \\ y_e \\ \varphi_e \end{bmatrix} + \begin{bmatrix} V_d \cos\varphi_e - V \\ V_d \sin\varphi_e \\ \omega_d - \omega \end{bmatrix}. \tag{3}$$

Transforming the original inputs to

$$V = V_d \cos\varphi_e - u_1,\quad \omega = \omega_d - u_2, \tag{4}$$

yields

$$\begin{bmatrix} \dot{x}_e \\ \dot{y}_e \\ \dot{\varphi}_e \end{bmatrix} = \begin{bmatrix} 0 & \omega_d & 0 \\ -\omega_d & 0 & 0 \\ 0 & 0 & 0 \end{bmatrix} \begin{bmatrix} x_e \\ y_e \\ \varphi_e \end{bmatrix} + \begin{bmatrix} 0 \\ \sin\varphi_e \\ 0 \end{bmatrix} V_d + \begin{bmatrix} 1 & 0 \\ 0 & 0 \\ 0 & 1 \end{bmatrix} \begin{bmatrix} u_1 \\ u_2 \end{bmatrix}. \tag{5}$$

New control inputs are defined as

$$u_1 = -k_1 x_e,\quad u_2 = -k_2\, sgn[V_d(t)]\, y_e - k_3 \varphi_e, \tag{6}$$

with k_1, k_2 and k_3 being control gains.

The Samson algorithm is not easy for numerical implementation due to control inputs u_i that have no physical interpretation. Backward transformation is needed to assess real input magnitudes and one needs to transform x, y and a heading angle φ to their errors and back, what may affect computing initialization. Also, linear and angular velocities directly resulting from reference motion need to be calculated.

A dynamics control model for a single robot is

$$\begin{gathered} M(q)\ddot{q} + C(q,\dot{q})\dot{q} + D(q) = E(q)\tau, \\ B_1(q)\dot{q} = 0. \end{gathered} \tag{7}$$

Applying static state feedback linearization yields

$$\begin{gathered} \dot{x}_1 = G(x_1)x_2, \\ \dot{x}_2 = u, \end{gathered} \tag{8}$$

The model (8) is taken for simulations for a MRF controlling a collective task mission.

3 Simulation case studies and experiments

Three simulation case studies and one experiment on a Pioneer robot are presented. The first two are for following a leader and an MRF reshaping during a mission. The third one is dynamics based control.

The four robot formation starts from positions shown in fig. 2. A desired motion is along a trajectory of a shape $x_d(t) = sin(0.1t)$ $y_d(t) = sin(0.1t)$ The followers start their motions every 6 seconds.

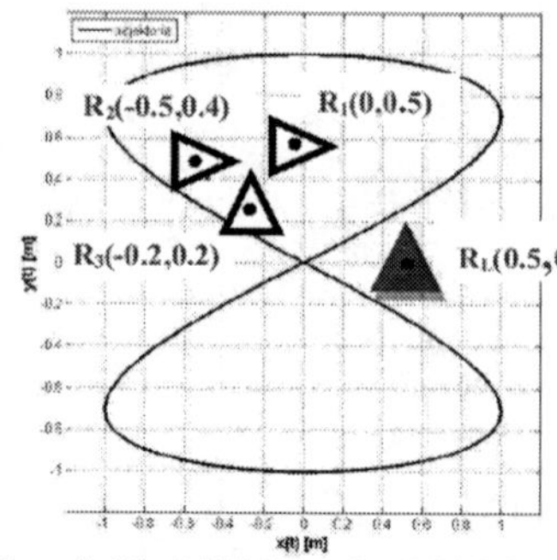

Figure 2. The MRF formation initial shape.

The simulation results, when robots follow a leader, are shown in fig. 3 for the input-output dynamics linearization and the Samson algorithm. In fig. 4 the analogous results are shown but this time a leader is a preceding robot. Notice the way robot 3 follows robot 2: it does the same „zig-zag" motion; the Samson algorithm relies upon the heading angle and angular velocity of the preceding robot. Also, robots controlled by the Samson algorithm reach about 3 times bigger linear velocities than controlled by the input-output linearization. This is due to control gains but they are adjusted to get some reasonable the convergence rate. Differences between angular velocities are smaller; 2 times bigger for the robot 3 comparing to the input-output linearization.

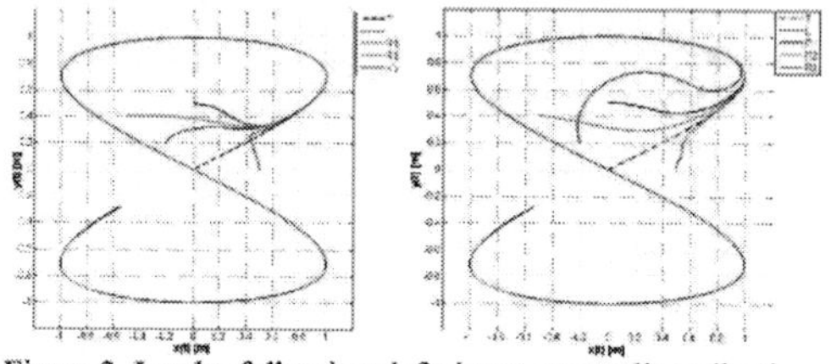

Figure 3. Leader following: left: input-output linearlization, right: Samson algorithm.

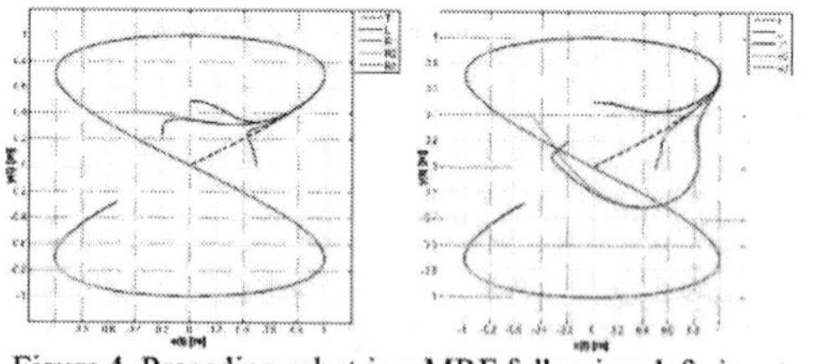

Figure 4. Preceding robot in a MRF following: left: input-output linearlization, right: Samson algorithm.

The reshaping scenario for the MRF is presented in fig. 5. The leader starts its motion, then the followers reach the desired trajectory, and the

DOI: 10.1051/matecconf/20168303004

command about the reshaping is given – form a triangle and continue following the leader.

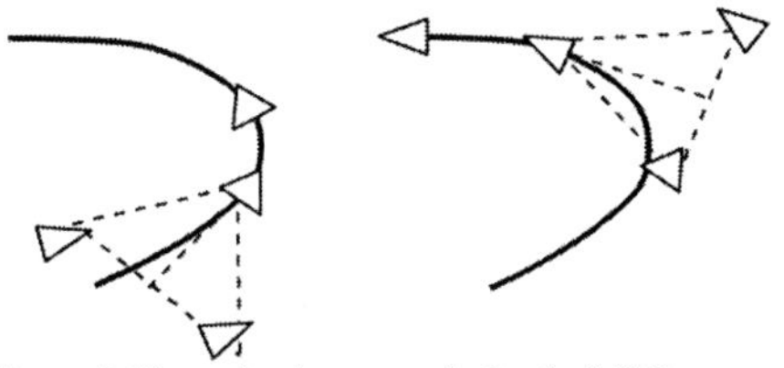

Figure 5. The reshaping scenario for the MRF.

Trajectory tracking results are shown in Fig. 6 for both kinematics based controllers.

The dynamics controller is based upon (7)-(8) and it is the Repetitive Control Law controller for a "zig-zag" reconnaissance mission. The simulation results for a sparse MRF are shown in Fig. 7.

An experiment with the application of the input-output linearizing controller was performed on a Pioneer 3DX robot (Fig. 1). The Pioneer robot is quite slow and it follows a desired trajectory quite well. It required however, completely different selection of control gains comparing to the ones adjusted in simulations studies. Also, tracking errors, a simulated one (in red) and a real one (in blue) are much different. This is of course a result of cumulating errors from all delays in a real robot software and hardware, and lacking dynamics.

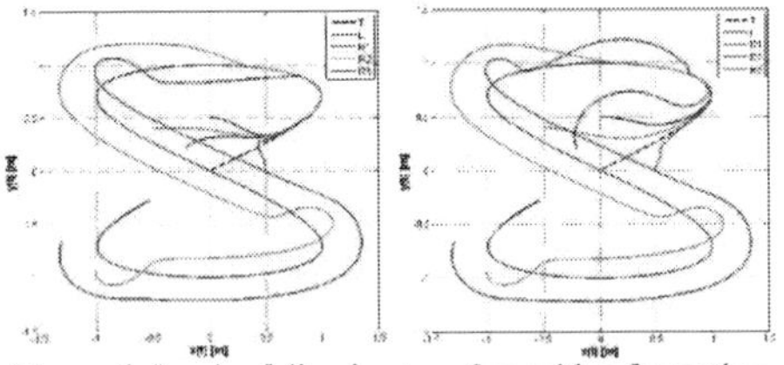

Figure 6. Leader following together with a formation reshaping: left: input-output linearlization, right: Samson algorithm.

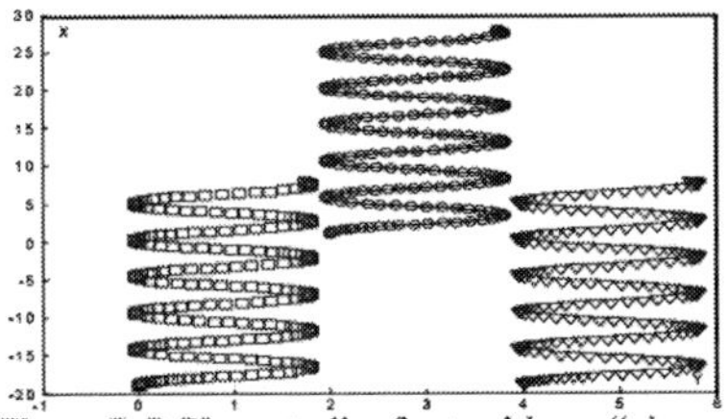

Figure 7. RCL controller for tracking a "zig-zag" reconnaissance mission for MRF.

Experimental results (in red) of a desired trajectory (in blue) tracking are shown in Fig. 8.

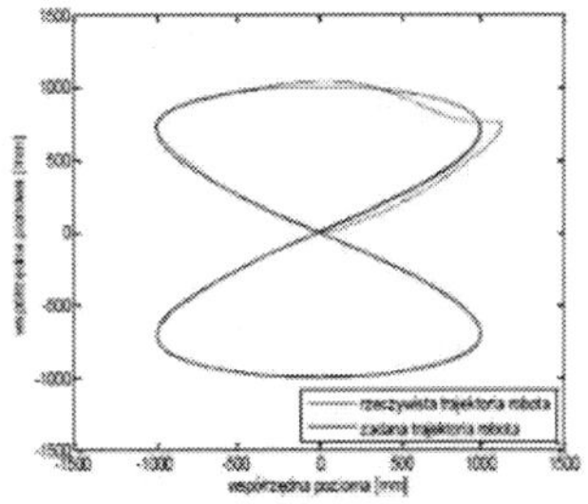

Figure 8. Trajectory tracking by Pioneer 3DX (in red).

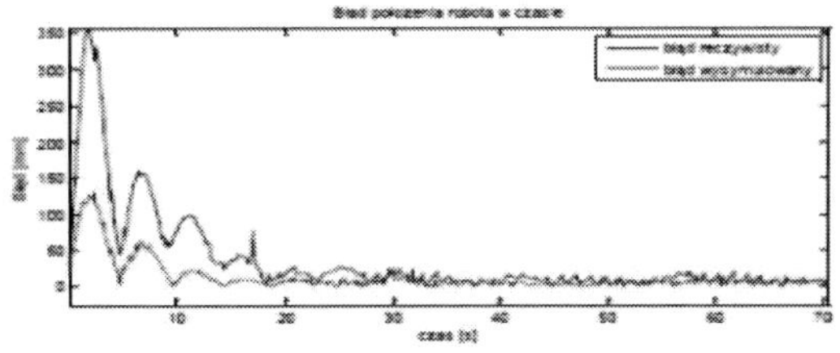

Figure 9. Position tracking errors: simulation – red, experiment – blue.

Conclusions

There are not many works verifying control algorithms for their further implementations. There are not many simulation studies for the control algorithms with regard of their applications to MRF. The future work is oriented to implementation of the algorithms to Pioneer robot.

References

1. Beard R.W., Lawton J., Hadaegh F.Y.: Coordination Architecture for Spacecraft Formation Control. IEEE Trans. Control Syst. Techn., vol. 9, no. 6, Nov. 2001, pp. 777–790.
2. Das A.K., Fierro R., Kumar V., Ostrowski J.P., Spletzer J., and Taylor C.J.: A Vision -Based Formation Control Framework. IEEE Trans. Robot. Automat., vol. 18, no. 5, 2002, pp. 813-825.
3. Desai J.P., Kumar V., Ostrowski P.: Modeling and Control of Formations of Nonholonomic Mobile Robots. IEEE Trans. Robot. Automat., vol.17, Dec. 2001, pp. 905–908.
4. Hou S.P., Cheah C.C., Slotine J.J.E.: Dynamic Region Following Formation Control for a Swarm of Robots, 2009 IEEE Int. Conf. Robot. Automat., Kobe, Japan, May 12-17, 2009, pp. 1929-1934.
5. Jarzębowska E.: Leader-Follower Tracking Control Design for Task-Based Missions. Int. J.

DOI: 10.1051/matecconf/20168303004

Vehicle Autonomous Syst., Vol. 9, No. 3/4, 2011, pp. 203-217.
6. Lawton J.R., Beard R.W., Young B.J.: A Decentralized Approach to Formation Maneuvers. IEEE Trans. Robot. Automat., vol. 19, no. 6, pp. 933–941, Dec. 2003.
7. Mas I., Li S., Acain J., Kitts C.: Entrapment/Escorting and Patrolling Missions in Multi-Robot Cluster Space Control. Proc. 2009 IEEE Int. Conf. Intelligent Robots and Systems, St. Louis, USA, October 2009, pp. 5855–5861.
8. Raghuwaiya K., Singh S., Sharma B., Vanualailai J.: Formation Types of a Flock of 1 - Trailer Mobile Robots. 7th IMT-GT Int. Conf. Mathematics, Statistics and its Applications, 2011, pp. 368-382.
9. Raghuwaiya K., Singh S., Vanualailai J.: Formation Control of Mobile Robots. World Academy of Science, Engineering and Technology 60, 2011, pp. 762-767.

MATEC Web of Conferences **83**, 03005 (2016)
CSNDD 2016
DOI: 10.1051/matecconf/20168303005

The Linear Stability of the Responses of Axially Moving Beams Supported by an Intermediate Spring

F. Kostekci[1]

[1]Mechanical Engineering Department, Hitit University, Corum, 19030, Turkey

Abstract. In the present paper, the stability investigation of the linear responses of axially travelling beams supported by an intermediate linear spring is performed numerically by means of searching linear complex eigenvalues. The boundary conditions and the dynamic equation of motion are obtained by means of the Extended Principle of Hamilton for the two-span Euler-Bernoulli beam. Assuming the harmonic solution of the governing differential equations of motion, the dispersion relation is derived. To have a nontrivial solution in conjunction with the dispersion relation, the frequency equation is obtained by letting the determinant of the matrix representing the coefficients of the equations of the boundary conditions be equal to zero. The aim of this contribution is to compute the complex values of the natural frequencies from the nonlinear frequency equation by means of numerical methods for the beam supported by an intermediate spring. The real and imaginary parts of the complex eigenvalues versus the constant axial transport speed curves are depicted in order to show the linear instability. Depending on the signs of the real and imaginary parts of the complex eigenvalues, the boundaries of the instability regions were observed.

1 Introduction

The industrial applications of moving materials classified as axially travelling stringlike, beamlike structures are numerous [1-4] such as magnetic tapes drives, band saws, power transmission belts, robot arms, serpentine belts and aerial cable tramways. The purpose of the current research paper is to investigate the linear instability of the beam. The dynamic equation of motion and

obtained utilizing the Extended Principle of Hamilton for the two-span Euler-Bernoulli beam. The complex mode method [5, 6] is used to obtain the nonlinear frequency equation. The boundaries of the instability regions were determined according to the signs of the real and imaginary parts of the eigenvalues.

2 Formulation of the beam problem

The axially moving continua with an intermediate spring support is illustrated schematically in Fig. 1.

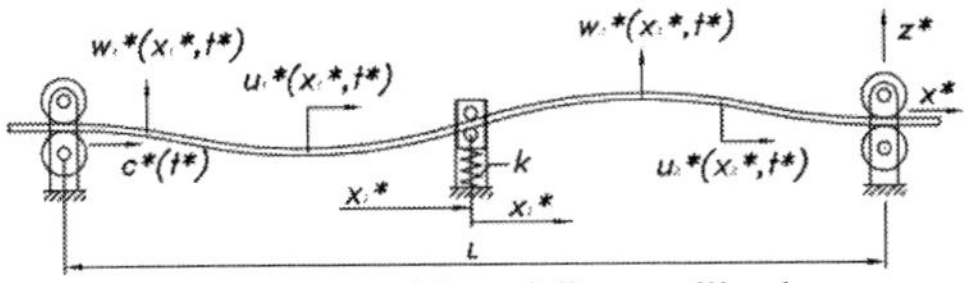

Fig. 1. Geometry of the axially travelling beam.

The equations of motion can be obtained by following the procedure presented in the reference [7] using the Extended Hamilton's Principle. The Euler-Bernoulli beam is moving axially at a slightly fluctuating harmonic velocity c^*. After writing the kinetic and potential energies of the system, the Lagrangian for the axially moving beam can be formulized through the Hamilton's Principle as shown in Eq. (1). In Fig. 1, w_1^*, w_2^* and u_1^* and u_2^* denote respectively the transverse and longitudinal displacements, c^* is the velocity of the axial transport of the beam, t^* denotes time, where the variables with asterisk denote dimensional ones. Properties of the beam include the beam length L, the cross-section area A, I is the cross-sectional inertia of the beam, the mass density ρ and E is the Young's modulus of the beam. k is the translational stiffness of the linear spring. The values of the spatial variable at the boundaries namely x_0^* and x_2^* should be written as 0 and L respectively. It is assumed that the beam has immovable boundaries at the outer ends. The assumption introduces nonlinearity because of the stretching of neutral fibres.

$$£ = (1/2)(\int_{x_0^*}^{x_1^*} \rho A((\dot{w}_1^* + w_1^{*\prime} c^*)^2 + (c^* + \dot{u}_1^* + u_1^{*\prime} c^*)^2)^2$$

$$dx^* + \int_{x_1^*}^{x_2^*} \rho A((\dot{w}_2^* + w_2^{*\prime} c^*)^2 + (c^* + \dot{u}_2^* + u_2^{*\prime} c^*)^2)^2 dx^*)$$

$$-(1/2)(\int_{x_0^*}^{x_1^*} EA(u_1^{*\prime} + (1/2)w_1^{*\prime 2})^2 dx^* + \int_{x_0^*}^{x_1^*} EIw_1^{*\prime\prime 2} dx^*$$

$$+\int_{x_0^*}^{x_1^*} P(u_1^{*\prime} + (1/2)w_1^{*\prime 2})^2 dx^* \quad (1)$$

$$+\int_{x_1^*}^{x_2^*} EA(u_2^{*\prime} + (1/2)w_2^{*\prime 2})^2 dx^* + \int_{x_1^*}^{x_2^*} EIw_2^{*\prime\prime 2} dx^*$$

$$+\int_{x_1^*}^{x_2^*} P(u_2^{*\prime} + (1/2)w_2^{*\prime 2})^2 dx^*$$

$$+(1/2)k w_1^{*2}(x_1^*))$$

where P is the total tensional force, which is composed of an applied artificial pre-tension and the tension caused by the extensional and flexural deformations of the centreline, in the axial direction.

2.1 Dimensionless formulation

For convenience, the equations and all accompanying boundary conditions are presented in dimensionless form for the two sides. The following non-dimensional variables and parameters introduced in Ref. [8] are employed.

$$w_1 = w_1^* / L, w_2 = w_2^* / L, u_1 = u_1^* / L, u_2 = u_2^* / L$$

$$x_0 = x_0^* / L, x_1 = x_1^* / L, x_2 = x_2^* / L, t = \gamma t^*$$

$$c = c^* / \sqrt{P / \rho A}, v_b^2 = EA / P, \bar{v}_f^2 = EI / PL^2$$

$$\gamma = (1/L)\sqrt{P / \rho A} \quad (2)$$

In addition to those given in Eq. 2, the dimensionless pan-wise variable, which represents a dimensionless location constant of the intermediate support and may take a value bigger than 0 and smaller than 1, is introduced: $x_1 = \eta$. Dimensionless parameter v_b presents the longitudinal rigidity parameter, and $\bar{v}_f$ denotes the flexural rigidity parameter. The axial velocity is converted to dimensionless quantity by dividing it with the critical velocity. Substituting Eq. 2 in the governing equation of the motion, performing required operations and adding damping term ($\bar{\mu}$) yield the dimensionless equations of motion and relevant non-dimensional boundary conditions.

$$\ddot{w}_1 + 2\dot{w}_1' c + w_1'\dot{v} + w_1''(c^2 - 1) + \bar{v}_f^2 w_1^{iv} +$$

$$\bar{\mu}(\dot{w}_1 + c\, w_1') = v_b^2 (1/2)(\int_0^\eta w_1'^2 dx + \int_\eta^1 w_2'^2 dx) w_1''$$

$$\ddot{w}_2 + 2\dot{w}_2' c + w_2'\dot{v} + w_2''(c^2 - 1) + \bar{v}_f^2 w_2^{iv} +$$

$$\bar{\mu}(\dot{w}_2 + c\, w_2') = v_b^2 (1/2)(\int_0^\eta w_1'^2 dx + \int_\eta^1 w_2'^2 dx) w_2'' \quad (3)$$

$$w_1(0,t) = 0, w_1'(0,t) = 0, w_2(1,t) = 0, w_2'(1,t) = 0,$$

$$w_1(\eta,t) = w_2(\eta,t), w_1'(\eta,t) = w_2'(\eta,t),$$

$$w_1''(\eta,t) = w_2''(\eta,t), w_1'''(\eta,t) - w_2'''(\eta,t) = \alpha_k\, w_1(\eta,t) \quad (4)$$

where the following abbreviations are introduced.

$$\alpha_k = k / (EI / L^3) \quad (5)$$

2.2 A systematic way

In the initial steps we tacitly used the method of multiple scales [9] to separate and obtain the linear parts of the equations in a systematic way [10]. The axial transport velocity varying about an arbitrary constant value, c_0, with a small amplitude fluctuation frequency is given as $c(t) = c_0 + \varepsilon\, c_1 \sin(\Omega t)$ where ε denotes a small variation of the velocity. The terms w_1 and w_2 representing displacements in Eq. 3 and 4 may be supposed in the form $w_1 = \sqrt{\varepsilon} y_1, w_2 = \sqrt{\varepsilon} y_2$. Thus, these terms satisfy that the longitudinal stiffness terms coming from nonlinear effects will stay in higher orders of expansions. Also, viscous damping term $\varepsilon\mu$ will be in ε order as $\bar{\mu} = \varepsilon\mu$. The displacement functions $y(x,t,\varepsilon)$ can be expressed in the series forms as follow

$$y_1(x,t,\varepsilon) = y_{11}(x,T_0,T_1) + \varepsilon\, y_{12}(x,T_0,T_2) + ...$$
$$y_2(x,t,\varepsilon) = y_{21}(x,T_0,T_1) + \varepsilon\, y_{22}(x,T_0,T_2) + ... \quad (6)$$

where $T_0 = t$, is the usual fast-time scale and $T_1 = t\varepsilon$ is the slow-time scales. The time derivatives are as follow:

$$d / dt = \varepsilon^0 D_0 + \varepsilon^1 D_1 + ...,$$

$$d^2 / dt^2 = \varepsilon^0 D_0^2 + 2\varepsilon^1 D_0 D_1 + ... \quad (7)$$

$$\bar{v}_f^2 = \varepsilon^0 v_f^2 \quad (8)$$

where $D_0 = \partial / \partial T_i$. Substituting expansions given with Eq. 6-8 into Eqs. 3 and 4, expanding outcomes and retaining terms up to $O(\varepsilon^0)$, result in the following equations of motion for each span. $O(\varepsilon^0)$:

$$D_0^2 y_{11} + 2c_0 D_0 y_{11}' + v_f^2 y_{11}^{iv} + (c_0^2 - 1) y_{11}'' = 0,$$

$$D_0^2 y_{21} + 2c_0 D_0 y_{21}' + v_f^2 y_{21}^{iv} + (c_0^2 - 1) y_{21}'' = 0 \quad (9)$$

2.3 Complex Mode Method

In this research, linear differential problem are taken into consideration, the nonlinear terms are not included in the subsequent sections. Assuming harmonic motion, Wickert and Mote [11] expressed the solution to Eq. (9) in the following span-wise characteristic functions:

$$y_{11}(x,T_0,T_1) = A(T_1)e^{i\omega T_0}W_1(x) + \overline{A}(T_1)e^{-i\omega T_0}\overline{W}_1(x),$$

$$y_{21}(x,T_0,T_1) = A(T_1)e^{i\omega T_0}W_2(x) + \overline{A}(T_1)e^{-i\omega T_0}\overline{W}_2(x) \quad (10)$$

where ω represents the natural frequency, A and $\overline{A}$ are complex amplitudes and their conjugates, respectively. By means of substituting Eqs. 10 into Eqs. 9, the equations of motion can be written for each span with appropriate non-dimensional end boundary conditions and compatibility conditions at the intermediate support.

$$v_f^2 W_1^{iv} + (c_0^2 - 1)W_1'' + 2i\omega c_0 W_1' - \omega^2 W_1 = 0,$$

$$v_f^2 W_2^{iv} + (c_0^2 - 1)W_2'' + 2i\omega c_0 W_2' - \omega^2 W_2 = 0 \quad (11)$$

$$W_1(0,t) = 0, W_1'(0,t) = 0, W_2(1,t) = 0, W_2'(1,t) = 0,$$

$$W_1(\eta,t) = W_2(\eta,t), W_1'(\eta,t) = W_2'(\eta,t),$$

$$W_1''(\eta,t) = W_2''(\eta,t), W_1'''(\eta,t) - W_2'''(\eta,t) = \alpha_k W_1(\eta,t) \quad (12)$$

where W_1 and W_2 is lateral displacement, (′) denotes differentiation with respect to the spatial variable (x), c_0 is the axial transport velocity, and v_f denotes the flexural rigidity of the beam. The deflections can be expressed in the following form [11]:

$$W_1 = d_1 e^{i\beta_5 x} + d_2 e^{i\beta_6 x} + d_3 e^{i\beta_7 x} + d_4 e^{i\beta_8 x},$$

$$W_2 = d_5 e^{i\beta_5 x} + d_6 e^{i\beta_6 x} + d_7 e^{i\beta_7 x} + d_8 e^{i\beta_8 x} \quad (13)$$

where $d_1 - d_8$ denote constants. These integration constants must be determined via the boundary conditions given with Equations 12. By substituting the functions given with Eqs. 13 into Eqs. 11, the dispersion relation, $v_f^2\beta^4 + (1-c_0^2)\beta^2 - 2\omega c_0\beta - \omega^2 = 0$, can be easily obtained. It is clear that there must be the relationships, $\beta_1 = \beta_5, \beta_2 = \beta_6, \beta_3 = \beta_7, \beta_4 = \beta_8$, among the eigenvalues. Using the boundary conditions given with Eqs. 13, the matrix of the coefficients can be obtained. We are interested in a sufficient condition for the nontriviality of the solution. The determinant of the matrix of the coefficients must vanish. Thus, the natural frequencies can be evaluated by means of Secant Method from this equality.

2.4. Stability

The beam loses its stability if the transport speed exceeds a certain value.

From the natural frequencies calculated in previous section, one can obtain eigenvalues $\mathrm{Re}(\lambda) + i\,\mathrm{Im}(\lambda)$ as complex parameter where $i\omega = \lambda$. When the real part of the eigenvalues λ is negative it indicates damping and consequently the variety of the amplitude. The imaginary parts of the eigenvalues λ characterize the natural frequencies. Depending on the signs of $\mathrm{Re}(\lambda)$ and $\mathrm{Im}(\lambda)$ the stability of the behaviour of the beam can be classified as [12]:

Stable: if $\mathrm{Re}(\lambda) \leq 0$ (14)

Unstable with static instability (divergence): $\mathrm{Re}(\lambda) > 0$ and $\mathrm{Im}(\lambda) = 0$ (15)

Unstable with dynamic instability (flutter): $\mathrm{Re}(\lambda) > 0$ and $\mathrm{Im}(\lambda) \neq 0$ (16)

3 Divergence and critical speeds

Critical speeds are readily derived from Eqs. (16) by inserting $\omega = 0$ [12]. Therefore, after assuming the solution function $W_1(x)$ and $W_2(x)$ as an exponetial function, a dispersion relation can be obtained. From the the roots of these equations, an artificial parameter can be evaluated as follow.

$$\bar{r} = ((c_0^2 - 1)/v_f^2)^{1/2} \quad (17)$$

Using the boundary conditions given with Eq. 17, the matrix of the coefficients of the solution function can be obtained. The determinant of the matrix must vanish. After inserting v_f into the determinant, $\bar{r}$ can be evaluated numerically from this determinant by means of the Newton-Raphson Method. By substituting the calculated values of $\bar{r}$ into Eq. (18), the divergence speeds can be evaluated. The lowest divergence speed is defined as the critical speed [12-14].

$$c_{div} = \pm\sqrt{1 + \bar{r}^2 v_f^2} \quad (18)$$

4 Numerical results

Our numerical investigations were conducted on the beam with the mechanical properties which are listed in Table 1. The flexural rigidity parameter was chosen as $v_f = 0.173$ [15].

Table 1. Mechanical properties of the system

The flexural rigidity parameter	$v_f = 0.173$
Spring-dependent coefficient	$\alpha_k = 50.118$
The location of the intermediate support	η

The first twelve speeds obtained from Eq. (18) for various η values are presented in Table 2.

Table 2. The speeds obtained from Eq. (18)

$\eta = 0.1$	$\eta = 0.2$	$\eta = 0.3$	$\eta = 0.4$	$\eta = 0.5$
1.32137	1.39434	1.49440	1.63663	1.43254
1.79028	1.94701	2.01820	2.44162	1.84855
2.32234	2.56129	2.15527	2.86930	2.36735
2.88283	2.84184	2.87443	3.29922	2.85387
3.45796	3.20197	3.61732	4.17721	3.39312
4.04148	3.85591	3.73183	5.06501	3.90310
4.63023	4.51736	4.37189	5.51258	4.44798
5.22244	5.18345	5.13297	5.95824	4.96857
5.47091	5.49883	5.50801	6.85596	5.51531
5.81705	5.85259	5.89805	7.75347	6.04171
6.41339	6.52385	6.66576	8.20440	6.58906
7.01102	7.19663	7.30159	8.65365	7.11905

The real and imaginary parts of the complex eigenvalues versus the constant axial transport speed curves are plotted in Fig. 2 in order to show the linear instability. Depending on the signs of the real and imaginary parts of the complex eigenvalues, the boundaries of the instability regions are observed in Fig. 2.

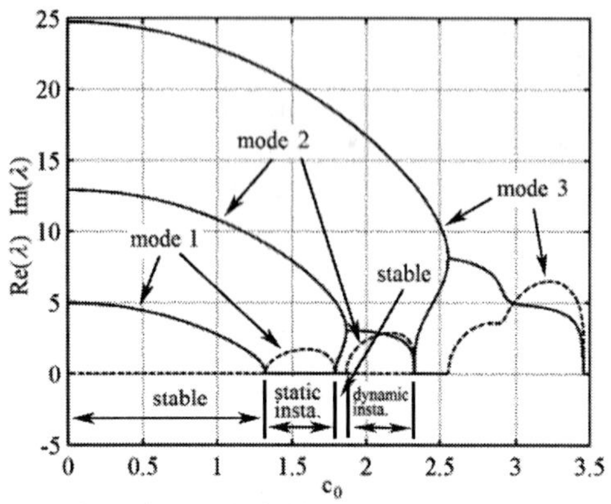

Fig. 2. The real parts (dashed line) and imaginary parts (solid line) of eigenvalues λ versus the transport speed for the first three modes

5 Summary and conclusions

The divergence speeds are analytically derived. The real and imaginary parts of the complex eigenvalues versus constant axial transport speed curves were calculated in order to show the linear instability regions. The signs of the real and imaginary parts of the eigenvalues are used to determine the boundaries of the instability regions.

Acknowledgments

This work is supported by the Scientific Research Project Unit of the Hitit University. (Project No. MUH19008.16.006). This support is gratefully acknowledged.

References

1. G. Chakraborty, A.K. Mallik, H. Hatwal, "Non-linear vibration of a travelling beam," International Journal of Non-Linear Mechanics, 4(**34**), pp.655-670, 1999.
2. K.Y. Sze, S.H. Chen, J.L. Huang, "The incremental harmonic balance method for nonlinear vibration of axially moving beams," Journal of Sound and Vibration, 3–5(**281**), pp.611-626, 2005.
3. L.H. Chen, W. Zhang, F.H. Yang, "Nonlinear dynamics of higher-dimensional system for an axially accelerating viscoelastic beam with in-plane and out-of-plane vibrations," Journal of Sound and Vibration, 25(**329**), pp.5321-5345, 2010.
4. M.H. Ghayesh, S. Balar, "Non-linear parametric vibration and stability analysis for two dynamic models of axially moving timoshenko beams," Applied Mathematical Modelling, 10(**34**), pp.2850-2859, 2010.
5. X.-D. Yang, W. Zhang, L.-Q. Chen, "Transverse vibrations and stability of axially traveling sandwich beam with soft core," Journal of Vibration and Acoustics, 5(**135**), pp.051013, 2013.
6. H.R. Öz, M. Pakdemirli, "Vibrations of an axially moving beam with time-dependent velocity," Journal of Sound and Vibration, 2(**227**), pp.239-257, 1999.
7. J.A. Wickert, "Non-linear vibration of a traveling tensioned beam," International Journal of Non-Linear Mechanics, 3(**27**), pp.503-517, 1992.
8. F. Köstekci, "Effects of flexural stiffness, axial velocity and internal support locations on translating beam's natural frequencies," Applied Mechanics and Materials, (**532**), pp.316-319, 2014.
9. A.H. Nayfeh, Introduction to perturbation techniques, Wiley, 1993.
10. M. Pakdemirli, A.H. Nayfeh, "Nonlinear vibrations of a beam-spring-mass system," Journal of Vibration and Acoustics, (**116**), pp.433-439, 1994.
11. J.A. Wickert, C.D. Mote, Jr., "Classical vibration analysis of axially moving continua," ASME Journal of Applied Mechanics, 57, pp.738–744, 1990.
12. U. Lee, Spectral element method in structural dynamics, 2009.
13. U. Lee, J. Kim, H. Oh, "Spectral analysis for the transverse vibration of an axially moving timoshenko beam," Journal of Sound and Vibration, 3-5(**271**), pp.685-703, 2004.
14. H. Oh, U. Lee, D.-H. Park, "Dynamics of an axially moving bernoulli-euler beam spectral element modeling and analysis," KSME International Journal, 3(**18**), pp.395-406, 2004.
15. M.H. Ghayesh, "Stability and bifurcations of an axially moving beam with an intermediate spring support," Nonlinear Dynamics, 1-2(**69**), pp.193-210, 2012.

MATEC Web of Conferences 83, 03006 (2016) DOI: 10.1051/matecconf/20168303006
CSNDD 2016

Identification of Nonlinearities in Joints of a Wing Structure

M.S.M. Sani[1,2] and H. Ouyang[3]

[1]Advanced Structural Integrity and Vibration Research (ASIVR), Faculty of Mechanical Engineering, Universiti Malaysia Pahang, 26600 Pekan, Pahang, Malaysia
[2]Automotive Engineering Centre, Universiti Malaysia Pahang, Malaysia
[3]Centre for Engineering Dynamics, School of Engineering, University of Liverpool, Liverpool L69 3GH, UK

Abstract. Nonlinear structural identification is essential in engineering. As new materials are being used and structures become slender and lighter, nonlinear behaviour of structures becomes more important. There have been many studies into the development and application of system identification methods for structural nonlinearity based on changes in natural frequencies, mode shapes and damping ratios. A great challenge is to identify nonlinearity in large structural systems. Much work has been undertaken in the development of nonlinear system identification methods (e.g. Hilbert Transform, NARMAX, and Proper Orthogonal Decomposition), however, it is arguable that most of these methods are cumbersome when applied to realistic large structures that contain mostly linear modes with some local nonlinearity (e.g. aircraft engine pylon attachment to a wing). In this paper, a multi-shaker force appropriation method is developed to determine the underlying linear and nonlinear structural properties through the use of the measurement and generation of restoring force surfaces. One undamped mode is excited in each multi-shaker test. Essentially, this technique is a derivative of the restoring surface method and involves a non-linear curve fitting performed in modal space. A reduced finite element model is established and its effectiveness in revealing the nonlinear characteristics of the system is discussed. The method is demonstrated through both numerical simulations and experiments on a simple jointed laboratory structure with seeded faults, which represents an engine pylon structure that consists of a rectangular wing with two stores suspended underneath.

1 Introduction

Linear identification in structural dynamics is matured and well established [1-3]. Nowadays, nonlinear identification is become a very popular area in structural dynamics. Most of engineering structures exhibit some degree of nonlinearity characteristics especially when the deformations are large [4-5]. Nonlinear systems have a range of behaviour not seen in linear vibrating systems. Furthermore, nonlinear dynamic analysis becomes very important for the identification of damage in structures. Detection, localisation and quantification of nonlinearity are very common in nonlinear structural dynamics area [6-7]. Five typical sources of nonlinearities in structural dynamics were as follows: geometric nonlinearity, inertia nonlinearity, material nonlinearity, damping dissipation and boundary conditions.

There have been many studies on the use of system identification methods to identify structural nonlinearity, which include changes in natural frequencies, mode shapes and damping ratios. Masri and Caughey [8] introduced the restoring forces surface (RFS) to identify nonlinearity in single-degree of freedom (SDOF) systems by exploiting Newton's 2nd law to directly measure restoring and dissipative forces in the system. RFS was extended by [9] to identify multi-degree of freedom (MDOF) systems by transforming the equations of motion from physical to modal coordinate space. RFS is more efficient than the Weiner-kernel approach in identifying nonlinear dynamic systems of the types considered. The parametric identification method by force state mapping technique was developed [10], which is similar to RFS. Dimitriadis and Cooper [11] attempted to identify MDOF systems using a variant of RFS method, which considers time response at similar amplitudes, and subsequently, constant nonlinear restoring forces could be achieved. Kerschen et al. [12] applied RFS method for two different cases: a symmetrical nonlinear beam with piecewise linear stiffness and an asymmetrical nonlinear beam with bilinear stiffness. The nonlinear identification method recently proposed [13-14] is based on the measured linear and nonlinear Frequency Response Functions (FRFs). The method is easy to implement and requires standard

[a] Corresponding author: mshahrir@ump.edu.my

DOI: 10.1051/matecconf/20168303006

testing methods. The data required is limited with measured linear and nonlinear FRFs. A method or procedure for the identification of non-linear single and multi-degree of freedom using restoring forces method with three types of nonlinearity was simulated by [15]. Noel et al. [16] demonstrated that the Restoring Force Surface (RFS) method can provide a reliable identification of a nonlinear spacecraft structure. The nonlinear component comprises an inertia wheel mounted on a support, the motion of which is constrained by eight elastomer plots and mechanical stops. Several adaptations to the RFS method are proposed, which include the elimination of kinematic constraints and the regularization of ill-conditioned inverse problems.

This paper presents the identification of nonlinear wing structure using combination force appropriation and restoring force method. Test structure in this paper represent the configuration wing structure with two stores which nonlinear pylon connection between them.

2 Finite Element (FE) Analysis of Wing Structure

Two engine pylons are designed and attached to a rectangular wing structure. The FE model of the wing structure is shown in Fig.1 which consists of 852 solid elements (CHEXA) and 40 plate elements (CQUAD4). There are 1824 nodes and free-free boundary condition is applied to this model.

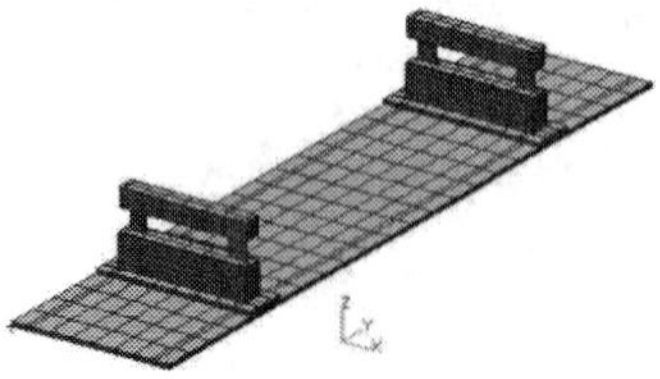

Fig. 1. FE Model of Wing Structure

NASTRAN code for normal mode analysis (SOL 103) is developed and used to compute natural frequencies and mode shape of wing structure. The results are shown in Table 1 and Fig. 2 respectively.

Table 1. FE Natural Frequencies of Wing Structure

Mode	Natural Frequencies (Hz)
1	15.25
2	18.40
3	33.10
4	78.10
5	134.80

M and **K** matrices are extracted from FE model by using matlab code of pch output file. However, **C** matrices can be developed using proportional damping (Rayleigh Method) is commonly used in nonlinear dynamics analysis. **M, K** and **C** matrices will be used in nonlinear modal model.

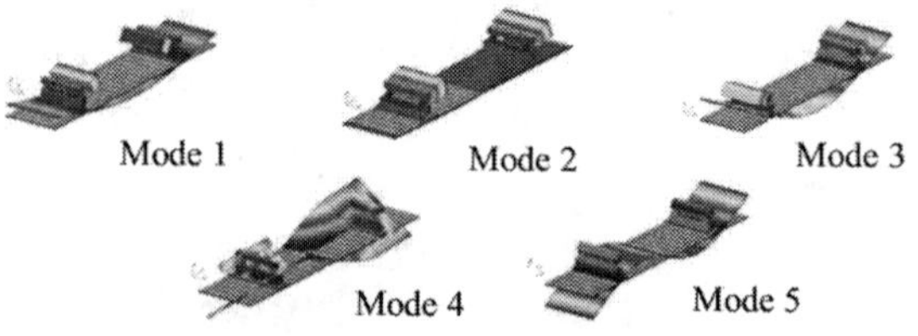

Fig. 2. FE Mode Shape of Wing Structure

3 Experimental Modal Analysis (EMA)

The overall test structure used in this project is intended to represent roughly the configuration of an aircraft wing having two under-wing stores with nonlinear pylon connection between them. Fig. 3 shows the arrangement of the wing model, supported from a frame through bungee cords in a free-free boundary configuration. Two Data Physics V4 shakers were attached to the center line of the wing via force transducers and driven by Data Physics power amplifiers. The natural frequencies, damping ratios, modal masses for the first five modes, extracted from FRF matrix using PolyMAX parameter estimation method are shown in Table 2.

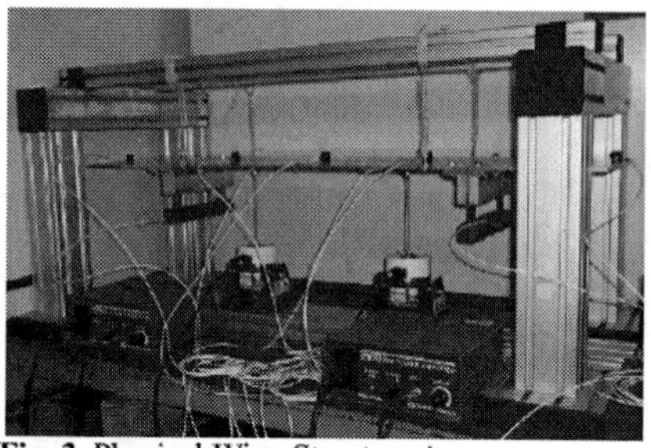

Fig. 3. Physical Wing Structure Arrangement

Table 2. Experimental frequencies, damping ratio and modal mass of overall structure of wing model by performing double shaker test.

Mode	Experimental Natural Frequencies (Hz)	Damping Ratio (%)	Modal Mass (kg)
1	15.09	0.76	2.15
2	18.40	0.91	2.44
3	32.33	0.32	1.66
4	78.01	0.09	0.99
5	135.08	0.27	1.17

4 Nonlinear Modal Model

The equations of motion of discretised structures in the physical space can be expressed as

$$\mathbf{M}\ddot{\mathbf{x}} + \mathbf{C}\dot{\mathbf{x}} + \mathbf{K}\mathbf{x} + \mathbf{g}_{nl}(\dot{\mathbf{x}}, \mathbf{x}) = \mathbf{f}(t) \quad (1)$$

where **M**, **C** and **K** are $n\times n$ mass, damping and stiffness matrices; $\mathbf{g}_{nl}$ is an $n\times n$ nonlinear stiffness matrix, $\mathbf{f}(t)$is applied nodal force vector and $\mathbf{x}(t)$ is the vector of physical displacements. The equations can be obtained for example, from finite element modelling of a structure. Transformation by $\mathbf{x} = \mathbf{\Phi p}$ leads to

$$\mathbf{\Phi^T M \Phi \ddot{p}} + \mathbf{\Phi^T C \Phi \dot{p}} + \mathbf{\Phi^T K \Phi p} + \mathbf{\Phi^T g}_{nl}(\mathbf{\Phi p}) = \mathbf{\Phi^T f}(t) \quad (2)$$

where $\mathbf{\Phi}$ is the modal vector matrix. By using orthogonality of the modes, equation (2) become

$$\bar{\mathbf{M}}\ddot{\mathbf{p}} + \bar{\mathbf{C}}\dot{\mathbf{p}} + \bar{\mathbf{K}}\mathbf{p} + \bar{\mathbf{g}}_{nl} = \bar{\mathbf{f}} \quad (3)$$

where $\bar{\mathbf{M}} = \mathbf{\Phi^T M \Phi} = [\bar{\mathrm{M}}_{rr}]$ and $\bar{\mathbf{K}} = \mathbf{\Phi^T K \Phi} = [\bar{\mathrm{K}}_{rr}]$ are diagonal matrices, and $\bar{\mathbf{g}}_{nl} = \mathbf{\Phi^T g}_{nl}$. If the structure has proportional damping, $\bar{\mathbf{C}} = \mathbf{\Phi^T C \Phi} = [\bar{\mathrm{C}}_{rr}]$ is also a diagonal matrix, and equation (3) reduces to

$$\bar{\mathrm{M}}_{rr}\ddot{\mathrm{p}}_r + \bar{\mathrm{C}}_{rr}\dot{\mathrm{p}}_r + \bar{\mathrm{K}}_{rr}\mathrm{p}_r + \bar{\mathrm{g}}_{nl,r} = \bar{\mathrm{f}}_r \quad (4)$$

where p_r is the rth modal displacement and other parameters in modal expression. Nonlinear terms, $\bar{\mathrm{g}}_{nl,r}$ refer to rth mode nonlinear restoring force and others mode allow for nonlinear cross-coupling terms.

5 Nonlinear Identification of Wing Structure

Polynomial basis functions of order 3 were used for the nonlinear models to retain the order of the model as low as possible. Static load test on two pylons and wing were performed. Equation (5) and (6) below show the nonlinearity in the pylons and wing structure contains cubic and quadratic terms. The curve for each pylon was fitted using a cubic polynomial and the equations were:

$$F_1 = (2.65 \times 10^8)x_1^3 + (2.76 \times 10^5)x_1^2 + (6.65 \times 10^3)x_1 - 0.05 \quad (5)$$

$$F_2 = (2.93 \times 10^8)x_2^3 + (3.87 \times 10^5)x_2^2 + (6.35 \times 10^3)x_2 + 0.45 \quad (6)$$

where F_1 is the load on the first pylon, F_2 is the load on the second pylon, x_1 is the deflection of first load and x_2 is the deflection of second load. From equation (5) and (6) it is shows that both assemblies pylon characteristic a slight symmetry, as the polynomial fits contain constant and second order terms.

Further investigation is dynamic test on overall wing structure using MIMO Normal mode test. This test was carried out on the unit to generate a freely decaying response. The existence of stiffness nonlinearity would be observed by a variation in natural frequency with amplitude and of a damping nonlinearity by changing damping ratio with amplitude. Damping effect on this study is very small and not affects the identification so much. No nonlinear damping was considered as the random excitation test showed little evidence of nonlinearity in the system. The finalized nonlinear curve fit was therefore of the form:

$$g_{nl,r} = f_r - m_r\ddot{p}_r - c_r\dot{p}_r - k_r p_r - A_{1,r}p_1^3 - B_{1,r}p_1^2p_2 - C_{1,r}p_1^2p_3 - D_{1,r}p_1p_2^2 - E_{1,r}p_1p_2p_3 - F_{1,r}p_1\dot{p}_3^2 G_{1,r}p_2^3 - H_{1,r}p_2^2p_3 - I_{1,r}p_2p_3^2 - J_{1,r}p_3^3 \quad (7)$$

for r=1,2,3, where the unknowns to be evaluated where c_r, k_r, $A_{1,r}$, $B_{1,r}$, $C_{1,r}$, $D_{1,r}$, $E_{1,r}$, $F_{1,r}$, $G_{1,r}$, $H_{1,r}$, $I_{1,r}$ and $J_{1,r}$. As the number of coefficient in this model is relatively small, the best model structure was chosen by implementing exhaustive search. Model structure selection is important part of nonlinear system identification method, because not every coefficient term should be included in the final model. In addition, some of the $A_{1,r}$ to $J_{1,r}$ coefficients may have value of 0. The resulting model is the complete optimum best curve fit that given particular sets of candidate terms and data. Table 3 shows the estimates of the interest linear parameters, c_r and k_r determined from the nonlinear curve fits and compare with those obtained from the PolyMAX modal estimation model. It can be seen that the parameters are small differences between two set of estimation. Furthermore, Table 4 shows nonlinear term coefficients identified for first three modes of overall wing structure. A zero value denotes that the corresponding term was not included in the final model.

Table 3: Comparison linear direct term coefficients identified by combination force appropriation and restoring method with PolyMAX.

Mode	Force Appropriation and Restoring Forces Method		PolyMAX (MIMO Normal Mode Test)	
	c_r, modal damping	k_r, modal stiffness	c_r, modal damping	k_r, modal stiffness
1	2.47	1.79E4	2.35	1.71E4
2	3.24	3.13E4	3.17	3.34E4
3	4.14	6.35E4	4.06	6.46E4

Table 4: Nonlinear term coefficients identified for first three modes

Coefficient	Mode 1	Mode 2	Mode 3
$A_{1,r}$	3.76E8	0	−2.56E8
$B_{1,r}$	−1.66E9	6.06E9	0
$C_{1,r}$	−7.8E8	0	1.58E9
$D_{1,r}$	3.45E9	0	0
$E_{1,r}$	0	−1.53E10	6.37E10
$F_{1,r}$	0	0	−2.42E9
$G_{1,r}$	0	1.33E9	0
$H_{1,r}$	0	1.87E9	0
$I_{1,r}$	0	2.32E10	1.42E9
$J_{1,r}$	−9.72E7	0	0

The final model structure in Equation (7), identified equation of motion for the overall wing structure can be written as:

$$2.15\ddot{p}_1 + 2.47\dot{p}_1 + 1.79 \times 10^4 p_1 + 3.76 \times 10^8 p_1^3 - 1.66 \times 10^9 p_1^2 p_2 - 7.87 \times 10^8 p_1^2 p_3 + 3.45 \times 10^9 p_1 p_2^2 - 9.72 \times 10^7 p_3^3 = f_1(t) \quad (8)$$

$$2.44\ddot{p}_2 + 3.24\dot{p}_2 + 3.13 \times 10^4 p_2 + 6.06 \times 10^9 p_1^2 p_2 - 1.53 \times 10^{10} p_1 p_2 p_3 + 1.33 \times 10^9 p_2^3 + 1.87 \times 10^9 p_2^2 p_3 + 2.32 \times 10^{10} p_2 p_3^2 = f_2(t) \tag{9}$$

$$\ddot{p}_3 + 4.14\dot{p}_3 + 6.35 \times 10^4 p_3 - 2.56 \times 10^8 p_1^3 + 1.58 \times 10^9 p_1^2 p_3 + 6.37 \times 10^{10} p_1 p_2 p_3 - 2.42 \times 10^9 p_1 p_3^2 + 1.42 \times 10^9 p_2 p_3^2 = f_3(t) \tag{10}$$

$$0.99\ddot{p}_4 + 2.95\dot{p}_4 + 2.21 \times 10^5 p_4 = f_4(t) \tag{11}$$

$$1.17\ddot{p}_5 + 5.35\dot{p}_5 + 7.19 \times 10^5 p_5 = f_5(t) \tag{12}$$

It should be noted that under appropriated excitation, the amplitudes of the cross coupled modal responses were about an order of magnitude lower than direct modal responses. Thus, all the cross coupled nonlinear terms in Equation (8) to (12) are at least an order of magnitude smaller than the direct nonlinear terms.

6 Conclusions

A nonlinear wing experimental structure was identified using the force appropriation and restoring forces method. In the pylon rotation degrees of freedom, discrete hardening stiffness nonlinearity was located. The results presented evidence nonlinear stiffness at higher amplitudes for the first three modes. The combination nonlinear identification methods succeeded to estimate good quality nonlinear modal models for the mode interest. It can be conclude that mode 1 and 2 showed hardening stiffness nonlinearities, mode 3 just showed very weak nonlinearity because of small degree of rotation at engine pylon structure. However, mode 4 and 5 behaved linearly for the wing structure.

Acknowledgement

The most work in this paper was conducted during the first author's PhD study at Liverpool sponsored by the Malaysia Ministry of Higher Education.

References

1. K. Worden, and G. R. Tomlinson, "Nonlinearity in structural dynamics: Detection, identification and modelling", University of Sheffield: Institute of Physics Publishing, (2001)
2. N. M. M. Maia, and J. M. M. Silva, "Theoretical and experimental modal analysis", Research Studies Press, Taunton, (1997)
3. D. J. Ewins, "Modal testing: theory and practice", Research Studies Press, Letchworth, (1995)
4. R. K. Dearson, "Discrete time dynamic model", Oxford University Press, (1994)
5. Ö. Arslan, M. Aykan , H. N. Özgüven, "Parametric identification os structural nonlinearities from measured frequency data", Mechanical System and Signal Processing, **vol. 25** , no. 11, pp 1112-1125, (2011)
6. G. Kerschen, K. Worden, A. F. Vakakis and J.C. Golinval, "Past, present and future of nonlinear system identification in structural dynamics", Mechanical System and Signal Processing, **vol. 20** , pp 505-592, (2006)
7. A. Hot, G. Kerschen, E. Foltête and S. Cogan, "Detection and quantification of non-linear structural behavior using principal component analysis", Mechanical System and Signal Processing, **vol. 26**, 104-116, (2012)
8. S. F. Masri, and T. K. Caughey, "A nonparametric identification technique for nonlinear dynamic problems", Journal of Applied Mechanics, **vol. 46**, pp. 433-447, (1979)
9. S. F. Masri, H. Sassi, and T. K. Caughey, "Nonparametric identification of nearly arbitrary nonlinear systems", Journal of Applied Mechanics, **vol. 49**, pp. 619-628,(1982)
10. E. F. Crawley, and A. C. Aubert, "Identification of nonlinear structural elements by force-state mapping", AIAA Journal, **vol. 24**, pp. 155-162, (1986)
11. G. Dimitriadis, and J. E. Cooper, "A method for the identification of non-linear multi-degree-of-freedom systems", Proceedings of the Institute of Mechanical Engineers, Part G 212, pp. 287-298, (1998)
12. G. Kerschen, J. C. Golinval and K. Worden, "Theoretical and experimental identification of a non-linear beam", Journal of Sound and Vibration, **vol. 244**, pp. 597-613, (2001)
13. M. Aykan, and H. N. Özgüven, "Parametric identification of nonlinearity from incomplete FRF data using describing function inversion", In: Proceedings of the SEM IMAC XXX conference, Jacsonville, **vol 3**, (2012)
14. M. Aykan, and H. N. Özgüven, "Identification of restoring force surfaces in Nonlinear MDOF systems from FRF data using nonlinearity matrix", Conference Proceedings of the Society for Experimental Mechanics Series, **vol. 35**, pp 65-76, (2013)
15 M.S.M. Sani, H. Ouyang, J.E. Cooper and C.K.E.N.C.K. Husin, "Smart Methodology of Stiffness Nonlinearity Identification Vibration System", 2nd International Conference on Mechanical Engineering Research 2013 (ICMER) (2013)
16. J. P. Noel, G. Kerschen, and A. Newerla, "Application of the Restoring Force Surface Method to a Real-life Spacecraft Structure", Conference Proceedings of the Society for Experimental Mechanics Series, pp 1-19, (2012).

Finite element model updating of natural fibre reinforced composite structure in structural dynamics

M.S.M. Sani[1,2,a], N.A.Z. Abdullah[1], S.N. Zahari[1], J.P. Siregar[3] and M.M. Rahman[3]

[1]Advanced Structural Integrity and Vibration Research (ASIVR), Faculty of Mechanical Engineering, Universiti Malaysia Pahang, 26600 Pekan, Pahang, Malaysia
[2]Automotive Engineering Centre, Universiti Malaysia Pahang, Malaysia
[3]Faculty of Mechanical Engineering, Universiti Malaysia Pahang, 26600 Pekan, Pahang, Malaysia

Abstract. Model updating is a process of making adjustment of certain parameters of finite element model in order to reduce discrepancy between analytical predictions of finite element (FE) and experimental results. Finite element model updating is considered as an important field of study as practical application of finite element method often shows discrepancy to the test result. The aim of this research is to perform model updating procedure on a composite structure as well as trying improving the presumed geometrical and material properties of tested composite structure in finite element prediction. The composite structure concerned in this study is a plate of reinforced kenaf fiber with epoxy. Modal properties (natural frequency, mode shapes, and damping ratio) of the kenaf fiber structure will be determined using both experimental modal analysis (EMA) and finite element analysis (FEA). In EMA, modal testing will be carried out using impact hammer test while normal mode analysis using FEA will be carried out using MSC. Nastran/Patran software. Correlation of the data will be carried out before optimizing the data from FEA. Several parameters will be considered and selected for the model updating procedure.

1 Introduction

The extensively use of composite structure across various field such as automotive, aerospace, naval and other high performance engineering application is due to their unique characteristics; light weight, high specific strength and stiffness, easy to fabricate and excellent thermal characteristics. This has aroused many researchers to study various aspect of their structural behaviour. Same with isotropic materials, composite materials also subjected to various type of damage, mostly crack and delamination. These resulting in local changes of stiffness of element thus the dynamic characteristic are modified. Therefore complete understanding of dynamic behaviour for composite structure should be not taken lightly. Dynamic behaviour in terms of natural frequency, damping ratio and mode shapes can be determine via modal analysis process either by numerical or experimental.

Finite element method is used for various type of structural analysis, although it is practical, discrepancy between the prediction results and the experimental result still exist. The reason this discrepancy exists is due to several factors. In order to reduce the discrepancy is by performing model updating procedure on the finite element model. Model updating is a technique of modifying the modeling assumptions and parameters until the correlation of analytical predictions and experimental results satisfies each other in practical requirement.

Abundant number of studies on model updating technique has been performed as model updating is a popular field of study in structural dynamics. Mottershead and Friswell have presented numerous findings and important issues regarding model updating topic [1-12] including topic of parameterization and regularization. Several concerns regarding this topic have been made in order to ensure the updated model has good physical meaning to the real structure and capable to carry out any prediction simulation and producing promising results.

Many approach of model updating procedure have been introduced by various authors in the past. For example, Esfandiari [13] in his study has demonstrated the utilization on frequency response function (FRF) in model updating aside from using the extracted modal data (natural frequencies, damping ratios and mode shapes) from the FRF obtained experimentally. On the other hand, in another study of model updating procedure explain the practice of making adjustment using modal data [14-18]. Although the above investigation examined the various approach of model updating, to the best of author's knowledge, there are quite a number of references in the literature systematically describe the

[a] Corresponding author: mshahrir@ump.edu.my

DOI: 10.1051/matecconf/20168303007

procedure of model updating on composite structure focusing on natural fibre. However, the study of model updating that focus on this type of fibre, which is kenaf fibre is quite a few. This was the motivation behind the present study. Accordingly, the present study attempts to demonstrate the finite element model updating procedure on selected composite material which is the kenaf fiber (hibiscus cannabinus, L. family Malvaceae) with different composition. This study will also attempt to improve the presumed geometrical and material properties of tested composite structure that was used beforehand in finite element prediction.

2 Kenaf fiber reinforced composite structure

In this investigation, the natural fiber, which is kenaf, was combined with the resin or matrix, which is epoxy, to complete the variation of composite material with different percentage of composition. Hand lay-up process was performed for the preparation of the composite material (Figure 1). The compositions of the composite structures were prepared as follows:

i. 5% kenaf + 95% epoxy
ii. 10% kenaf + 90% epoxy
iii. 15% kenaf + 85% epoxy

To start with, both kenaf and epoxy were weighted using the weighing scales with the correct measurement. Then, kenaf was mixed with the polyester and stirred together to obtain the new composite material. The mixture was poured into glass mold and paint roller was used to consolidate the mixture thoroughly wetting the reinforcement and eliminate the entrapped air. The composite material was left for 24 hours until it became harden before the composite was separated with the mold.

Fig. 1. Full process for preparing the specimen.

3 Numerical Analysis

Normal mode analysis was performed on the finite element (FE) model of composite plate via MSC Nastran software to obtain modal data of the composite plate structure for each composition. The FE model of all the composite plates ware modelled using four-node shell elements (CQUAD4) built in MSC Patran software. The composite plates consist of 24 element and 35 nodes. The geometrical and material properties as stated in Table 1 were assigned to the FE model of the composite structure. The value of Young's Modulus of the composite with each composition, were obtained and calculated based on the tensile test carried out on other specimen of the same composition of kenaf fibres. The free-free boundary condition was applied to the structure. Then, normal mode analysis (SOL 103) was executed to compute modal data of the composite structure. The first five natural frequencies obtained throughout the FE analysis are shown in Table 2. The density of the composites plates for each composition was calculated based on general rule of mixture formula for density (refer to Equation 1).

$$\rho_c = \rho_f \upsilon_f + \rho_m \upsilon_m \quad (1)$$

where ρ_c, ρ_f and ρ_m is the density of the composite, fiber and matrix respectively while υ_f and υ_m is the volume percentage of fiber and matrix respectively.

Table 1. Assigned geometrical and material properties of the composite plate structure.

Properties	Value according to kenaf composition		
	5%	10%	15%
Young's Modulus (GPa)	1.55	1.7	1.8
Density (kg/m^3)	1390	1380	1370
Poisson Ratio	0.291	0.291	0.291
Thickness (m)	0.003	0.003	0.003

4 Experimental Modal Analyses (EMA)

Modal testing or EMA was used to extract modal parameters of the composite structures experimentally. Impact hammer test was used as the excitation method in this EMA. All the composite plate was divided into 24 grids point to represent the point where FRF will be measured. A tri-axial accelerometer was used to measure the acceleration of the structures when external force was applied to them by using impact hammer. The composite plates were supported by elastic sponge during testing in order to simulate the free-free boundary condition as assigned in numerical analysis.

DasyLab software was used in order to measure the input signal from the impact hammer and the acceleration of the plates from the accelerometer. All the signals were converted into frequency response function (FRF) which was later processed in post-processing software in order to obtain modal data of all tested composite plates. The modal data obtained were correlated with the modal data that was gathered previously using finite element analysis (FEA) so as to ensure that the selected modes were paired correctly. Comparison of natural frequencies obtained experimentally and through FEA was shown in Table 2. Based on the table, it is apparent that there are significant value of error between data gathered through EMA and FEA. The presupposition that can be done regarding this matter is that presume properties used for the composite structure in FEA was inexact.

5 Model Updating

Model updating, as stated in previously in this paper, is an effective method that can be used to improve correlation between finite element model and experimental results. This can be done by changing the modelling parameters that have been assigned beforehand as an assumption value. By using data gathered through experiment, the assumption that was assigned can be corrected into more accurate value. Therefore, more precise numerical model with more trusted properties can be computed.

The optimization algorithm built in MSC Nastran was used. The objective function, J, which was used for error minimization in the updating process, is stated in Equation 2. The prediction of the modal data was put for error reduction in this updating procedure. The operation was carried out until convergence was achieved when the error function between two data was small enough.

$$J = \Sigma^n j_{=1} (\lambda_j / \lambda_j^{exp} - 1)^2 \qquad (2)$$

As shown in Equation 1, λ_j is the jth prediction of eigenvalue from FE model while λ_j^{exp} is the jth eigenvalue prediction obtained experimentally. It is prominent that this equation is only considered as valid for use if the experimental eigenvalue and analytical eigenvalues are paired correctly. Comparison of natural frequencies values of initial FE results and the model updating results is given in Table 2 as well. As shown in the table, significant difference exists between natural frequency value before and after model updating was performed. Percentage of errors of data before updating and post updating are also lessening satisfactorily.

On the other hand, sensitivity analysis was done in order to select the updating parameters to ensure that only sensitive or significant parameters are chosen for model updating [16]. Formulation of sensitivity analysis is shown as follows:

$$\omega^2_{n+1} = \omega^2_n + S_i\,(\delta\theta) \qquad (3)$$

$$S_i = \delta\omega^2/\,\delta\theta = u_i^T\,(\delta K/\,\delta\theta - \omega^2\,\delta M/\,\delta\theta)\;u_i \qquad (4)$$

In Equation 3, S_i is a sensitivity matrix at ith iteration, which represents the rates of change of the structural eigenvalues ω^2 with respect to changes in parameter $\delta\theta$, which can be expressed as in Equation 4.

Two parameters that were picked for this study are the Young's modulus of the kenaf and epoxy composite and the thickness of the composite plate structure. The value of You ng's modulus was allowed to vary from 1.549 GPa to 2.184 GPa for plate of 5% kenaf composition, 1.69 GPa to 2.184 GPa for plate of 10% kenaf composition, and 1.79 GPa to 2.184 GPa for plate of 15% kenaf composition which are the range value of the composite material with the tested composition [17]. At the same time, the thickness value was allowed to vary in small range from its initial presumes value since the thickness of the composite plate can hardly diverse much all over the plate structure due to its slight unevenness on lower surface.In addition, by performing the sensitivity analysis [18-20] on the selected parameters, the actual properties of the composite structure can be computed and the presumed value of the composite was corrected. Table 3 shows the changes of the updated parameter properties from their initial values.

Table 2. Comparison of natural frequency values between initial results and updated results.

Mode	EMA natural frequency (Hz)			Initial FE results						Model updating FE results					
				Natural frequency (Hz)			Error (%)			Frequency (Hz)			Error (%)		
	5% kenaf	10% kenaf	15% kenaf	5% kenaf	10% kenaf	15% kenaf	5% kenaf	10% kenaf	15% kenaf	5% kenaf	10% kenaf	15% kenaf	5% kenaf	10% kenaf	15% kenaf
1	37.0	38.9	39.3	33.98	35.72	36.89	8.16	8.17	6.13	37.72	38.56	38.55	1.95	0.87	1.91
2	40.4	42.0	44.0	38.18	40.13	41.45	5.50	4.45	5.80	42.39	43.33	43.32	4.93	3.17	1.55
3	87.1	89.4	90.3	71.91	75.58	78.06	17.44	15.46	13.55	79.84	81.61	81.58	8.34	8.71	9.66
4	90.3	92.7	93.0	75.29	79.13	81.72	16.62	14.64	12.13	83.56	85.42	85.39	7.46	7.85	8.18
5	101.0	104.0	107.0	87.92	92.40	95.43	12.95	11.15	10.81	97.58	99.75	99.71	3.39	4.09	6.81

Table 3. Change of updated parameter properties from their initial values.

Parameter	Kenaf composition	Initial value (I)	Updated value (II)	Change (%) |(II-I)/I|
Young's modulus, E (GPa)	5%	1.55	1.76	13.55
	10%	1.70	1.83	7.65
	15%	1.80	1.87	3.89
Thickness, t (m)	5%	0.0030	0.0031	3.33
	10%	0.0030	0.0031	3.33

	15%	0.0030	0.0031	3.33

6 Conclusions

This study was carried out in order to perform modal based model updating on composite structure as well as to improve the presumed geometrical and material properties of the tested composite structure in FE prediction. The updating procedure is observed as parameter identification, which intends to bring the finite element prediction to be as closely as possible to the actual test subject.

The initial FE prediction of the composite plate has shown quite diversion from the experimental findings, mostly caused by uncertain value of properties assigned on the structure. However, the uncertain properties values have successfully improved after updating procedure. Furthermore, the percentage of error for those two sets of data (FE and experiments) reduced satisfactory after updating. The most sensitive parameters for updating the composite plate also have been identified as well as computing the updated value for each of the updating parameters. Obviously, the updated value for those parameters showed dissimilarity when compared to the values that was assigned initially on the structure in FE.

The evidence from this study testifies that model updating technique or also called model calibration is proven to be a good method in reducing divergences. Concurrently, SOL200 is proven to be an effective algorithm for finding the sensitivity of parameters. More information on model updating technique would help us to establish a greater degree of accuracy on this matter. If the investigation is to be moved forward, a better understanding of model updating technique and parameterization needs to be developed.

Acknowledgement

The authors would like to express their gratitude to the Malaysia Ministry of Higher Education (MOHE) for providing the FRGS fund for this study.

References

1. H. Ahmadian, J. E. Mottershead, and M. I. Friswell, "Physical Realization of Generic-Element Parameters in Model Updating," *Journal of Vibration and Acoustics*, **vol. 124**. p. 628, (2002)
2. H. H. Khodaparast, J. E. Mottershead, and K. J. Badcock, "Interval model updating with irreducible uncertainty using the Kriging predictor," *Mech. Syst. Signal Process.*, **vol. 25**, no. 4, pp. 1204–1226, (May 2011)
3. M. Palmonella, M. I. Friswell, J. E. Mottershead, and A. W. Lees, "Finite element models of spot welds in structural dynamics: Review and updating," *Computers and Structures*, **vol. 83**. pp. 648–661, (2005)
4. J. E. Mottershead, M. I. Friswell, G. H. T. Ng, and J. A. Brandon, "Geometric Parameters for Finite Element Model Updating of Joints and Constraints," *Mech. Syst. Signal Process.*, **vol. 10**, pp. 171–182, (1996)
5. J. E. Mottershead, M. Link, and M. I. Friswell, "The sensitivity method in finite element model updating: A tutorial," *Mech. Syst. Signal Process.*, **vol. 25**, no. 7, pp. 2275–2296, (Oct. 2011)
6. J. E. Mottershead and M. I. Friswell, "Model Updating In Structural Dynamics: A Survey," *J. Sound Vib.*, **vol. 167**, pp. 347–375, (1993)
7. W. Wang, J. E. Mottershead, C. M. Sebastian, and E. A. Patterson, "Shape features and finite element model updating from full-field strain data," *Int. J. Solids Struct.*, **vol. 48**, no. 11–12, pp. 1644–1657, (Jun. 2011)
8. C. Mares, M. I. Friswell, and J. E. Mottershead, "Model updating using robust estimation," *Mech. Syst. Signal Process.*, **vol. 16**, no. 1, pp. 169–183, (Jan. 2002)
9. M. I. Friswell and J. E. Mottershead, *Finite element model updating in structural dynamics*, **vol. 38**. (1995)
10. M. I. Friswell, J. E. Mottershead, and H. Ahmadian, "Finite-element model updating using experimental test data: parametrization and regularization," *Philos. Trans. R. Soc. London, Ser. A (Mathematical, Phys. Eng. Sci.*, **vol. 359**, pp. 169–186,(2001)
11. C. Mares, J. E. Mottershead, and M. I. Friswell, "Natural-Frequency Errors and Using Physical Reasoning," *J. Vib. Acoust.*, **vol. 17**, no. 1, pp. 39–46, (Jan. 2003)
12. W. Wang, J. E. Mottershead, A. Ihle, T. Siebert, and H. Reinhard Schubach, "Finite element model updating from full-field vibration measurement using digital image correlation," *J. Sound Vib.*, **vol. 330**, no. 8, pp. 1599–1620, (Apr. 2011)
13. A. Esfandiari, F. Bakhtiari-Nejad, M. Sanayei, and A. Rahai, "Structural finite element model updating using transfer function data," *Comput. Struct.*, **vol. 88**, no. 1–2, pp. 54–64, (Jan. 2010)
14. H.-P. Chen and T. S. Maung, "Regularised finit e element model updating using measured incomplete modal data," *J. Sound Vib.*, **vol. 333**, no. 21, pp. 5566–5582, (Oct. 2014)
15. E. Jacquelin, S. Adhikari, and M. I. Friswell, "A second-moment approach for direct probabilistic model updating in structural dynamics," *Mech. Syst. Signal Process.*, **vol. 29**, pp. 262–283, May 2012.
16. N. Husain and A. Snaylam, "FE Model Updating for Damage Detection–Application to a Welded Structure," *Key Eng. ...*, **vol. 414**, no. 2009, (2009)
17. R. Bow and M. Selection, "The Investigation of Mechanical Properties on Natural Fiber Composites for Recurve Bow Material Selection," **vol. 9**, no. 25, pp. 27–31, (2015)
18. N. Abdullah, et al. A review on model updating in structural dynamics. in IOP Conference Series: Materials Science and Engineering. **vol. 100** 01 (2015) doi:10.1088/1757-899X/100/1/012015

Application of the Dahl friction model in the dynamics analysis of grab cranes

A. Urbaś

University of Bielsko-Biala, Department of Mechanics, 43-309 Bielsko-Biala, Willowa 2, Poland

Abstract. The paper presents dynamics analysis of grab cranes. The cranes can be built out of any number of links, however, only selected links have their own drive that is considered in the flexible form. The cranes are mounted to the ground by means of any number of flexible supports. Joint coordinates and homogeneous transformation matrices are used to describe the cranes' geometry. Equations of motion are derived using Lagrange equations. Friction phenomenon is taken into account in all joints. The joint forces and torques used to calculate friction torques in the revolute joints and friction forces in the prismatic joints are determined using the recursive Newton-Euler algorithm. Models of revolute and prismatic joints are worked out. Friction coefficients are defined using the Dahl friction model. The results of numerical calculations present the influence of friction on the driving torques, force and behaviour of the load.

Nomenclature

g – acceleration of gravity

i – symbol of support

p – symbol of link (joint, drive)

$e^{(i)}_{s,\alpha}\big|_{\alpha\in\{x,y,z\}}$ – elongation of support i in α direction

$l^{(p)}$ – length of link p

$l^{(i)}_{s,\alpha,0}, l^{(i)}_{s,\alpha}\big|_{\alpha\in\{x,y,z\}}$ – initial and current length of support i in α direction

$m^{(p)}$ – mass of link p

$\tilde{n}^{(p)}_{dof}$ – number of generalised coordinates describing the motion of link p with respect to link $p-1$, $\tilde{n}^{(p)}_{dof} \le 6$

$n^{(p)}_{dof}$ – number of generalised coordinates describing the motion of link p with respect to reference system $\{0\}$

n_{dr} – number of drives

n_l – number of links

n_s – number of supports

$s^{(i)}_{s,\alpha}, d^{(i)}_{s,\alpha}\big|_{\alpha\in\{x,y,z\}}$ – stiffness and damping coefficients of support i in α direction

$e^{(i)}_{s,\alpha}\big|_{\alpha\in\{x,y,z\}}$ – elongation of support i in α direction

$s^{(p)}_{dr}, d^{(p)}_{dr}$ – stiffness and damping coefficients of drive p

$\tilde{\mathbf{q}}^{(p)}$ – vector of generalised coordinates describing the motion of link p with respect to link $p-1$

$\mathbf{q}^{(p)}$ – vector of generalised coordinates describing the motion of link p with respect to reference system $\{0\}$

$\mathbf{r}^{(p)}_{C^{(p)}}$ – vector of position of point $C^{(p)}$ defined in the local coordinate system of link p

$\mathbf{H}^{(p)}$ – pseudo-inertia matrix of link p

$\tilde{\mathbf{T}}^{(p)}$ – homogeneous transformation matrix from the local coordinate system of link p to the system of link $p-1$

$\mathbf{T}^{(p)}$ – homogeneous transformation matrix from the local coordinate system of link p to reference system $\{0\}$,

$$\mathbf{T}^{(p)}_i = \frac{\partial \mathbf{T}^{(p)}}{\partial q^{(p)}_i}, \quad \mathbf{T}^{(p)}_{i,j} = \frac{\partial^2 \mathbf{T}^{(p)}}{\partial q^{(p)}_i \partial q^{(p)}_j}$$

Friction parameters

i – parameter that defines the shape of the material curve

$\mu^{(p)}_\alpha\big|_{\alpha\in\{A,B,C\}}, \mu^{(p)}$ – friction coefficient in the revolute and prismatic joint

$\mu^{(p)}_{\alpha,0}\big|_{\alpha\in\{A,B,C\}}, \mu^{(p)}_0$ – maximum value of the friction coefficient in the revolute and prismatic joint

$\sigma^{(p)}_\alpha\big|_{\alpha\in\{A,B,C\}}, \sigma^{(p)}$ – contact stiffness coefficient of the revolute and prismatic joint

DOI: 10.1051/matecconf/20168303008

1 Introduction

Grab cranes belong to the group of handling equipment that is used in various industries, and in particular in the forest industry. A versatile analysis of these cranes is conducted by means of computer methods of modelling and simulation. For this purpose, engineers use commercial programs or, less often, authors' programs which are cheaper and dedicated for specific applications. However, the essential advantage of creating their own programs is primarily the ability to test alternative and sometimes new methods of modelling (e.g. flexibility of links, friction) or to solve equations of motion.

The mathematical model for dynamics analysis of a grab crane [1-4] built with eight links is presented in the paper. Flexibility of the support system and drives is taken into account. The influence of friction in the joints on the driving torques and force is analysed.

2 Mathematical model of the grab crane

As an example, the dynamics analysis of a grab crane built out of eight links, of which only five have their own drive with driving torque/force $\mathbf{t}_{dr}^{(p)}\Big|_{p=2,3,4,5,8}$ (Fig. 1), is presented. The crane is sited by means of eight flexible supports.

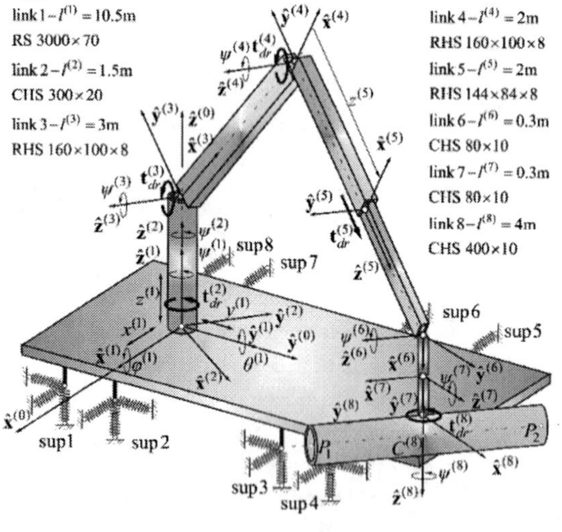

Fig. 1. Model of the grab crane.

The motion of link p of the crane is described by the vector of joint coordinates:

$$\mathbf{q}^{(p)}\Big|_{p=1,\dots,n_l} = \left[\mathbf{q}^{(p-1)^T} \quad \tilde{\mathbf{q}}^{(p)^T}\right]^T, \tag{1}$$

where: $\mathbf{q}^{(0)} = \varnothing$,

$\tilde{\mathbf{q}}^{(1)} = \left[x^{(1)} \quad y^{(1)} \quad z^{(1)} \quad \psi^{(1)} \quad \theta^{(1)} \quad \varphi^{(1)}\right]^T$, $\tilde{\mathbf{q}}^{(2)} = \left[\psi^{(2)}\right]$,

$\tilde{\mathbf{q}}^{(3)} = \left[\psi^{(3)}\right]$, $\tilde{\mathbf{q}}^{(4)} = \left[\psi^{(4)}\right]$, $\tilde{\mathbf{q}}^{(5)} = \left[z^{(5)}\right]$, $\tilde{\mathbf{q}}^{(6)} = \left[\psi^{(6)}\right]$,

$\tilde{\mathbf{q}}^{(7)} = \left[\psi^{(7)}\right]$, $\tilde{\mathbf{q}}^{(8)} = \left[\psi^{(8)}\right]$.

The Denavit-Hartenberg notation [5] is used to describe the geometry of the crane. The homogeneous transformation matrices from the local coordinate systems to the reference system are determined as follows:

$$\mathbf{T}^{(p)}\Big|_{p=1,\dots,n_l} = \mathbf{T}^{(p-1)}\tilde{\mathbf{T}}^{(p)}, \tag{2}$$

where: $\mathbf{T}^{(0)} = \mathbf{I}$,

$$\tilde{\mathbf{T}}^{(1)} = \begin{bmatrix} c\psi^{(1)}c\theta^{(1)} & c\psi^{(1)}s\theta^{(1)}s\varphi^{(1)} - s\psi^{(1)}c\varphi^{(1)} & c\psi^{(1)}s\theta^{(1)}c\varphi^{(1)} + s\psi^{(1)}s\varphi^{(1)} & x^{(1)} \\ s\psi^{(1)}c\theta^{(1)} & s\psi^{(1)}s\theta^{(1)}s\varphi^{(1)} + c\psi^{(1)}c\varphi^{(1)} & s\psi^{(1)}s\theta^{(1)}c\varphi^{(1)} - c\psi^{(1)}s\varphi^{(1)} & y^{(1)} \\ -s\theta^{(1)} & c\theta^{(1)}s\varphi^{(1)} & c\theta^{(1)}c\varphi^{(1)} & z^{(1)} \\ 0 & 0 & 0 & 1 \end{bmatrix} \text{ – nonlinear model,}$$

$$\tilde{\mathbf{T}}^{(1)} = \begin{bmatrix} 1 & -\psi^{(1)} & \theta^{(1)} & x^{(1)} \\ \psi^{(1)} & 1 & -\varphi^{(1)} & y^{(1)} \\ -\theta^{(1)} & \varphi^{(1)} & 1 & z^{(1)} \\ 0 & 0 & 0 & 1 \end{bmatrix} \text{ – linear model,}$$

$$\tilde{\mathbf{T}}^{(2)} = \begin{bmatrix} c\psi^{(2)} & -s\psi^{(2)} & 0 & 0 \\ s\psi^{(2)} & c\psi^{(2)} & 0 & 0 \\ 0 & 0 & 1 & 0 \\ 0 & 0 & 0 & 1 \end{bmatrix},$$

$$\tilde{\mathbf{T}}^{(3)} = \begin{bmatrix} c\psi^{(3)} & -s\psi^{(3)} & 0 & 0 \\ 0 & 0 & -1 & 0 \\ s\psi^{(3)} & c\psi^{(3)} & 0 & l^{(2)} \\ 0 & 0 & 0 & 1 \end{bmatrix},$$

$$\tilde{\mathbf{T}}^{(4)} = \begin{bmatrix} c\psi^{(4)} & -s\psi^{(4)} & 0 & l^{(3)} \\ s\psi^{(4)} & c\psi^{(4)} & 0 & 0 \\ 0 & 0 & 1 & 0 \\ 0 & 0 & 0 & 1 \end{bmatrix},$$

$$\tilde{\mathbf{T}}^{(5)} = \begin{bmatrix} 1 & 0 & 0 & 0 \\ 0 & 0 & -1 & -z^{(5)} \\ 0 & 1 & 0 & 0 \\ 0 & 0 & 0 & 1 \end{bmatrix},$$

$$\tilde{\mathbf{T}}^{(6)} = \begin{bmatrix} c\psi^{(6)} & -s\psi^{(6)} & 0 & 0 \\ 0 & 0 & 1 & 0 \\ -s\psi^{(6)} & -c\psi^{(6)} & 0 & l^{(5)} \\ 0 & 0 & 0 & 1 \end{bmatrix},$$

$$\tilde{\mathbf{T}}^{(8)} = \begin{bmatrix} c\psi^{(8)} & -s\psi^{(8)} & 0 & 0 \\ 0 & 0 & 1 & l^{(7)} \\ -s\psi^{(8)} & -c\psi^{(8)} & 0 & 0 \\ 0 & 0 & 0 & 1 \end{bmatrix}.$$

In order to determine friction torques in the revolute joints and friction forces in the prismatic joints, models of these joints were developed (Fig. 2).

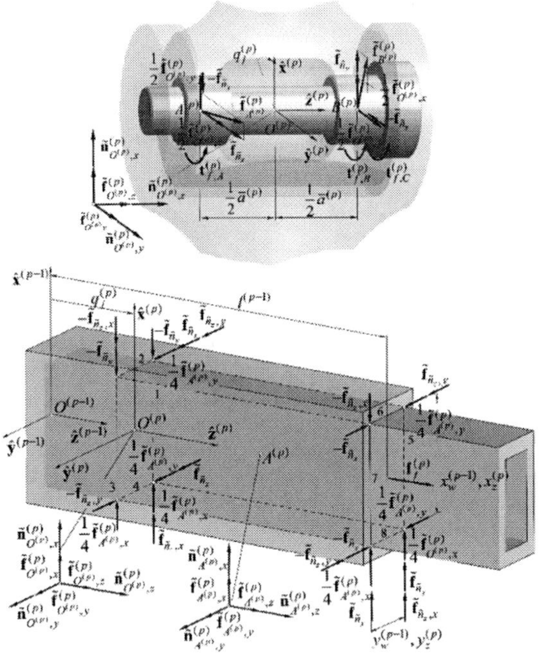

Fig. 2. Models of revolute and prismatic joints.

In order to determine the joint forces $\tilde{\mathbf{f}}^{(p)}_{O^{(p)}}$ and torques $\tilde{\mathbf{n}}^{(p)}_{O^{(p)}}$, and then the values of friction torques and forces, the recursive Newton-Euler algorithm [5] is used.
The values of the friction coefficients are defined using the Dahl friction model [6], described by the following form:

$$\dot{\mu}\Big|_{\substack{\mu=\mu_\alpha^{(p)}\big|_{\alpha\in\{A,B,C\}} \\ \mu=\mu^{(p)}}} = \sigma \dot{q}_j^{(p)} \left(1-\frac{\mu}{\mu_0}\operatorname{sgn}\left(\dot{q}_j^{(p)}\right)\right)^i \times \times\operatorname{sgn}\left(1-\frac{\mu}{\mu_0}\operatorname{sgn}\left(\dot{q}_j^{(p)}\right)\right)\Bigg|_{\substack{\mu_0=\mu_{\alpha,0}^{(p)},\ \sigma=\sigma_\alpha^{(p)}\big|_{\alpha\in\{A,B,C\}} \\ \mu_0=\mu_0^{(p)},\ \sigma=\sigma^{(p)}}} \tag{3}$$

The equations of motion are derived using the Lagrange equations and the algorithms taken from paper [7]:

$$\mathbf{A}\ddot{\mathbf{q}} = \mathbf{e} + \mathbf{f}_s + \mathbf{t}_{dr} - \mathbf{s}_f , \tag{4}$$

where: $$\mathbf{A} = \begin{bmatrix} \mathbf{A}_{1,1} & \cdots & \mathbf{A}_{1,j} & \cdots & \mathbf{A}_{1,n_l} \\ \vdots & & \vdots & & \vdots \\ \mathbf{A}_{i,1} & \cdots & \mathbf{A}_{i,j} & \cdots & \mathbf{A}_{i,n_l} \\ \vdots & & \vdots & & \vdots \\ \mathbf{A}_{n_l,1} & \cdots & \mathbf{A}_{n_l,j} & \cdots & \mathbf{A}_{n_l,n_l} \end{bmatrix},$$

$$\mathbf{A}_{i,j} = \sum_{l=\max\{i,j\}}^{n_l} \tilde{\mathbf{A}}_{i,j}^{(l)} , \ \tilde{\mathbf{A}}_{i,j}^{(p)}\Big|_{i,j=1,\dots,p} = \left(\tilde{a}^{(p)}_{n_{dof}^{(i-1)}+k,\,n_{dof}^{(j-1)}+l}\right)_{\substack{k=1,\dots,\tilde{n}_{dof}^{(i)} \\ l=1,\dots,\tilde{n}_{dof}^{(j)}}} ,$$

$$\tilde{a}_{i,j}^{(p)} = \operatorname{tr}\left\{\mathbf{T}_i^{(p)}\mathbf{H}^{(p)}\mathbf{T}_j^{(p)^T}\right\},$$

$$\mathbf{e} = \begin{bmatrix} \mathbf{e}_1 \\ \vdots \\ \mathbf{e}_i \\ \vdots \\ \mathbf{e}_{n_l} \end{bmatrix}, \ \mathbf{e}_i = -\sum_{l=i}^{n_l}\left(\tilde{\mathbf{h}}_i^{(l)} + \tilde{\mathbf{g}}_i^{(l)}\right),$$

$$\tilde{\mathbf{h}}_i^{(p)}\Big|_{i=1,\dots,p} = \left(\tilde{h}^{(p)}_{n_{dof}^{(i-1)}+k}\right)_{k=1,\dots,\tilde{n}_{dof}^{(p)}},$$

$$\tilde{h}_i^{(p)} = \sum_{m=1}^{n_{dof}^{(p)}}\sum_{n=m}^{n_{dof}^{(p)}} \operatorname{tr}\left\{\mathbf{T}_i^{(p)}\mathbf{H}^{(p)}\mathbf{T}_{m,n}^{(p)^T}\right\}\dot{q}_m^{(p)}\dot{q}_n^{(p)},$$

$$\tilde{\mathbf{g}}_i^{(p)}\Big|_{i=1,\dots,p} = \left(\tilde{g}^{(p)}_{n_{dof}^{(i-1)}+k}\right)_{k=1,\dots,\tilde{n}_{dof}^{(p)}}, \ \tilde{g}_i^{(p)} = m^{(p)}g\mathbf{j}_3\mathbf{T}_i^{(p)}\tilde{\mathbf{r}}^{(p)}_{C^{(p)}},$$

$$\mathbf{j}_3 = \begin{bmatrix} 0 & 0 & 1 & 0 \end{bmatrix},$$

$$\mathbf{f}_s = \left[-\left(\frac{\partial E_{p,s}}{\partial \tilde{\mathbf{q}}^{(1)}} + \frac{\partial R_s}{\partial \dot{\tilde{\mathbf{q}}}^{(1)}}\right)^T \ \ \mathbf{0}\right]^T,$$

$$E_{p,s} = \frac{1}{2}\sum_{i=1}^{n_s}\sum_{\alpha\in\{x,y,z\}} s_{s,\alpha}^{(i)}\left(e_{s,\alpha}^{(i)}\right)^2,$$

$$R_s = \frac{1}{2}\sum_{i=1}^{n_s}\sum_{\alpha\in\{x,y,z\}} d_{s,\alpha}^{(i)}\left(\dot{e}_{s,\alpha}^{(i)}\right)^2,$$

$$\mathbf{t}_{dr} = \begin{bmatrix} \mathbf{0} & t_{dr}^{(2)} & t_{dr}^{(3)} & t_{dr}^{(4)} & t_{dr}^{(5)} & 0 & 0 & t_{dr}^{(8)} \end{bmatrix}^T,$$

$$t_{dr}^{(p)} = -\left(\frac{\partial E_{p,dr}^{(p)}}{\partial q_j^{(p)}} + \frac{\partial R_{dr}^{(p)}}{\partial \dot{q}_j^{(p)}}\right),$$

$$E_{p,dr}^{(p)} = \frac{1}{2}s_{dr}^{(p)}\left(q_{dr}^{(p)} - q_j^{(p)}\right)^2, \ R_{dr}^{(p)} = \frac{1}{2}d_{dr}^{(p)}\left(\dot{q}_{dr}^{(p)} - \dot{q}_j^{(p)}\right)^2,$$

$$\mathbf{s}_f = \begin{bmatrix} \mathbf{0} & t_f^{(2)} & t_f^{(3)} & t_f^{(4)} & f_f^{(5)} & t_f^{(6)} & t_f^{(7)} & t_f^{(8)} \end{bmatrix}^T,$$

$$t_f^{(p)} = t_{f,A}^{(p)} + t_{f,B}^{(p)} + t_{f,C}^{(p)},$$

$$t_{f,A}^{(p)} = \frac{1}{2}\mu_A^{(p)}\tilde{f}^{(p)}_{A^{(p)}}d_A^{(p)}, \ t_{f,B}^{(p)} = \frac{1}{2}\mu_B^{(c,p)}\tilde{f}^{(p)}_{B^{(p)}}d_B^{(p)},$$

$$t_{f,C}^{(p)} = \frac{1}{3}\mu_C^{(p)}\left|\tilde{f}^{(p)}_{O^{(p)},z}\right|\frac{d_C^{(p)^3} - d_A^{(p)^3}}{d_C^{(p)^2} - d_A^{(p)^2}},$$

$$f_f^{(p)} = \mu^{(p)}\tilde{f}^{(p)}, \ \tilde{f}^{(p)} = \sum_{i=1}^{8}\tilde{f}_i^{(p)}.$$

The equations of motion are integrated using the Runge-Kutta method of the fourth order with a fixed step equal to 10^{-4}.

DOI: 10.1051/matecconf/20168303008

3 Results of the calculations

The influence of friction on the courses of the driving torques in the revolute joints and driving force in the prismatic joint are presented in Fig. 4.

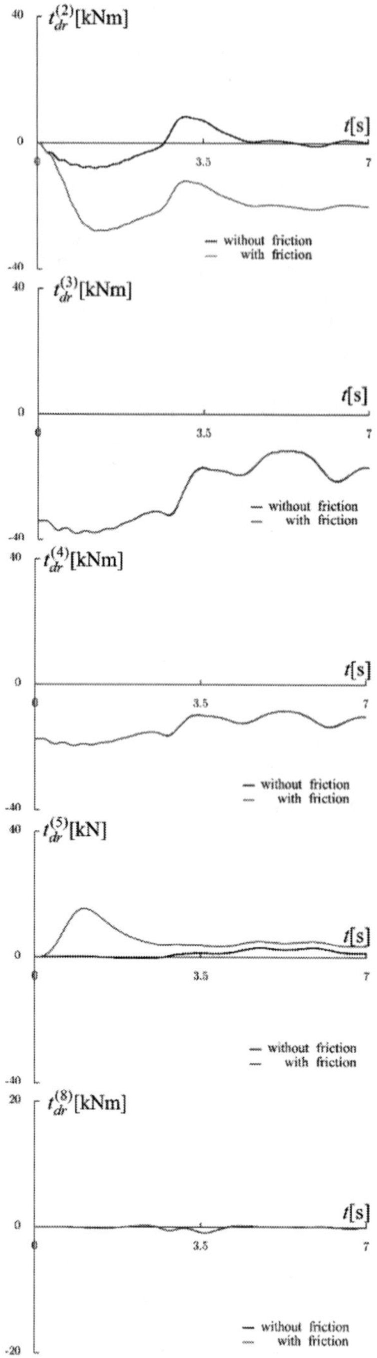

Fig. 4. Courses of driving torques and force.

After analysing the results, a significant influence can be noticed of the friction on the driving torque in joint 2 and the driving force in joint 5. These influences are not significant in the other drives.

The trajectories of selected points of load determined in $\hat{\mathbf{x}}^{(0)}\hat{\mathbf{y}}^{(0)}$ are presented in Fig. 5.

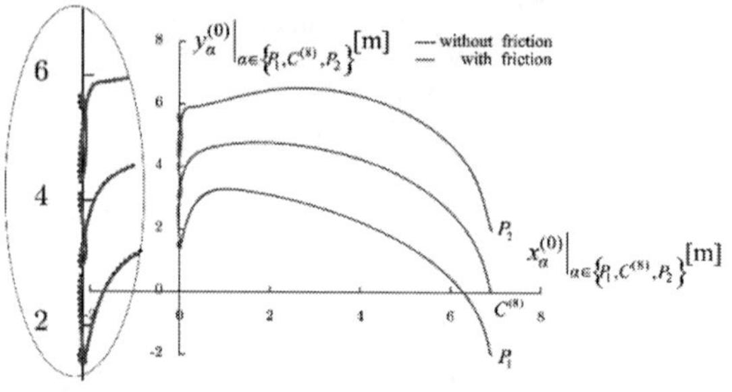

Fig. 5. Trajectories of selected points of the load.

An analysis of the trajectories allows to observe that friction does not significantly affect the behaviour of the load.

4 Conclusions

The dynamics analysis of a grab crane is presented in the paper. In formulating the mathematical model of the crane, the following are taken into account: flexibility of the supports, drives and friction in the joints, which is modelled using the Dahl friction model. The influence of these quantities on the driving torques and force, and also the behaviour of the load, are analysed. In the author's opinion, the model presented here can be useful in the process of design.

References

1. E. Papadopoulos, S. Sarkar, On the dynamic modeling of an articulated electrohydraulic forestry machine in Proc. of the 1996 AIAA Forum on Advanced Developments in Space Robotics, WI, 1-2 August 1996.
2. A. Urbaś, Analysis of flexibility of the support and its influence on dynamics of the grab crane, Latin American Journal of Solids and Structures **10**(1) (2013).
3. B. Posiadała, P. Waryś, D. Cekus, M. Tomala, The dynamics of the forest crane during the load carrying, International Journal of Structural Stability and Dynamics, **13**(7) (2013). doi: 10.1142/S0219455413400130
4. P. La Hera and D. O. Morales, Model-based development of control systems for forestry cranes, Journal of Control and Science and Engineering, vol. 2015, no. ID 256951, doi:10.1155/2015/256951.
5. J. J. Craig, Introduction to robotics. Mechanics and control. Addison-Wesley Publishing Company, Inc., (1989)
6. P. R. Dahl, Solid friction damping of mechanical vibrations. AIAA Journal, **14**(12) (1976)
7. E. I. Jurevič (ed.), Dynamics of robot control, Nauka, Moscow (1984), in Russian

Effects of a slow harmonic displacement on an Atomic Force Microscope system under Lennard-Jones forces

Morad Khadraoui, Faouzi Lakrad[a], and Mohamed Belhaq

Laboratory of Renewable Energy and Dynamics of Systems, Faculty of Sciences Aïn Chock, University Hassan II-Casablanca, Casablanca, Morocco

Abstract. We focus in this paper on the modeling and dynamical analysis of a tapping mode atomic force microscopy (AFM). The microbeam is subjected to a low frequency harmonic displacement of its base and to the Lennard-Jones (LJ) forces at its free end. Static and modal analysis are performed for various gaps between the tip of the microbeam and a sample. The Galerkin method is employed to reduce the equations of motion to a fast-slow dynamical system. We show that the dynamics of the AFM system is governed by the contact and the noncontact invariant slow manifolds. The tapping mode is triggered via two saddle-node bifurcations of these manifolds. Moreover, the contact time is computed and the effects of the base motion amplitude and the initial gap are discussed.

1 Introduction

The Atomic Force Microscope (AFM) is a scanning probe microscope that is used as a nano-scale tool for manipulation and characterization in nanosciences [1]. It can be used in a broad spectrum of applications such as imaging, nanolithography, electronics, chemical and biological analysis [2,3]. It is based mainly on a vibrating microcantilever with a nano-scale tip that interacts with a sample surface via intermolecular forces [4]. Indeed, understanding AFM vibrations is central to the correct interpretation of the AFM outputs. In the present paper, a Lennard-Jones (LJ) force [4] is used to model the tip-surface highly nonlinear interactions.

Several studies investigated the effects of resonant harmonic external and/or parametric forcings on an AFM subjected to the LJ forces [5–9] in the contact, noncontact and tapping modes. This latter was the dominant imaging mode for most scanning probe microscopes during the last decade [2]. It is based mainly on resonant excitations of the cantilever with a feedback loop keeping the cantilever vibration amplitude constant. This mode minimizes the shear forces, present in the contact mode, that can be destructive to the tip and samples. It overcomes some deficiencies of the noncontact mode by improving the resolution and enabling the measurement of mechanical properties [10]. However, operating the cantilever near its resonances can likely cause complex dynamics due to the nonlinearities. As a remedy nonresonant tapping mode techniques are used. Thus, AFM with low frequency excitation compared to the fundamental natural frequency of the cantilever is used, for instance, in the pulsed-force mode AFM [11] and the peak-force AFM [12]. Indeed, the low frequency excitation leads to lower the tapping force causing limited tip-sample contact areas and minimizing the loss of resolution.

The present work is focused on the dynamics of an AFM system under LJ forces and a very slow harmonic base displacement. Consequently, the system can be viewed as a fast-slow system with two time scales dynamics: one ruled by natural frequencies of the system and the other by the low frequency of the base displacement. Solutions of the system follow the stable invariant slow manifolds in large regions of the phase space. For more informations on the fast-slow systems see for instance [13, 14]. In fact, two stable invariant slow manifolds coexist, one corresponds to the contact mode and the other to the noncontact mode. These two stable slow manifolds undergo dynamic saddle-node bifurcations (through the collision with an unstable slow manifold) when the amplitude of the base displacement is varied. These dynamic bifurcations rule the contact time between the tip and the sample and they determine the operational mode of the AFM: contact, noncontact and tapping modes, respectively.

This work uses a continuous model of the AFM system and it can be viewed as a continuation of a previous paper by Lakrad [15] where a lumped mass model was used. The present paper is organized as follows: in section 2 we derive a continuous nonlinear model of the microcantilever under the LJ force and the base displacement using the Hamilton principle. Then, the static configurations and the corresponding natural frequencies and mode shapes are investigated in sections 3. In section 4, the Galerkin method is used to reduce the equations of motion to a fast-slow system. Then, the slow invariant manifolds are computed and the conditions of existence of the tapping mode and the contact time are determined.

2 Mathematical modelling

The classical beam theory based on the Euler-Bernoulli assumptions is used to develop a continuous model of an AFM probe, of length L, operating in air. As shown in Fig. 1, Z denotes the tip/sample separation distance in the reference configuration and $w(x,t) + y(t)$ indicates the to-

[a] e-mail: f.lakrad@fsac.ac.ma

DOI: 10.1051/matecconf/20168304001

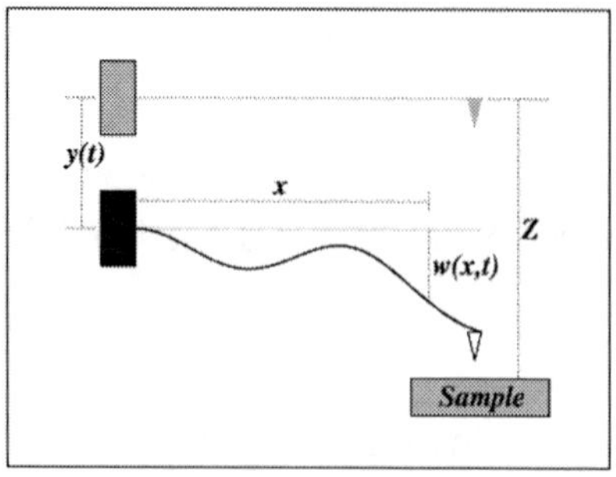

Fig. 1. A schematic of the AFM system

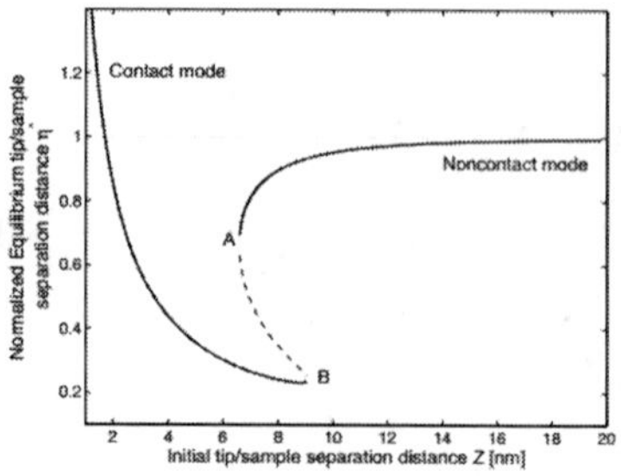

Fig. 2. Variation of the normalized tip/sample distance η^* versus the initial separation Z

tal deflection of the microcantilever, where $w(x,t)$ is the deflection of the microcantilever relative to a non-inertial reference frame attached to the base. The base excitation from a dither piezo is assumed to be a vertical harmonic displacement; that is, $y(t) = Y^* \cos(\Omega t)$.

The cantilever-tip-sample interaction is modelled by a LJ force F_{LJ} between a sphere of radius R and a flat surface [4,7]

$$F_{LJ} = \frac{A_1 R}{180(Z - w(L,t) - y(t))^8} - \frac{A_2 R}{6(Z - w(L,t) - y(t))^2} \quad (1)$$

where A_1 and A_2 are the Hamaker constants for the repulsive and attractive potentials, respectively. The attractive part of the LJ force corresponds to the van der Waals force [4]. In spite of its simplicity, this interaction model captures generic properties present in the near-field interactions [9]. In the present work, the LJ force F_{LJ} is assumed to be the unique source of nonlinearity of the system.

Using the Hamilton principle, the nondimensional form of the equation of motion and the associated boundary conditions are given by

$$\ddot{W}(X,\tau) + W''''(X,\tau) + c\dot{W}(X,\tau) = -\ddot{Y}_b(\tau) - c\dot{Y}_b(\tau) \quad (2)$$

$$W(0,\tau) = 0, \qquad W'(0,\tau) = 0, \qquad W''(1,\tau) = 0 \quad (3)$$

and

$$W'''(1,\tau) = \frac{-\Gamma_1}{(1 - W(1,\tau) - Y_b(\tau))^8} - \frac{\Gamma_2}{(1 - W(1,\tau) - Y_b(\tau))^2} \quad (4)$$

where the dots and the primes denote derivatives with respect to the nondimensional time $\tau = \omega t$ and the nondimensional space $X = x/L$, respectively. The displacements are normalized with respect to the initial gap Z: $W(x,t) = w(x,t)/Z$ and $Y_b(t) = y(t)/Z = Y\cos(\Omega t)$. The other quantities are given by

$$\omega^2 = \frac{EI}{\rho A L^4}; \quad c = \frac{c^*}{\rho A \omega}; \quad \Gamma_1 = \frac{-A_1 R L^3}{180 Z^9 EI}; \quad \Gamma_2 = \frac{A_2 R L^3}{6 Z^3 EI}$$

in which E, I, A and ρ are, respectively, the cantilever modulus of elasticity, second moment of area, cross-section area, and material density. In addition, c denotes the damping coefficient. The nondimensional instantaneous tip/sample separation is $\eta(1,\tau) = 1 - W(1,\tau) - Y_b(\tau)$.

Numerical applications are carried out for the case of the interaction of a soft monocrystalline silicon microcantilever with the (111) reactive face of a flat silicon sample [9]. Indeed, such a soft cantilever can be used to minimize the tapping force and sample penetration, especially for soft samples. Moreover, in order to operate in the tapping mode it is recommended that the sample should have a low adhesion to the surface. All the key system parameters of the AFM system are listed in table 1.

Table 1. Properties and dimensions of the studied system [9]

Descriptions	Symbol	Values
Length	L	449 μm
Width	b	46 μm
Thickness	h	1.7 μm
Tip radius	R	150 nm
Material density	ρ	2330 kg.m^{-3}
Elastic modulus	E	176 GPa
Damping ratio	ξ	0.01
Hamaker (rep.)	A_1	1.3596×10^{-70} J m^6
Hamaker (att.)	A_2	1.865×10^{-19} J

3 Static and modal analyses

In this section we solve the linearized undamped eigenvalue problem associated to Eqs. (2)-(4). For that, we first calculate static deflections $W^*(X)$ obtained from Eqs. (2)-(4) by dropping the time derivatives and the base excitation. The static tip/sample separation distance η^* is defined by $\eta^* = 1 - W^*(1)$, and is given by solving a ninth order algebraic equation.

In Fig. 2, we show the variation of η^* versus the initial tip/sample separation Z. We can distinguish two regions: the first region corresponds to $Z < Z^* = 1.7$nm where the system is repulsive since $\eta^* > 1$. The second region corresponds to $Z > Z^*$ which is attractive (since $\eta^* < 1$). In this second region, for 6.6 nm < Z < 9.09 nm, there is coexistence of three static solutions with two stable and one unstable.This region of bistability is located between the points A and B where saddle-node bifurcations occur. The lower and upper stable equilibria correspond, respectively, to the contact and noncontact configurations. Moreover, for high values of Z the effect of the attractive van der Waals force becomes weaker till it asymptotically vanishes and consequently $\eta^* \to 1$.

We compute the natural frequencies and mode shapes around a chosen static deflection $W^*(X)$ by setting

$$W(X,\tau) = W^*(X) + V(X,\tau) \qquad (5)$$

Substituting Eq.(5) into Eqs. (2)-(4), dropping the damping and excitation terms and linearizing the outcome for small $V(X,\tau)$, then the free vibration of the cantilever is given by the following set of equations

$$\ddot{V}(X,\tau) + V^{''''}(X,\tau) = 0 \qquad (6)$$

The boundary conditions are

$$V(0,\tau) = 0, \qquad V^{'}(0,\tau) = 0, \qquad V^{''}(1,\tau) = 0 \quad (7)$$

and

$$V^{'''}(1,\tau) = -[\frac{8\Gamma_1}{(1-W^*(1))^9} + \frac{2\Gamma_2}{(1-W^*(1))^3}]V(1,\tau) \quad (8)$$

Using separation of variables, the nth mode shape is given by

$$\Phi_n(X) = a_n[\cos(\lambda_n X) - \cosh(\lambda_n X) - \frac{\cos(\lambda_n) + \cosh(\lambda_n)}{\sin(\lambda_n) + \sinh(\lambda_n)}(\sin(\lambda_n X) - \sinh(\lambda_n X))] \quad (9)$$

where a_n is computed by using the orthonormality condition of the modes, the nth natural frequencies coefficients λ_n are zeros of the following algebraic equation

$$\lambda_n^3 = \beta\frac{[\cos(\lambda_n)\sinh(\lambda_n) - \sin(\lambda_n)\cosh(\lambda_n)]}{[1 + \cos(\lambda_n)\cosh(\lambda_n)]}; \quad (10)$$

$$\text{where} \quad \beta = -[\frac{8\Gamma_1}{(1-W^*(1))^9} + \frac{2\Gamma_2}{(1-W^*(1))^3}]$$

The dimensional natural frequencies of the micro-cantilever are given by $\omega_n = \frac{\lambda_n^2}{L^2}\sqrt{\frac{EI}{\rho A}}$.

Fig. 3 displays the variations of the first two natural frequencies λ_1 and λ_2 with the initial tip/sample separation Z, for three different cases: the free-end microcantilever and the AFM microcantilever operating in the contact and non-contact states. For $Z < 6.6$nm, the AFM is in the contact mode, the frequencies are larger than in the case of the free end microcantilever because of the repulsive interactions. They are decreasing with increasing initial gap Z. In the bistable region Z ∈]6.6nm, 9.09nm[, there is birth of a noncontact natural mode and an unstable mode (dashed red lines) through a saddle-node bifurcation. The noncontact mode is below the free-end mode and tends to it for increasing Z. The contact mode and the unstable mode disappear through a saddle-node bifurcation. For Z ∈]9.09nm, +∞[i.e., the noncontact mode, increasing Z leads natural frequencies to tend toward the free-end natural frequencies. Moreover, the differences with the free end case frequencies are decreasing with increasing the order of the mode.

The two first mode shapes of the microcantilever under the LJ force are depicted in Fig. 4 for Z = 5nm belonging to the monostable contact mode. These mode shapes are compared to the free-end mode shapes. It is observed that the first two mode shapes about the contact equilibrium are significantly different from those of the free-end case. On the other hand, the mode shapes about the noncontact static equilibrium are almost the same as those of the free-end microcantilever.

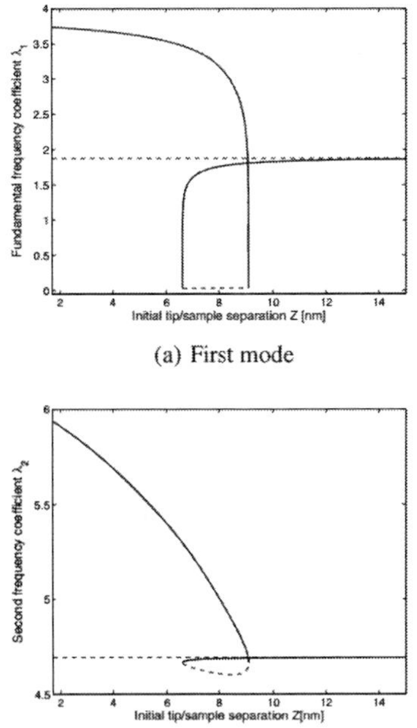

(a) First mode

(b) Second mode

Fig. 3. Variation of the first two frequency coefficients λ_1 and λ_2, solution of Eq.(10), with the initial separation Z. The free-end microcantilever (dashed grey lines), the contact mode (blue), the noncontact mode (black) and the unstable mode (dashed red lines)

4 Effects of the slow harmonic base motion

In this section we will show that, due to the slow harmonic excitation of the base, the dynamics of the microcantilever can be reduced to the dynamics of the contact and the noncontact slow invariant manifolds. Moreover, based on the time spent on the attracting contact slow invariant manifold, the contact time during the tapping mode could be computed and controlled by the amplitude of the base displacement and the initial gap.

We use the Galerkin procedure to reduce the partial differential equation (2) and the associated boundary conditions into a set of nonlinear ordinary differential equations. In what follows only the first mode will be considered, thus

$$W(X,\tau) = W^*(X) + q_1(\tau)\Phi_1(X) \qquad (11)$$

Then, the extended Hamilton principle and the orthonormality property of the mode shapes (9) are applied to obtain the following discretized equation

$$\begin{aligned} &q_1^{''} + \xi q_1^{'} + [\lambda_1^4 + \Phi_1^2(1)(\frac{8\Gamma_1}{(1-W^*(1))^9} + \frac{2\Gamma_2}{(1-W^*(1))^3})]q_1 \\ &+ \Phi_1(1)[\frac{\Gamma_1}{(1-W^*(1))^8} + \frac{\Gamma_2}{(1-W^*(1))^2}] \\ &- \frac{\Gamma_1\Phi_1(1)}{(1-W^*(1) - q_1\Phi_1(1) - Y_b(\tau))^8} \\ &- \frac{\Gamma_2\Phi_1(1)}{(1-W^*(1) - q_1\Phi_1(1) - Y_b(\tau))^2} = -\alpha_1(cY_b^{'}(\tau) + Y_b^{''}(\tau)) \end{aligned} \qquad (12)$$

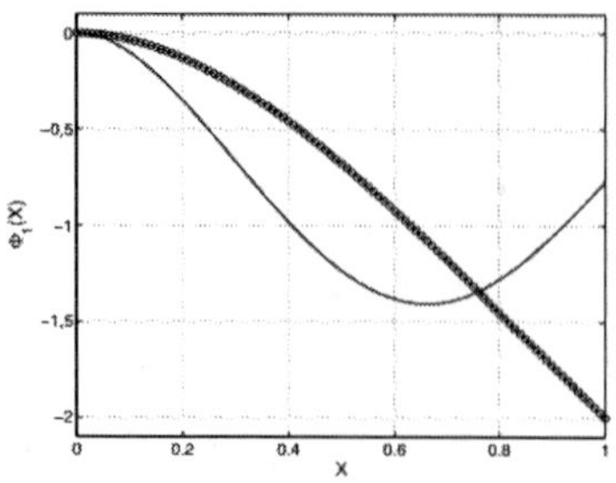

(a) Mode shape 1

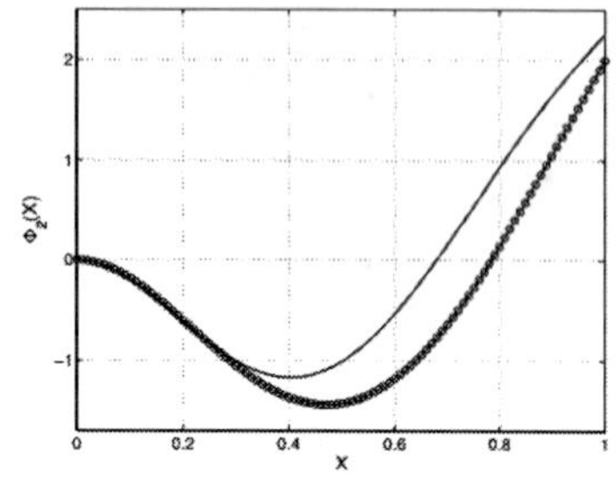

(b) Mode shape 2

Fig. 4. The first two modes shapes in the contact mode for $Z = 5$ nm. The blue line corresponds to the microcantilever with LJ force and the line with circles corresponds to the free end microcantilever

where the primes denote derivatives with respect to τ and the base motion $Y_b(\tau) = Y\cos(\epsilon\tau)$ with $\epsilon = \Omega/\omega << 1$. Furthermore,

$$\xi = \frac{c^*}{\rho A}; \qquad \alpha_1 = \int_0^1 \Phi_1(X)dX. \qquad (13)$$

4.1 Invariant slow manifolds

Equation (12) can be transformed to a fast-slow system by introducing the slow time scale $T = \epsilon\tau$. Following Fenichel's results [13] there exist stable slow invariant manifolds $\mathcal{M}$ such that all orbits, starting in some region of the phase space, will reach them and the dynamics of the system is reduced to the dynamics in $\mathcal{M}$. A first order approximation of the invariant slow manifolds, corresponding to $\epsilon = 0$ in the fast-slow system, is given by the following algebraic equation

$$\begin{aligned}
&-[\lambda_1^4 + \Phi_1^2(1)(\frac{8\Gamma_1}{(1-W^*(1))^9} + \frac{2\Gamma_2}{(1-W^*(1))^3})]q_1(T) \\
&- [\frac{\Gamma_1\Phi_1(1)}{(1-W^*(1))^8} + \frac{\Gamma_2\Phi_1(1)}{(1-W^*(1))^2}] \\
&+ \frac{\Gamma_1\Phi_1(1)}{(1-W^*(1)-q_1(T)\Phi_1(1)-Y\cos(T))^8} \\
&+ \frac{\Gamma_2\Phi_1(1)}{(1-W^*(1)-q_1(T)\Phi_1(1)-Y\cos(T))^2} = 0 \qquad (14)
\end{aligned}$$

Hence, the approximation of $\mathcal{M}$ is given by the graph of a function of the generalized coordinate $q_1(T)$ in terms of

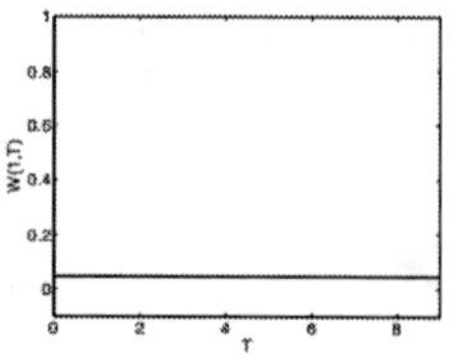

(a) $Y = 0$

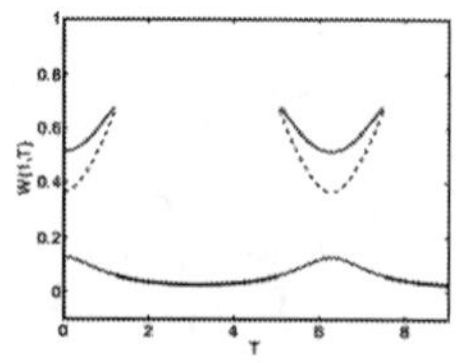

(b) $Y = 0.3$

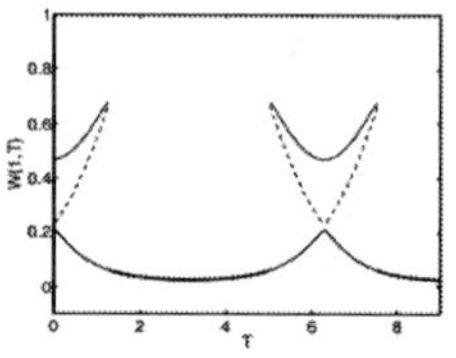

(c) $Y = 0.345$

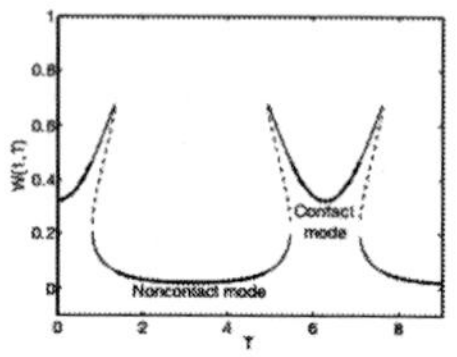

(d) $Y = 0.5$

Fig. 5. Graphs of the invariant slow manifolds, given by Eq.(14), for $Z = 10$nm and for various amplitude of the base Y.

T and the parameters of the system. In Fig. 5 are shown the graphs of the slow manifolds versus the slow time T, for various values of the amplitude Y and for the initial gap $Z = 10$nm. The stable manifolds are plotted with continuous lines and unstable ones with dashed lines. When only one stable manifold exists, it is plotted with a thick continuous line. The lower and the upper stable solutions correspond to the noncontact and contact slow manifolds, respectively. For $Y = 0$ i.e., no base motion, the slow manifold corresponds to the noncontact static equilibrium which is independent of T. For $Y = 0.3$, a reverse saddle-node bifurcation gives birth to two slow manifolds and the noncontact manifold is the visited solution. Then, for $Y = 0.3452$ a second saddle-node bifurcation is taking place since the noncontact slow manifold collides with the unstable manifold. Subsequently, for $Y = 0.5$ the contact and the noncontact slow manifolds are both visited during a period of the base displacement. Physically, this corresponds to the tapping mode and geometrically to a periodic burster. Figure 6 confirms that numerical solutions of Eq.(12) follow the slow invariant manifolds according to the scenario described in Fig 5.

4.2 Contact time

The contact time t_c is the time that the tip is interacting repulsively with the sample [16]. It depends on the mechanical properties of the sample and the tapping amplitude and frequency. In our case, the tapping frequency is equal to the base excitation frequency and is very low than the fundamental frequency of the microcantilever. Indeed, the tapping mode in our case is governed by saddle-node bifurcations of slow manifolds, consequently the contact duration can be computed as the interval of time starting when the contact slow manifold is attracting and ending when it is repelling. In Fig. 7 are shown the normalized contact time versus the amplitude of the base motion Y for various Z that are initially in the noncontact mode i.e., $t_c = 0$. The contact time is normalized by the period of the slow base displacement. Indeed, increasing Z requires higher amplitude Y to trigger the tapping mode. Moreover, a nontrivial

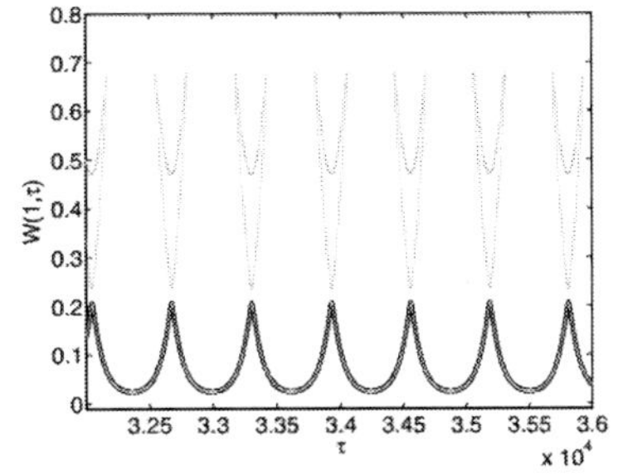

(a) $Y = 0.345$

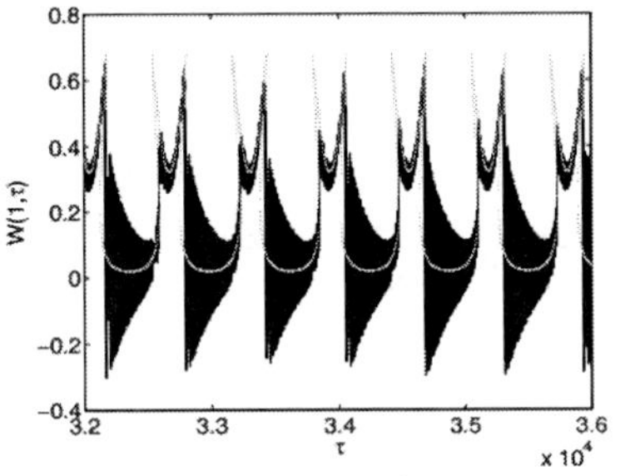

(b) $Y = 0.5$

Fig. 6. Numerical solutions of Eq.(12), for Z = 10nm, in black and the approximated slow manifolds in grey

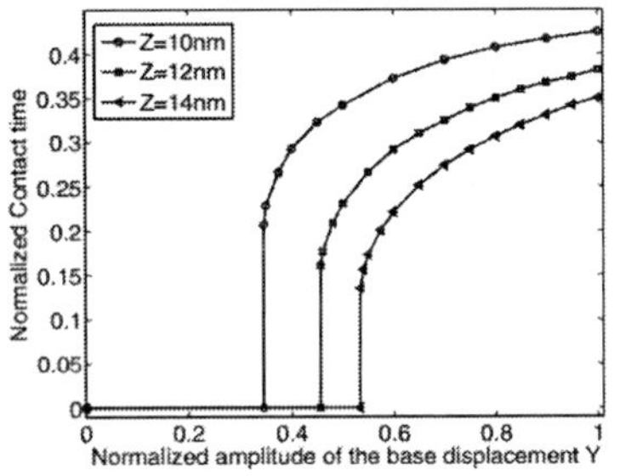

Fig. 7. Normalized contact time versus Y for various initial gaps

time contact takes place through a saddle-node bifurcation that causes a jump from zero. This jump is decreasing by increasing the initial gap Z. Furthermore, the normalized contact time tends asymptotically to 0.5 for higher amplitudes Y, which means that the tip will spend 50% of the time in contact with the sample.

5 Conclusion

In this paper, we developed a mathematical model of an AFM microbeam subjected to a slow harmonic base motion and Lennard-Jones forces. We investigated the static contact and noncontact configurations and the associated natural frequencies and modes shapes. It was shown that the fundamental natural frequency near the contact mode is the most affected by the intermolecular interactions. Then, a one mode Galerkin method was employed and the equations of motion were transformed to a fast-slow dynamical system. It was found that the contact and the noncontact slow manifolds govern the dynamics of the AFM. Thus, we showed that the tapping mode is triggered via two saddle-node bifurcations of these slow manifolds. Moreover, the contact time between the tip and the sample was computed and the influences of the base motion amplitude and the initial gap discussed.
As a continuation of the present work the effects of higher modes is under investigation. The use of other types of interaction forces between the tip and the sample is planned.

References

1. Binnig G, Quate CF, Gerber CH. Atomic force microscope. *Phys Rev Lett* **56**(9):930-934, 1986.
2. Jena BP, Heinrich Hörber JK. *Force microscopy. Applications in biology and medicine*. Wiley-Liss; 2006.
3. Eaton P, West P. *Atomic force microscopy*. Oxford university press; 2010.
4. Israelachvili J. *Intermolecular and surface forces*. 3rd edn. Academic Press; 2011.
5. Kirrou I, Belhaq M. Contact stiffness modulation in contact-mode atomic force microscopy. *International Journal of Non-Linear Mechanics* **55**: 102-109, 2013.
6. Garcia R, Prez R. Dynamic atomic force microscopy methods. *Surface science reports* **47**: 197-301, 2002.
7. Ashhab M, Salapaka MV, Dahleh M, Mezic I. Melnikov-based dynamical analysis of microcantilevers in scanning probe microscopy. *Nonlinear Dyn* **20**: 197-220, 1999.
8. Basso M, Giarrk L , Dahleh M, Mezic I. Complex dynamics in a harmonically excited Lennard-Jones oscillator: microcantilever-sample interaction in scanning probe microscopes. *ASME J Dynam Syst Meas Control* **122**: 240-5, 2000.
9. Rützel S, Lee S, Raman A. Nonlinear dynamics of atomic-force-microscope probes driven in Lennard-Jones potentials. *Proc. R. Soc. Lond. A* **459**: 1925-1948, 2003.
10. Butt HJ, Cappella B, Kappl M. Force measurements with the atomic force microscope: technique, interpretation and applications. *Surface Science Reports* **59**: 1-152, 2005.
11. Rosa-Zeiser A, Weilandt E, Hild S, Marti O. The simultaneous measurement of elastic, electrostatic and adhesive properties by scanning force microscopy: pulsed-force mode operation. *Meas. Sci. Technol.* **8**: 1333-1338, 1997.
12. Pittenger B, Erina N, Su C. Quantitative mechanical property mapping at the nanoscale with PeakForce QNM. Bruker Application Note 128, 2010.
13. Verhulst F. Singular perturbation methods for slow-fast dynamics. *Nonlinear Dyn* **50**: 747-753, 2007.
14. Lakrad F, Schiehlen W. Effects of a low frequency parametric excitation. *Chaos Solitons & Fractals* **22**: 1149-1164, 2004.
15. Lakrad F. Invariant slow manifolds of an Atomic Force Microscope system under the effects of Lennard-Jones forces and a slow harmonic base motion. *Commun. Nonlinear Sci Numer Simulat* **32**, 49-62, 2016.
16. Tamayo J, García R. Deformation, Contact Time, and Phase Contrast in Tapping Mode Scanning Force Microscopy. *Langmuir* **12**: 4430-4435, 1996.

Modeling and parametric analysis of a piezoelectric flexoelectric nanoactuator

Sourour Baroudi[1,a], Ahmed Jemai[1], and Fehmi Najar[1]

Applied Mechanics and Systems Research Laboratory, Tunisia Polytechnic School, BP 743, La Marsa 2078, University of Carthage, Tunisia.

Abstract. With the development of nanotechnology, nanoactuators have recently re-stimulated a surge of scientific interests in research communities. One of the interesting transduction mechanisms that showed high efficiency at the nanoscale was flexoelectricity. In fact, the flexoelectric effect in dielectric solids couples polarization and strain gradient, rather than polarization and strain for piezoelectricity, to convert mechanical stimulus into electricity and vice cersa. The objective of the current work is to develop a complete comprehensive electromechanical model of a nanobeam whose for piezoelectrically-actuated nanocantilever sensor in which both the flexoelectricity and piezoelectricity effects will be tzken into consideration. Starting from the enthalpy density function, the Hamilton's principle is applied to drive the governing coupled equations with appropriate boundary conditions. Then, we investigate the free vibration of the mechanism by formulating the eigenvalue problem associated with the coupled partial differential equations. Using the Galerkin procedure we develop both the static and dynamic of our structure . The results show that a certain aspect ratio flexoelectric effect significantly increases the performance of the nanoactuator.

1 Introduction

Flexoelectricity, referring to the spontaneous polarization in response to strain gradient, is a universal effect in all classes of dielectric materials even in the centrosymmetric crystals. A rational physical interpretation of this phenomenon is the local breaking of inversion symmetry of the material structure caused by the strain gradient, rendering the formation of dipole moments and thus the induced polarization[2].

Recently, Zubko et al. [3], Nguyen et al. [4], Yudin and Tagantsev [5] have conducted a thorough and comprehensive review elaborating the fundamentals of the flexoelectricity in solids, its consequences in the physical properties of nano-scale systems and the potential applications of this electromechanical phenomenon. Moreover,at the nanoscale, where large strain gradients are expected, the flexoelectric effect becomes more appreciable. Thus the size-dependent flexoelectricity may contribute significantly to the electromechanical coupling of piezoelectric materials at the nanoscale and it is necessary to incorporate such an effect when investigating the static and dynamic behaviors of those nanostructured materials. Indeed, a comprehensive theoretical framework incorporating the flexoelectricity, surface effects and electrostatic force was developed by Hu and Shen [9].

Then, the size-dependent electroelastic responses and dynamic behaviors of one-dimensional piezoelectric nanostructures were investigated by Yan and Jiang in their works [11] and[12]. Recently,[13] was particularly interested by the possibility of using the phenomenon of flexoelectricity for energy harvesting and showed that flexoelectricity based energy harvesting can be a viable alternative to piezoelectrics.In his work, Qian adopted a linear form of the electric potential and considered only the strain gradient along the beam thickness direction.

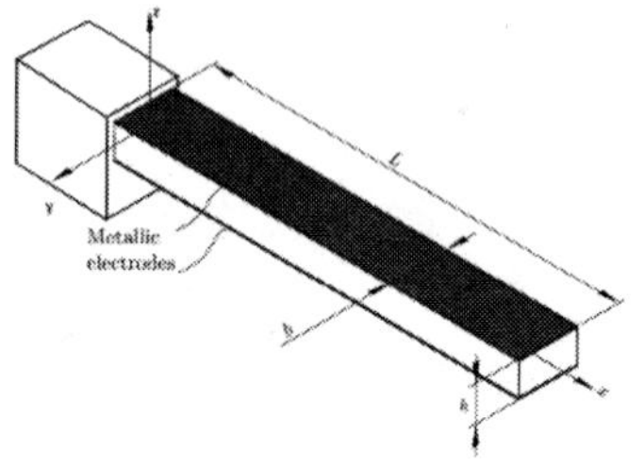

Fig. 1. Geometry of the beam

Here we aim to study the electroelastic responses and dynamic behaviors of a nanoactuator in which both the flexoelectricity and piezoelectricity effects are considered. We propose to take into account the effect of the self generated electrical potential.

2 Problem Formulation

We consider a nanobeam with length L, width b and thickness h under an electric potential V. It is modeled using the Euler-Bernoulli beam theory. Thus the displacement vector **u** of an arbitrary point of the nanobeam is given by

$$\begin{aligned}\mathbf{u} &= -z\frac{\partial w(x,t)}{\partial x}\hat{e}_x + w(x,t)\hat{e}_z \\ &= -zw'(x,t)\hat{e}_x + w(x,t)\hat{e}_z\end{aligned} \tag{1}$$

where $w(x,t)$ represent the transverse displacement component, $(\hat{\mathbf{e}}_x, \hat{\mathbf{e}}_y, \hat{\mathbf{e}}_z)$ is the rectangular orthonormal Cartesian

[a] e-mail: sourour.baroudi@gmail.com

basis,t is the time and the prime corresponds to derivative with respect to x. From this displacement field, the only non-zero strain component can be expressed as $\varepsilon_{xx} = -zw''$.Therefore, the non-zero strain gradient components are $\varepsilon_{xx,x} = -zw'''$ and $\varepsilon_{xx,z} = -w''$. The polarization density field within the cantilever beam is considered to be along the two axis: x-axis and z-axis, as follows:

$$\mathbf{P} = P_x(x,z,t)\hat{e}_x + P_z(x,z,t)\hat{e}_z \tag{2}$$

We assume that the potential ϕ is approximated by the following formula [14]

$$\phi(x,z,t) = \varphi(x,t)\cos(\frac{\pi}{h}z) + \frac{z+h/2}{h}v(t)(H(x) - H(x-L)) \tag{3}$$

where $v(t)$ is the voltage across the electrodes.

Refering to [8], given the above assumptions, and setting $c_{1111} = c_{11}$, $d_{111} = d_{11}$, $d_{113} = d_{13}$, $f_{1111} = f_{11}$, $f_{1133} = f_{14}$, and $g_{111111} = g_{113113} = g_{11}$, the expression of the internal energy density W^L can be written as:

$$\begin{aligned} W^L = &\tfrac{1}{2}a_{11}P_x^2 + \tfrac{1}{2}a_{33}P_z^2 + \tfrac{1}{2}c_{11}\varepsilon_{xx}^2 + d_{11}\varepsilon_{xx}P_x \\ &+ d_{13}\varepsilon_{xx}P_z + f_{11}\varepsilon_{xx,x}P_x + f_{14}\varepsilon_{xx,z}P_z \\ &+ \tfrac{1}{2}g_{11}\varepsilon_{xx,x}^2 + \tfrac{1}{2}g_{11}\varepsilon_{xx,z}^2 \end{aligned} \tag{4}$$

We obtain then the explicit expressions of the nonzero components of the stress tensor $\mathbf{T}$,the higher order stress tensor $\tilde{\mathbf{T}}$ and the electric filed $\mathbf{E}$ as follows:

$$\begin{aligned} T_{11} &= c_{11}\varepsilon_{xx} + d_{11}P_x + d_{13}P_z \\ \tilde{T}_{111} &= f_{11}P_x + g_{11}\varepsilon_{xx,x} \\ \tilde{T}_{113} &= f_{14}P_z + g_{11}\varepsilon_{xx,z} \\ -\mathrm{E}_1 &= a_{11}P_x + d_{11}\varepsilon_{xx} + f_{11}\varepsilon_{xx,x} \\ -\mathrm{E}_3 &= a_{33}P_z + d_{13}\varepsilon_{xx} + f_{14}\varepsilon_{xx,z} \end{aligned} \tag{5}$$

To derive the governing electroelastic equations and the boundary conditions of the system, we apply the Hamilton principle as follows:

$$\int_{t_1}^{t_2}\left(\delta K - \int_V \delta H\, dV + \delta W_{nc}\right) dt = 0 \tag{6}$$

where K is the kinetic energy, H is the enthalpy density and W_{nc} is the work done by the external loads. According to Toupin[8], the enthalpy density of perfectly elastic dielectrics is defined as:

$$H = W^L - \tfrac{1}{2}\varepsilon\phi_{,i}^2 + P_i\phi_{,i} \quad i = x,y,z \tag{7}$$

where W^L is the internal energy density, $\varepsilon = 8.854 \times 10^{-12}$ $CV^{-1}m^{-1}$ is the permittivity of the vacuum or the air and the index following a coma in the subscript corresponds to the derivative with respect to the ith-direction. Then the variation of the electric enthalpy density is given by:

$$\begin{aligned} \delta H = &\delta W^L + (P_x - \varepsilon\phi_{,x})\delta\phi_{,x} \\ &+ (P_z - \varepsilon\phi_{,z})\delta\phi_{,z} + \phi_{,x}\delta P_x + \phi_{,z}\delta P_z \\ = &\left(-zT_{11} - \tilde{T}_{113}\right)\delta w'' - z\tilde{T}_{111}\delta w''' \\ &+ (\phi_{,x} - E_1)\delta P_x + (\phi_{,z} - E_3)\delta P_z \\ &+ (P_x - \varepsilon\phi_{,x})\delta\phi_{,x} + (P_z - \varepsilon\phi_{,z})\delta\phi_{,z} \end{aligned} \tag{8}$$

The expression of the variation of the kinetic energy is reduced to

$$\delta K = -\int_0^L m_0\ddot{w}\,\delta w dx \tag{9}$$

where where m_0 is the principal inertia and equal to $m_0 = \rho A$ The work denoted by W_{nc} is given as

$$\delta W_{nc} = \int_V F\delta w dV \tag{10}$$

where F is the external body force Inserting the expressions of K, H and W_{nc} in (8) , setting the quantities δw, δP_x, δP_z and $\delta\phi$ as arbitrary and integrating over the volum the obtained equations, gives the following nondimensional coupled differential equations:

$$\begin{aligned} \ddot{w} = &-w^{iv} + \alpha_0 w^{vi} + \varphi'' + \alpha_1 v(t)\left(\delta'''(x) - \delta'''(x-1)\right) \\ &- \alpha_2 v(t)\left(\delta''(x) - \delta''(x-1)\right) - \alpha_3 v(t)\left(\delta'(x) - \delta'(x-1)\right) \\ &+ \alpha_4 F \end{aligned} \tag{11}$$

$$\varphi'' = \beta_0\varphi + \beta_1 w'' - \beta_2 v(t)\left(\delta'(x) - \delta'(x-1)\right) \tag{12}$$

where the terms τ and φ_0 are introduced and given as

$$\tau = \sqrt{\frac{L^4 m_0}{K_1^*}} \quad \text{and} \quad \varphi_0 = \frac{\pi K_1^* a_{33}}{2bL^2 d_{13}} \tag{13}$$

The nondimensional coeffients are given as follows

$$\begin{aligned} \alpha_0 &= \frac{K_2^*}{L^2 K_1^*}, \alpha_1 = \frac{f_{11}J}{K_1^* h^2 a_{11}}, \alpha_2 = \frac{Ld_{11}J}{K_1^* h^2 a_{11}}, \\ \alpha_3 &= \frac{L^2 f_{14}A}{K_1^* a_{33}h^2}, \alpha_4 = \frac{L^4 A}{K_1^* h}, \beta_0 = \frac{L^2\pi^2\varepsilon_{r33}}{h^2\varepsilon_{r11}}, \\ \beta_1 &= \frac{d_{13}A\pi}{2a_{33}\varepsilon_{r11}b\varphi_0}, \beta_2 = \frac{A\pi}{4bh\varphi_0} \end{aligned} \tag{14}$$

where the quantities ε_{r11}, ε_{r33}, K_1^*, K_2^* and K_3^* are defined as follows: $\varepsilon_{r11} = \left(\frac{1}{a_{11}} + \varepsilon\right)$, $\varepsilon_{r33} = \left(\frac{1}{a_{33}} + \varepsilon\right)$,
$K_1^* = -\left(-c_{11} + \frac{d_{11}^2}{a_{11}} + \frac{d_{13}^2}{a_{11}}\right)J - \left(\frac{f_{14}^2}{a_{33}} - g_{11}\right)A$,
$K_2^* = \left(-\frac{f_{11}^2}{a_{11}} + g_{11}\right)$ and $K_3^* = \frac{f_{11}d_{11}}{a_{11}}J$.
$A = bh$ and $J = \frac{bh^3}{12}$ is the beam moment of inertia.
Associated with the following normalized boundary conditions

$$\begin{cases} w(0,t) = 0 \\ w'(0,t) = 0 \\ w''(0,t) = 0 \\ \varphi(0,t) = 0 \end{cases} \quad \text{and} \quad \begin{cases} w''(1,t) - \alpha_0 w^{iv}(1,t) - \varphi(1,t) = 0 \\ w'''(1,t) - \alpha_0 w^{v}(1,t) - \varphi'(1,t) = 0 \\ -w'''(1,t) + \gamma_0 w^{iv}(1,t) = 0 \\ \varphi(1,t) = 0 \end{cases} \tag{15}$$

where $\gamma_0 = \frac{K_2^*}{K_3^* L}$

3 Free vibration analysis

We propose to resolve the eigenvalue problem of the obtained system of coupled differential equations considering the free vibration condition, ie $v(t) = 0$ and $F = 0$. Let's first write

$$w(x,t) = W(x)e^{i\omega t}, \text{ and } \varphi(x,t) = \psi(x)e^{i\omega t} \tag{16}$$

DOI: 10.1051/matecconf/20168304002

Table 1. The first three nondimensional natural frequencies

ω_1	ω_2	ω_3
3.5208	22.0731	61.7811

In such way the system of coupled equations becomes

$$-\omega^2 W = -W^{iv} + \alpha_0 W^{vi} + \psi'' \qquad (17)$$

$$\psi'' = \beta_0 \psi + \beta_1 W'' \qquad (18)$$

For reason of simplicity, we propose to resolve the eigenvalue problem of the above system of coupled equations by neglecting the sixth order of W. Writing W and ψ in exponetial form as : $W(x) = \bar{W}e^{\lambda x}$ and $\psi(x) = \bar{\psi}e^{\lambda x}$, the general solution of W and ψ can be wwritten as follows:

$$W(x) = \sum_{i=1}^{6} C_i e^{\lambda_i x}, \text{ and } \psi(x) = \sum_{i=1}^{6} C_i D_i e^{\lambda_i x} \qquad (19)$$

where $(\lambda_i)_{(i=1..6)}$ are the six eigen values of the eigen value problem considered and $(D_i)_{(i=1..6)} = \frac{\beta_1 \lambda_i^2}{\lambda_i^2 - \beta_0}$. The circular eigenfrequencies ω_n for each mode can be obtained by resolving the following equation

$$Det\left([\mathbf{J}_0]\right) = 0 \qquad (20)$$

where the matrix $\mathbf{J}_0$ is given as

$$[\mathbf{J}_0] = \begin{bmatrix} 1 & 1 & 1 & 1 & 1 & 1 \\ \lambda_1 & \lambda_2 & \lambda_3 & \lambda_4 & \lambda_5 & \lambda_6 \\ \lambda_1^2 e^{\lambda_1} & \lambda_2^2 e^{\lambda_2} & \lambda_3^2 e^{\lambda_3} & \lambda_4^2 e^{\lambda_4} & \lambda_5^2 e^{\lambda_5} & \lambda_6^2 e^{\lambda_6} \\ \Lambda_1 e^{\lambda_1} & \Lambda_2 e^{\lambda_2} & \Lambda_3 e^{\lambda_3} & \Lambda_4 e^{\lambda_4} & \Lambda_5 e^{\lambda_5} & \Lambda_6 e^{\lambda_6} \\ D_1 & D_2 & D_3 & D_4 & D_5 & D_6 \\ D_1 e^{\lambda_1} & D_2 e^{\lambda_2} & D_3 e^{\lambda_3} & D_4 e^{\lambda_4} & D_5 e^{\lambda_5} & D_6 e^{\lambda_6} \end{bmatrix} \qquad (21)$$

where$(\Lambda_i)_{(i=1..6)} = -\lambda_i^3 + D_i \lambda_i$.
Considering the general solution of W and ψ (19), applying the boundary conditions appearing in (15) and setting C_1 as arbitrary constant, the constants C_2 to C_6 can be found. For numerfical results, $BaTiO_3$ is chosen as the model material.Its material properties are obtained from [15]. Considering the plane stress condition, we calculate the material constants as follows: the elastic constant $c_{11} = 131GPa$,the dielectric constants $a_{11} = 0.897 \times 10^8 Vm/C$, $a_{33} = 0.788{\times}10^8 Vm/C$, the piezoelectric constants $d_{11} = -18.52 \times 10^8 V/m$, $d_{13} = 1.87 \times 10^8 V/m$.The mass density is $\rho = 6.02 \times 10^3 kg/m^3$. We refer to [16] to deduce the flexoelectric constant $f_{11} = -0.013V$. Following Yan et al. [11, 12] we set the fexoelectric oefficient $f_{14} = 5V$. Refering to [13, 15], we approximate $\sqrt{g_{11}/c_{11}}$ with the characterestic nonlocal elastic length scale noted by l_0 which is chosen to be equal to $3nm$.

Table 1 gives the first three non-dimensional natural frequencies of piezoelectric flexoelectric cantilever nanobeam. Figures 1 and 2 plot the first three mode shapes of respectively the deflection W and the potential ψ.

4 Static response

For reason of comparison, we deduced the static response or our structure when assuming that the applied external force is a constant and concentrated load F at the free end of the beam, In order to aply the Galerkin methode, we write

$$w(x) = \sum_{i=1}^{N_i} Q_i W_i(x), \text{ and } \varphi(x) = \sum_{j=1}^{N_j} U_j \psi_j(x) \qquad (22)$$

Substituting the series representation of w and φ in the corresponding system of differential coupled equations and applying the orthogonality condition gives the two uncoupled equations in terms of the two amplitudes $(Q_i)_{(i=1..N_i)}$ and $(U_j)_{(j=1..N_j)}$.

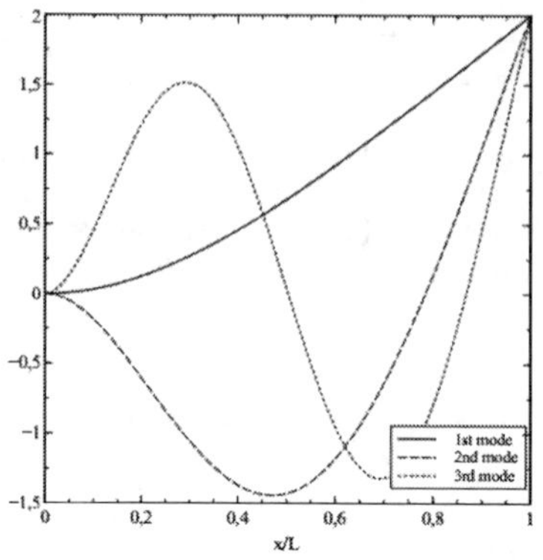

Fig. 2. First three deflection modes

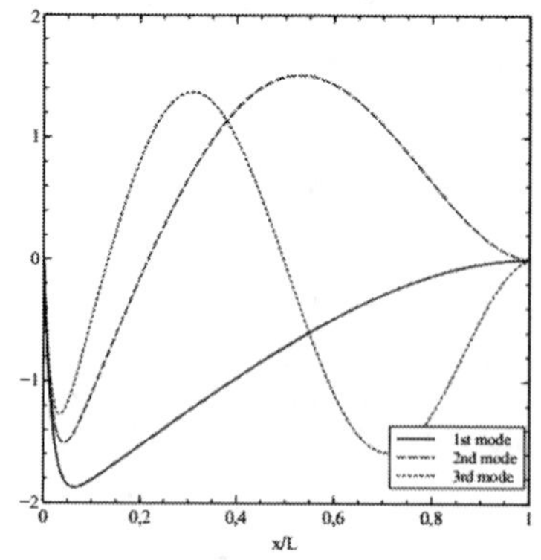

Fig. 3. First three potential modes

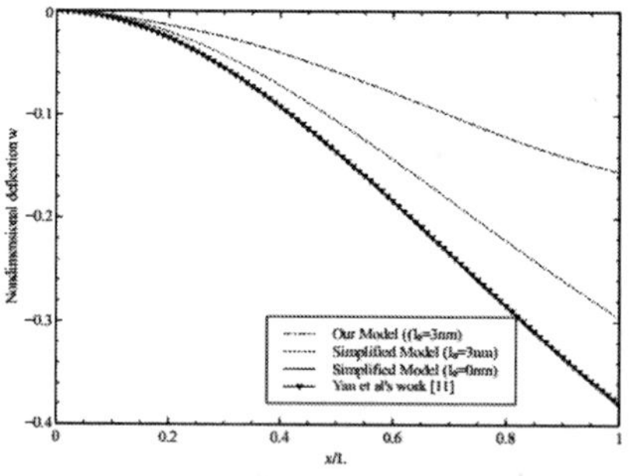

Fig. 4. Normalized dfelction variation in comparison with previous work [11] (V=-0.1, L/h=20, h=20nm, F=-1nN)

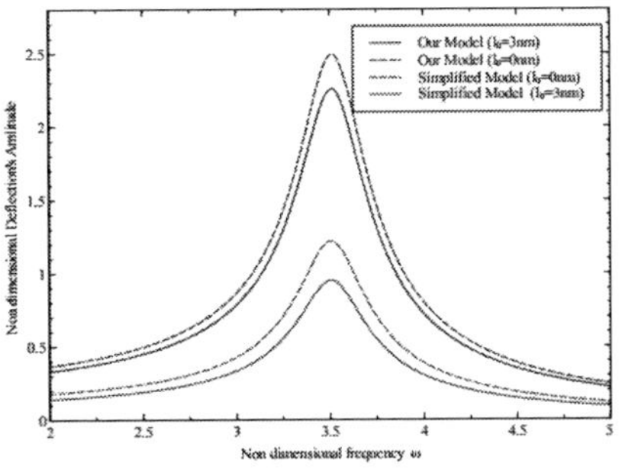

Fig. 5. Comparison of nondimensional dynamic dfelction's amplitude for different models (V=-0.1,L/h=20,h=20nm)

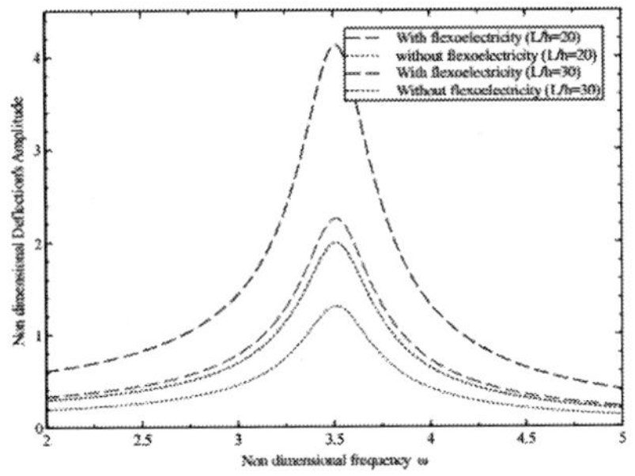

Fig. 6. Normalized dynamic dfelction for different aspect ratios L/h (V=-0.1, h=20nm)

For numerical results, we consider that the nanobeam is loaded with a concentrated force $F = 1nN$ and an electric potential $V = -0.1$. The dimensions of the nanobeam are chosen as follows: $L/h = 30$, $b = h$.

Figure 4 represents the normalized dfelction in comparison with Yan et al's work [11] showing a good convenince when setting $l_0 = 0$ for our corresponding simplified model (meaning considering a one dimensional model and a linear distribution of the electrical potential ϕ, i.e $\varphi(x) = 0$).

5 Dynamic response

Here we consider the time variation of the two variables $w(x,t)$ and $\varphi(x,t)$ as follows

$$w(x,t) = \sum_{i=1}^{N_i} q_i(t)W_i(x), \text{ and } \varphi(x,t) = \sum_{j=1}^{N_j} u_j(t)\psi_j(x) \quad (23)$$

Introdcing the diagonal damping matrix **D** defined as $D_n = \frac{\omega_n}{Q}$ where Q is the quality factor chosen equal to 10 and considering a harmonic behavior for the generalized coordinate of displacement **q(t)**, the generalized coordinate of the potential **u(t)**, the force $\bar{\mathbf{f}}$ and the voltage v(t), as follows $\mathbf{q}(t) = \mathbf{Q}e^{j\omega t}$, $\mathbf{u}(t) = \mathbf{U}e^{j\omega t}$ and $v(t) = Ve^{j\omega t}$, we deduce the dynamic response of our system in terms of the amplitude of the displacement $\mathbf{q}(t)$ and the amplitude of the electric potential $\mathbf{u}(t)$.

We compare our model's results with that obtained for the simplified one in figure 5. The discrepancy between the two models indicates that more accurate and exact results are obtained thanks to our complete model and shows the neccessity of taking into consideration the current assumptions and adopted approach. Figure 6 represents the normalized dynamic dfelction for different aspect ratios L/h and a thickness equal to 20×10^{-9}. It is clear that the the response of the nanoactuator is well accentuated for slender beam ($L/h = 30$ for example) when taking into consideration its flexoelectric coupling effect.

6 Conclusion

In this paper, a two dimensional flexoelectric piezoelectric cantilever nanobeam is modeled to investigate the static and dynamic response of a nanoactuator. Starting with the extended linear theory of piezoelectricity and applying the Hamilton principle, the governing system of coupled differential equations is derived. An eigenvalue problem is estalished to study the free vibration of the model. Finally, using the Galerkin method, both the static bending and the dynamic respnse are studied and compared to previous and simplified works. Our results show the considerable effect when rather considering our complete model, meaning when taking into account the two dimnesional aspect of the structure and considering the non linear bahaviour of the the electric potential. We illustrate also that the flexoelectric coupling effect can be responsible for increasing the size-dependent properties of dielectric materials at the nanoscale.

References

1. J. S. Yang, Appl. Mech. Rev **59**, (2006) 335-345.
2. R. Maranganti, N. D. Sharma and P. Sharma, P., Phys. Rev. **B 74**, (2009) 014110.
3. P. Zubko, G. Catalan and A. K. Tagantsev, Annu. Rev. Mater. **43**, (2013) 387-421.
4. T. D. Nguyen, S. Mao, Y. W. Yeh, P. K. Purohit and M. C. McAlpine, Adv. Mater. **25**, (2013) 946-974.
5. P. V. Yudin and A. K. Tagantsev, Nanotechnology **24**, (2013) 432001.
6. M. S. Majdoub, P. Sharma, and T. Çağin, Physical Review B **78.12**, (2008) 121407.
7. X. Liang, S. Hu and S. Shen, Smart Mater. Struct. **23**, (2013) 035020.
8. R. A. Toupin, J. Rational Mech. Anal. **5.6**, (1956) 849-915.
9. S. L. Hu and S. P. Shen, Science China Physics, Mechanics and Astronomy **53**, (2010) 1497-1504.
10. C. C. Liu, S. L. Hu and S. P. Shen, Smart Mater. Struct. **21**, (2012) 115024.
11. Z. Yan and L. Y. Jiang, J. Appl. Phys. **113**, (2013) 194102.
12. Z. Yan and L. Y. Jiang, J. Appl. Phys. **46**, (2013) 355502.
13. Q. Deng, M. Kammoun, A. Erturk, P. Sharma, Int. J. Solids and Struct. **51**, (2014) 3218-3225.
14. Q. Wang, Engineering Structures **24**, (2002) 199-205.
15. A. K. Tagantsev and A. S. Yurkov, J. Appl. Phys. **112**, (2012) 044103.
16. R. Maranganti and P. Sharma, Physical Review B **80**, (2009) 054109.

MATEC Web of Conferences **83**, 04003 (2016) DOI: 10.1051/matecconf/20168304003
CSNDD 2016

Analysis of new actuation methods for capacitive shunt micro switchs

S Ben Sassi[1,a], M E Khater[2], E M Abdel-Rahman[3], and F Najar[1]

[1] Applied Mechanics and Systems Research Laboratory, Tunisia Polytechnic School, University of Carthage, B.P.743, La Marsa 2078, Tunisia
[2] Department of Mechanical Engineering, KFUPM, Dhahran, Saudi Arabia
[3] Department of Systems Design Engineering , University of Waterloo, Waterloo, ON, Canada

Abstract. This work investigates the use of new actuation methods in capacitive shunt micro switches. We formulate the coupled electromechanical problem by taking into account the fringing effects and nonlinearities due to mid-plane stretching. Static analysis is undertaken using the Differential Quadrature Method (DQM) to obtain the pull in voltage which is verified by means of the Finite Element Method (FEM). Based on Galerkin approximation, a single degree of freedom dynamic model is developed and limit-cycle solutions are calculated using the Finite Difference Method (FDM). In addition to the harmonic waveform signal, we apply novel actuation waveform signals to simulate the frequency-response. We show that, biased signals, using a square wave signal reduces significantly the pull-in voltage compared to the triangular and harmonic signal . Finally, these results are validated experimentally.

1 Introduction

MEMS technology is becoming an attractive technology thanks to an expanded applications domain and remarkable operational advantages [1,2]. RF MEMS switches are among the most important micro devices targeting a combination of high reliability and very low cost. Their role consists of performing circuit ON/OFF switching through a mechanical deflection. The driving force can be electrostatic, piezoelectric [3], electromagnetic [4] or a combination of them [5].

Electrostatic actuation is the most widespread method thanks to its simplicity and its near-zero power consumption. But a constant obstacle is the high actuation voltage required when designing MEMS switches. Reducing the pull-in voltage is essential not only for CMOS circuit integration but also to enhance the device lifetime. Researches have employed various methods to reduce the actuation voltage. Pacheco et al [6] designed a low-voltage RF MEMS switch by inserting folded suspensions above the micro beam. Balaraman et al [7] fabricated a low cost and actuation voltage RF micro switch using copper. Taking advantage of residual stress to induce buckling and bending, Chu et al [8] designed a high-isolation and low actuation voltage RF micro switch. Khater et al [9] demonstrated 60 % voltage reduction by changing the actuation mode from static to dynamic pull-in. Samaali et al [10] studied the dynamic behavior of a Single-Pole Double-Throw (SPDT) micro switch. They showed that the predicted pull in voltage using square signal is lower than that predicted for a harmonic wave.

In this work, we aim at studying the dynamic response of a micro switch under harmonic, square and triangular signal waveforms. Toward this end, we formulate the electromechanical problem of an electrostatic shunt micro switch.

[a] e-mail: `sara.bensassi@gmail.com`

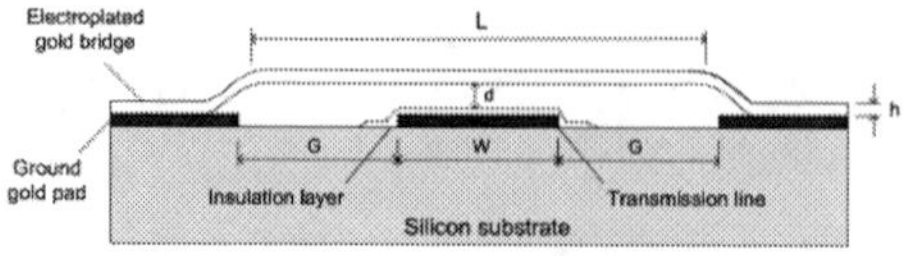

Fig. 1. A schematic diagram of the micro switch

We solve the static equation of equilibrium and calculate the static pull-in voltage using the Differential Quadrature Method (DQM). Then, we develop a Reduced Order Model (ROM) based on a one-mode Galerkin approximation and use the Finite Difference Method (FDM) to obtain the micro switch limit-cycles. Experimental tests are performed to validate the ROM. Finally, we study the impact of harmonic, square and triangle waveform signals on the micro switch numerically and experimentally and prove the superiority of square wave signal actuation.

2 Problem Formulation

The micro switch is one of the most popular devices in the telecommunications field. The switch under study is made of L long doubly-clamped micro beam fixed to a coplanar waveguide (CPW) as shown in Figure 1 [9]. The transmission line has a width W and is covered by a thin dielectric layer. Accounting for third order geometric nonlinearities and modeling the micro beam using the Euler-Bernoulli beam theory [11], the equation of motion that describes the beam deflection $\hat{w}(\hat{x}, \hat{t})$ is given by:

$$\rho A \frac{\partial^2 \hat{w}}{\partial \hat{t}^2} + (\hat{c} + \hat{c}_s)\frac{\partial \hat{w}}{\partial \hat{t}} + EI\frac{\partial^4 \hat{w}}{\partial \hat{x}^4} = \frac{\varepsilon b_e V^2}{2(d-\hat{w})^2} S(\hat{x}) + \frac{\partial^2 \hat{w}}{\partial \hat{x}^2}\Big[\hat{N} + \frac{EA}{2L}\int_0^L \Big(\frac{\partial \hat{w}}{\partial \hat{x}}\Big)^2 dx\Big] \quad (1)$$

subject to two sets of fixed-end boundary conditions described by:

$$\hat{w}(0,\hat{t}) = \hat{w}(L,\hat{t}) = 0, \qquad \frac{\partial \hat{w}}{\partial \hat{x}}(0,\hat{t}) = \frac{\partial \hat{w}}{\partial \hat{x}}(L,\hat{t}) = 0 \quad (2)$$

Viscous damping is represented by $\hat{c} = \frac{\hat{\omega}}{Q}$ where $\hat{\omega}$ is the fundamental natural frequency of the micro switch and Q is its quality factor. Squeeze film damping is represented by [12]:

$$\hat{c}_s = \frac{\hat{\mu}\hat{b}^3}{(1+6K_n)(d-\hat{w})^3} \quad (3)$$

where $\hat{\mu}$ is air viscosity and K_n the Knudsen number. The axial force in the beam due to residual stresses is represented by $\hat{N}$ and ε is air permittivity.

The micro switch is actuated via the transmission line lying under the beam center only. In such a case, where the two 'electrodes' are not equal in length, it is important the fringing electrostatic field must be taken into account when evaluating the electrostatic force [13]. To represent the non-overlapping lower electrode (transmission line), we adopt the function $S(\hat{x})$ given by:

$$S(\hat{x}) = \begin{cases} 1 \text{ if } G \le \hat{x} \le L-G \\ 0 \text{ elsewhere} \end{cases} \quad (4)$$

The effects of the fringing field are taken into account through the deployment of an effective beam width b_e [12]:

$$b_e = (1 + 0.65\frac{d-\hat{w}}{b})\hat{b} \quad (5)$$

For convenience in implementation of DQM discretization, we rewrite Equations (1) and (2) in non-dimensional form as:

$$\frac{\partial^2 w}{\partial t^2} + (c+c_s)\frac{\partial w}{\partial t} + \frac{\partial^4 w}{\partial x^4} = \frac{\partial^2 w}{\partial x^2}\Big[N + \alpha_1 \int_0^1 \Big(\frac{\partial w}{\partial x}\Big)^2 dx\Big] + \alpha_2 \frac{V^2}{(1-w)^2}\Big(1 + 0.65\frac{1-w}{b}\Big)S(x) \quad (6)$$

subject to the boundary conditions:

$$w(0,t) = w(1,t) = 0, \qquad \frac{\partial w}{\partial x}(0,t) = \frac{\partial w}{\partial x}(1,t) = 0 \quad (7)$$

where:

$$x = \frac{\hat{x}}{L},\ w = \frac{\hat{w}}{d},\ t = \frac{\hat{t}}{T},\ \omega = \frac{\hat{\omega}}{T},\ b = \frac{\hat{b}}{d}$$
$$c = \frac{\hat{c}L^4}{EIT},\ c_s = \frac{\mu}{(1+6K_n)(1-w)^3},\ \mu = \frac{\hat{\mu}\hat{b}^3L^4}{EIT}, \quad (8)$$
$$\alpha_1 = 6\frac{d^2}{h^2},\ \alpha_2 = \frac{6\varepsilon L^4}{Eh^3d^3},\ N = \frac{\hat{N}L^2}{EI}$$

3 Static Response and Pull-in Voltage

The first step in studying of the micro switch is to determine its static response and predict its pull in voltage, which defines the actuation limit for electrostatic MEMS. To this end, we employ the DQM which consists of discretizing the domain using nonuniform grid points, then expressing the displacement function (beam deflection) and its derivatives as weighted linear combinations of the beam displacement at the grid nodes. We adopt the Chebyshev-Gauss-Lobatto grid distribution scheme thanks to its accuracy compared to grid distributions [14]. The derivatives of the displacement function are then approximated by:

$$\Big[\frac{\partial w^r(x)}{\partial x^r}\Big]_{x=x_i} = \sum_{j=1}^{n} A_{ij}^{(r)} w_j \quad (9)$$

Where the weighting coefficients are determined using the Lagrange interpolation polynomials. They are given by:

$$A_{ij}^{(1)} = \frac{\prod_{v=1;v\neq i}^{n}(x_i - x_v)}{(x_i - x_j)\prod_{v=1;v\neq j}^{n}(x_j - x_v)},\quad i,j = 1,2,...,n,\ j \neq i \quad (10)$$

$$A_{ij}^{(r)} = r\Big(A_{ij}^{r-1}A_{ij}^{1} - \frac{A_{ij}^{r-1}}{(x_i - x_j)}\Big),\quad i,j = 1,2,...,n;\ j \neq i \quad (11)$$

$$A_{ij}^{(r)} = -\sum_{v=1;v\neq i}^{n} A_{iv}^{(r)},\quad i = j = 1,2,...,n \quad (12)$$

Due to non-overlapping of the micro structure capacitor plates, two of the grid points must correspond to the edges of the lower electrode. To achieve that, we discretize the beam into 49 nodes. The micro switch dimensions are: $L = 225\,\mu m$, $\hat{b} = 20\,\mu m$, and $h = 1.7\,\mu m$. The initial capacitor gap is $d = 1.7\,\mu m$. It is fabricated from polysilicon with material properties $E = 80\,\text{GPa}$ and $\rho = 9320\,\text{kg/m}^3$ and residual stress of $\sigma = 17\,\text{GPa}$.

In addition to solving the static version of Equation (6) using the DQM, we performed a FEM analysis of the micro switch using the commercial software ANSYS. We compare the FEM results to those of the DQM to validate it. Table 1 shows the calculated static pull-in voltage using DQM, FEM and published results for the same micro switch in [9]. The comparison shows good agreement between DQM and FEM despite the large number of grid points.

Table 1. Comparison of static pull-in voltage values

DQM	FEM	Experiment [9]
64 V	66.45 V	68.5 V

4 Dynamic Response of the Micro Switch

To solve the equation of motion, Equation 6, we utilize an innovative ROM based on Galerkin decomposition and FDM. The novelty in this method is to adopt a discretized mode shape calculated using 49 grid points in the DQM [15]. We take advantage of the discretized mode shape $\phi(x_i)$ in the Galerkin approximation of the equation of motion. This allows us to reduce the number of modes retained in the Galerkin expansion and therefore, to reduce considerably the computational time of simulations.

Writing the dynamic displacement as

$$w(x,t) = q(t)\phi(x)$$

DOI: 10.1051/matecconf/20168304003

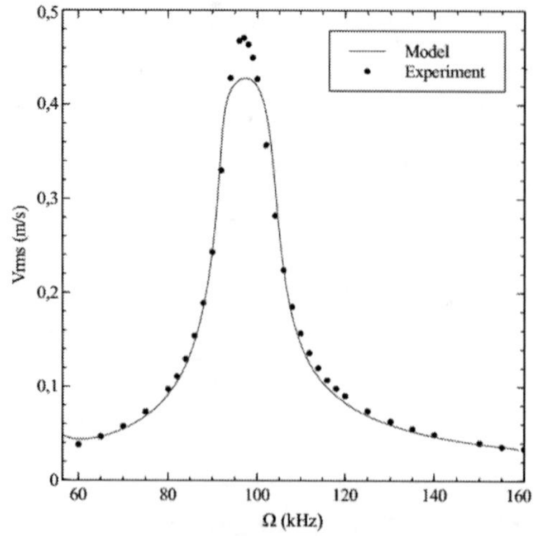

Fig. 2. Frequency-response curves of the micro switch mid-point velocity when actuated by a harmonic signal with V_{dc} = 23 V, V_{ac} = 23.28 V

where $q(t)$ is the modal coordinate and integrating the resulting equation using Newton-Cotes rule, we end up with the following reduced order model:

$$\begin{aligned} &\ddot{q} + \dot{q}(c + c_s) + q\sum_{i=1}^{n} C_i\phi_i \sum_{j=1}^{n} A_{ij}^{(4)}\phi_j \\ &+ \alpha_1 q^3 \sum_{i=1}^{n} C_i\phi_i \sum_{j=1}^{n} A_{ij}^{(2)}\phi_j \sum_{k=1}^{n} C_k\phi_k \sum_{v=1}^{n} A_{kv}^{(2)}\phi_v \\ &- Nq\sum_{i=1}^{n} C_i\phi_i \sum_{j=1}^{n} A_{ij}^{(2)}\phi_j = \alpha_2\Big(\sum_{i=1}^{n} C_i\phi_i \frac{V^2}{(1-q\phi_i)^2}\Big) \\ &\times \Big(1 + 0.65\frac{(1-q\phi_i)}{b/d}\Big)S_i \end{aligned} \tag{13}$$

where we write $\phi_i = \phi(x_i)$, the approximate value of the cord length integral is given by:

$$C_i = \int_0^1 \prod_{v=1;v\neq i}^{n} \frac{x - \zeta_v}{\zeta_i - \zeta_v} dx \tag{14}$$

and ζ_v are the grid points used in the DQM. The function:

$$S_i = \begin{cases} 1 & \frac{n+1}{2} - 4 \leq i \leq \frac{n+1}{2} + 4 \\ 0 & \text{elsewhere} \end{cases} \tag{15}$$

is a discretized unit-step function representing the size of the lower electrode compared to the micro switch span. We use a two-step finite difference scheme to rewrite the time derivatives in Equation (13) and 150 time steps to decompose one orbit. The resulting linear algebraic system is then solved to obtain the limit cycle corresponding to the operating conditions.

To investigate the validity of the proposed ROM, we perform experiments on the shunt micro switch using a harmonic signal in the form:

$$V(t) = 23 + 23.28\cos(\Omega t)\ (\text{V})$$

Figure 2 displays good agreement between the micro switch frequency-response curves of the midpoint RMS velocity

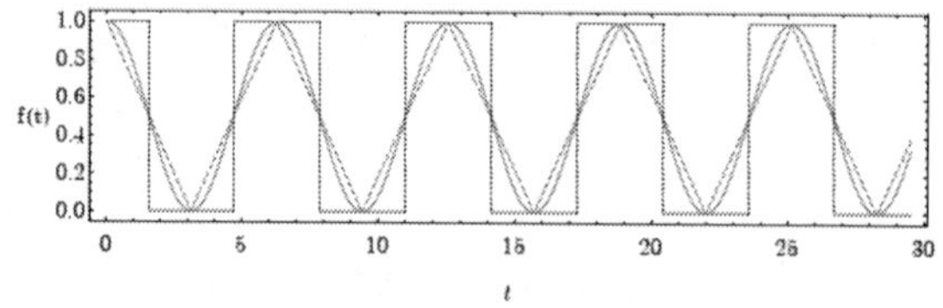

Fig. 3. Actuation waveforms: square (solid thin), harmonic (solid thick) and triangular (dashed)

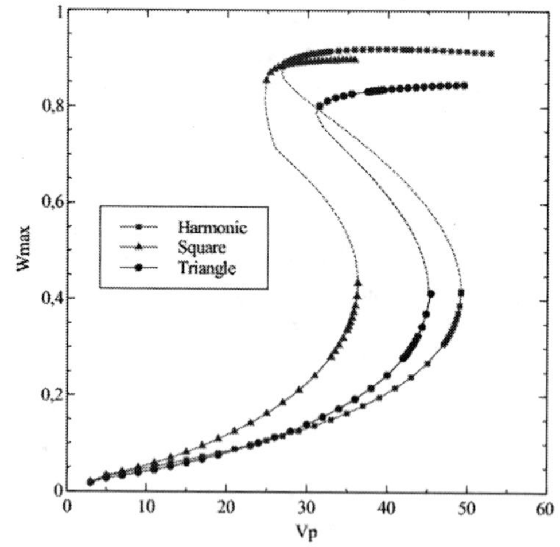

Fig. 4. The force-response curves under square (green), harmonic (red) and triangular (blue) actuation waveforms

obtained from the ROM and the experiment. We conclude that adopting a one-mode approximation is enough to converge to the solution.

Next, we compare among the dynamic response of the micro switch under the three biased square, harmonic, and triangular waveforms plotted in Figure 3. The normalized maximum (mid-point) deflection of the micro switch is shown in Figure 4 as a function of the peak voltage V_p of the harmonic, square and triangular voltage signals for $\Omega = 70\,kHz$. The bias voltage is set to $V_{DC} = 20V$ for all three signals. In each case, dynamic pull in occurs due to a homoclinic bifurcation observable in the phase space.

We remark that the lowest dynamic pull-in voltage is attained at $V_p = 20 + 36.3\,g(\Omega t)$ using the square function $g(t)$ while the highest one is attained at $V_p = 20 + 49.2\cos(\Omega t)$ for a harmonic signal. The dynamic pull-in for triangular signals expressed mathematically by the function $h(t)$ is obtained at $V_p = 20 + 45.2\,h(\Omega t)$. We conclude that a square actuation voltage signal is a more efficient approach to the actuation of micro switches than harmonic or triangular waveforms since it reduces considerably the dynamic pull in voltage required to switch the state of (open and close) MEMS shunt switches.

5 Conclusion

In this work, we carried out numerical and experimental studies of a shunt micro switch. First, we studied the static response of the micro switch under DC voltage using the DQM to solve the equation of equilibrium. We validated our model by comparing the predicted static pull-in voltage to that predicted using an FEM model and previous experimental results. Then, we built a ROM based on one-term modal decomposition and FDM to obtain the limit-cycle of the micro switch for three waveforms: harmonic, square, and triangular. Comparison of simulations and experimental findings shows good agreement and validates our model. Besides this, we demonstrate that the micro switch responds differently to the three signals. In fact, it was shown that for a biased signals, a square waveform reduces considerably the actuation voltage compared with harmonic and triangular waveforms. This is an important insight in the design of low voltage MEMS shunt switches.

References

1. Yadav, R., Yadav, R., Nehra, V., & Rangara, K. J. Key Features, Application & Design Tools.
2. KOBAYASHI, S., & KAWAI, H. (2004). A Capacitive RF MEMS Shunt Switch. In International Workshop on SiP/SoC Integration of MEMS and Passive Components with RF-ICs, Chiba.
3. Park, J. H., Lee, H. C., Park, Y. H., Kim, Y. D., Ji, C. H., Bu, J., & Nam, H. J. (2006). A fully wafer-level packaged RF MEMS switch with low actuation voltage using a piezoelectric actuator. Journal of Micromechanics and Microengineering, 16(11), 2281.
4. Gatzen, H. H. (2012, June). Magnetic micro and nano actuator systems. In Meeting Abstracts (No. 48, pp. 3409-3409). The Electrochemical Society.
5. Bahrami, M. N., Yousefi-Koma, A., & Raeisifard, H. (2014). Modeling and nonlinear analysis of a micro-switch under electrostatic and piezoelectric excitations with curvature and piezoelectric nonlinearities. Journal of Mechanical Science and Technology, 28(1), 263-272.
6. Pacheco, S. P., Katehi, L. P., & Nguyen, C. C. (2000, June). Design of low actuation voltage RF MEMS switch. In Microwave Symposium Digest. 2000 IEEE MTT-S International (Vol. 1, pp. 165-168). IEEE.
7. Balaraman, D., Bhattacharya, S. K., Ayazi, F., & Papapolymerou, J. (2002, June). Low-cost low actuation voltage copper RF MEMS switches. In Microwave Symposium Digest, 2002 IEEE MTT-S International (Vol. 2, pp. 1225-1228). IEEE.
8. Chu, C. H., Shih, W. P., Chung, S. Y., Tsai, H. C., Shing, T. K., & Chang, P. Z. (2007). A low actuation voltage electrostatic actuator for RF MEMS switch applications. Journal of micromechanics and microengineering, 17(8), 1649.
9. Khater, M. E., Vummidi, K., Abdel-Rahman, E. M., Nayfeh, A. H., & Raman, S. (2011). Dynamic actuation methods for capacitive MEMS shunt switches. Journal of Micromechanics and Microengineering, 21(3), 035009.
10. Samaali, H., Najar, F., Ouni, B., & Choura, S. (2015). MEMS SPDT microswitch with low actuation voltage for RF applications. Microelectronics International, 32(2), 55-62.
11. Younis, M. I. (2011). MEMS linear and nonlinear statics and dynamics (Vol. 20). Springer Science & Business Media.
12. Osterberg, P. M., & Senturia, S. D. (1997). M-TEST: a test chip for MEMS material property measurement using electrostatically actuated test structures. Microelectromechanical Systems, Journal of, 6(2), 107-118.
13. Snow, M. G., & Bajaj, A. K. (2015). Uncertainty quantification analysis of the dynamics of an electrostatically actuated microelectromechanical switch model. Journal of Sound and Vibration, 349, 375-388.
14. Bert, C. W., & Malik, M. (1996). Differential quadrature method in computational mechanics: a review. Applied Mechanics Reviews, 49(1), 1-28.
15. Sassi, S. B., & Najar, F. (2015). Novel Reduced Order Model for Electrically Actuated Microbeam-Based MEMS. In Design and Modeling of Mechanical Systems-II (pp. 513-520). Springer International Publishing.

Nonlocal modeling of a Carbon Nanotube actuated by an electrostatic force

Hassen Ouakad[1,a], Fehmi Najar[2], and Sami El-Borgi[3,2]

[1] Mechanical Eng. Dep., King Fahd University of Petroleum and Minerals, P.O. Box 31261, Dhahran, Saudi Arabia.
[2] Applied Mechanics and Systems Research Lab., Tunisia Polytechnic School, University of Carthage, La Marsa, Tunisia.
[3] Mechanical Eng. Program, Texas A&M University at Qatar, Eng. Building, P.O. Box 23874, Education City, Doha, Qatar.

Abstract. Carbon nanotubes (CNTs) are promising mechanical structures at the nano-scale which have attracted increasing attention due to their amazing mechanical, chemical, thermal, and electrical properties. To take into account size dependence of such small sized structures, the use of nonlocal continuum theory is proposed where intrinsic length scales is taken into account. Based on the Eringen theory, a nonlinear nonlocal model of a clamped-clamped CNT is developed in this study. Static and free vibration responses are simulated and analyzed. The main objective of this work is to study the influence of CNT size and length scale parameter on the static and free vibration response to better understand their effect on the general behavior of the CNT. It has been found that the nonlocal effect can largely influence the performance of the CNT and change qualitatively its nonlinear response.

1 Introduction

Carbon nanotubes (CNTs) are viewed as promising mechanical structures in the nano-scale which have attracted the engineers and researchers in this field mostly due to their amazing mechanical, chemical, thermal, and electrical properties. Many engineering applications have reported the possible use of CNTs as building block for novel nanoscale devices [1–3]. Therefore, studying properly the structural behaviors of CNTs under various conditions is a fundamental problem in any nano?scale investigation. For example, many conducted experiments on CNT structures acknowledged their size dependent behavior in the nanoscale.

On the other hand, the conventional continuum mechanics, adopted and assumed in many previous investigations [4, 5], fails to predict the size dependence of such small-sized structures due to lacking of intrinsic length scales. Therefore, only recently, several higher-order elasticity theories have been presented to develop size–dependent continuum models. In what follows is a brief summary of selected studies in which nonlocal elasticity dierential model was used to model size dependence in CNTs based nano–structures. The outcomes of the below investigations are important in mechanical design considerations of devices that use CNTs as main building structures.

Numerous studies investigated the linear free and vibration and wave propagation of CNTs [6–9]. Thongyothee et al. [6] investigated the free vibration problem of CNTs including the effect of nonlocal elasticity to study the effect of their chirality and various boundary conditions. Ansari et al. [7] investigated the free vibration of double-walled carbon nanotubes (DWCNTs) using the Eringen's nonlocal elastic model along with the Donnell shell model. They accounted for the van der Waals forces between the inner and outer nanotubes. Lim and Yang [8] discussed the physics and understanding of nonlocal nanoscale wave propagation in CNTs based on nonlocal elastic stress field theory. In this regards, they developed an analytical nonlocal shear deformable nanobeam model based on the variational principle for wave propagation in CNTs. Further works focused on the nonlocal effect on the buckling characteristics of CNTs considering linear theory [9]. In a more recent work, Kiani [10] developed several novel models, all based on the nonlocal stress theory, to study the lateral buckling of groups of vertically aligned single-walled carbon nanotubes (SWCNTs). In the model, he accounted for the effect of vdW forces along with the nonlocal Rayleigh beam theory.

The previous concisely summarized works are all assuming linear problems. Those who included nonlinearity in either the geometry or the actuating force for CNTs are few [11–17]. To mention some, Hosseini-Ara et al. [11] proposed analytical solutions based on nonlocal Timoshenko kinematics, strain gradient approach and some variational methods to derive the higher-order boundary conditions as well as governing for the sake of investigating the buckling characteristics of CNTs. Karlicic et al. [12] analyzed the free flexural vibration and buckling SWCNT under compressive axial loading. Ansari et al. [13] studied numerically the torsional vibration behaviors of SWCNTs. They assumed an Euler?Bernoulli model while including the material length scale parameters through the strain gradient elasticity theory as to capture the CNT material size dependent effect. Fakhrabadi et al. [14] presented an investigation into the free and forced nonlinear vibration of CNTs under step electrostatic actuation based on the modified couple stress theory. Askari et al. [15] accounted for Eringen nonlocal elasticity and reformulated the classical Euler-Bernoulli beam theories to carry out the nonlinear free vibration analysis of a SWCNT. Mehdipour et al. [16] investigated the use electrostatically actuated clamped-free CNT with an attached mass as a bio-mass sensor. Ke et al. [17] investigated the nonlinear free vibration problem

[a] e-mail: houakad@kfupm.edu.sa

 DOI: 10.1051/matecconf/20168304004

Table 1: Geometrical and physical parameters.

d(nm)	L(nm)	$\hat{R}$(nm)	h(nm)	E(TPa)
500	3000	1	0.34	1.2

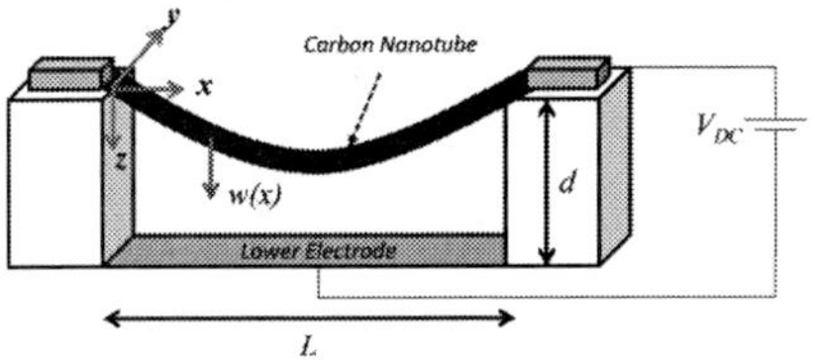

Fig. 1: 3D schematic of an electrostatically actuated straight clamped-clamped CNT actuator.

of functionally graded nanocomposite beams reinforced by SWCNTs considering the Timoshenko beam theory and von Karman geometric nonlinearity

2 Problem Formulation

In this section, we formulate the governing equation of an electrostatically actuated CNT actuator, shown in Fig. 1. The nanotube is triggered by an electrode underneath it with a gap width d. It is modeled as a hollow cylinder Euler-Bernoulli beam of radius $\hat{R}$, shell thickness h, and length L. It has a cross-sectional area A and an area moment of inertia I. The nanotube is assumed to have a Young's modulus E. The CNT is considered here to be straight without slack (initial curvature). Following [18], we calculate the the nonlocal equation describing the in-plane bending deflection $w(x)$ of the clamped-clamped CNT actuator. As in [18] the Eringen's theory is adopted.

$$\ddot{w} + w^{iv} - \mu_0 \ddot{w}'' = V_{DC}^2 \left(\Gamma_e(w) - \mu_0 \Gamma_e''(w)\right) - \alpha_1 \left(w - \mu_0 w''\right)'' \int_0^1 w w'' dx \quad (1)$$

The boundary conditions are

$$w(0) = 0, \quad w(1) = 0, \quad w'(0) = 0, \quad w'(1) = 0 \quad (2)$$

where the longitudinal coordinates x is normalized with respect to the length l. $w(x)$ is the nondimesional transverse deflection toward the substrate and V_{DC} is the applied DC voltage amplitude.

Also the function describing the variation of the electrostatic force is given by

$$\Gamma_e(w) = \frac{\alpha_2}{\sqrt{(1-w)(1-w+2R)}\left(\cosh^{-1}\left(1+\frac{1-w}{R}\right)\right)^2}$$

The nondimensional coefficients are defined by

$$\alpha_1 = \frac{Ad^2}{2I}, \quad \alpha_2 = \frac{\pi\epsilon_0 L^4}{EId^2}, \quad R = \frac{\hat{R}}{d}, \quad \mu_0 = \frac{\hat{\mu}_0}{L^2}$$

where $A = \pi(\hat{R} + \frac{h}{2})^2 - \pi(\hat{R} - \frac{h}{2})^2$ is the cross-section area, $I = \frac{\pi}{4}((\hat{R} + \frac{h}{2})^4 - (\hat{R} - \frac{h}{2})^4)$ is the moment of inertia, ϵ_0 is the air permittivity and $\hat{\mu}_0$ is the nonlocal parameter.

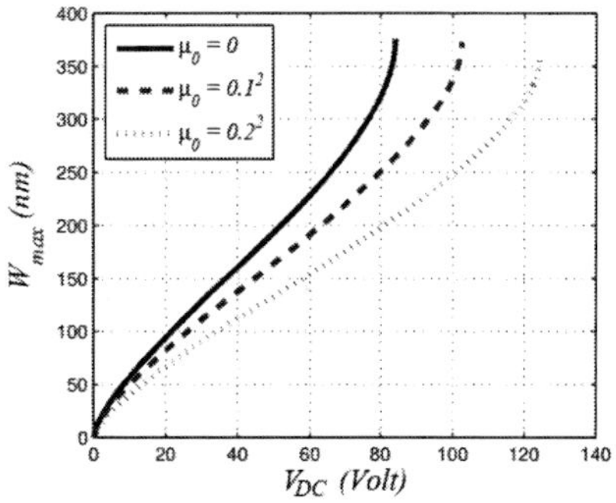

Fig. 2: Variation of the CNT maximum static deflection with the DC voltage and for different values of the nonlocal parameter.

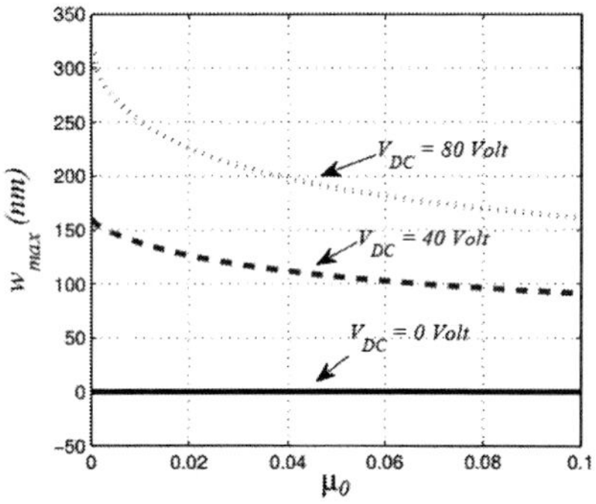

Fig. 3: Variation of the CNT maximum static deflection with the nonlocal parameter and for different values of the DC voltage.

3 Results

3.1 Static Response

We drop time dependent terms in Eq. (1) and end up with a nonlinear ODE describing the static deflection of the CNT. We use the Differential Quadrature Method (DQM) to solve for the equilibrium positions (see [18] for details of the DQM formulation). Fig. 2 shows the maximum static deflection w_{max} at midpoint as a function of the applied DC voltage and for different values of the nondimensional nonlocal parameter μ_0. It is worth noting here that the classical theory corresponds to the case where $\mu_0 = 0$. It is clear from Fig. 2 that for a fixed DC voltage, increasing the value of the nonlocal parameter has a tendency to decrease the maximum deflection indicating that the CNT beam becomes stiffer due essentially to variation of the effective nonlinear stiffness.

Similarly in Fig. 3 illustrates the variation of the maximum static deflection w_{max} at midpoint in terms of the nondimensional nonlocal parameter μ_0 and for different values of the applied DC voltage. It is evident that the effect of the nonlocal parameter becomes more pronounced for increasing values of the applied DC voltage.

 DOI: 10.1051/matecconf/20168304004

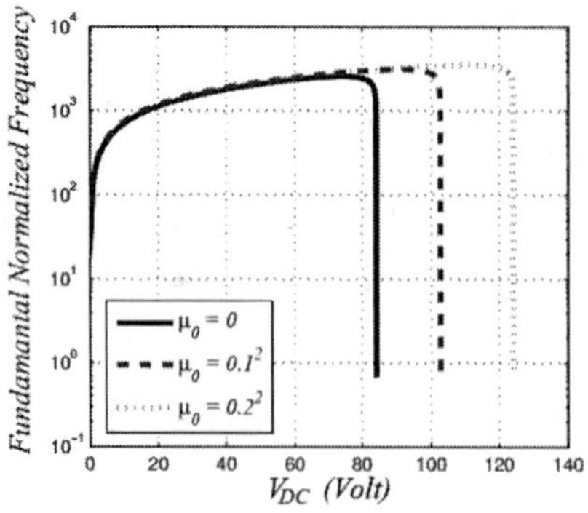

Fig. 4: Variation of the CNT fundamental normalized frequency with the DC voltage and for different values of the nonlocal parameter.

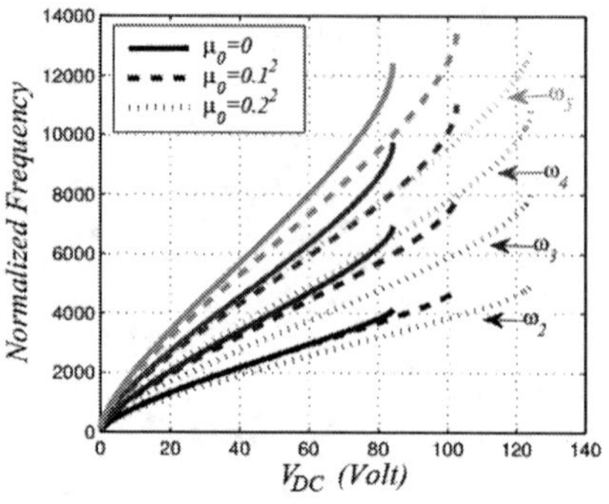

Fig. 5: Variation of the CNT normalized higher-order frequencies with the DC voltage and for different values of the nonlocal parameter.

3.2 Linear Natural Frequencies

We start by linearizing Eq. (1) around the stable equilibrium position obtained by solving the static response of the CNT. Therefore, the associated linear eigenvalue problem is solved for the natural frequencies of the CNT using the DQM [18]. As a result, the linear nonlocal effect in both midplane stretching (cubic nonlinearity) and the actuating electric force (quadratic nonlinearity) are taken into account.

Fig. 4 depicts the variation of the fundamental normalized frequency of the CNT beam, for different values of the applied DC voltage and the nonlocal parameter. The fundamental frequency clearly increases for increasing DC voltage before the occurrence of pull-in. Increasing the values of the nonlocal parameter allows a slight increase in the frequency and a further growth of the pull-in voltage. This indicates that the effect of the nonlocal parameters has a tendency to make the CNT beam more rigid, which is the same conclusion reached in the static analysis.

For higher natural frequencies and different values of the nonlocal parameter, we show in Fig. 5 the influence of the applied DC voltage. As expected, the frequencies increase for high values of the DC voltage. Furthermore, the nonlocal parameter tends to reduce the frequencies and this effect is more significant in higher modes. To understand further this effect, the first five natural frequencies are plotted for increasing values of the nonlocal parameter in Fig. 6 and 7 for fixed values of the DC voltage of 0 and 40 volts, respectively. It is clear from the case of 0 volt DC voltage (Fig. 6) that all frequencies are reduced for increasing values of the nonlocal parameter indicating that the beam becomes more flexible which is the opposite conclusion than that of the static analysis.

We can see from Fig. 6 that without DC load, the CNT is not stretched and then the nonlocal effect will be of softening type. Whereas, if we increase the DC load to a nonzero value (see Fig. 7) the frequencies start to increase, mainly due to a hardening cubic effect of the nonlocal parameter, and then the quadratic effect will jump as the dominant for higher values of the nonlocal parameter. Adding to that the effect of the rotary inertia and because of its minus sign visible in Fig. 6, the effective stiffness of the CNT is further reduced.

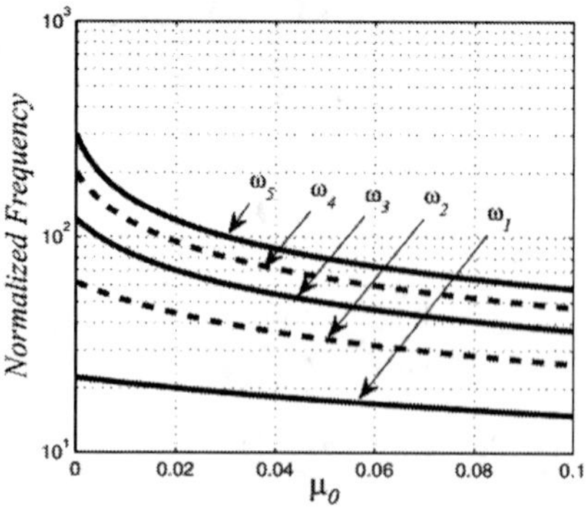

Fig. 6: Variation of the CNT normalized higher-order frequencies with the nonlocal parameter and for zero DC voltage.

For the cases of 40 volts DC voltage (Fig. 7), the same conclusion is reached except that for the fundamental frequency where an increase is first observed due to the influence of the quadratic nonlinearity of the electrostatic force. Then, when the μ_0 is higher than 0.02 a decrease of the fundamental frequency is obtained. It is also interesting to note in Fig. 7 the cross-over of the first two natural frequencies for a value of the nonlocal parameter equal to 0.06, the second mode of vibration becomes the fundamental mode.

Fig. 9 illustrates the effect of the gap width to CNT radius ratio on the pull-in voltage for a fixed beam length to radius ratio equal to 3000nm and for different values of the nonlocal parameter. It can be observed that for small values of the gap width to CNT radius ratios, the effect of the nonlocal parameter is insignificant. However, this effect becomes significant for larger values of the ratio indicating that the pull-in voltage increases for increasing values of the nonlocal parameter.

Finally, Fig. 10 depicts the effect of the beam length to radius ratio on the pull-in voltage for a fixed the gap width to CNT radius ratio equal to 500nm and for different values of the nonlocal parameter. It can be concluded that for large values of the beam length to radius ratio, the effect of the nonlocal parameter is insignificant. On the other hand, this effect becomes pronounced for small values of the ratio

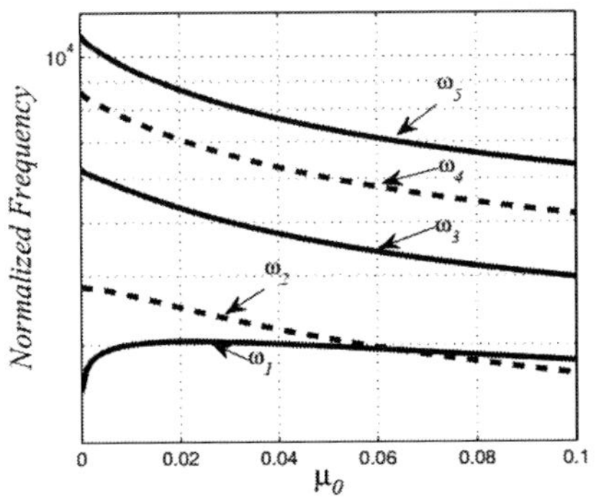

Fig. 7: Variation of the CNT normalized higher-order frequencies with the nonlocal parameter and for 80 Volt DC voltage.

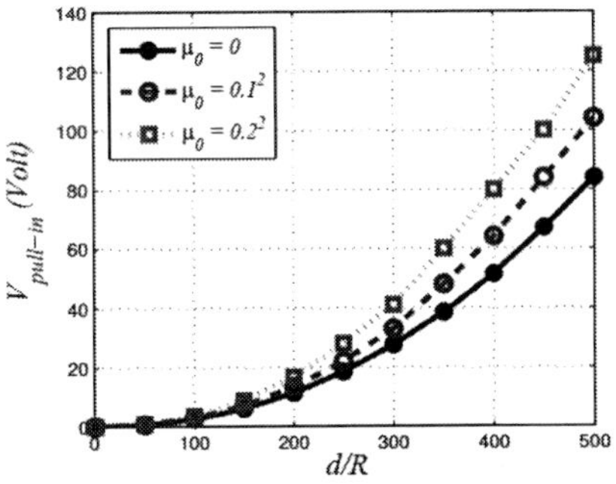

Fig. 8: Variation of the CNT pull-in voltage of the for different gap size to CNT radius ratio and for L=3000 nm.

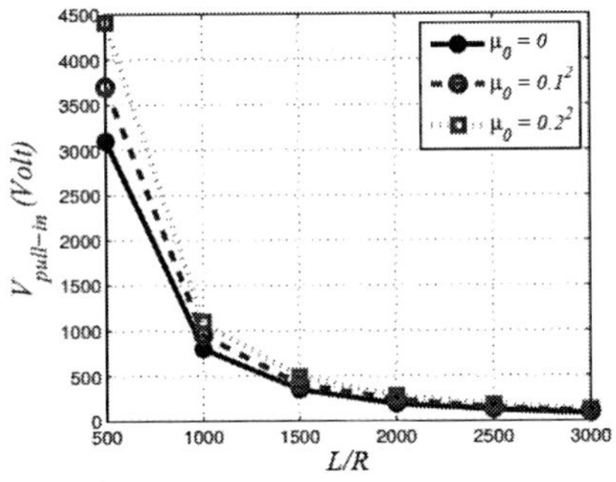

Fig. 9: Variation of the CNT pull-in voltage of the for different CNT length to radius ratio and for d=500 nm.

indicating that the pull-in voltage increases for increasing values of the nonlocal parameter.

4 Conclusions

Carbon nanotubes (CNT) are mechanical structures depicting interesting behavior at the nano-scale. However, size dependence of such small sized structures can influence their responses to external excitations. The nonlocal continuum theory has been used to take into account intrinsic length scales of the CNT. Based on the Eringen theory, a nonlinear nonlocal model of a clamped-clamped CNT was developed. We used the Differential Quadrature Method to solve the derived Static and free vibration equations.

The influence of CNT size and length scale parameter on the static response has been analyzed. It has been shown that an increase of the nonlocal parameter increases the effective forcing on the CNT which implicate an apparent stiffness reduction reducing the static response and enlarging the pull-in voltage.

For free vibration responses linearized eigenvalue problem around the static solution has been solved to calculate the natural frequencies of the CNT, it has been found that the nonlocal effect reduces all natural frequencies up to the fifth mode when no external DC voltage is applied. If the linear par t of the electrostatic force is taken into account, the fundamental frequency is first increased under the hardening behavior of the quadratic nonlinearity of the electrostatic force, then it start to decrease again following the general trend of the other frequencies.

References

1. V. Sazonova, Y. Yaish, H. Üstünel, D. Roundy, T. A. Arias, and P. L. McEuen, Nature **431**, (2004) 284–287.
2. H. Üstünel, D. Roundy, and T. A. Arias, Nano Letters **5**, (2005) 523–526.
3. Zhao, C., Song, Y., Ren, J., Qu, X., Biomaterials **30**, (2009) 1739–1745.
4. H. M. Ouakad, M. I. Younis, J. Computational and Nonlinear Dynamics **5**, (2010) 011009.
5. W. G. Conley, A. Raman, C. M. Krousgrill, S. Mohammadi, Nano Letters **8**, (2008) 1590–1595.
6. C. Thongyothee, S. Chucheepsakul, T. Li, In Advanced Materials Research **747**, (2013) 257–260.
7. Ansari, R., H. Rouhi, and B. Arash. IJST **37**, 2 (2013) 91–105.
8. Lim, C. W., and Y. Yang. J. Mechanics of Materials and Structures **5**, 3 (2010) 459–476.

y

9. Xu, Xiao-jian, Zi-chen Deng. Applied Mathematics and Mechanics **35**, (2014) 1115–1128.
10. Kiani, Keivan. Acta Mechanica **225**, 12 (2014) 3569–3589.
11. Hosseini-Ara, R., H. R. Mirdamadi, H. Khademyzadeh, R. Mostolizadeh. J. of the Brazilian Society of Mechanical Sciences and Engineering **34**, 2 (2012) 126–134.
12. D. Karlicic, P. Kozic, R. Pavlovic, J. Theoretical and Applied Mechanics **53**, (2015) 217–233.
13. R. Ansari, R. Gholami, S. Ajori. J. Vibration and Acoustics **135**, 5 (2013) 051016.
14. M. M. S. Fakhrabadi, A. Rastgoo, M. T. Ahmadian, and M. M. Mashhadi, Acta Mechanica **225**, 6 (2014) 1523–1535.
15. H. Askari, D. Zhang, E. Esmailzadeh, In Nanotechnology (IEEE-NANO), (2013) 545–548.
16. I. Mehdipour, A. Erfani-Moghadam, C. Mehdipour. Current Applied Physics **13**, 7 (2013) 1463–1469.
17. L-L Ke, J. Yang, S. Kitipornchai, Composite Structures **92**, 3 (2010) 676–683.
18. F. Najar, S. El-Borgi, JN Reddy, K. Mrabet, Composite Structures **120**, (2015) 117–128.

MATEC Web of Conferences **83**, 04005 (2016) DOI: 10.1051/matecconf/20168304005
CSNDD 2016

Investigation of the nonlinear static and dynamic behaviour of rectangular microplates under electrostatic actuation

S. Saghir, M.I. Younis
4700 King Abdullah University of Science and Technology (KAUST), Thuwal 23955-6900, Saudi Arabia

Abstract. We present an investigation of the static and dynamic behavior of the nonlinear von-Karman plates when actuated by the nonlinear electrostatic forces. The investigation is based on a reduced order model developed using the Galerkin method, which rely on modeshapes and in-plane shape functions extracted using a finite element method. In this study, a fully clamped microplate is considered. We investigate the static behavior and the results are validated by comparison with the results calculated by a finite element model. The forced-vibration response of the plate is then investigated when the plate is excited by a harmonic AC load superimposed to a DC load. The dynamic behavior is examined near the primary resonance. The microplate shows a strong hardening behavior due to the cubic nonlinearity of mid-plane stretching. However, the behavior switches to softening as the DC load is increased.

1. Introduction

Micro-Electro Mechanical Systems (MEMS) commonly use electrically actuated flexible micro-structures, such as microbeams and microplates [1, 2]. Examples include pressure, mass and gas sensors, micropumps, microjets, micromirror, and microphones [3, 4]. Modeling accurately the mechanical behavior of such structures under the applied electric force is required to predict the response prior to the experimental testing and actual use of the device. Accurate models can guide the design engineer through the design process; reducing the design time on one hand and on the other hand can help to improve the existing devices. It is common to study the mechanical behavior of MEMS using linear theory [2]; which is applicable only for small deflections. Since in MEMS, structures often undergo large deflection, linear theory becomes inaccurate. Common modeling approaches include lumped mass models and Finite Element Method (FEM) [4, 5]. Lumped mass models give rough estimate of the response only. FEM based software tools are accurate but computationally expensive, especially when it comes to study the nonlinear dynamic behavior. On the other hand Reduced Order Models (ROM) based on the Galerkin approach have got popularity during the last decades because of their accuracy and low computational cost [6-10]. They have the capability to reveal the effect of different design parameters very conveniently.
A common way to model the plate structures in MEMS devices is using the linear plate theory [11-13], which is correct when out of plane deflection is small. But in cases when the transverse deflection is comparable to the thickness of the plate, a strong geometric nonlinearity is present due to mid-plane stretching making the predications of linear theories erroneous.
Most of the works reported in literature, have been mainly focused on the static and linear vibration behavior of microplates. Almost no work has been presented about the nonlinear behavior of microplates when actuated by large electrostatic loading or when undergoing large motion. Understanding such behaviors is fundamental to the development of the next generation microplates-based MEMS devices.
The objective of this paper is to investigate the nonlinear forced vibration behavior of electrically actuated fully clamped microplates. Towards this, we develop a reduced order model based on the von Karman equations of the plate, in which all three equations of the plate motion are discretized using the Galerkin method. For validation purpose, we compare the static results computed by the reduced order model with the results calculated using the FEM software COMSOL [14]. Further the reduced order model is employed to investigate the dynamic behavior of the plates under large vibration amplitude.

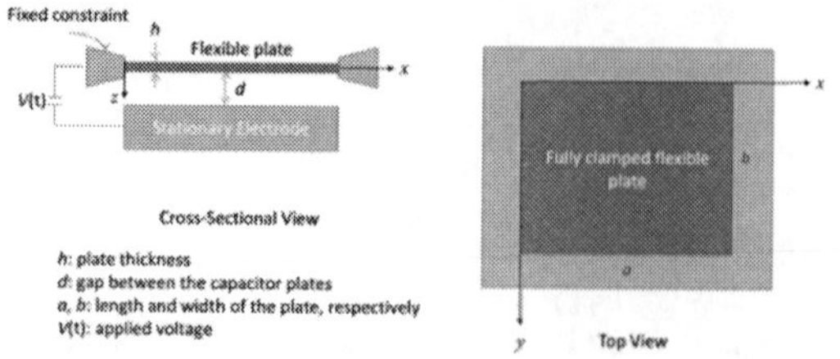

Fig. 1: A schematic diagram of an electrically actuated fully clamped microplate.

2. Problem Formulation

We consider a fully clamped rectangular microplate as shown in figure 1 and adopt the von-Karman equations of motion [15, 16]. We ignore the in-plane inertia since the in-plane natural frequencies are much higher than the transverse natural frequencies. Hence, the in-plane inertia has negligible effect on the transverse motion. Dropping the in plane inertia terms, the governing equations of the plate motion can be written as

$$\frac{\partial^2 u}{\partial x^2}+\frac{1}{2}(1+\upsilon)\frac{\partial^2 v}{\partial x\partial y}+\frac{1}{2}(1-\upsilon)\frac{\partial^2 u}{\partial y^2}+\frac{\partial w}{\partial x}\frac{\partial^2 w}{\partial x^2}+\frac{1}{2}(1+\upsilon)\frac{\partial w}{\partial y}\frac{\partial^2 w}{\partial x\partial y}+\frac{1}{2}(1-\upsilon)\frac{\partial w}{\partial x}\frac{\partial^2 w}{\partial y^2}=0 \quad (1)$$

$$\frac{\partial^2 v}{\partial y^2}+\frac{1}{2}(1+\upsilon)\frac{\partial^2 u}{\partial x\partial y}+\frac{1}{2}(1-\upsilon)\frac{\partial^2 v}{\partial x^2}+\frac{\partial w}{\partial y}\frac{\partial^2 w}{\partial y^2}+\frac{1}{2}(1+\upsilon)\frac{\partial w}{\partial x}\frac{\partial^2 w}{\partial x\partial y}+\frac{1}{2}(1-\upsilon)\frac{\partial w}{\partial y}\frac{\partial^2 w}{\partial x^2}=0 \quad (2)$$

$$\begin{aligned}&\frac{h^2}{12}\nabla^4 w+\frac{\rho(1-\upsilon^2)}{Eh}\frac{\partial^2 w}{\partial t^2}+\frac{c(1-\upsilon^2)}{Eh}\frac{\partial w}{\partial t}=\frac{\varepsilon(1-\upsilon^2)V(t)^2}{2Eh(d-w)^2}\\&+\frac{(1-\upsilon^2)}{Eh}\left(N_{xx}\frac{\partial^2 w}{\partial x^2}+2N_{xy}\frac{\partial^2 w}{\partial x\partial y}+N_{yy}\frac{\partial^2 w}{\partial y^2}\right)\\&+\frac{\partial u}{\partial x}\left(\frac{\partial^2 w}{\partial x^2}+\upsilon\frac{\partial^2 w}{\partial y^2}\right)+\frac{\partial v}{\partial y}\left(\upsilon\frac{\partial^2 w}{\partial x^2}+\frac{\partial^2 w}{\partial y^2}\right)\\&+(1-\upsilon)\left(\frac{\partial u}{\partial y}+\frac{\partial v}{\partial x}\right)\frac{\partial^2 w}{\partial x\partial y}+\frac{1}{2}\left(\frac{\partial w}{\partial x}\right)^2\left(\frac{\partial^2 w}{\partial x^2}+\upsilon\frac{\partial^2 w}{\partial y^2}\right)\\&+\frac{1}{2}\left(\frac{\partial w}{\partial y}\right)^2\left(\upsilon\frac{\partial^2 w}{\partial x^2}+\frac{\partial^2 w}{\partial y^2}\right)+(1-\upsilon)\left(\frac{\partial w}{\partial x}\frac{\partial w}{\partial y}\frac{\partial^2 w}{\partial x\partial y}\right)\end{aligned} \quad (3)$$

where u, v and w are the displacements along the x, y and z direction, respectively, ∇^4 is the bi-harmonic operator, $\nabla^4=\frac{\partial^4}{\partial x^4}+2\frac{\partial^4}{\partial x^2\partial y^2}+\frac{\partial^4}{\partial y^4}$, ρ is the mass density per unit area, c is the damping coefficient, and υ and E are Poisson's ratio and modulus of elasticity, respectively. The first term on the right side of equation (3) i.e. $\frac{\varepsilon(1-\upsilon^2)V(t)^2}{2Eh(d-w)^2}$ is the applied electric pressure in the transverse direction while N_{ij} is the applied force per unit length on the edge perpendicular to i^{th} axis in the jth direction.

For convenience, we introduce the non-dimensional variables (denoted by hats);

$$\hat{x}=\frac{x}{a},\ \hat{y}=\frac{y}{b},\ \hat{w}=\frac{w}{d},\ \hat{u}=\frac{au}{d^2},\ \hat{v}=\frac{av}{d^2},\ \hat{t}=\frac{t}{T} \quad (4)$$

Substituting equation (4) into equations (1)-(3) and dropping the hats for convenience we get the following equations:

$$\frac{\partial^2 u}{\partial x^2}+\frac{(1+\upsilon)}{2\alpha}\frac{\partial^2 v}{\partial x\partial y}+\frac{(1-\upsilon)}{2\alpha^2}\frac{\partial^2 u}{\partial y^2}+\frac{\partial w}{\partial x}\frac{\partial^2 w}{\partial x^2}+\frac{(1+\upsilon)}{2\alpha^2}\frac{\partial w}{\partial y}\frac{\partial^2 w}{\partial x\partial y}+\frac{(1-\upsilon)}{2\alpha^2}\frac{\partial w}{\partial x}\frac{\partial^2 w}{\partial y^2}=0 \quad (5)$$

$$\frac{\partial^2 v}{\partial y^2}+\frac{(1+\upsilon)\alpha}{2}\frac{\partial^2 u}{\partial x\partial y}+\frac{(1-\upsilon)\alpha^2}{2}\frac{\partial^2 v}{\partial x^2}+\frac{1}{\alpha}\frac{\partial w}{\partial y}\frac{\partial^2 w}{\partial y^2}+\frac{(1+\upsilon)\alpha}{2}\frac{\partial w}{\partial x}\frac{\partial^2 w}{\partial x\partial y}+\frac{(1-\upsilon)}{2}\frac{\partial w}{\partial y}\frac{\partial^2 w}{\partial x^2}=0 \quad (6)$$

$$\begin{aligned}&\frac{\partial^4 w}{\partial x^4}+\frac{2}{\alpha^2}\frac{\partial^4 w}{\partial x^2\partial y^2}+\frac{1}{\alpha^4}\frac{\partial^4 w}{\partial y^4}+\frac{\partial^2 w}{\partial t^2}+\hat{c}\frac{\partial w}{\partial t}=\alpha_2\frac{V(t)^2}{(1-w)^2}\\&+3\alpha_0^2\left\{\hat{N}_{xx}\frac{\partial^2 w}{\partial x^2}+\frac{2}{\alpha}\hat{N}_{xy}\frac{\partial^2 w}{\partial x\partial y}+\frac{1}{\alpha^2}\hat{N}_{yy}\frac{\partial^2 w}{\partial y^2}\right\}\\&+12\alpha_1^2\left\{\begin{aligned}&\frac{\partial u}{\partial x}\left(\frac{\partial^2 w}{\partial x^2}+\frac{\upsilon}{\alpha^2}\frac{\partial^2 w}{\partial y^2}\right)+\frac{1}{\alpha}\frac{\partial v}{\partial y}\left(\upsilon\frac{\partial^2 w}{\partial x^2}+\frac{1}{\alpha^2}\frac{\partial^2 w}{\partial y^2}\right)\\&+(1-\upsilon)\left(\frac{1}{\alpha}\frac{\partial u}{\partial y}+\frac{\partial v}{\partial x}\right)\frac{1}{\alpha}\frac{\partial^2 w}{\partial x\partial y}\end{aligned}\right\}\\&+12\alpha_1^2\left\{\begin{aligned}&\frac{1}{2}\left(\frac{\partial w}{\partial x}\right)^2\left(\frac{\partial^2 w}{\partial x^2}+\frac{\upsilon}{\alpha^2}\frac{\partial^2 w}{\partial y^2}\right)\\&+\frac{1}{2\alpha^2}\left(\frac{\partial w}{\partial y}\right)^2\left(\upsilon\frac{\partial^2 w}{\partial x^2}+\frac{1}{\alpha^2}\frac{\partial^2 w}{\partial y^2}\right)+\frac{(1-\upsilon)}{\alpha^2}\left(\frac{\partial w}{\partial x}\frac{\partial w}{\partial y}\frac{\partial^2 w}{\partial x\partial y}\right)\end{aligned}\right\}\end{aligned}$$

(7)

The parameters appearing in equations (5)-(7) are

$$\alpha=\frac{b}{a},\ \alpha_0=\frac{a}{h},\ \alpha_1=\frac{d}{h},\ \alpha_2=\frac{6(1-\upsilon^2)}{Eh^3d^3}\varepsilon a^4,\ \hat{c}=\frac{ca^4}{TD}$$

and $\hat{N}_{ij}=\frac{(1-\upsilon^2)}{Eh}N_{ij}$ (8)

where $T=\sqrt{\frac{\rho a^4}{D}}$ and $D=\frac{Eh^3}{12(1-\nu^2)}$.

3. Reduced Order Model

To develop the reduced order model for the governing equations (5)-(7), we assume the solution for the transverse and in-plane displacements as follows

$$w=\sum_{i=1}^{n}q_i(t)\phi_i(x,y) \quad (9)$$

$$u=q_u(t)\psi_u(x,y) \quad (10)$$

$$v=q_v(t)\psi_v(x,y) \quad (11)$$

where $\phi_i(x,y)$ are the transversal linear modeshapes of the plate and $q_i(t)$ are the unknown time dependent coefficients, $\psi_u(x,y)$ and $\psi_v(x,y)$ are the shape functions for the in-plane displacements u and v, while $q_u(t)$ and $q_v(t)$ are the corresponding time dependent unknown coefficients.

Since the linear eigenvalue problem governing the transversal linear modeshapes of the fully clamped plate cannot be solved analytically, we resort to the finite element method for this purpose. We use the commercial software COMSOL [14] to obtain the modeshapes of the plate. The in-plane displacement shape functions are also obtained by FEM while the plate is deflected by a uniform transverse pressure.

To treat the electric force term we multiply equation (7) by $(1-w)^2$ so that electric force term is represented exactly [6, 17]. Substituting equations (9)-(11) into equations (5)-(7), multiplying equation (5) and (6) by $\psi_u(x,y)$ and $\psi_v(x,y)$, respectively, and equation (7) by $\phi_i(x,y)$, and integrating over the plate domain, we get a system of differential algebraic equations (DAEs) for the time dependent coefficients. Equations (5) and (6) generate algebraic equations while equation (7) results in ordinary differential equations (ODEs). The resulting system of DAEs is solved for the unknown time dependent coefficients $q_i(t)$, $q_u(t)$ and $q_v(t)$. Toward this, we solve equations (5) and (6) for $q_u(t)$ and $q_v(t)$ in terms of $q_i(t)$ and then substitute the results into equation (7). This equation is then integrated in time using Runge Kutta method. These coefficients are substituted back into equations (9)-(11) to get the displacements w, u and v.

4. Static Analysis

To calculate the static deflection of the microplate under a DC load, we drop the time derivatives in the reduced order model and the time dependent unknown coefficients $q_i(t)$, $q_u(t)$, and $q_v(t)$ are replaced by constant coefficients q_i, q_u, and q_v. This results in a system of nonlinear algebraic equations, which is numerically solved for q_i, q_u and q_v. Then equation (9) is used to find the transversal deflection.

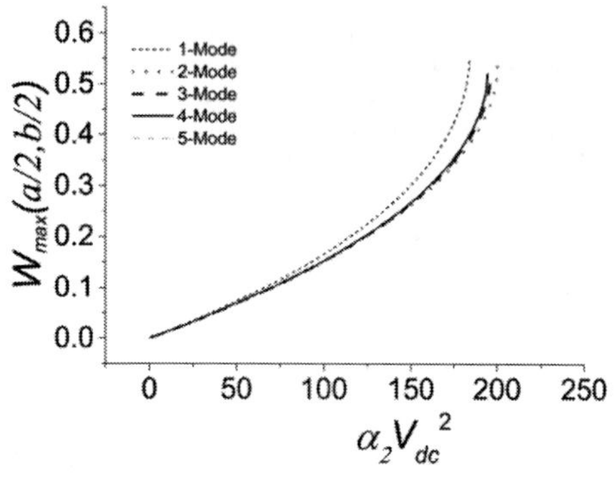

Fig. 2: Convergence of the static response with the number of transverse modes retained in the reduced order model. Variation of the maximum non-dimensional deflection W_{max} with the electrostatic voltage parameter $\alpha_2 V_{dc}^2$ when $\alpha = 1$ and $\alpha_1 = 1$.

First we study convergence of the static results with the number of transverse modes, $\phi_i(x,y)$ retained in the reduced order model. Figure 2 shows the stable solution, the non-dimensional deflection W_{max} at the center of plate against various values of $\alpha_2 V_{dc}^2$. We note that the static results converge by retaining at least four transverse modes in the model. One can note that the deflection curve is limited by the pull-in instability, where the slope of the curve approaches infinity.

Next we validate the reduced order model by comparing its results with the results obtained by FEM software COMSOL [14] for different aspect ratios α. Figure 3 shows the maximum deflection W_{max} at the center of the microplate calculated by both models. We use rectangular microplates of four different aspect ratios of $\alpha = 1$, $\alpha = 1.1$, $\alpha = 1.2$, and $\alpha = 1.5$. Dimensional specifications of the electrically actuated microplates are given in Table 1, while material properties are; $E = 153\text{GPa}$ and $\upsilon = 0.23$.

Table 1: Dimensional specifications of electrically actuated microplates used in the model validation.

Aspect ratio $\alpha = b/a$	Length $a(\mu m)$	Width $b(\mu m)$	Thickness $h(\mu m)$	Capacitor gap $d(\mu m)$
1	300	300	2	2
1.1	300	330	2	2
1.2	300	360	2	2
1.5	300	450	2	2

Results are shown for various values of V_{dc}, until the pull in instability, for $\alpha = 1.5$. It shows excellent agreement among the results calculated by the ROM model and the FEM model.

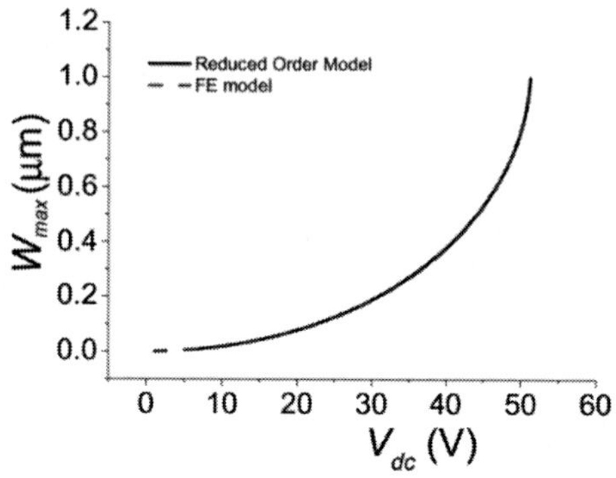

Fig. 3: Comparison of the maximum deflection W_{max} at the center of the microplate, calculated by the reduced order with the results obtained from FEM for various values of V_{dc} until the pull-in instability, for a rectangular microplate of aspect ratio $\alpha = 1.5$.

5. Dynamic Analysis

In this section we investigate the dynamic response of the square microplate in the neighborhood of primary, resonances of the fundamental mode. We analyze the dynamics of the microplate by generating frequency response curves. The Runge Kutta method is used to perform long time numerical integration to solve the system of DAEs. The stable steady solution is captured after making sure that the transient response is no longer contributing to the response. Figure 4 shows the time history of the dynamic response of the microplate when actuated at $V_{dc} = 3\text{V}$ and $V_{ac} = 0.1\text{V}$ and $\Omega = 35.1$. Figure 4a depicts the diminishing transient response, while figure

4b shows the stable steady state response. We use steady-state responses similar to figure 4b to construct the frequency response curves.

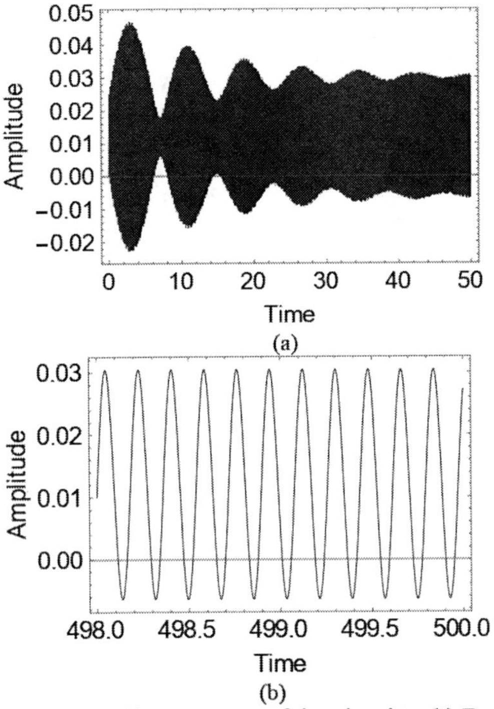

Fig. 4: Time history response of the microplate. (a) Transient response. (b) Steady state response.

To investigate the dynamic response at primary resonance, we fix $\alpha = 1$, $\alpha_1 = 1$, $\alpha_2 = 1$ and assume that the in-plane external forces are zero i.e. $N_{xx} = N_{xy} = N_{yy} = 0$. With the above parameters the static pull-in voltage for the microplate is near 14 Volts. We investigate the nonlinear dynamic behavior of the microplate near primary resonance, a small harmonic V_{ac} is superimposed to a V_{dc} i.e. $V(t) = V_{dc} + V_{ac}Sin(\Omega t)$. Figure 5 shows the nonlinear response of the microplate when actuated at $V_{dc} = 3\text{V}$ for various values of V_{ac} while the quality factor is fixed at $Q = 250$ near primary resonance, which for the linear plate is near 36. Our choice of a constant value of Q means that we assumed negligible effect of squeeze-film damping, which reasonable assumption is assuming that the microplate is placed inside a vacuum chamber and is operated at reduced pressure. Otherwise, squeeze-film damping can have strong effect on the dynamics of microplates and needs to be modeled using Reynolds equation and a structural-fluidic model [17, 18].

The microplate exhibits strong hardening effect due to the cubic nonlinearity, which comes into play due to mid-plane stretching. The hysteresis in the curves is captured by performing forward and backward frequency sweeps. Nonlinear resonance peaks occur near $\Omega = 36.15$ for $V_{ac} = 0.1\text{V}$ while it is at $\Omega = 36.6$ for $V_{ac} = 0.3\text{V}$ and $\Omega = 36.9$ for $V_{ac} = 0.5\text{V}$. Further we notice that there exist multiple stable solutions over some range of frequency and amplitude jumps from higher to lower or lower to higher values depending on the type of frequency sweep.

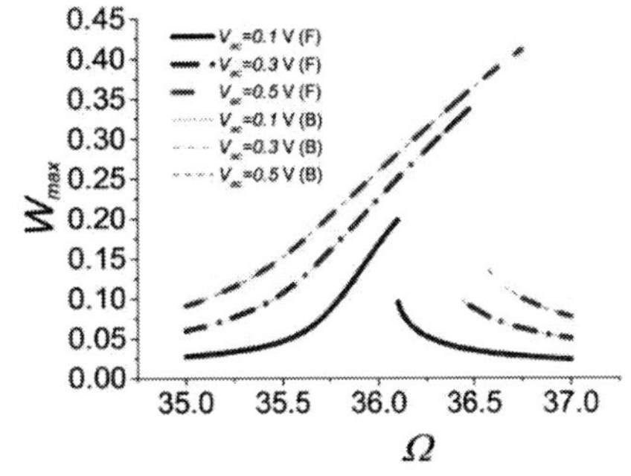

Fig. 5: Maximum non-dimensional deflection $W_{\max}$ at the center of the microplate against the actuating frequency Ω when actuated at $V_{dc} = 3\text{V}$ and various values of V_{ac} while quality factor $Q = 250$; (F) Forward frequency sweep, (B)Backward frequency sweep.

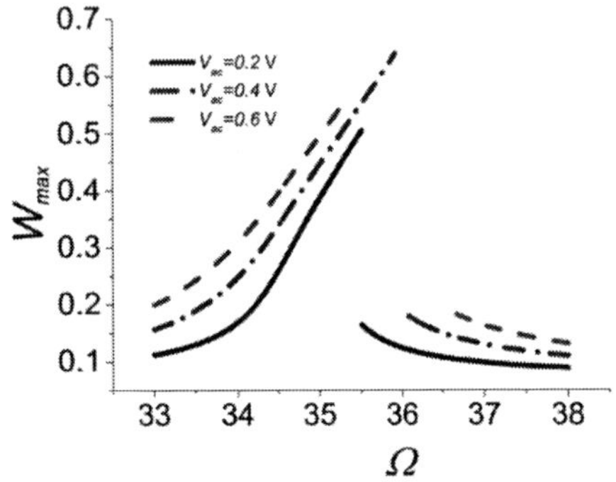

Fig. 6: Maximum non-dimensional deflection $W_{\max}$ of the microplate against the actuating frequency Ω when actuated at $V_{dc} = 7\text{V}$ and various values of V_{ac} while quality factor $Q = 250$.

Figure 6 shows the frequency response curves of the microplate when actuated at $V_{dc} = 7\text{V}$ for various values of V_{ac} with a quality factor $Q = 250$. An overlap of stable solutions exists when actuated at $V_{ac} = 0.2\text{V}$ contrary to the responses at $V_{ac} = 0.4\text{V}$ and $V_{ac} = 0.6\text{V}$, respectively. A gap between the two stable solutions starts to emerge in the case of $V_{ac} = 0.4\text{V}$ by the forward and backward frequency sweeps. This gap might indicate that the microplate becomes unstable at that actuating voltage and pulls on the stationary electrode. This kind of instability is called dynamic pull-in instability, which usually occurs at a lower DC load superimposed to a small harmonic load [17, 19-21]. Another possibility exists, especially in the case of $V_{ac} = 0.4\text{V}$, that this gap is created due to the numerical divergence of the time integration scheme, due to its inability to find suitable initial conditions that lead to a stable periodic orbit. In other words, this might indicate highly fractal behavior, which usually gets stronger as the system approaches the dynamic pull-in regime [20]. To confirm if this divergence is due to fractal behavior or due to pull-in exactly, one should resort to other numerical techniques to find periodic motions, such as shooting and finite difference methods as well as basin of attraction analysis [17, 21]. The response on the other hand for the case of $V_{ac} = 0.6\text{V}$ is most likely an indication of a pull-in band, since as reported in [21], further increase in V_{ac} widens the pull-in band gap between the stable solutions and makes the upper stable branches terminated at lower values [21].

6. Conclusions

In this article we developed a reduced order model for the investigation of the static as well as the dynamic behavior of electrically actuated rectangular microplates. First convergence of the static results with the number of modeshapes retained in the ROM has been studied. We found that four modes are sufficient for convergence. The dynamic behavior of the microplate has been investigated near primary resonance using long time numerical integration. We captured the stable solutions using forward and backward frequency sweeps. The microplate actuated near primary resonance shows a strong hardening behavior due to the cubic nonlinearity, which comes into play due to mid-plane stretching. Increasing V_{ac} further widens the gap between the two stable solutions captured by forward and backward frequency sweeps and pull-in instability occurs at a lower vibration amplitude.

References

1. Chuang, W.C., H.L. Lee, P.Z. Chang, and Y.C. Hu, *Review on the modeling of electrostatic MEMS.* Sensors, 2010. **10**(6): p. 6149-6171.
2. Batra, R.C., M. Porfiri, and D. Spinello, *Review of modeling electrostatically actuated microelectromechanical systems.* Smart Materials and Structures, 2007. **16**(6): p. R23.
3. Tong, P. and W. Huang, *Large Deflection of Thin Plates in Pressure Sensor Applications.* Journal of Applied Mechanics, 2002. **69**(6): p. 785-789.
4. Ng, T.Y., T.Y. Jiang, H. Li, K.Y. Lam, and J.N. Reddy, *A coupled field study on the non-linear dynamic characteristics of an electrostatic micropump.* Journal of Sound and Vibration, 2004. **273**(4–5): p. 989-1006.
5. Pursula, A., P. Råback, S. Lähteenmäki, and J. Lahdenperä, *Coupled FEM simulations of accelerometers including nonlinear gas damping with comparison to measurements.* Journal of Micromechanics and Microengineering, 2006. **16**(11): p. 2345-2354.
6. Younis, M.I., E.M. Abdel-Rahman, and A. Nayfeh, *A reduced-order model for electrically actuated microbeam-based MEMS.* Microelectromechanical Systems, Journal of, 2003. **12**(5): p. 672-680.
7. Zhao, X., E.M. Abdel-Rahman, and A.H. Nayfeh, *A reduced-order model for electrically actuated microplates.* Journal of Micromechanics and Microengineering, 2004. **14**(7): p. 900-906.
8. Vogl, G.W. and A.H. Nayfeh, *A reduced-order model for electrically actuated clamped circular plates.* Journal of Micromechanics and Microengineering, 2005. **15**(4): p. 684-690.
9. Batra, R.C., M. Porfiri, and D. Spinello, *Reduced-order models for microelectromechanical rectangular and circular plates incorporating the Casimir force.* International Journal of Solids and Structures, 2008. **45**(11–12): p. 3558-3583.
10. Nayfeh, A.H., M.I. Younis, and E.M. Abdel-Rahman, *Reduced-order models for MEMS applications.* Nonlinear Dynamics, 2005. **41**(1-3): p. 211-236.
11. Machauf, A., Y. Nemirovsky, and U. Dinnar, *A membrane micropump electrostatically actuated across the working fluid.* Journal of Micromechanics and Microengineering, 2005. **15**(12): p. 2309-2316.
12. Chao, P.C., C.-W. Chiu, and C. Tsai, *A novel method to predict the pull-in voltage in a closed form for micro-plates actuated by a distributed electrostatic force.* Journal of Micromechanics and Microengineering, 2006. **16**(5): p. 986-998.
13. Nayfeh, A.H. and M.I. Younis, *A new approach to the modeling and simulation of flexible microstructures under the effect of squeeze-film damping.* Journal of Micromechanics and Microengineering, 2004. **14**(2): p. 170-181.
14. COMSOL, http://www.comsol.com/.
15. Nayfeh, A.H. and D.T. Mook, *Nonlinear oscillations.* 2008: John Wiley & Sons.
16. Nayfeh, A.H. and P.F. Pai, *Linear and nonlinear structural mechanics.* 2008: John Wiley & Sons.
17. Younis, M.I., *MEMS Linear and Nonlinear Statics and Dynamics: Mems Linear and Nonlinear Statics and Dynamics.* Vol. 20. 2011: Springer.
18. Younis, M.I. and A.H. Nayfeh, *Simulation of squeeze-film damping of microplates actuated by large electrostatic load.* Journal of Computational and Nonlinear Dynamics, 2007. **2**(3): p. 232-241.
19. Alsaleem, F.M., M.I. Younis, and H.M. Ouakad, *On the nonlinear resonances and dynamic pull-in of electrostatically actuated resonators.* Journal of Micromechanics and Microengineering, 2009. **19**(4): p. 045013.
20. Alsaleem, F.M., M. Younis, and L. Ruzziconi, *An experimental and theoretical investigation of dynamic pull-in in MEMS resonators actuated electrostatically.* Microelectromechanical Systems, Journal of, 2010. **19**(4): p. 794-806.
21. Nayfeh, A.H., M.I. Younis, and E.M. Abdel-Rahman, *Dynamic pull-in phenomenon in MEMS resonators.* Nonlinear Dynamics, 2007. **48**(1-2): p. 153-163.

Identification of the operating parameters for the mechanical system using EMD algorithm

J. Jonak[1], A. Machrowska[1], J. Podgórski[2], and J. Bęc[2]

[1]Department of Machine Design and Mechatronics, Faculty of Mechanical Engineering, Lublin University of Technology, Poland
[2]Department of Structural Mechanics, Faculty of Civil Engineering and Architecture, Lublin University of Technology, Poland

Abstract. The subject of paper was a research on mechanical system with pneumatic coupling, consisting of a compressor turbine (T_I) and of so-called free driving turbine (T_{II}).The purpose of the tests is to determine the applicability of Empirical Mode Decomposition (EMD) algorithm to identify operating parameters of the mechanical system, which is necessary to build a system for structure condition monitoring with use of vibroacoustic method. There has been assumed possibility of easy separation of working frequencies of these components on order to estimate amplitudes generated by them.

1 Introduction

Identification of the operating parameters for the mechanical systems is crucial for the construction and development of monitoring and diagnosis systems.
Each of the elements of the monitored object is characterized by a number of frequency bands. Knowledge of these bands determines the correctness of the technical condition assessment. One of the most effective methods of appraisal provides the vibration analysis.
Input signal obtained as a result of measurements may constitute a nonlinear and non-stationary data set, resulting in significant limitations in case of application of commonly known analytical methods. The most of the conventional diagnostic methods are based on the assumption of operating parameters stationarity, thus they tend to be unreliable in case of operating parameters changes. As an interesting example of the difficulties associated with the identification of operating parameters, which occur during vibration analysis (performed by conventional techniques) we can indicate fluctuating load conditions [1] or changes in rotational speed of mechanical system components [2].
As a result of numerous literature reports with reference to the possibility of processing strongly nonlinear and non- stationary vibration signals, it was decided to test the EMD algorithm for recorded vibration signals. The object of research was the mechanical system, whose essential components were two turbines (T_I and T_{II} turbine) coupled pneumatically.

2 EMD algorithm

Adaptive technique Empirical Mode Decomposition (EMD) is a method of time-frequency analysis of multicomponent, nonlinear and non-stationary signals [3].
The basic assumption of this algorithm is to decompose the signal into a finite number of mode oscillations imbedded in the input signal called Intrinsic Mode Functions (IMFs). EMD analyses have found a wide range of applications [3-7].
EMD algorithm is based on the process of sifting the IMFs from given input data set. IMF obtaining is related to local extrema identification. The first phase of this method includes identification of all local extremes. Found local maxima are interpolated by cubic spline in order to build the upper signal envelope. The same procedure is used for the determination of the lower envelope (Figure 1).

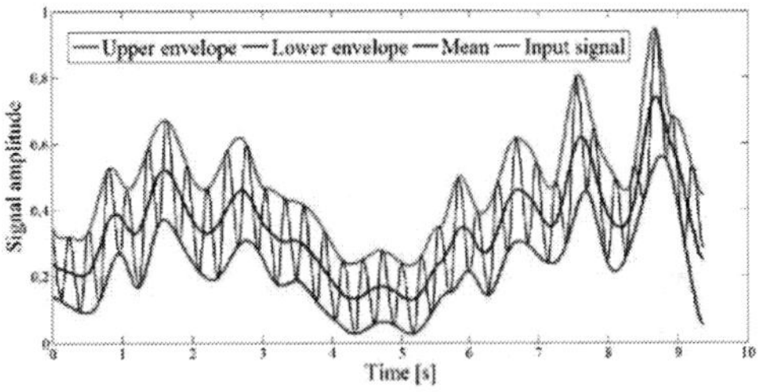

Fig. 1. EMD algorithm: input signal envelopes building

Subsequently their mean $m_1(t)$ is calculated at each point and the difference between input data $x(t)$ and already obtained $m_1(t)$ is determined.

$$h_1(t) = x(t) - m_1(t) \quad (1)$$

According to Huang [8], function $h_1(t)$ is an IMF, when it satisfies two conditions: in the whole data set, the number of extremes and the number of zero crossings must be either equal or differ at most by one, at any point the mean value $m_1(t)$ of the envelope defined by the local maxima, and the envelope defined by the local minima is zero.
Sifting procedure is repeated k times for each of the resulting difference $h_k(t)$, often referred to prototype mode or shortly: proto-IMF [9, 10].

$$h_n(t) = h_{n-1}(t) - m_n(t) \quad (2)$$

Process can be stopped when the condition of equality of the zero crossings and extremes number is fulfilled. Then it is determined a residual signal $r_n(t)$, described as:

$$r_n(t) = x(t) - \sum c_i(t), \quad (3)$$

where $c_i(t)$ is a i-th Intrinsic Mode Function and n expresses the largest number of mode oscillations possible to obtain.

3 Tests and results

3.1 Signal preprocessing

There have been recorded time series of acceleration for the analyzed mechanical system. These ones have been numerically elaborated (including integration), in order to conduct the velocity analysis. A typical course, at sampling rate of 12 kHz, for the received vibration velocity signal is presented in Figure 2.

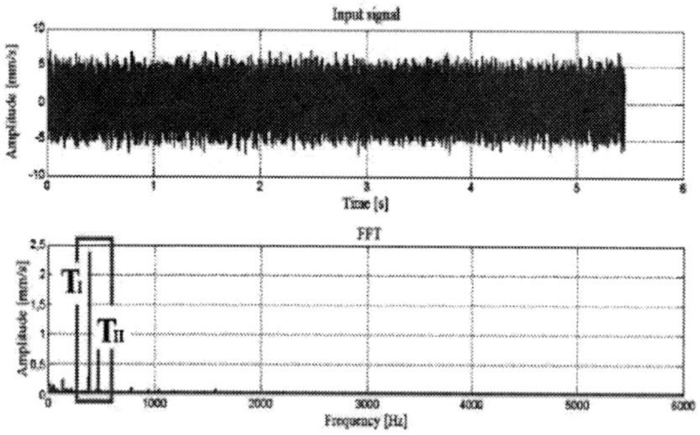

Fig. 2. Time series and spectral analysis of the vibration velocity

3.2 Results of EMD algorithm

As the result of using the EMD algorithm, the distribution of the signal at 15 IMFs and the residual signal has been found and shown in Figure 3.

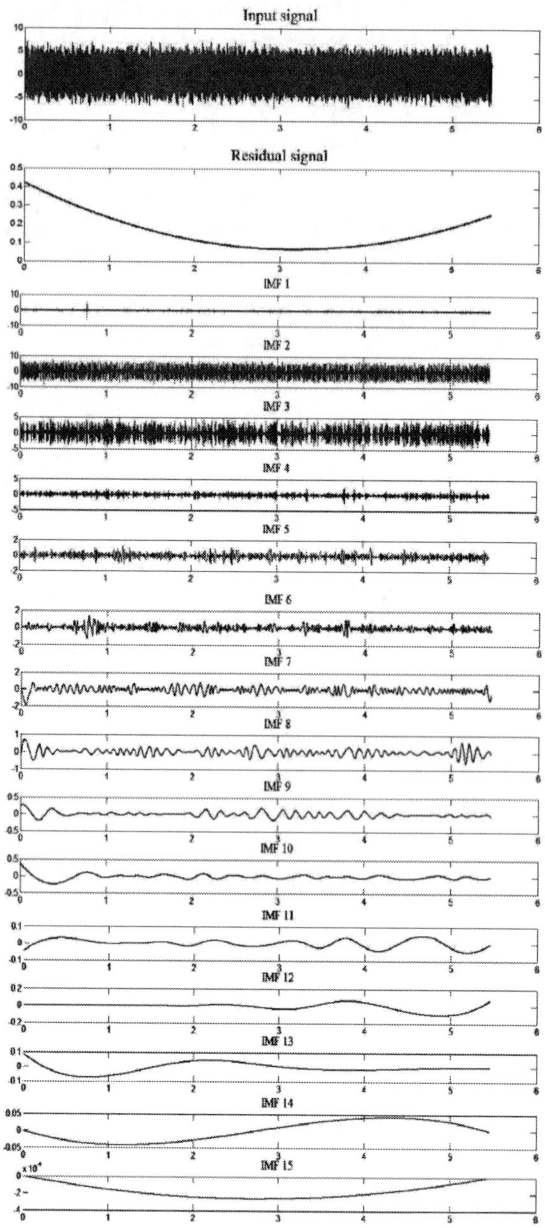

Fig. 3. Distribution of the analyzed signal at 15 IMF

In order to verify the results of decomposition for each of the previously extracted IMF functions, FFT (Fast Fourier Transform) analysis was carried out. Research has shown the occurrence of the energy "leakage" effects between IMF 2 and IMF 3 (Figure 4).

DOI: 10.1051/matecconf/20168305001

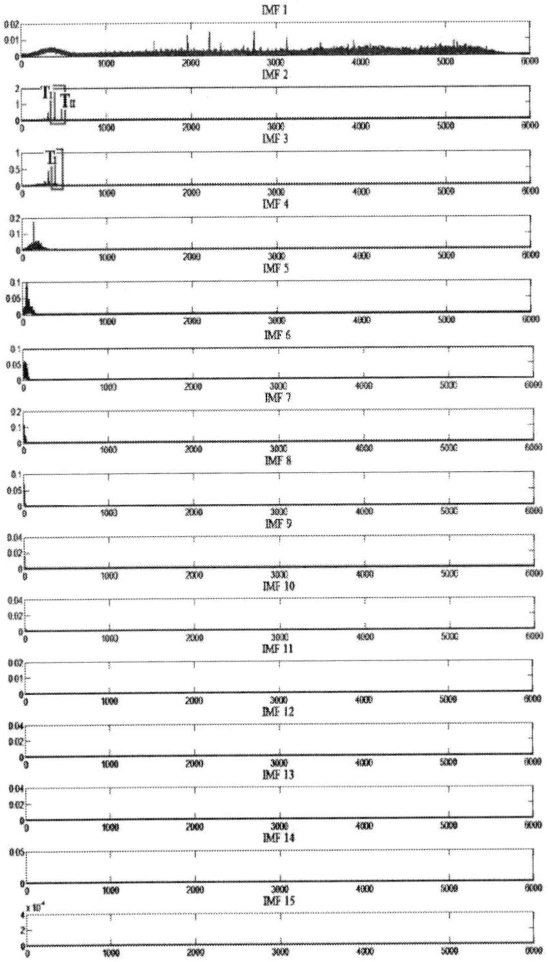

Fig. 4. FFT analysis of the IMF

3.3 EMD limitations

Within the framework of EMD algorithm, analytical test has shown that the procedure has some limitations. The first is limitation of the decomposition for the single data channel. The second is bounded resolution arising from the fact of performing the so-called "dyadic" filtration. Another one is observed "energy leakage" phenomenon. As the result, there is no successful separation of the characteristic frequencies for T_I and T_{II} turbines.

4 Conclusions

Characteristic for the analyzed mechanical system is the existence of significant angular velocity and amplitude fluctuations. This feature is observed especially for examined object with turbines being the basic components. In such a case, there is a considerable variability of amplitude ratios and rotational frequencies. Equally important is the variability of relative location of characteristic frequencies.

Further analyses showed that the ratio of the T_I and T_{II} frequencies (fT_I/ fT_{II}) was 1,22 with amplitude ratio (AT_I/ AT_{II}) equal to 0,29. According to the Feldman nomogram (Figure 5) [11], such a parameter set covers an area, where it is impossible to clearly separate signal components.

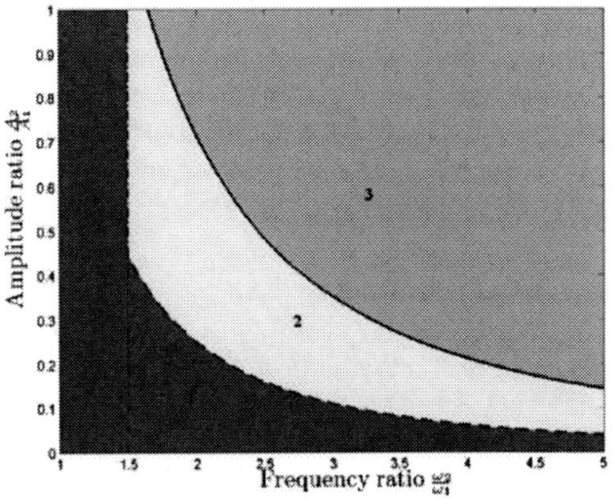

Fig. 5. Summary of the EMD ranges for harmonics separation: (1) impossible decomposition for very close frequency harmonics and small amplitude ratio; (2) decomposition requires several sifting iterations for close frequency harmonics; (3) single iteration separation for distant frequency harmonics and large amplitude ratio [11]

Further research is needed with the focus on the using of the selected sifting levels combinations (e.g. IMF2 and IMF3 band).
Mentioned combination can enable separation T_I and T_{II} signals as well as allow accurate estimation of vibration amplitude values of turbine components in the considered mechanical system.

References

1. C. J. Stander, P.S. Heyns, W. Schoombie, *Using vibration monitoring for local fault detection on gears operating under fluctuating load conditions*, Mechanical Systems and Signal Processing **16** 6 (2002)
2. M. Zhao, J. Lin, X. Xu, Y. Lei, *Tacholess envelope order analysis and its application to fault detection of rolling element bearings with varying speeds*, Sensors **13** 8 (2013)
3. H. Li, Y. Zhang, H. Zheng, *Bearing fault detection and diagnosis based on order tracking and Teager-Huang Transform*, Journal of Mechanical Science and Technology, **24** 3 (2010)
4. R. Ricci, P. Pennacchi, *Diagnostics of gear faults based on EMD and automatic selection of intrinsic mode functions*, Mechanical Systems and Signal Processing (2011)
5. J. Dybała, R. Zimroz, *Application of Empirical Mode Decomposition for impulsive signal extraction to detect bearing damage- industrial case study*, Condition Monitoring of Machinery in Non-Stationary Operations, Springer Berlin Heidelberg (2012)

6. P. Gaur, R.B. Pachori, H. Wang, G. Prasad, *An empirical mode decomposition based filtering method for classification of motor- imagery EEG signals for enhancing brain- computer interface*, Neural Networks (IJCNN), 2015 International Joint Conference on, IEE, **1-7** (2015)
7. Y. Yuan, C. Sun, M. Li, S.S. Choi, Q. Li, *Determination of optimal supercapacitor- lead- acid battery energy storage capacity for smoothing wind power using empirical mode decomposition and neural network*, Electric Power Systems Research, **127** (2015)
8. N.E. Huang, Z. Shen, S.R. Long, M.C. Wu, H.H. Shih, Q. Zheng, H.H. Liu, *The empirical mode decomposition and the Hilbert spectrum for nonlinear and non- stationary time series analysis*, Proceedings of the Royal Society of London A: Mathematical Physical and Engineering Sciences, **454** 1971, The Royal Society (1998)
9. J. Dybała, R. Zimroz, *Rolling bearing diagnosing method based on empirical mode decomposition of machine vibration signal*, Applied Acoustics **77** (2014)
10. D. Yang, H. Li, Y. Hu, J. Zhao, H. Xiao, Y. Lan, *Vibration condition monitoring system for wind turbine bearings based on noise suppression with multi- point data fusion*, Renewable Energy **92** (2016)
11. M. Feldman, *Analytical basics of the EMD: Two harmonics decomposition*, Mechanical Systems and Signal Processing, **23** 7 (2009)

MATEC Web of Conferences **83**, 05002 (2016)
CSNDD 2016
DOI: 10.1051/matecconf/20168305002

Wind tunnel study of axial force variation in guys of flexible mast models

J. Bęc[1], E. Błazik-Borowa[1] and T. Lipecki[1]

[1]Lublin University of Technology, Department of Structure Mechanics, Nadbystrzycka 38d St., 20-618 Lublin, Poland

Abstract. Measurements of the axial force variation in guys of mast model subjected to wind load of three different wind structures are presented in the paper. The measurements have been carried out in the wind tunnel of Wind Engineering Laboratory at Cracow University of Techn ology. The subjects of the research are two guyed masts of triangular and square shaft cross-sections. Mean values of forces and standard deviations as functions of wind parameters and wind attack angle have been calculated. Significant influence of wind structure on the forces in guys has been confirmed as the result. Additionally there has been noticed possibility of mast vibrations at critical velocities.

1 Introduction

Guyed masts are used as supporting structures for radio and television antennas and mobile phone transmitters. These media are still being developed and therefore we can expect more and more masts to be built. Problems of loads evaluation for guyed masts, static and dynamic analyses of such structures, as well as design methods are still up-to-date scientific problems.

The paper presents a study of axial force variation in guys of flexible mast models under the wind load, and indicates that the structure of the wind has significant effect on their values. This issue and others (see [1, 2, 3]) related to the wind action on masts, are very important for designers of such structures.

2 Description of the research methods

The measurements were carried out in the wind tunnel of Wind Engineering Laboratory at Cracow University of Technology. The research was conducted on mast models with a height of 1.0 m, with two levels of guys and two types of shaft cross-sections: square and triangular (Fig.1). The models of masts were rotated during the tests, for the mast of square cross-section, the following angles of wind attack α (the angle between the plane formed by the guy and shaft axis, and the vertical windward plane) were used: 0°, 45°, 90°, 135° and 180°, while for the mast of triangular cross-section the α angles were: 0°, 30°, 60°, 90°, 120°, 150° and 180°.

The measurements were performed at 16 wind speed levels equally spread in the range from 0 m/s to 20 m/s. Three wind structures, described in detail in [3, 4], were used as the inflow condition (Fig. 2). The measurements of forces in guys were made with use of HBM Hottinger S2/100 force transducer with the range of measured forces: -100 ÷ +100 N.

a)

b)

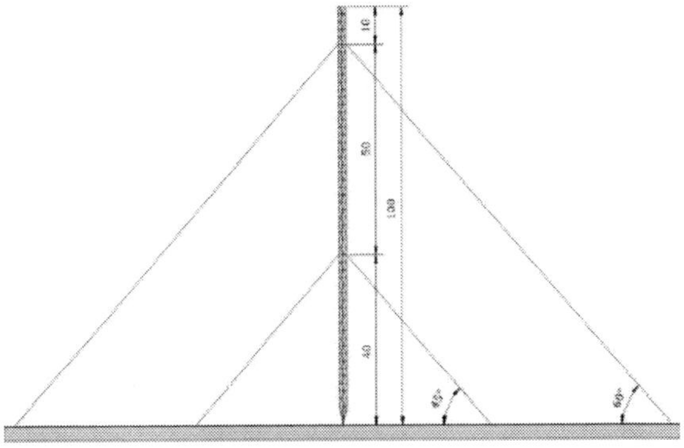

Fig. 1. Geometry: a) view of the model in the wind tunnel, b) model dimensions [cm].

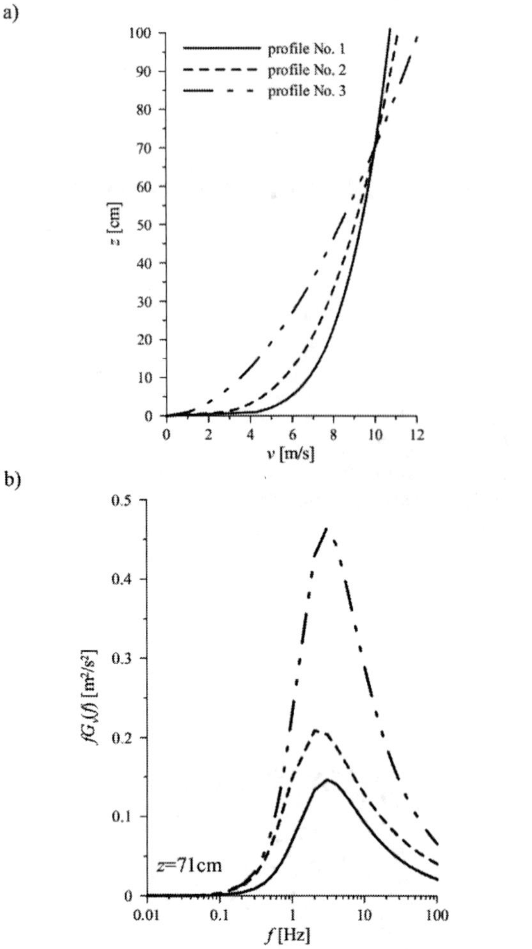

Fig. 2. Wind parameters: a) mean wind speed profile, b) power spectral density of wind; where v – velocity, f – frequency, G - power spectral density, z – height.

3 Results

Time series of the axial forces in guys caused by the wind were obtained as the result of the measurements. Exemplary measurement results for the triangular shaft mast at v=10 m/s have been shown in Fig. 3. The presented results do not show the total value of axial force but only the difference between the tensional force value F_t and the initial force in a guy F_o. The positive value (F=F_o-F_t>0) means that the force is diminished, while the negative value (F=F_o-F_t<0) stands for the increasing force.

Mean values (Fig. 4) and standard deviations (Fig. 5) were calculated for each time series in relation to the wind attack angle and mean wind speed. Figures 6, 7 and 8 were put down in order to analyze the wind structure influence on the forces in guys. Influence of mean wind speed profiles on the mean values and standard deviations of forces in guys at the mean wind speed of v=10m/s at measured the height 70 cm can be observed at these graphs.

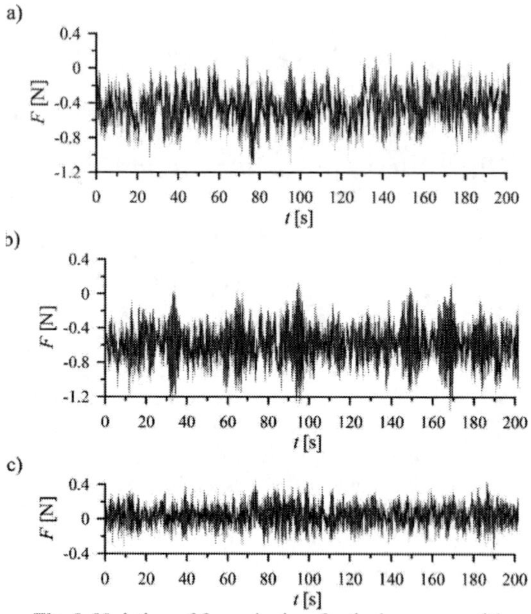

Fig. 3. Variation of forces in time for the lower guy of the square mast at the wind profile No. 3 at v = 10 m/s for: a) α =0°, b) α = 45°, b) α = 90°

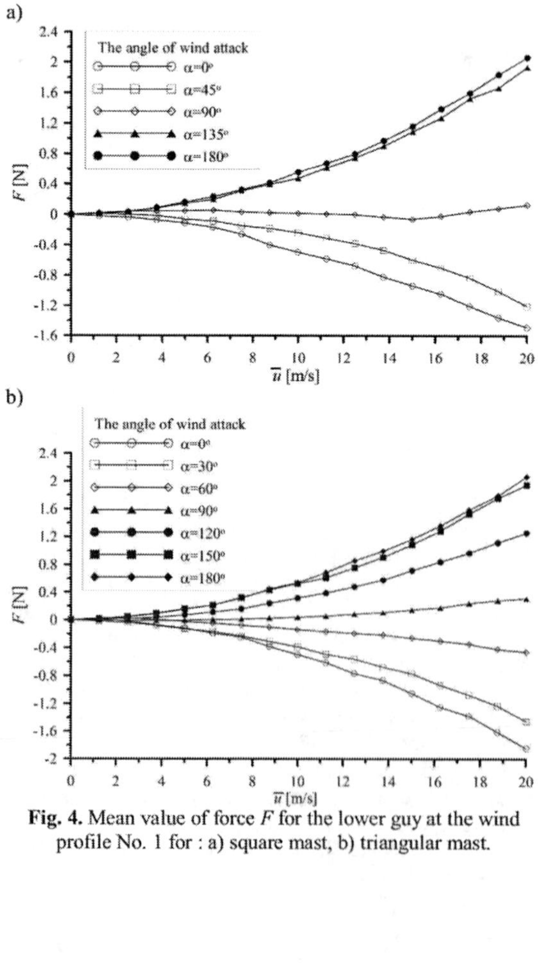

Fig. 4. Mean value of force F for the lower guy at the wind profile No. 1 for : a) square mast, b) triangular mast.

DOI: 10.1051/matecconf/20168305002

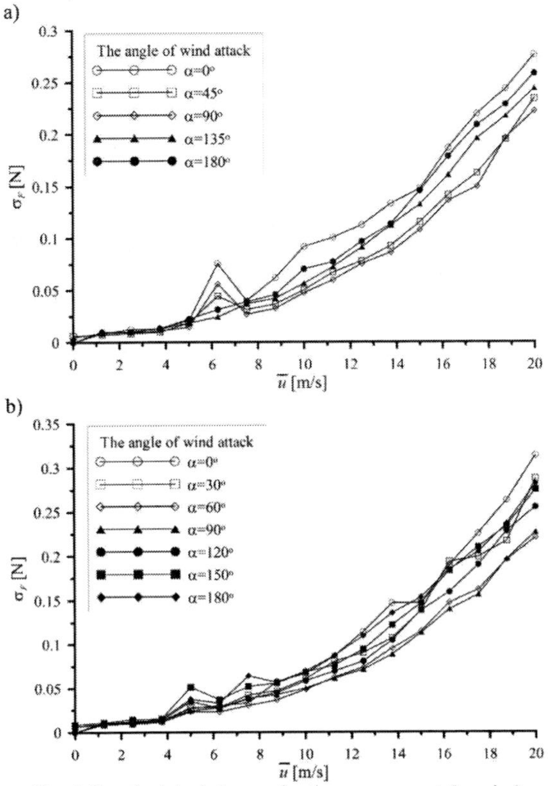

Fig. 5. Standard deviation σ_F for the upper guy at the wind profile No. 1: a) square mast, b) triangular mast.

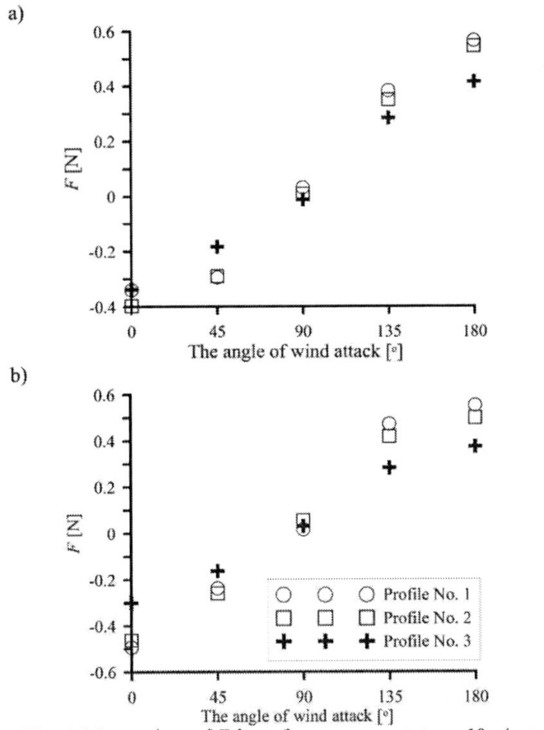

Fig. 6. Mean values of F force for square mast at v = 10m/s at different angles of wind attack: a) upper level of guys, b) lower level of guys.

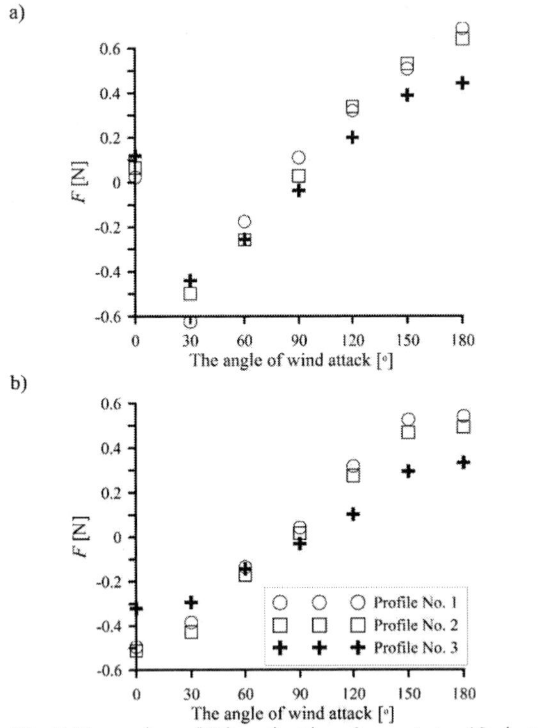

Fig. 7. Mean values of F force for triangular mast at v=10m/s at different angles of wind attack: a) upper level of guys, b) lower level of guys.

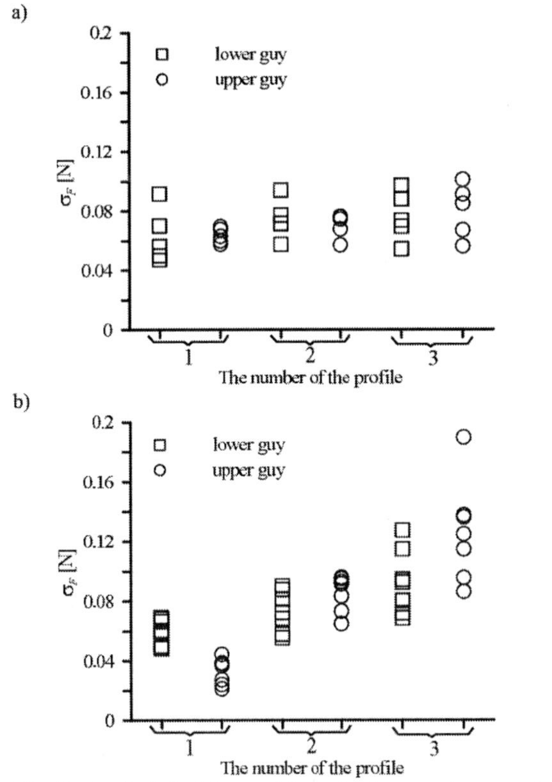

Fig. 8. Standard deviations of the axial force σ_F in guys at v = 10 m/s: a) square mast, b) triangular mast.

DOI: 10.1051/matecconf/20168305002

Under the action of wind the forces in windward guys increase. If we rotate the mast, differences between forces F_o and F_t decrease until the guy is in the plane perpendicular to the direction of flow. At this setting the force is small in comparison to other positions. When the angle of wind attack α is greater than 90° the difference between forces F_o and F_t is positive (force in the guy is smaller than the initial tension) and its value increases to achieve the biggest value at $\alpha = 180°$.

There are peaks of standard deviations noticeable at the graphs in Fig. 5 at some mean values of wind speeds. They are more evident at higher numbers of wind profiles, which means higher values of turbulence. The peaks are observed in graphs referring to all guys of mast, nevertheless if the guys are located in the planes along or across the mean wind. This means that the top of the mast is moving in both directions. This last conclusion means that the force variation is generated by the vibration of the mast shaft, rather than the vibrations of guys only. Increasing vibration amplitudes which follow the increase in turbulence means that the mast vibrations are caused by buffeting.

There are changes of guy forces at different wind profiles shown in Figures 6 and 7. It can be noticed, that the higher number of the wind profile and higher turbulence intensity give smaller changes in guy forces. The difference of forces in windward and leeward guys may reach even up-to 40 %.

The wind structure influences the standard deviations, as well. This can be seen in Fig. 8. In both cases of the square shaft cross-section, as well as the triangular one, the higher values of turbulence intensity generate higher values of standard deviations of guy forces. However, it is much more visible in case of the triangular shaft, than the square one.

4 Conclusions

The study shows that the structure of the wind has a great influence on the value of wind load on guys of masts. The differences in the mean values of normal forces in guys, exposed to the different wind profiles, are up to 40% for the lowest values of the force. The standard deviations for the normal forces also depend on the structure of the wind. The higher peak value at the abscissa in the graph of power spectral density produces greater maximum value of the standard deviation of the normal force. The study also shows that the values of forces in guys are mainly influenced by the work of the mast itself. There is much smaller influence on the axial forces of the wind forces applied directly to the guys. This is confirmed by the fact that if the guys are set in the plane perpendicular to the mean wind direction, axial forces are negligibly small when compared to ones obtained at other rotations.

In this study, there has been observed critical velocity, which produces mast vibrations and as the result, higher amplitudes of variation of axial forces in guys. This phenomenon requires further research mostly on the description of the origin of the critical velocity and the ways to avoid this occurring.

References

1. P. Harikrishna, A. Annadurai, S. Gomathinayaga m, N. Lakshmanan, Full scale measurements of the structural response of a 50 m guyed mast under wind loading, Engineering Structures 25 (2003) 859–867.
2. S. Fabre, M. Stickland, T. Scanlon, A. Oldroyd, D. Kindler, F. Quail, Measurement and simulation of the flow field around the FINO 3 triangular lattice meteorological mast, J. WindEng.Ind.Aerodyn.130 (2014) 99–107.
3. J. Bęc, T. Lipecki, E. Błazik-Borowa, Pomiary drgań aeroelastycznych modeli masztów z odciągami w tunelu aerodynamicznym, Budownictwo i Architektura 12 (1) (2013) 211–218 (in Polish)
4. J. Bęc, T. Lipecki, E. Błazik-Borowa, Research on wind structure in the wind tunnel of Wind Engineering Laboratory of Cracow University of Technology, Journal of Physics: Conference Series 318 (2011), doi: 10.1088/1742-6596/318/7/072003

Efficient asymptotic basis to reduce the forced dynamic problem of viscoelastic sandwich plates

F. Boumediene[1] , E.M. Daya[2,3] , J.M. Cadou[4] , L. Duigou[4]

[1] Faculté de Génie Mécanique & Génie des Procédés, USTHB, BP 32, El Alia, 16111 Bab Ezzouar, Alger, Algeria,
[2] Lab d'Etude des Microstructures et Mécanique des Matériaux, UMR CNRS 7239, Université de Lorraine, Metz, France
[3] Lab of Excellence on Design of Alloy Metals for low-mAss Structures 'DAMAS', University of Lorraine, Metz, France.
[4] Institut de Recherche Dupuy de Lôme, FRE CNRS 3744, IRDL, F-56100 Lorient, France

Abstract. In this paper, an efficient asymptotic basis is proposed to reduce the forced dynamic problem of viscoelastic sandwich plates. The numerical resolution is based on the asymptotic numerical method (ANM) and finite element method (FEM). Numerical tests have been performed in the case of sandwich plates with Young's modulus variable with respect to the frequency. The comparison of the results obtained in the reduced order model with those given in the full order model shows both a good agreement and a significant reduction in computational cost.

1 Introduction

Viscoelastic sandwich materials are often used as a passive solution to reduce vibrations and noise in many domains. To increase the damping in structures, the viscoelastic material is often sandwiched between identical elastic layers. The dynamic analysis of this type of structures subjected to a harmonic load excitation **F**, using the finite element method, can be written in the following form [1]:

$$[\mathbf{K_0} + E(\omega)\mathbf{K_v} - \omega^2\mathbf{M}]\mathbf{U} = \mathbf{F} \tag{1}$$

where ω is the vibration circular frequency, $\mathbf{U}$ is the complex displacement vector of dimension [*ND*]; $\mathbf{K_0}$ and $\mathbf{K_v}$ are real constant stiffness matrices of dimension [*ND*×*ND*], $E(\omega)$ is the complex Young's modulus of the viscoelastic core, **M** is the global mass matrix of dimension [*ND*×*ND*] and **F** is the vector of the excitation amplitude of dimension [*ND*].

Resolution of this type of problem has remained relevant to researchers until now, because of its difficulty. The problem is that all resolution methods need triangulation of Jacobean matrix which is of complex type. In addition, if the displacement vector is written with separating the real and imaginary parts as follows: $\mathbf{U} = \{\mathbf{U^R}, \mathbf{U^I}\}$; the equation (1) will have two time the number of degrees of freedom (*dof*) of equivalent real problem. Then resolution could lead to high computational cost in the case of large scale structures.

Recently, we have proposed a new basis to reduce this type of problem in the cases of beams [2] and plates with constant Young's modulus [3]. In this paper, this reduction basis is improved to be applied in the case of sandwich structure with Young's modulus dependent on the frequency. For the numerical computation, structures are discretized using shell element with eight nodes and eight degrees of freedom by node [5]. The reduction method is combined to the asymptotic numerical method (ANM) [4] to compute the response of viscoelastic sandwich plate.

2 Asymptotic numerical method (ANM)

The dynamic governing equation can be written in the following form:

$$\mathbf{R(X)} = \mathbf{0} \tag{2}$$

where **R** is the residual vector of dimension [*ND*], the vector $\mathbf{X} = (\mathbf{U}, \omega, E)$ contains all the unknowns of the problem.

The resolution of equation (2), using ANM, are widely explained in different papers [2,4]. Here we explain only the principle.

Within the ANM we can solve the residual equation (2) in a step by step manner, where the unknown

$\mathbf{X}$ is sought in the form of truncated power series with respect to a path parameter " a " [2]:

$$\mathbf{X}(a) = \mathbf{X_0} + \sum_{p=1}^{n} a^p \mathbf{X_p} \quad (3)$$

where $\mathbf{X_0}$ is a regular initial solution of equation (2), $\mathbf{X_p}$ is the unknown, at each order p (p = 1, n) to be computed and n is the truncation order.

Series (3) are inserted in equations (2). Then, from the identifications of the like powers of 'a', a set of recurrent linear problems is obtained at each order p as follows:

$$\mathbf{K_t X_p} = \mathbf{F_p^{nl}} \quad (4)$$

where $\mathbf{K_t}$ denotes the Jacobian matrix at the initial point $\mathbf{X_0}$ and $\mathbf{F_p^{nl}}$ is the second member vector at order p. The second member is known since it depends only on known parameters $\mathbf{X_i}$ computed in precedent orders (i = 1, p-1).

The limit of validity of series (3) can be computed automatically using the tolerance ε. Then, resolution of one step needs one matrix triangulation and resolution of p linear systems (4). This leads to a significant computational time. For that, we propose to use reduction model.

3 Reduction model

To build the projection matrix, one ANM step without reduction is realized and vectors at each order $\mathbf{U_i}$ issued from series (3) are saved to build the basis $\mathfrak{R}$. After that, the computation is carried out in a reduced form. The reduction consists of projecting the unknown vector onto a small basis as follows:

$$\mathbf{U} = \mathfrak{R}.\mathbf{u} \quad (5)$$

where $\mathfrak{R}$ is the projection matrix of dimension $[ND \times nd]$ and $\mathbf{u}$ is the reduced vector of dimension $[nd]$.

Two reduction algorithms are tested:

- In the first one (noted RED1), the reduction is applied directly to equation (1): After replacing the vector $\mathbf{U}$, by its expression (5), and multiplying the left side by $\mathfrak{R}^\mathbf{T}$; the governing equation to resolve becomes smaller. Its dimension is now $[nd]$ with $(nd \ll ND)$:

$$[\mathbf{k_0} + E(\omega)\mathbf{k_v} - \omega^2 \mathbf{m}]\mathbf{u} = \mathbf{f} \quad (6)$$

where $\mathbf{k_0} = \mathfrak{R}^\mathbf{T}\mathbf{K_0}\mathfrak{R}$, $\mathbf{k_v} = \mathfrak{R}^\mathbf{T}\mathbf{K_v}\mathfrak{R}$, $\mathbf{m} = \mathfrak{R}^\mathbf{T}\mathbf{M}\mathfrak{R}$ and $\mathbf{f} = \mathfrak{R}^\mathbf{T}\mathbf{F}$

Then, the solution of the reduced problem (6) is performed by the ANM but only reduced matrices and vectors are considered (Figure 2). This procedure works well in the case of beam [2].

- In the second one (noted RED2), the reduction is applied to resolve only equation (4) in the application of the ANM to the problem. Then only matrix decomposition is done in the reduced form. The second member is computed in a complete order (Figure 2). After that, we reduce it to resolve the problem (4) in a reduced forme, as follow:

$$\mathbf{k_t x_p} = \mathbf{f_p^{nl}} \quad (7)$$

where $\mathbf{k_t} = \mathfrak{R}^\mathbf{T}\mathbf{K_t}\mathfrak{R}$ and $\mathbf{f_p^{nl}} = \mathfrak{R}^\mathbf{T}\mathbf{F_p^{nl}}$
This procedure was applied to a viscoelastic material with a constant Young's modulus [5].

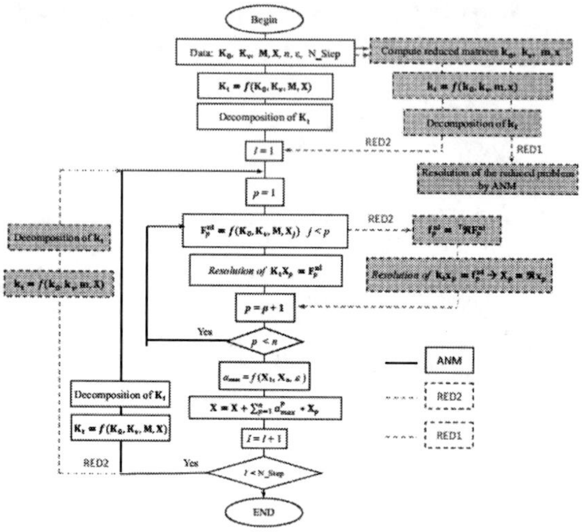

Fig. 1 Resolution chart

4 Numerical examples

The first plate (plate 1) has two clamped edges with dimensions (348×304.8×3.175 mm^3) (Figure 2). Its characteristics are given in Table 1. The Young's modulus is supposed variable with respect to the frequency following the generalized Maxwell model:

$$E(\omega) = k_0 + \eta_0.i\omega + \sum_{j=1}^{N_{max}} i\omega / \left(\frac{i\omega}{k_j} + \frac{1}{\eta_j}\right) \quad (8)$$

where k_j and η_j are material Maxwell constants (given experimentally) and 'i' is the complex number ($i^2 = -1$).

The plate is discretized into 173 nodes (ND =2768 d.o.f, real and imaginary parts). The transverse excitation force (F=100N) is applied to the center of the plate.

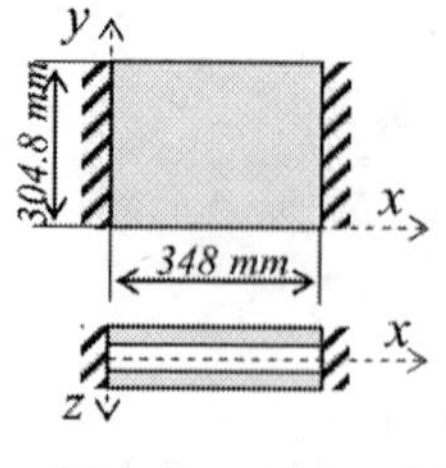

Fig. 2. Clamped-free-clamped-free plate (Plate 1)

Table 1. Material properties

Material properties	Elastic layers	Viscoelastic layer
Young's modulus [N/m^2]	0,21.10^{12}	0,2722.10^7

Poisson's ratio	0,3	0,44
Density [kg/m³]	7800	1200
Thickness [mm]	1,524	0,127

Results from the full order model (using ANM without reduction) and those from the reduced order model with basis containing twenty vectors (*nd*=40 real and imaginary parts), are superposed for comparison (Figure 3). To confirm the validity of our program, results are compared with those computed directly by inversing the tangent matrix at each frequency as follows:

$$\mathbf{U}(\omega) = [\mathbf{K_0} + E(\omega)\mathbf{K_v} - \omega^2\mathbf{M}]^{-1}.\mathbf{F} \quad (9)$$

Figure 3, represents the magnitude of the transverse displacement evolution with respect to frequency at the load position. Figure 4, presents the residual value evolution in a logarithm scale where the residual value is computed as follows:

$$RES = \|\mathbf{R(X)}\|/\|\mathbf{K_0}.\mathbf{U}\| \quad (10)$$

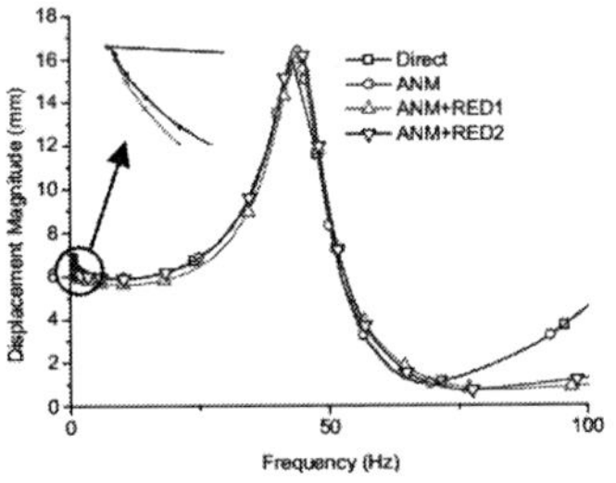

Fig. 3. Transverse displacement magnitude (Plate 1)

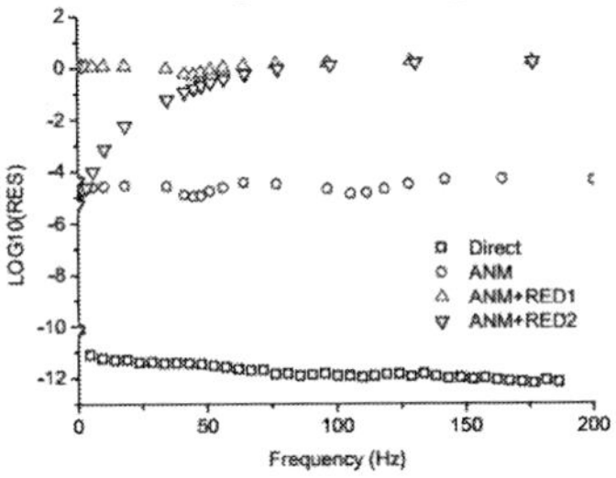

Fig. 4. Residual value evolution (Plate 1)

These figures show good agreement between results with the 2nd strategy (RED2) and ANM results around the first mode until a frequency equal to 50Hz. However, the first strategy (RED1) gives a large error. From study of the computational time, we found that the matrix triangulation consumes more time in one step. For example, the resolution step time is equal to (4.84E-02) the decomposition time for a problem with (2768 dof) and 0.004 the decomposition time for a problem with (9104 dof)

Then avoiding triangulation of full order matrices is sufficient to reduce significantly the computational time. Therfore, we prefer keeping the second strategy which consists of reducing only the linear equations at each order *p*. But we also remark, that this strategy needs a basis updating.

To control the solution accuracy, the basis is updated if the residual value is greater than the tolerance (1.E-4), we proceed as follows:

1. recalculate a new exact solution (9)
2. do one ANM step in a full order model,
3. update the basis.
4. resolution in reduced form and verify residual value. If it is greater than the tolerance return to 1.

Following this procedure, we found good result (Figure 5) but five update steps are needed (Figure 6). Then, this requires five computations in full size model. So to decrease this number, it is possible to correct the solution at each order to avoid error accumulation. This is the objective of the next section.

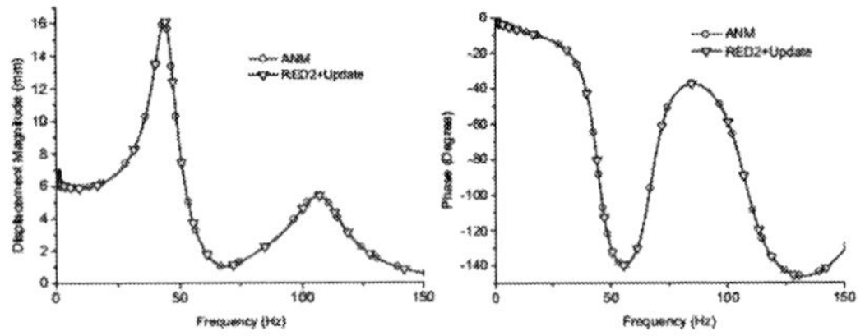

(a) magnitude (b) Phase

Fig. 5. Transverse displacement with basis update (Plate 1)

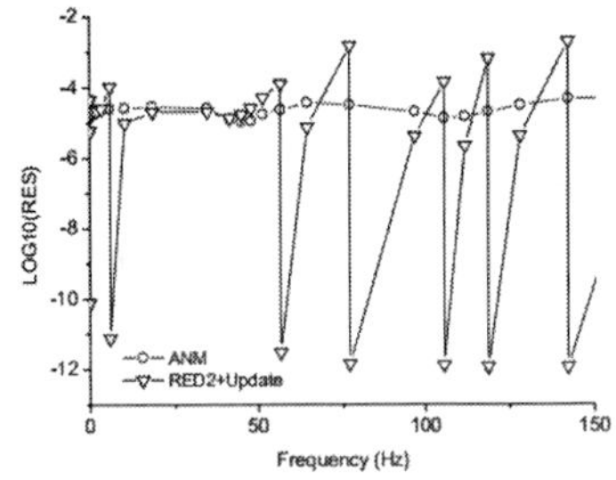

Fig. 6. Residual value evolution with basis update (Plate 1)

6 Reduction-correction techniques

In this section, we present the method used to improve the validity of the reduced solution. We consider the reduced solution, $\mathrm{X_p^{red}}$, as already known at the order of truncation p. This vector is issued from resolution of the problem (7). The solution of the full-size problem at the order p can be defined by:

$$\mathbf{X_p} = \mathbf{X_p^{red}} + \Delta\mathbf{X_p} \quad (11)$$

where $\Delta\mathbf{X_p}$ is the error.

We replace this expression in equation (4):

$$\mathbf{K_t}(\mathbf{X_p^{red}} + \Delta\mathbf{X_p}) = \mathbf{F^{nd}} \quad (12)$$

Then, the error $\Delta\mathbf{X_p}$ can be computed as follows (13):

$$\mathbf{K}_t(\Delta \mathbf{X}_p) = \mathbf{F}^{nd} - \mathbf{K}_t(\mathbf{X}_p^{red}) \tag{13}$$

Instead of taking the true matrix $\mathbf{K}_t$, we use the factorized matrix used in the first full-size computation where the basis has been defined. This procedure can be repeated until an accepted residual value is obtained. With this procedure, we success to find all the solution curves of figures 5 and 6 with only three basis updates (Figure 7).

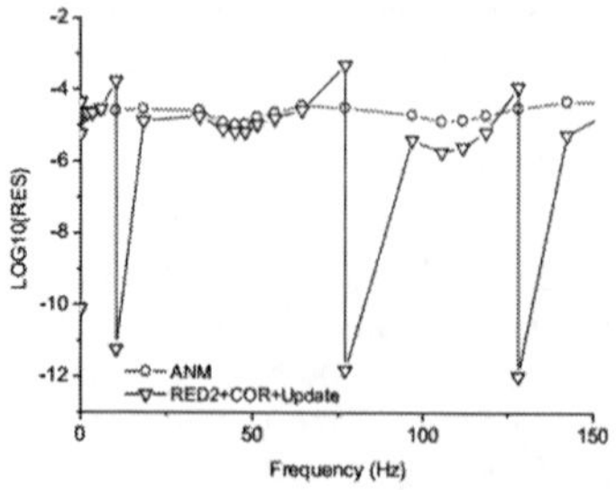

Fig. 7. Residual value evolution with correction step and basis update (Plate 1)

The method is now applied to simply supported plate (figure 8). Geometrical and material characteristics are the same as the plate 1. We remark the good agreement between the Full order model (FOM) and the reduced order model (ROM) results (Figure 9) with only three basis updates (Figure 10).

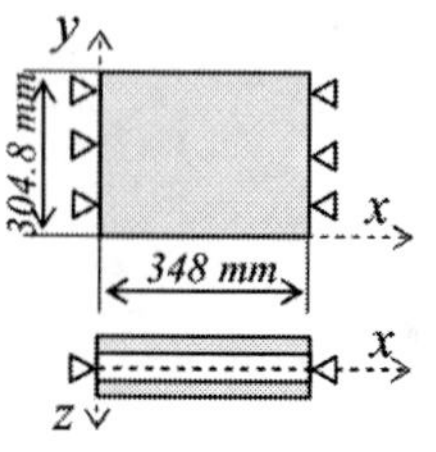

Fig. 8. Plate 2

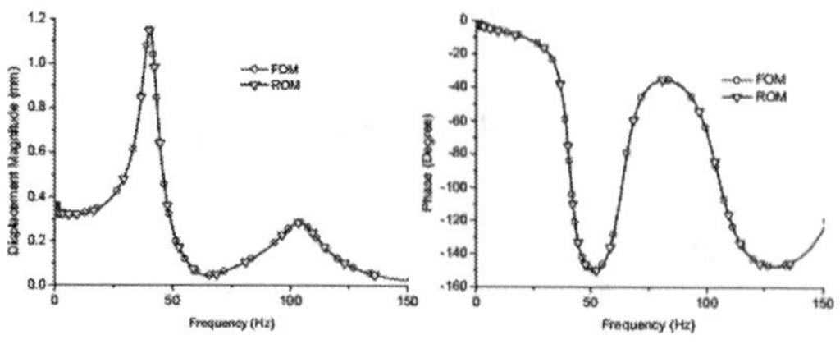

(a) magnitude (b) Phase

Fig. 9. Transverse displacement with correction step and basis update (Plate 2)

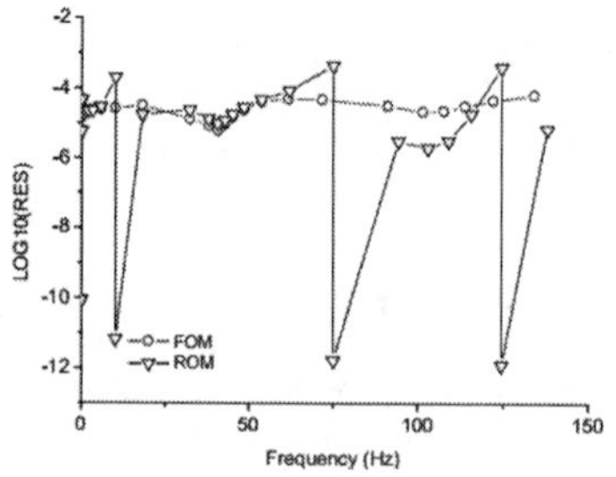

Fig. 10. Residual value evolution with correction step and basis update (plate 2)

8 Conclusion

Many tests have been performed on plates with different boundary conditions and viscoelastic properties. All examples treated confirmed that the proposed reduced basis is efficient and reproduces perfectly the results issued from the full order model. In addition, a significant computational gain is obtained by using the reduction-correction procedure, up to 80% with considering the time needed to build the basis and to compute the reduced initial vectors and matrices.

References

1. E.M. Daya, M. Potier-Ferry. A numerical method for nonlinear eigenvalue problems application to vibrations of viscoelastic structures. *Computers and Structures* **79**, 533-541 (2001)
2. F. Boumediene E.M Daya, J.M Cadou & L. Duigou. Forced harmonic response of viscoelastic sandwich beams by a reduction method. *Mechanics of Advanced Materials and Structures*. **23**(11), 1290-1299 (2016)
3. J.M. Cadou, F. Boumediene, Y. Guevel, G. Girault, L. Duigou, E.M. Daya, M. Potier-Ferry. A high order reduction-correction method for Hopf bifurcation in fluids and for viscoelastic vibration. *Computational Mechanics* **57**, 305–324 (2016)
4. F. Abdoun, L. Azrar, E.M. Daya, M. Potier -Ferry. Forced harmonic response of viscoelastic structures by an asymptotic numerical method. *Computers and Structures* **87**, 91–100 (2009)
5. L. Duigou, E.M Daya, M. Potier-Ferry. Iterative algorithms for nonlinear eigenvalue problems. Application to vibrations of viscoelastic shells. *Computer Methods in Applied Mechanics and Engineering* **192**, 1323–1335 (2003)

Low cost metamodel for robust design of periodic nonlinear coupled micro-systems

K. Chikhaoui[1,2], N. Bouhaddi[1], N. Kacem[1], M. Guedri[2], and M. Soula[2]

[1] FEMTO-ST Institute, UMR 6174, Department of Applied Mechanics, University of Franche-Comté, UBFC, 24 Rue de l'Epitaphe, 25000 Besançon, France
[2] National High School of Engineers of Tunis (ENSIT), University of Tunis, Tunis, Tunisia

Abstract. To achieve robust design, in presence of uncertainty, nonlinearity and structural periodicity, a metamodel combining the Latin Hypercube Sampling (LHS) method for uncertainty propagation and an enriched Craig-Bampton Component Mode Synthesis approach (CB-CMS) for model reduction is proposed. Its application to predict the time responses of a stochastic periodic nonlinear micro-system proves its efficiency in terms of accuracy and reduction of computational cost.

1 Introduction

To design large-scale systems incorporating parametric uncertainties with an ever increasing performance, it is inevitable to apply uncertainty quantification and propagation methods to evaluate the influence of the randomness of the structural input parameters on the dynamic behavior of a system. In a probabilistic framework [1], the most frequently-applied stochastic methods are the sample-based methods which consist on a succession of deterministic simulations corresponding to a set of random variables modeling the parametric uncertainties. The Monte Carlo (MC) [2] and the Latin Hypercube Sampling (LHS) [3] methods are the most well-known sample-based methods. Nevertheless, in spite of the flexibility or their implementation, these methods require a great number of samples for a reasonable accuracy and consequently a prohibitive computational cost. Therefore, it is interesting to combine them with reduction methods. In presence of uncertainties and localized nonlinearities, standard reduction methods are no longer efficient since they do not contain any information about these structural modifications. To overcome this issue, the transformation basis used in the Component Mode Synthesis (CMS) approach to reduce each component of the structure can be enriched by taking into account stochastic and nonlinear effects. This advantage enables considering some components including uncertainties and/or localized nonlinearities subject to be reduced independently of the others. Moreover, when considering the case of periodic structures, the CMS approach simplifies the analysis by considering each cell as a component and permits subsequently to attain more important computing time gain. In this work, we focus on the well-known Craig-Bampton CMS (CB-CMS) [4]. In the literature, enriching the Ritz basis with static contribution of neglected eigenvectors was proposed in [5] for accurately evaluating frequency response of modified structures. Masson [6] extends the standard reduction basis by a set of optimized static residual vectors both for direct reductions and CMS approaches. Bouazizi et al. [7] associated the equivalent linearization method with a reduction basis enriched by static residual vectors accounting for parametric modifications and localized nonlinearities. In [8] the CB transformation is enriched with a set of static residual vectors calculated using a priori knowledge of the initial component designs. In [9], to take into account geometrical nonlinearities, the CMS method is enriched by adding the most important modal derivatives, describing second order nonlinear contributions of vibration modes perturbed with the shapes of some others.

Furthermore, combining reduction methods and uncertainty propagation methods was successfully applied in [10] by propagating uncertainties using perturbation techniques at the component level or the global level in a CB-CMS framework. In [11], the CMS is combined with the stochastic spectral finite element method to compute frequency linear structures responses. In [12], the CMS methods are coupled to an adaptive polynomial chaos expansion in order to investigate the frequency transfer functions for large finite element systems with linear and nonlinear stochastic parameters. A robust design optimization is obtained in [13] by combining multi-objective optimization techniques, reduced order method for cheap function evaluations and the probabilistic collocation method for uncertainty quantification.

The emphasis of this paper is to develop a metamodel which is robust against uncertainties and nonlinearities and which takes advantage of the structural periodicity. Its efficiency is evaluated in the light of the proposed numerical application which consists on the computation of the time responses of a periodic coupled micro-beams structure containing stochastic parameters and localized nonlinearities. The robustness of the metamodel is evaluated in terms of accuracy and computational time reduction ratio with respect to the reference solutions computed using the LHS method implemented on the full finite element model.

2 Proposed metamodel

The sample-based LHS method [3] for uncertainty propagation consists on computing N successive deterministic responses $\{y(\xi^{(n)}), n = 1, \dots, N\}$ according to a set of N samples of random variables. The choice of the LHS method is due to its reduced computing time with respect to the very time-consuming MC method. In fact, a great number of samples are typically required in MC to achieve good accuracy while partitioning the variability space into regions of equal probability and picking up one sampling point in each region allow reducing the number of samples needed for the LHS method implementation.

In this paper, the proposed metamodel consists on combining the LHS method and a model reduction method by projecting the responses on the robust enriched basis defined hereafter.

In general case, analyzing the time domain behavior of a nonlinear mechanical system consists on solving the differential equation of motion

$$\begin{cases} M\ddot{y} + B\dot{y} + Ky + f_{NL}(y, \dot{y}) = f_{ext} \\ y(0) = y_0, \dot{y}(0) = \dot{y}_0 \text{ initial conditions} \end{cases} \quad (1)$$

where K, M and B stand for the stiffness, mass and damping matrices of the system, f_{NL} the nonlinear forces vector and f_{ext} the exciting forces vector.

Moving from full finite element model to reduced order model requires projecting the time response on a reduction basis, considering the variable transforms $y(t) = Tq(t)$ to modal coordinates. Hence, the equation of motion (1) becomes

$$T^T M T\ddot{q}(t) + T^T B T\dot{q}(t) + T^T K T q + T^T f_{NL}(q, T\dot{q}) = T^T f_{ext} \quad (2)$$

Solving the equation (2) in the time domain is done using the nonlinear Newmark time integration scheme and the Newton-Raphson prediction-correction algorithm [14, 15].

In deterministic linear case, the reduced model is obtained using standard CB-CMS of the linear system. In this case, the CB transformation (CBT) is defined, for a component k, as

$$y^k = \begin{Bmatrix} y_j \\ y_i \end{Bmatrix}^k = \begin{bmatrix} I_{jj} & 0 \\ \psi_{ij} & \varphi_i \end{bmatrix}^k \begin{Bmatrix} q_j \\ q_c \end{Bmatrix}^k = T_{CB}{}^k q^k \quad (3)$$

where ${y_i}^k$ are the interior coordinates reduced to modal coordinates ${q_c}^k$, ${y_j}^k$ are the junction coordinates, ${\psi_{ij}}^k = -({K_{ii}}^k)^{-1} {K_{ij}}^k$ is the static subbasis which contains the constraints modes, I_{jj} is the identity matrix, and ${\varphi_i}^k$ is the truncated basis containing the first n_r normal modes of the corresponding component.

Nevertheless, a standard CBT basis does not contain any information about uncertainties and nonlinearities. Therefore, adding complementary subbases ΔT^k to enrich ${T_{CB}}^k$ is necessary. The ΔT^k subbases contain a well selected static residual vectors calculated according to the type of enrichment. Once the complementary subbases ${\Delta T_{NL}}^k$, ${\Delta T_S}^k$ and ${\Delta T_E}^k$, corresponding respectively to the nonlinear, stochastic and external excitation enrichments are generated, the enriched CBT (ECBT) ${T_{ECB}}^k$, for each component k, takes the form

$$T_{ECB}{}^k = [T_{CB} \;\vdots\; \Delta T_{NL} \;\vdots\; \Delta T_S \;\vdots\; \Delta T_E]^k$$

$$= \begin{bmatrix} I_{jj} & 0 & \vdots & 0 & \vdots & 0 & \vdots & 0 \\ \psi_{ij} & \varphi_i & \vdots & \Delta T_{NL} & \vdots & \Delta T_S & \vdots & \Delta T_E \end{bmatrix}^k \quad (4)$$

This transformation matrix contains n_j^k junction dofs, n_r^k retained normal modes and n_e^k enriching static residual vectors.

3 Numerical simulations

The proposed stochastic periodic nonlinear micro-system, Figure 1, is composed of twenty identical coupled micro-beams submitted to transversal vibration. The discretization of each micro-beam into 20 elements leads to a FEM of 800 degrees of freedom (dofs). The micro-beams, which geometrical and mechanical parameters are listed in Table1, are coupled using twenty localized linear dampers and linear and nonlinear springs. Each beam i is submitted to a localized excitation force $P_{fi}(N) = 10^{-6} cos(2\pi f_1 t)$ where $f_1 = 565.53 \times 10^3 Hz$ is the first eigenfrequency.

The time responses evaluation is done in the time interval [0 - 5× 10^{-5} s], divided into steps of $10^{-8} s$, in which the stationary regime is already attained.

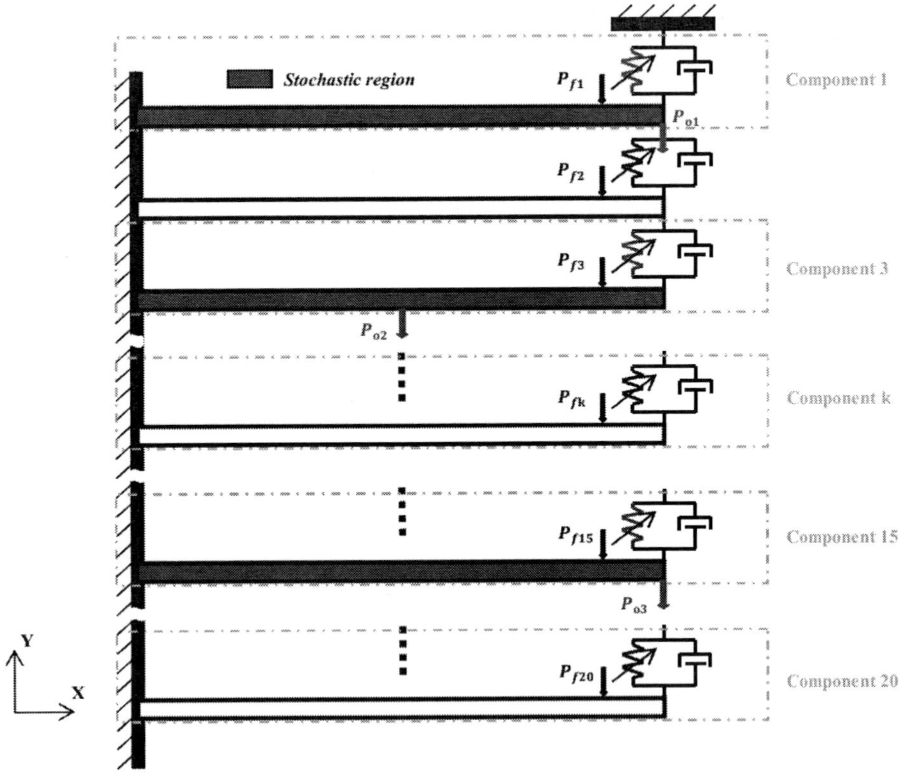

Fig. 1. Model of periodic nonlinear coupled micro-system.

Table 1. Geometrical and mechanical parameters of the micro-system.

b (μm)	h (μm)	L_b (μm)	E_0 (GPa)	ρ_0 ($kg.m^{-3}$)	c ($N.s.m^{-1}$)	k_0 ($N.m^{-1}$)	k_{NL} ($N.m^{-3}$)
3	1	50	169	2330	10^{-5}	10	10^{14}

Being in the case of a periodic structure, to apply the CB-CMS method, we propose to consider each set of coupling elements (localized linear damper, linear and nonlinear springs) and a micro-beam as a component (Fig. 1). Three stochastic zones are considered including the first, the third and the fifteenth components. In these zones, the Young modulus of the micro-beam and the linear coupling stiffness are supposed to be the uncertain parameters such as

$$p = p_0(1 + \delta_p \xi_p) \tag{5}$$

where p refers to the parameter and ξ_p are two types of random variables of uniform and lognormal probability distributions modeling respectively the Young Modulus and the linear stiffness uncertainties. $\delta_p = 10\%$ are the considered dispersions.

The (1-MAC) matrix, Figure 2, shows the effect of the uncertainties of the associated linear structure (MAC for Modal Assurance Criterion). This matrix compares the normal modes of the deterministic model with the mean of stochastic modes computed for each random variable of the stochastic model computed using the LHS method on the full finite element model. If (1-MAC) = 0, the compared modes are consistent. If (1-MAC) = 1, no correlation between these modes occurs.

The responses are presented in terms of velocities in order to illustrate the nonlinear effect. Figures 3-5 show the time responses of three observation degrees of freedom (dofs) obtained when applying the proposed metamodel (LHS-ECBT) and the reference LHS method implemented on the full finite element model (LHS-REF).

To compare the obtained results in terms of accuracy and time consuming, the Central Processing Unit (CPU) time and the time indicators E, T and D^2 (Table 2) are calculated. These time indicators, also called energy criteria (Hemez and Doebling, 2003), are calculated using the time moments

$$\mathcal{M}_i = \int_{-\infty}^{+\infty} (t - t_s)^i y(t)^2 dt \tag{6}$$

where $y(t)$ is the time response, i the order of the moment and t_s the temporal shift chosen in our case as

$t_s = 0$. In fact, $E = \mathcal{M}_0$ is the total energy of the response, $T = \mathcal{M}_1/\mathcal{M}_0$ the central time and $D^2 = \mathcal{M}_2/\mathcal{M}_0 - (\mathcal{M}_1/\mathcal{M}_0)^2$ the root mean square duration, E permits to evaluate the accuracy of the response in term of amplitude while T and D^2 allow it in term of periodicity.

As shown in Table 2, a model reduction ratio of 90% (reduced model of 82 dofs) and a computational time gain of 55% can be achieved when implementing the metamodel compared to the reference model. The accuracy errors are less than 2.5% and 0.14% in terms of amplitude and periodicity respectively. To reduce these errors, other vectors must be added in the transformation basis.

It is important to note that taking into account the periodicity of the structure permits to simplify the implementation of the CB-CMS method by taking the unit cell as a component.

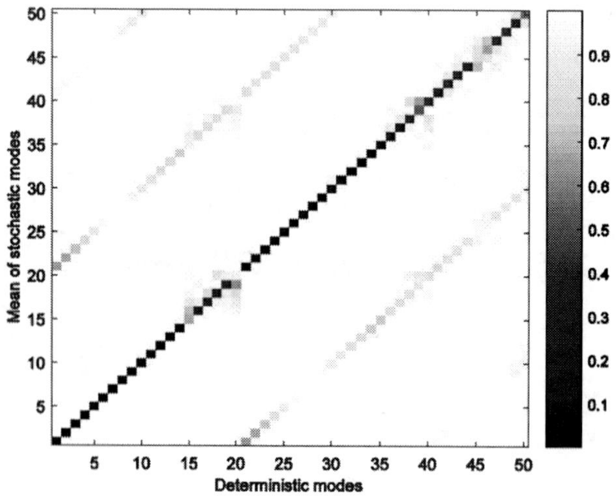

Fig. 2. (1-MAC) matrix illustrating the comparison of deterministic normal modes and the means of the stochastic modes.

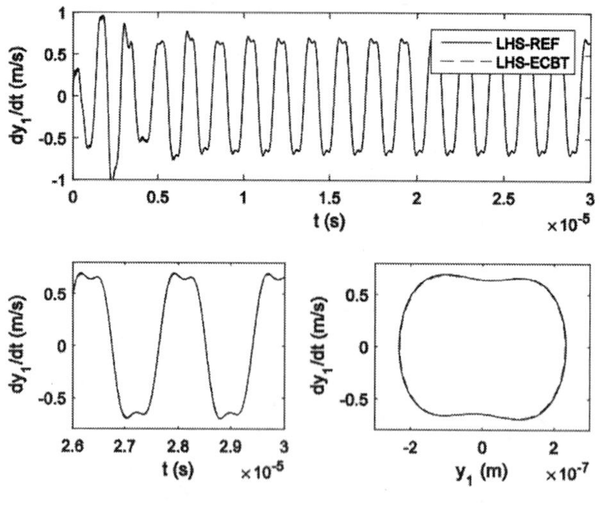

Fig. 3. Mean of the stochastic velocities and phase diagrams, at the observation dofs Po_1, computed using LHS-REF and LHS-ECBT methods.

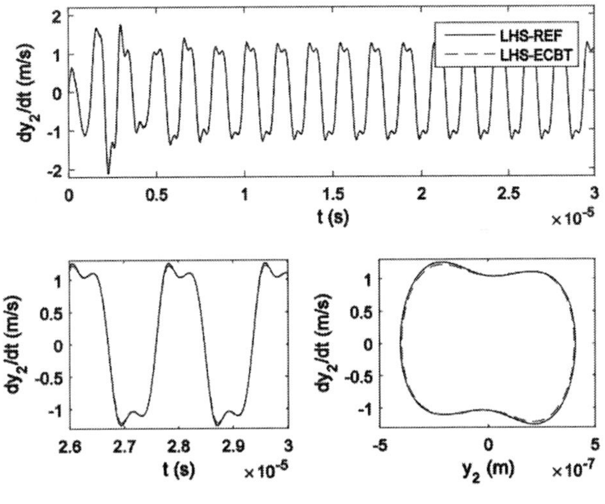

Fig. 4. Mean of the stochastic velocities and phase diagrams, at the observation dofs Po_2, computed using LHS-REF and LHS-ECBT methods.

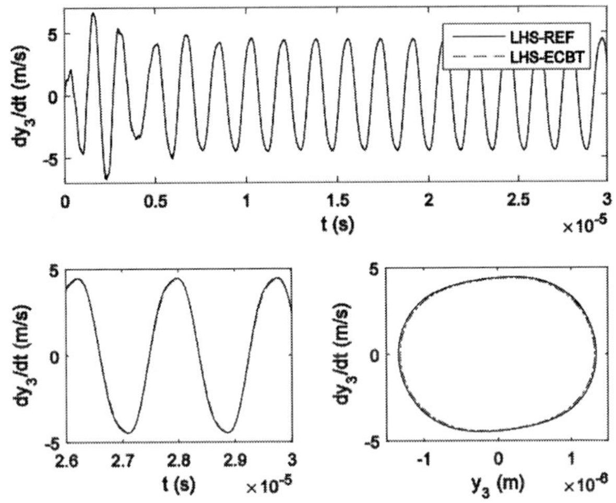

Fig. 5. Mean of the stochastic velocities and phase diagrams, at the observation dofs Po_3, computed using LHS-REF and LHS-ECBT methods.

Table 2. Model size, time indicators and CPU time.

Methods	**Model size (dofs)**	**Time indicators (error %)**			**CPU (%)**
		E	**T**	**D²**	
LHS-REF	800	-	-	-	100
LHS-ECBT	82	2.50	0.07	0.14	45.1

DOI: 10.1051/matecconf/20168305004

4 Conclusion

This paper proposed a metamodel combining the LHS method for uncertainty propagation and the CB-CMS for model reduction. In the light of the results of the presented application, the proposed metamodel permits to approximate the structure behavior with a reasonable computational cost and without a significant loss of accuracy, with respect to the reference method and so proves to be a powerful tool for robust design of periodic nonlinear micro-systems.

References

1. C. Soize, Eur J Comp Mech **19**: 25-40 (2009)
2. R. -Y. Rubinstein, *John Wiley & Sons* (1981)
3. J. C. Helton, F. J. Davis, Reliab Eng Sys Safe **81**: 23-69, (2003)
4. Jr. R. R. Craig, M.C.C. Bampton, *AIAA J* **6** (7): 1303-1319, (1968)
5. E. Balmès, Mech Syst Signal Pr **10** (4): 381-394, (1996)
6. G. Masson, B. Ait Brik, S. Cogan, N. Bouhaddi, *J Sound Vib* 296 (4-5): 845-860, (2006)
7. M. L. Bouazizi, M. Guedri, N. Bouhaddi, Mech Syst Signal Pr **20**: 131-157, (2006)
8. D. A. Perdahcıoglu, M. H. M. Ellenbroek, A. De Boer, *ICSV16*, July 5-9, Krakow, Poland, (2009)
9. F. Wenneker, P. Tiso, *In M. Allen et al. (Eds) Dynamics of Coupled Structures* **1**: 157-165, (2014)
10. L. Hinke, F. Dohnal, B. R. Mace, T. P. Waters, N. S. Ferguson, J Sound Vib **324**: 161-178, (2009)
11. M. Guedri, N. Bouhaddi, R. Majed, J Sound Vib **297**: 123-45, (2006)
12. D. Sarsri, L. Azrar, A. Jebbouri, A. El Hami, Comput Struct **89**: 346-356, (2011)
13. R. S. Motta, S. M. B. Afonso, P. B. Lyra, R. B. Willmersdorf, Eng Comput, 32 (2): 258-288, (2015)
14. M. Gérardin, D. Rixen, *John Wiley & Sons*, (1997)
15. S. Krenk, *Cambridge University Press*, (2009)

DOI: 10.1051/matecconf/20168305005

Non-linear models for Rotor-AMB system drop

C. Jarroux[1,2], R. Dufour[1], J. Mahfoud[1], B. Defoy[2], T. Alban[2] and A. Delgado[3]

[1]Université de Lyon, CNRS, INSA-Lyon, LaMCoS UMR5259, 69621 Villeurbanne, France
[2]Thermodyn, GE Oil and Gas, 71200 Le Creusot, France
[3]GE Global Research Center, NY 12309 Niskayuna, USA

Abstract. The proposed investigation deals with the assessment of several models for the prediction of the dynamic behaviour of turbomachinery supported by Active Magnetic Bearing when drop events occur leading the rotor to drop onto its Touch-Down Bearings (TDBs), which are emergency rolling element bearings in series with a ribbon damper. The rotor exhibits a non-linear transient motion where the amount of damping provided by the ribbon damper is a key parameter for avoiding dangerous behaviours. The ribbon damper is modelled successively with the Masing, the generalized Dahl and the Kelvin-Voigt models. Usually, the TDB is modelled as a viscous damper, used here as reference for comparisons. The originality of this work lies in the new modellings of that ribbon, based on dry-friction theory, where parameters are benchmarked to experimental results. The three models presented are successively integrated in the TDB model followed by rotor drop simulations. Comparisons are done in terms of rotor drop dynamics and show that the transient responses predicted with the Masing and generalized Dahl models are similar.

1 Introduction

Rotating machines supported by Active Magnetic Bearings (AMBs) lead to frictionless system with reduced energy losses. However, in some particular events, the AMB power supply may fail leading to a non-linear transient behaviour of the horizontal rotor mainly subject mass unbalance, contact and gravity forces. The resulting dynamic behaviour has been studied over the last twenty years [1-6]. The rotor, forced by gravity, drops onto its Touch-Down Bearings (TDBs), ball-bearings in this case, providing a back-up pivot linkage and protecting the AMBs. The rotor can follow several types of trajectory after impact. Its dynamic behaviour depends mostly on the amount of damping in the system, mainly provided by the ribbon damper, a corrugated steel foil fitted between the outer-race of the TDB and the casing [5, 6], see figure 1.

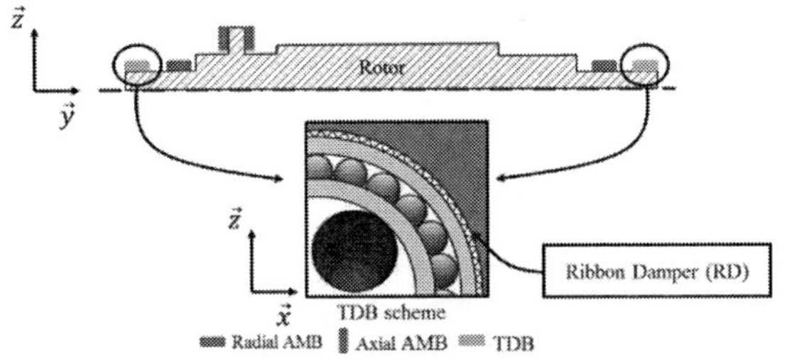

Fig. 1. Schematic position of AMBs and TDBs

With the purpose of enhancing rotor drop dynamic predictions, measured force-deflection loops were performed to evaluate the dynamic behaviour of the ribbon damper. The latter is related to dry-friction phenomenon rather than viscous damping, as it is usually assumed. The first part of this work describes the test-rig and experimental results. In the following section, the Masing and generalized Dahl models, both dry-friction models, are described and adjusted to fit experimental results. The Kelvin-Voigt model, considered as the reference, is used with provided data. Then, an eight degrees of freedom (DoF) model is established, comprising 3-DoF rigid rotor and 5-DoF TDB models. Finally, the influence of the three TDB models on the rotor drop dynamics is investigated.

2 Experiments

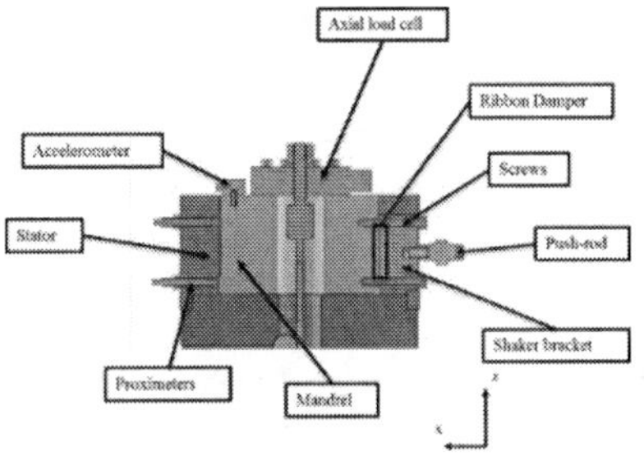

Fig. 2. Vertical cross-section of the ribbon damper test-rig

DOI: 10.1051/matecconf/20168305005

2.1 Test-rig

The objective of the experiments was to measure the restoring forces generated by the preloaded corrugated foil (ribbon damper). Figure 2 shows a cross-section of the test-rig comprising a stationary housing and a movable mandrel connected to two electro-hydraulic shakers (90 deg. apart) via threaded rods instrumented with load cells. The ribbon is positioned between the housing and the mandrel. The housing includes Eddy-current sensors to measure relative motion of the mandrel while an accelerometer atop the mandrel measures acceleration.

The mandrel was excited with single frequency excitations along a single axis ranging from 20 Hz to 200 Hz. Once the mandrel reached steady state, 0.5 s of signal was recorded using a sampling frequency of 8192 Hz. The ribbon reaction force is obtained by subtracting the mandrel inertial force from the shaker input force. The experimental results indicate the restoring force is a hysteresis force-deflection loop quasi-symmetric regarding the zero origin, which denotes that the compression and traction phases are rather similar. In addition, sharp corners at the change of the velocity indicate the dynamic behavior of the ribbon damper should be numerically described by dry-friction models.

2.2 Dry-friction phenomenon

The dry-friction damping induces a highly non-linear force which is able to flatten the frequency response of dynamic systems. This peak flattening could be due to the out-of-phase between the friction force and the displacement generated by the stick-slip phenomenon, as explained in [7]. Thus, the damping provided by the ribbon when a rotor drops onto its TDBs would come mainly from the alternation of stick and slip motion of each bump of the ribbon, rubbing along the inner-race and housing surfaces, generating this force-deflection out-of-phase. Two main categories of friction model can be employed: macroslip and microslip models. Macroslip models generate homogenous dynamics meaning that everything slips or everything sticks at the same time, while in the microslip model mixed configurations appear. In the case of the ribbon, the large imposed displacements generate brutal stick-slip transition while this is not absolutely true for small imposed displacements. The next section deals with the description of the ribbon damper models.

3 Ribbon damper models

3.1 The Masing model

A first approach is the modelling of the ribbon damper with the Masing macroslip model, used in [7] in parallel with a purely elastic restitution force, see equation (1). The contact parameters are assessed thanks to the experimental force-deflection loops, plotted in dotted lines in figure 4. The stick stiffness k_1 so-called contact stiffness, corresponds to the scenario where all the ribbon bumps are stuck along the inner-race and housing surfaces, resulting in a high global stiffness. Once the slipping threshold μF_N is reached, all the bumps slip and a change of slope occurs in the hysteretic loop. The resulting stiffness so-called the slipping stiffness k_2 is much smaller than the contact stiffness. The rubbing device z records the last position before each change of dynamic behaviour.

$$f_{rd} = k_2 u + f_{nl}\left(z, u, \frac{du}{dt}\right) \tag{1}$$

$$\begin{cases} if\ |k_1(u-z)| > \mu F_N \rightarrow \begin{cases} f_{nl} = sign\left(\frac{du}{dt}\right)\mu F_N \\ z = u - sign\left(\frac{du}{dt}\right)\frac{\mu F_N}{k_1} \end{cases} \\ else \begin{cases} f_{nl} = sign\left(\frac{du}{dt}\right) k_1 (u-z) \\ z = z \end{cases} \end{cases} \tag{2}$$

3.2 The generalized Dahl model

The Dahl model, generalized in [8] and used in [9], is a dry-friction restoring force model, describes by a non-linear first Order Differential Equation (ODE), detailed in equation (3). The advantage of this model is that it can adopt any shape of loop.

$$\frac{df_{rd}}{dt} = \beta \frac{du}{dt}\left(h - f_{rd}\, sign\left(\frac{du}{dt}\right)\right) \tag{3}$$

$$h = \frac{1}{2}\left((h_u + h_l)\, sign\left(\frac{du}{dt}\right) + (h_u - h_l)\right) \tag{4}$$

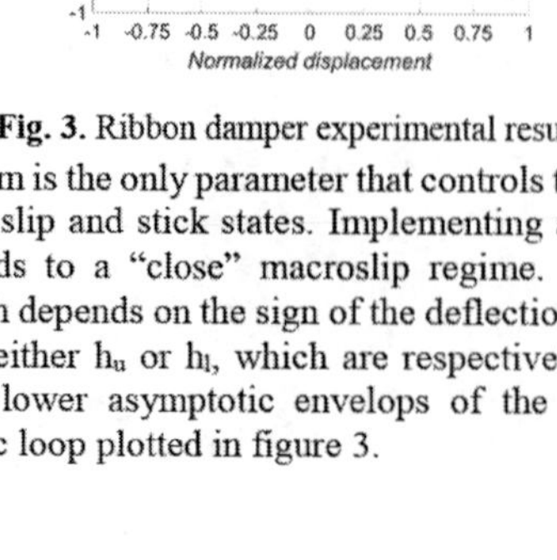

Fig. 3. Ribbon damper experimental results

The β term is the only parameter that controls the transition between slip and stick states. Implementing a high value of β leads to a "close" macroslip regime. The current envelop h depends on the sign of the deflection velocity to become either h_u or h_l, which are respectively the upper and the lower asymptotic envelops of the quasi static hysteretic loop plotted in figure 3.

DOI: 10.1051/matecconf/20168305005

As these envelops are curved fitted, one can observe that the hysteretic behavior is characterized by a linear force-deflection relation from -0.8 to 0.8, thus indicating the ribbon exhibits a hardening-hardening non-linear behavior for the higher deflections.

3.3 The Kelvin-Voigt model

Classically employed in rotor drop dynamic investigations, the linear spring-damper so-called Kelvin-Voigt model, is described in equation (5). The stiffness is provided by the manufacturer and the damping, measured with logarithmic decrement technique, is set to 0.16:

$$f_{rd} = c_{rd}\frac{du}{dt} + k_{rd}u \tag{5}$$

The displacements measured with the Eddy-current sensors as well as their time derivatives are the input for the three developed models described by equations (1), (3) and (5). The figure 4 compares the predicted (solid lines) and measured (dotted lines) force-deflection loops. Results are normalized with respect to the ribbon damper clearance. Both dry-friction models provide accurate predictions and are capable of reproducing the phenomenon. On the other hand, the Kelvin-Voigt model, representing only the slip stiffness without including stick-slip phenomenon, significantly underpredicts the amount of damping associated with the area defined by the force-deflection loop (dissipated energy).

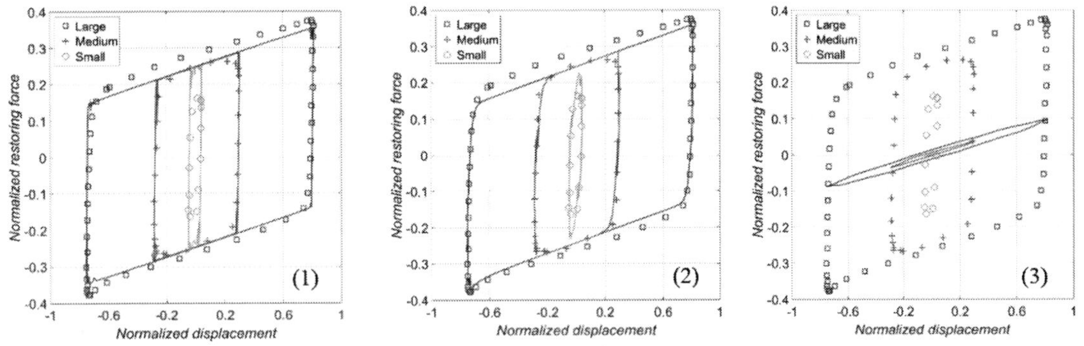

Fig. 4. Force-deflection loop with the Masing model (1), the generalized Dahl model (2) and the Kelvin-Voigt model (3)

4 Rotor – TDB system

A classical 5-DoF TDB model [3-6], sketched in figure 5, is used to describe the rotor drop dynamics. The rotor model consists in a rigid rotating mass with 2 lateral DoF and 1 angular DoF. AMBs are considered having constant stiffness k_{amb} and damping c_{amb} coefficients at a constant rotating speed. The normal contact forces f_{crir} is modelled with a vibro-impact model based on Hunt and Crossley [10] derived from the Hertz theory for inelastic collisions. The inner-race rotor interface friction force is described by a Coulomb model smoothed with the arctangent law, also used in [11]. The TDB angular dynamics is taken into account by an equivalent polar inertia, a resistive torque and the TDB is driven in rotation by the Coulomb interface friction force. The bearing restoring forces are implemented using a quadratic law. The pre-loaded ribbon provides a f_{rd} force. Finally, f_{cor}, called hard-stop, is the contact force between the outer-race and the housing, modeled like f_{crir}. This force appears when the ribbon is fully crushed.

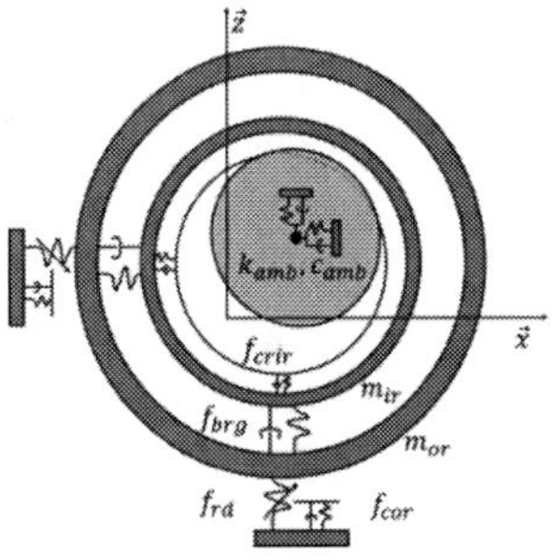

Fig. 5. Schematic presentation of the TDB model

4.1 Drop simulations

The shaft rotates at 6 500 rpm and is subject to mass unbalance forces, based on API 617 standard. When the steady-states regime is reached, the AMBs are shut-down and the gravity is applied along the z-axis. The rotor is decelerated by the interface friction force while the TDB accelerates. Time transient simulation of 3 s are computed using the 5th order Runge-Kutta algorithm with typical time-step for non-linear analysis.

4.2 Results

Rotor drop orbits are plotted in figure 6. Results are normalized with respect to the rotor-TDB clearance. Predictions with the dry-friction models are similar. Drop

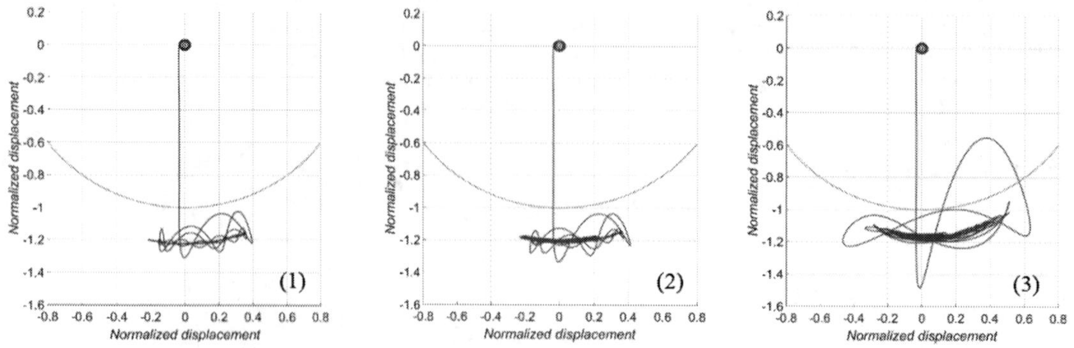

Fig. 6. Rotor orbits with the Masing model (1), the generalized Dahl model (2) and the Kelvin-Voigt model (3)

rebounds are flattened with both the Masing and the generalized Dahl models meaning that the dry-friction generates a considerable amount of damping. In contrast, rotor drop predictions with the Kelvin-Voigt model exhibit rebounds. This is in agreement with observations previously made based on figure 3. For both the dry-friction model, the rotor drop energy is quite fully absorbed during the first rebound.

5 Conclusion

The rotor drop dynamics on three different ribbon damper model was numerically investigated. The reference model was the commonly used Kelvin-Voigt model based on provided data. The two others were the macroslip Masing model and the microslip generalized Dahl model, both based on dry-friction theory and tuned with harmonic tests performed on the ribbon. These last two models were able to reproduce accurately the real dynamic behaviour of the ribbon under harmonic excitations. Both of them provided similar results concerning the rotor drop dynamics. A considerable amount of damping was generated and drop rebounds were flattened. However, rebounds are usually observed in case of rotor drop events as exhibited by the Kelvin-Voigt model predictions. The ribbon damper may have a different dynamic behaviour when subject to shock or harmonic excitations. Shock tests should be performed on that ribbon to validate this assumption.

Acknowledgement

This work is a part of Clément JARROUX PhD performed with the support of CIFRE funding N0 2013/1376. The authors are grateful to French National Agency ANRT.

References

[1] A. Gelin, "Etude théorique et expérimentale des comportements dynamiques permanents et transitoires de rotors sur paliers magnétiques," PhD Thesis, INSA de Lyon, Lyon, 1991.

[2] M. O. T. Cole, P. S. Keogh, and C. R. Burrows, "The Dynamic Behavior of a Rolling Element Auxiliary Bearing Following Rotor Impact," Journal of Tribology, vol. 124, pp. 406-413, 2001.

[3] G. Sun, "Rotor drop and following thermal growth simulations using detailed auxiliary bearing and damper models," Journal of Sound and Vibration, vol. 289, pp. 334-359, 2006.

[4] J. G. Lee and A. Palazzolo, "Catcher Bearing Life Prediction Using a Rainflow Counting Approach," Journal of Tribology, vol. 134, pp. 031101-031101, 2012.

[5] J. Wilkes, J. Moore, D. Ransom, and G. Vannini, "An improved catcher bearing model and an explanation of the forward whirl / whip phenomenon observed in AMB transient drop experiments," Journal of Engineering for Gas Turbines and Power, vol. 136, pp. 504-515, 2014.

[6] C. Jarroux, R. Dufour, J. Mahfoud, B. Defoy, and T. Alban, "Parametric analysis of a rigid rotor drop onto touchdown bearings," in Colloquium 573 – Coupling and Nonlinear interactions in Rotating Machinery, Lyon, 2015.

[7] B. Al Sayed, E. Chatelet, S. Baguet, and G. Jacquet-Richardet, "Dissipated energy and boundary condition effects associated to dry friction on the dynamics of vibrating structures," Mechanism and Machine Theory, vol. 46, pp. 479-491, 2011.

[8] A. Al Majid and R. Dufour, "Formulation of a Hysteretic Restoring Force Model. Application to Vibration Isolation," Nonlinear Dynamics, vol. 27, pp. 69-85, 2002/01/01 2002.

[9] A. Al Majid and R. Dufour, "Harmonic response of a structure mounted on an isolator modelled with a hysteretic operator: experiments and prediction," Journal of Sound and Vibration, vol. 277, pp. 391-403, 2004.

[10] K. H. Hunt and F. R. E. Crossley, "Coefficient of Restitution Interpreted as Damping in Vibroimpact," Journal of Applied Mechanics, vol. 42, pp. 440-445, 1975.

[11] C. Duran, L. Manin, M. A. Andrianoely, C. Bordegaray, F. Battle, and R. Dufour, "Effect of rotor-stator contact on the mass unbalance response," in IFToMM International Conference on Rotor Dynamics, Milan, 2014.

DOI: 10.1051/matecconf/20168305006

Metamodel for nonlinear dynamic response analysis of damaged laminated composites

S. Mahmoudi[a], Frédérique Trivaudey, N. Bouhaddi, and M.L. Boubakar

FEMTO-ST Institute, UMR 6174, Department of Applied Mechanics, University of Franche-Comté, UBFC, 24 rue de l'Épitaphe, 25000 Besançon, France

Abstract. Damage affects negatively the safety of the structure and can lead to failure. Thus, it is recommended to use structural health monitoring techniques in order to detect, localize and quantify damage. The main aim of the current work is the development of a numerical metamodel to investigate the dynamic behavior of damaged composite structures. Hence, a metamodelling for damage prediction and dynamic behavior analysis of laminate composite structures is proposed, wherein the stress state in the structure is used as indicative parameters and artificial neural networks as a learning tool.

1 Introduction

Damage in composite structures can be classified in three forms, matrix cracking, fibre fracture and delamination. These forms of damage may not be visible from surface, but they may have a significant effect on the structure stiffness, its strength and its life-time. Hence, the damage assessment is the subject of many studies in the literature, especially with regard to techniques of detection, quantification and localization of the damage. Kachanov [1] proposed a first meso-modelling of damage which is introduced as the degree of reduction in the structure resistance and the loss of rigidity. For laminated composite structures, fibre rupture generally occurs in the final phase of layer rupture whereas matrix cracking is considered as the first damage mechanism and presents a common defect in laminated composites.

Over the last decades, several methods of damage detection in composite materials were developed to analyse the damage mechanism and as a result to ensure the reliability and safety of service structures. Most of these methods are costly and they may need to take the structure out of service. Indeed, in some cases, the identification of damage can be considered as an inverse problem where the damage is determined with respect to its effects and the solution become not obvious to get. Thus, for reasons of simplicity, advanced identification techniques based on artificial intelligence methods such as artificial neural networks are increasingly used due to their excellent pattern recognition capability.

Artificial neural networks (ANNs) are defined as computational models whose design is schematically inspired from basic operating concept of biological neurons [2]. The key strength of ANNs is generalisation. Once the relationship between the input and output variables is established via the training process, the ANNs model can be able to accurately deal with inputs which were not present in the training set. Indeed, they are generally considered as a tool for pattern recognition or classification, modelling and prediction problem. And, they represent a promising metamodelling technique, especially for data sets having non-linear relationships as the case of damage detection and quantification problem. Several researchers used ANNs to detect, localize, and quantify damage in homogeneous and composite structures. Sahin et al. [3] proposed a damage detection algorithm using a combination of changes in eigenfrequencies and curvature mode shapes as inputs to ANNs for damage detection and prediction of its severity. Zhang et al. [4] used numerical data from a finite element model of a delaminated composite beam and eigenfrequencies from modal testing as inputs for ANNs in order to predict the interface, lengthwise location and size of delamination. Xu et al. [5] proposed an adaptive multiplayer perceptron (MLP) for cracks detection in anisotropic laminated plates where the only displacement response on the plate surface is provided to the MLP as input. A literature reviews in which the ANNs are used in the damage detection is provided in [6, 7].

In this paper, the dynamic behaviour is expressed through elasticity coupled with damage using a phenomenological macro-model for cracked composites structures made of polymer reinforced with long glass fibres. The damage is fully described by a single scalar variable whose evolution law is expressed through the maximum dissipation principle [8, 9]. Then, taking into account the nonlinearity induced by damage and using the classical linear Kirchhoff-Love theory of plates, the resulting nonlinear problem is solved in time domain. Several numerical simulations have been performed to generate a dataset consisting with stress and damage states for various combinations of layer orientation and applied load. These data have been used to train a feed-forward neural network till the network learns to an acceptable level of accuracy. The trained ANN has been tested to predict the damage from the input stress state. The established ANN can learn effectively about the damage location and severity present in the composite structure and can predict reasonably well when tested with unknown data set. This approach provides a quick response for damage level prediction in online applications reducing significantly the computational costs.

[a] e-mail: saber.mahmoudi@femto-st.fr

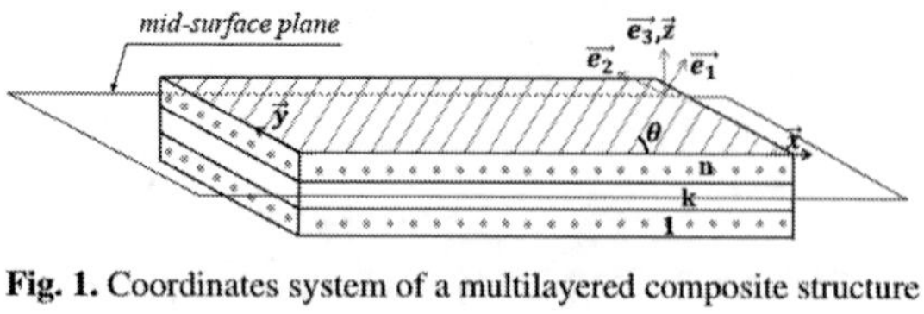

Fig. 1. Coordinates system of a multilayered composite structure

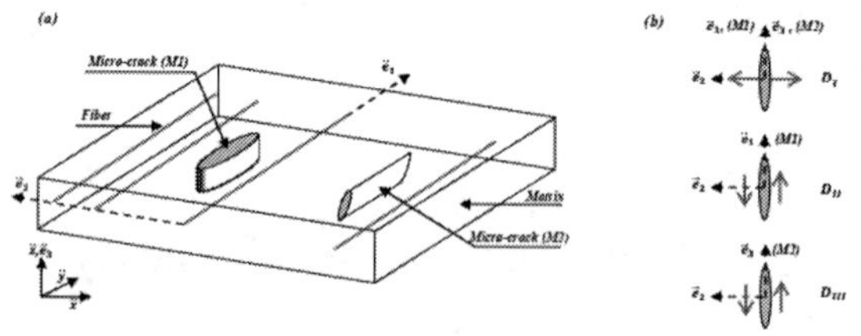

Fig. 2. Thin layer of polymer reinforced with long glass fibres: (a) orientation of the micro-cracks in the matrix, (b) micro-cracks opening modes [8]

2 Damaged dynamic behaviour

2.1 Finite element model

The thin composite structure, whose thickness is constant h, is composed of n layers having different fibre orientations θ. The global coordinate system $(\mathbf{x}, \mathbf{y}, \mathbf{z})$ is chosen such as the plan $(\mathbf{x}, \mathbf{y})$ is located in the mid-plane of the structure and the $\mathbf{z}$ is oriented in the thickness direction as shown in figure 1. Assuming that there is no stress in the $\mathbf{z}$ direction and according to the First-order Shear Deformation Theory, the displacement field can be written as follows:

$$\left\{\underline{u}(x,y,z,t)\right\} = \left\{\begin{array}{c} u(x,y,z,t) \\ v(x,y,z,t) \\ w(x,y,z,t) \end{array}\right\} + z\left\{\begin{array}{c} \varphi_y(x,y,t) \\ -\varphi_x(x,y,t) \\ 0 \end{array}\right\} \quad (1)$$

The problem is discretized by finite elements method where the quadrangular element Serendip $Q8$ with eight nodes is used to mesh the 2D model. Then, using the variational Hamilton principle, the elementary stiffness and mass matrices are carried out for each ply. The global mass $[M]$ and stiffness $[K]$ matrices are then obtained by assembling the elementary contributions all over the mesh elements. The damping is assumed proportional such as $[B] = \alpha_1[M] + \alpha_2[K]$. When the damage occurs, the stiffness will depend on the the damage state. This allows to rewrite it as $[K(D)]$ and the dynamic equation of motion is expressed as follow:

$$[M]\ddot{u}(t) + [B]\dot{u}(t) + [K(D)]u(t) = F(t) \quad (2)$$

where $u(t)$ is the dynamic displacement vector and $F(t)$ is the external load vector. The last equation depicts a nonlinear behaviour of the composite structure due to the material nonlinearity induced by the nonlinear evolution of the stiffness during the load.

2.2 Bilateral damage model

Generally, the matrix cracking is considered as the first damage mechanism and results in a common defect in composites. It evolves and grows leading mainly to the degradation of the structure stiffness through the decrease of the transverse elastic Young's modulus. The investigated structures in this paper are made of a thin layer of polymer reinforced with long glass fibres oriented with respect to $\mathbf{e}_1$ direction as depicted in figure 2(a) where the micro-cracks are parallel to the fibre direction. Boubakar et al [8] deduced, through experimental analysis, that the damage in the polymer matrix depends on the micro-cracks opening modes $(M1)$ and $(M2)$ as shown in figure 2(b).
Three parameters D_I, D_{II} and D_{III} are introduced to describe the damage effects such as:

$$D_I = \frac{\Delta E_2}{E_2} \quad ; \quad D_{II} = \frac{\Delta G_{12}}{G_{12}} \quad ; \quad D_{III} = \frac{\Delta G_{23}}{G_{23}} \quad (3)$$

A damage matrix $[H(D)]$ is fully expressed by a single scalar variable D representing the relative reduction of the transverse Young's modulus [9]:

$$\begin{array}{l} H_{22} = \dfrac{D_I}{1-D_I}S_{22} = \dfrac{D}{1-D}S_{22} \\ H_{44} = \dfrac{D_{II}}{1-D_{II}}S_{44} = \dfrac{D}{\sqrt{1-D}}\sqrt{S_{11}S_{22}} \\ H_{66} = \dfrac{D_{III}}{1-D_{III}}S_{66} = \dfrac{D}{\sqrt{1-D}}S_{22} \\ Others \quad H_{ij} = 0 \end{array} \quad (4)$$

where S_{ii} depict the flexibility components of $[S]$ of the undamaged material. The $[H(D)]$ is added to the flexibility matrix to describe the damaged elastic behavior of the composite structure.

$$\varepsilon = [S + H(D)] : \sigma \quad (5)$$

where σ is the stress tensor. A threshold function $\bar{Y}$ is defined in order to record the damage level and adapt the threshold of damage activation to the damage level. Thereby, the following charge function is established:

$$f_d = Y - \bar{Y} = \frac{1}{2}\sigma^T\left(\frac{\partial H}{\partial D}\right)\sigma - (Y_c + qD^p) \quad (6)$$

where the constants Y_c, q and p are damage parameters. $Y = \frac{1}{2}\sigma^T\left(\frac{\partial H}{\partial D}\right)\sigma$ is the thermodynamic force associated to the damage and it is obtained using the thermodynamics of irreversible processes.

3 Metamodel approach

The damage prediction procedure is fully numerical and is divided into two main steps: the first one consists in the offline simulations where several numerical simulations using the finite element model are carried out. The second one is the application of the ANNs to estimate the damage level and its location.

3.1 Damage simulation

The implicit time integration scheme of Newmark for nonlinear dynamic problems is improved and adapted in order to solve the equation (2). For each time step, the displacement field is known and the effective and predicted stress tensors are calculated at each integration Gauss point with Newton-Raphson method. Then the damage level is obtained through the charge function in equation (6). If $f_d \leq 0$, the last value of damage is kept as the damage level. But, if f_d tends to be positive, damage occurs and the damage increment will be obtained using the Newton-Raphson method. After getting the damage level, the stiffness matrix in equation (2) is updated to take into account the change in material properties.

3.2 Artificial neural networks metamodel

ANNs are composed of a numerous neurons arranged in layers and connected together by weights. Each neuron communicates with each other to process probabilistic or deterministic information. The full model for damage computing described in the section 3.1 is used in offline simulations in order to get enough data able to obtain a high accuracy in severity and location prediction of the damage. Since the damage mechanism depends on the orthotropic stresses $(\sigma_{22}, \sigma_{12}, \sigma_{23})$, these latter are chosen as inputs of the ANNs and the corresponding damage level D is the ANNs target. Different damage scenarios are simulated using a three layered clamped-free composite beam oriented as $(90/0/90)$, $(67/0/-67)$ and $(45/0/-45)$. Each structure is subjected to a sufficient load able to get fully damaged zones. After carrying out these simulations, some Gauss points are selected where the three stresses $(\sigma_{22}, \sigma_{12}, \sigma_{23})$ and corresponding damage level are extrated and thereby a database is made. Then the obtained samples of the database are normalized in the interval $[-1, 1]$ to make the training process faster and to ensure the uniform distribution of the samples between the network inputs and the targets. Once the input and output sets are organised, the ANNs is now ready to be trained. The organised database is randomly divided into training, validation and test ratios. During the training process, the predicted ANNs output (*i.e: using the ANNs model*) is compared to the given target (*i.e: full model*) and the weights are adjusted until achieving an optimal set of weights that minimizes the mean square error of the entire training data set. The validation set is used to control the over fitting and finally, the performance of the fully trained ANNs is evaluated with the test set. To summarize, the offline simulations are used to obtain the relationship between the state of stress and the damage level. This relationship is introduced in online simulations instead of the Newton-Raphson method to reduce the computational costs. The solving procedure is more detailed in [10]

4 Numerical Results

Several numerical simulations have been performed in order to highlight how the stress state can be used as a damage indicator in terms of location and severity. The mechanical properties of the considered laminated structure are given in table 1. The investigated structures are clamped at $x = 0$ and free at $x = L$ and they have the same geometric data as $(0.3, 0.03, 0.001)\,m$ in $(\mathbf{x}, \mathbf{y}, \mathbf{z})$ directions, respectively. A distributed impact load is applied along the free side in $\mathbf{x}$ direction with F as magnitude and $\tau = 1\,ms$ as a duration. Firstly, to validate the proposed ANNs metamodel, the results provided by the ANNs are compared to

Table 1. Mechanical properties of the laminated structure

Elastic modulus E_1 ($\mathbf{e}_1$) (MPa)	45680
Elastic modulus E_2 ($\mathbf{e}_2$) (MPa)	16470
Shear modulus G_{12} (MPa)	6760
Poisson ratio ν_{12}	0.34
Poisson ratio ν_{23}	0.34
Y_c (MPa)	0.0027
q (MPa)	1.246
p	0.816

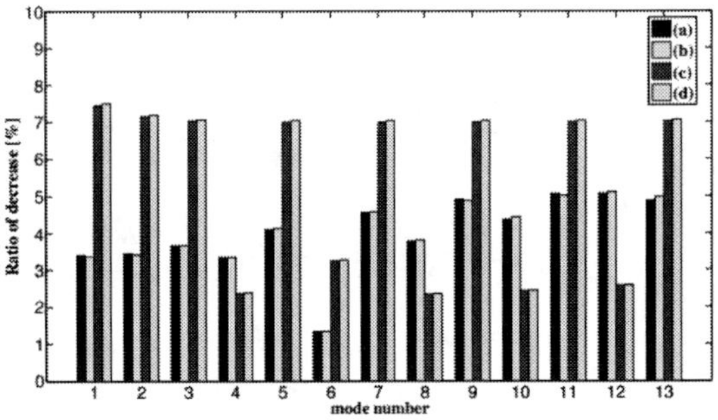

Fig. 3. Damage severity performance: (a) and (b) are the decrease ratios obtained via the full damage model and the ANNs metamodel, of the (45/0/ − 45) laminated beam, (c) and (d) are the same corresponding ratios of the (90/0/90) laminated beam

Fig. 4. Damage state, obtained via the full model, per ply of the $(45/0/-45)$ beam where $F = 2.28\,10^5 N/m$

those given by the full damage model. Thus, the ANNs method is applied to the structures used in the offline simulations. The decrease ratio of eigenfrequencies obtained via the two methods is used to ensure the performance of ANNs in terms of damage severity. This ratio is defined as $R_d = 100\|\frac{f_0 - f_D}{f_0}\|$. As shown in figure 3, the ANNs metamodel predict the damage severity with a good agreement comparing to the full model. Then, comparing the figure 5 to figure 4 and the figure 7 to figure 6 where (a), (b) and (c) are the upper, middle and the lower layers respectively, the ANNs demonstrate their ability to perfectly estimate the spatial damage levels in the layered beams with high accuracy.

Now, the designed ANNs is applied to a new layered beam. This latter has the same dimensions of the beams used previously but the layer orientations $(60/0/60)$ are not present is the the trainig data set. An impact distributed load is applied to the free side of this beam where $F = 1.52\,10^5 N/m$. As shown in figure 8, the used ANNs permit to obtain the eigenfrequencies shift with a good agreement with those given by the full damage model. Then, the damage states

DOI: 10.1051/mateconf/20168305006

Fig. 5. ANNs estimed damage state per ply of the $(45/0/-45)$ beam where $F = 2.28\ 10^5 N/m$

Fig. 6. Damage state, obtained via the full model, per ply of the $(90/0/90)$ beam where $F = 1.2\ 10^5 N/m$

Fig. 7. ANNs estimed damage state per ply of the $(90/0/90)$ beam where $F = 1.2\ 10^5 N/m$

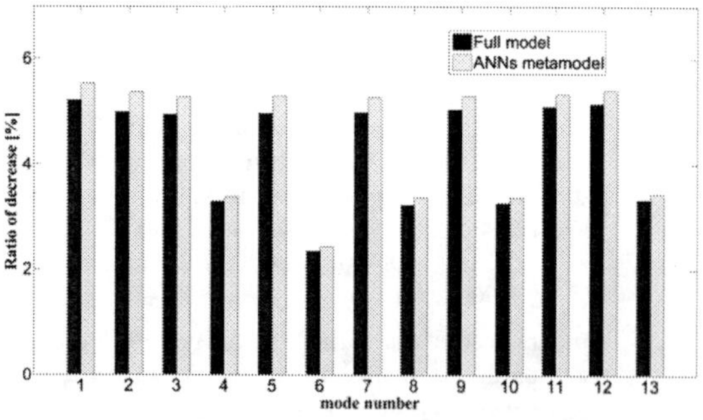

Fig. 8. Damage severity performance in terms of the decrease ratios obtained via the full damage model and the ANNs metamodel of the $(60/0/-60)$ layered beam

Fig. 9. Damage state, obtained via the full model, per ply of the $(60/0/-60)$ layered beam

Fig. 10. ANNs estimed damage state per ply of the $(60/0/-60)$ layered beam

obtained through the full model and the ANNs metamodel are used to verify the damage dispersion in each layer. As shown in figures 9 and 10, ANNs localize the damage and estimate its level in each layer with high accuracy. In addition, ANNs metamodel reduces significantly the computational cost in the online simulations. Hence, once the relationship between the stresses and the corresponding damage level is obtained using the offline simulations, it is introduced in online simulations instead of the Newton-Raphson method to compute the damage. The damage estimation becomes faster reducing about 66% of the computational cost as shown in table 2.

Table 2. Computational costs (CPU time [*hour*]) of the numerical simulations using the full model and the ANNs metamodel

	$(45/0/-45)$	$(60/0/-60)$	$(90/0/90)$
Full model	77.5	72.4	67.8
ANNs (online)	29	27.6	22.8
Reduction ratio [%]	**62.6**	**62**	**66.4**

5 Conclusion

This paper proposes a named stress-based method which describes the damage evolution in the laminate structures made of polymer reinforced with long glass fibers. The only considered damage is the matrix micro-cracking and is fully described by a single scalar variable. The damage depends on the stress state which can be used as a damage indicator in terms of location and severity. Hence, an ANNs metamodel for severity and location prediction of damage in composite structures is proposed and investigated using features extracted from stress-based analysis data as inputs and the corresponding damage level as target. Numerical simulations validate the proposed metamodel and have shown a high accuracy in terms of damage evolution, severity and localization. Moreover, the ANNs metamodel permits to reduce significantly the computational costs. The proposed method is flexible in use since only the stress state is required to predict the damage level. Thereby, this method can be used in life-time estimations and monitoring strategies of composite structures.

References

1. L. M. Kachanov, Izv. Akad. Nauk. S.S.R. Otd. Tech. Nauk. **8**, (1958) 26-31
2. E. Alpaydin, *Introduction to Machine Learning* (The MIT Press, 2010 2nd Edition)
3. M. Sahin, R. A. Shenoi, Journal of Mechanical Engineering Science **217 (6)**, (2003) 661-676
4. Z. Zhang, K. Shankar, T. Ray, E. V. Morozov, M. Tahtali, Composite Structures **102**, (2013) 226-236.
5. Y. G. Xu, G. R. Liu, Z. P. Wu, X. M. Huang, International Journal of Solids and Structures **38 (32)**, (2001) 5625-5645
6. H. Sohn, C. R. Farrar, F. M. Hemez, D. D. Shunk, D. W. Stinemates, B. R. Nadler, J. J. Czarnecki, *A review of structural health monitoring literature: 1996-2001* (Los Alamos National Laboratory, Los Alamos, 2004)
7. K. Waldron, A. Ghoshal, M. Schulz, M. Sundaresan, F. Ferguson, P. Pai, J. Chung, Finite Elements in Analysis and Design **38 (3)**, (2002) 193-226.
8. M. L. Boubakar, F. Trivaudey, D. Perreux, L. Vang, Composite structures **58 (2)**, (2002) 271-286.
9. D. Perreux, C. Oytana, Composites Engineering **3**, (1993) 115-122.
10. S. Mahmoudi, F. Trivaudey, N. Bouhaddi, Finite Elements in Analysis and Design, **119**, (2016) 1-14.

Vibration Based Methods For Damage Detection In Structures

E. Manoach[1], J. Warminski[2], L. Kloda[2], A. Teter [2]

[1]Institute of Mechanics, Bulgarian Academy of Sciences, Bulgaria
[2] Department of Applied Mechanics, Lublin University of Technology, Lublin, Poland

Abstract. Vibration based damage detection methods are among the most popular and promising approaches for health monitoring of structures. In this work a critical review of different methods for damage detection methods of structures is presented. The theoretical bases of the most popular methods based on the changes in the modal properties of the structures are deduced. The review includes the modal displacements, the mode shape slopes, the modal curvatures and the strain energy methods. The efficiency of all these methods is compared by using a finite element analysis of intact and damaged beams. The methods are tested experimentally by using a scanning laser vibrometer to measure the modal properties of specially prepared composite beams with defects. All this methods are compared with the damage detection method based on the analysis of the Poincaré maps of the motion of the structures). Conclusions concerning the advantages and the applicability of the considered methods are deduced.

1 Introduction

Vibration based methods are among the popular and very important methods for structural health monitoring (SHM) and damage detection (DD). In this work only a small part of the vibrational based damage detection methods are considered. The easiest and the most popular from the vibration based methods are the methods based on the modal analysis of the structure. The earliest technique for DD was based on analyzing the eigen-frequencies of the system. The natural frequencies are easily measurable with high accuracy and are usually less contaminated by experimental noise. It is well known, however, that the natural frequencies are not very sensitive to damages. It was shown by many authors that significant damage may cause very small changes in natural frequencies, particularly for larger structures, and these changes may go undetected due to measurement or processing errors. [1-3]. The natural frequencies are global parameters of the systems and often they are weakly affected by local faults

Developing of the experimental techniques – contactless method with sensors or specially Scanning Laser Doppler Vibrometers (SLDV) makes possible to measure the modal displacements in large numbers of nodes. This fact extended a lot the possibility to use extensively the modal based damage detection methods.

In this work some basic modal based methods for damage detection and localization will be presented briefly. Time domain method based on modal measurements will be also deduced. Then, all listed methods will be tested numerically and experimentally and conclusions about their applicability will be invoked.

2 Modal based methods

In this work are reviewed some well know vibration based methods. The detailed analysis of the most of them can be found in the excellent review articles [4-7]

2.1 Modal displacements

Mode shapes are found to be quite sensitive to damage and are able to directly provide damage location information [8]. Moreover, the mode shapes are less sensitive to environmental effects, such as temperature, than natural frequencies [9, 10].
The simplest method based on the mode shapes is the so-called modal displacements method (MD) [11].
It seeks for a maximal difference in the measured displacements of modes of damaged and healthy structures:

$$\Delta W_i = \sum_{j=1}^{N} \left| w_{ij}^* - w_{ij} \right| \tag{1}$$

In Eqn (1) w_{ij} is the displacements of i-[th] mode at j-th node of the intact beam and w_{ij}^* is the same for damaged beam. Often, it is accepted N=1, i.e. only first mode is used for DD purpose.

2.2. Modal slopes square

Another simple method is the based on the modal slopes. It estimates the absolute differences between the squared modal slopes (the first derivative of modal shapes) of the damaged and healthy structures

$$\Delta W_i'^2 = \sum_{j=1}^{N} \left| w_{ij}'^{*2} - w_{ij}'^2 \right| \qquad (2)$$

This method is also more often used considering only the first mode , i.e. N=1.

2.3. Modal curvatures and modal curvatures square

The modal curvature method (MC) is widely used because it shows a good sensitivity to damage [12]. The methods have a few drawbacks such as the high noise sensitivity due to curvature calculations and no explicit relationship between damage and damage features [13]. By using SLDV, however, it is expected that the efficiency of the method to be very good.

The method can be expressed as:

$$\Delta W_i'' = \sum_{j=1}^{N} \left| w_{ij}''^{*} - w_{ij}'' \right| \qquad (3)$$

In [10] it is shown that the MC based methods are much more promising than the MD based methods, especially when curvatures are directly measured from the strain mode shape.

Some authors [13] mention that together with the peak of MC at the damage location, some smaller peaks at different undamaged locations can appear, especially for the higher modes. This can cause confusion in a practical application in which one does not known in advance the location of faults. A variation of modal curvature method is the modal curvatures square method which some authors [11] have used in order to enforce the differences in the curvatures of the intact and damaged structure.

2.4. Modal strain energy methods

The modal strain energy method **(MSEM)** was developed by Stubbs and Kim [14] and [15] to locate damage in structures. It was developed for Bernoulli -Euler beams. Later it has been extended by Cornwell et. al. in [16] for plate-lake structures. The method will be presented briefly here for the case of Timoshenko beam where the strain energy consists of two parts – due to bending and to shear. For location j on the beam this change in the ith mode strain energy healthy and damaged beams can be presented as:

$$U_{ij} = \frac{1}{2}\int_{a_j}^{a_{j+1}} \left\{ (EI)_j \left(\frac{\partial^2 w_i}{\partial x^2} \right)^2 + k\frac{(Ebh)_j}{2(1+\nu)} \left(\frac{\partial w_i}{\partial x} - \psi_i \right)^2 \right\} dx ;$$

$$U_{ij}^* = \frac{1}{2}\int_{a_j}^{a_{j+1}} \left\{ (EI)_j^* \left(\frac{\partial^2 w_i^*}{\partial x^2} \right)^2 + k\frac{(Ebh)_j}{2(1+\nu)} \left(\frac{\partial w_i^*}{\partial x} - \psi_i^* \right)^2 \right\} dx , \qquad (4a,b)$$

where i is the mode number, j is the element number U_{ij} is the strain energy of ith mode for the jth element . With * are denoted the variables for the damaged beam and ψ_i and ψ_i^* are the rotational angles of the cross-section of i^{-th} elements for healthy and damaged beams. It can be shown that the following relation can be obtained:

$$\frac{(EI)_j^*}{(EI)_j} = \frac{J_{ij}/U}{J_{ij}^*/U^*} = \frac{\gamma_j^*}{\gamma_j} = f_j ; \gamma_j^* = J_{ij}U^* , \gamma_j = J_{ij}^*U \qquad (5)$$

where

$$J_{ij}^* = \int_{a_j}^{a_{j+1}} \left\{ \left(\frac{\partial^2 w_i^*}{\partial x^2} \right)^2 + k\frac{6}{h^2(1+\nu)} \left(\frac{\partial w_i^*}{\partial x} - \psi_i^* \right)^2 \right\} dx$$

$$J_{ij} = \int_{a_j}^{a_{j+1}} \left\{ \left(\frac{\partial^2 w_i}{\partial x^2} \right)^2 + k\frac{6}{h^2(1+\nu)} \left(\frac{\partial w_i}{\partial x} - \psi_i \right)^2 \right\} dx$$

The strain energy ratio f_j potentially provides indications for damage location as well as its extent. It is expected to be equal to one over the undamaged regions and different than one over a damaged region. It is well known that damage affects the strain energy when it is located close to the regions of maximum strain energy. Therefore, the damage index f_j is affected by the frequency of excitation. For this reason it is reasonable to use the cumulative strain energy ratio:

$$f_m = \frac{1}{N}\sum_{j=1}^{N} f_{j_m} \qquad (6)$$

This approach including strain energy due to bending and shear could be considered as a small extension of the method developed in [14], if the angular rotation of ψ_i could be measured experimentally. This, however, is not an easy task. Therefore, in this work we have checked the influence of the second term in the strain energy (due to shear strains) only numerically.

3. Forced response methods

A big group of SHM methods are based on the analysis of the measured forced response of the structure. This is actually the most popular alternative for in-operation machinery monitoring. Most of these methods are trying to extract some features from the nonlinear forced response of the structures which are sensitive to the damage.

3.1. Poincare map based method

In [17] and [18] a method for damage detection and location on the base of the analysis of the Poincaré map of the forced response of the structure was developed. It was tested numerically and experimentally for plates and beams. The intention of the authors of this study is to check the method effectivness and to extent it if only the modal properties of the damaged and healthy structures are known.

According to this method a damage index can be constructed in the following way:

$$I_i^d = C\left(S_i^u - S_i^d \right) , \qquad (7)$$

where

DOI: 10.1051/matecconf/20168305007

$$S_i^u = \sum_{j=1}^{N_p-1} \sqrt{\left(w_{i,j+1}^u - w_{i,j}^u\right)^2 + \left(\dot{w}_{i,j+1}^u - \dot{w}_{i,j}^u\right)^2} \qquad \text{(8 a-b)}$$

$$S_i^d = \sum_{j=1}^{N_p-1} \sqrt{\left(w_{i,j+1}^d - w_{i,j}^d\right)^2 + \left(\dot{w}_{i,j+1}^d - \dot{w}_{i,j}^d\right)^2}$$

and C is a constant. This constant could be connected with the number of the nodes in the Poincaré map, for example $C=1/N_p$ where, $i=1,2...N_{nodes}$, N_{node} is the number of nodes, N_p is the number of points on the Poincaré map and $(w_{ij}^u, \dot{w}_{ij}^u)$ and $(w_{ij}^d, \dot{w}_{ij}^d)$ denotes the j th point on the Poincaré maps of the undamaged and the damaged states of i-th node, respectively.

This damage index represents the relative difference between the lengths of two curves formed by connecting the dots on the Poincaré maps in i-th node for the non-damaged and the damaged structure. This difference is accepted as a measure for the global change (during of the total period of vibration) in the dynamic behaviour of the damaged structure in comparison with the undamaged one. The assumption is that if the function $I^D(x)$ has maximum and it is strongly concave in the vicinity of the maximum, then the structure has damage and the nodes close to the maximal value of the function will represent the damaged area.

3.2. Obtaining the time series from the modal data

In this way, if we know from the experimental measurements the natural frequencies and modes of vibrations ω_n and w_n , following a simple numerical procedure we obtain time series describing the forced response of the beam subjected to artificially applied harmonic loading $p(x,t)=p_0 sin(\omega_e t)$, where p_0 is the amplitude of the artificial excitation and ω_e is the excitation frequency (see, for example [3]). In order to check these formulas for DD and to check the applicability of a few of the popular and easy modal methods for damage detections, numerical and experimental testing were performed.

4. Numerical results

A beam with length l=0.44 m , thickness h=0.0033 m and width b=0.0181 m was considered.

The material properties of the beam are E=4.6^{10} Pa, ρ=2032kg/m^3 and ν=0.4 . They correspond of the effective properties of composite cross ply laminated beam made of 10 layers glass-epoxy composite. The beam was modeled as a Timoshenko beam by a finite element program. The length of the beam was divided to 80 linear beam elements (81 nodes) with equal length.

Three cases of reductions of beam thickness (cuts) were considered:

a) Reduction with 0.125 mm (Cut1)
b) Reduction with 0.25 mm (Cut 2)
c) Reduction with 0.33 mm (Cut 3)

All reductions were made in elements 49 and 50 (between nodes 48-50).

For the intact beam and the beams with 3 different depth of the cuts were calculated the first seven eigen frequencies and normal modes. The introduced minimal damages lead to very small changes in the natural frequencies of the beam. The decrease of the natural frequencies in the damaged beams does not exceeds 1 % (not shown here).

The computed damage indexes (DI) based on modal displacements (Eq. (1)) for the case a) – cut with depth 0.125 mm are shown in figure 1. Calculations were performed for N=1. Despite of the fact that the thickness in damaged elements is reduced by only 3.78 %, the damage and its location is very well predicted in all cases (N=1,2,3). The inclusion of more modes in the computations enlarge the value of the DI (figure 2). The next criterion, mode shape slope squared, is demonstrated in figure 3. For this criterion the results from the computations of beam with Cut 2 are shown. As can be seen the location of the fault is well predicted, but there are two peaks in the damage location. These two peaks appeared in several cases which have been tested. This fact can provoke confusion in the potential experiment. The inclusion of more modes in the criteria leads to increasing the value of DI but the figures remain the same.

In the next figure (figure 4) the computed modal curvatures for each node of the beam are shown. It is clearly seen that the fault is predicted very well. The increased number of modes in the formula (3) leads to increasing the value of the peak of the curve. Modal curvatures square method gives even more sharp demonstration of the damage location (not shown here).

The application of strain energy method for damage detection is shown in figure 5. The FE program allows obtaining of modes of vibrations w_n and angular rotations ψ_n which allows checking the method's performance for the cases of the total energy and for the case of the bending energy only. Three deductions could be made from this figure: 1)The damage is extremely well predicted for both cases; 2) The influence of the shear deformation is obviously very strong for this kind of damage and therefore the consideration of the total energy gives much better results for damage identification. It must be noted that for different damages (delamination) which we have tested, the inclusion of the term due to shear stresses in strain energy is not essential. This calculations are not shown here.; 3) The method could be used to estimate the severity of the damage because the undamaged case corresponds to value $DI_{energy}\approx 1$.

Then a forced response of the beam was simulated. The beams were subjected to harmonic force with amplitude P = 20 Pa and excitation frequency ω_e =523.5 rad/s. The forced responses of the intact and damaged beams were computed , Poincaré maps were constructed and damage index (Eq. (8)) was computed. The results for the beams with three different damages (Cut1, Cut2 and Cut3) are plotted in figure 6. In all three cases the damage and its location are predicted very well, in spite of the fact that the picks are not very sharp. These low grade slopes of the Poincaré map DI are typical for this method and do not decrease its ability to predict damage.

5. Experimental setup

The natural frequencies and mode shapes of free vibrations were obtained by using scanned laser vibrometer Polytec PSV-500 and mini SmartShaker K2007E01. The beam was manually discretized by 41 grid points. The structure of fiber glass generates internal scattering which leads to high levels of noise. That is why the reflective labels were stuck on each of measured points. The first and last points have been set at the clamped ends. 41 markers were located along the beam length at equal distance from each other. (Fig. 7a).
The specimens are composite beams prepared from 10-layer glass-epoxy composite. The measurements were made for the intact beam. After that measurement of the intact beam 3 different cuts were made consecutively. The cut was made on the entire width of the beam (Fig. 8).
After the introduction of each cut modes shape and frequencies were measured by SLDV. This technique allow us to avoid the influence of non-perfect boundary conditions.

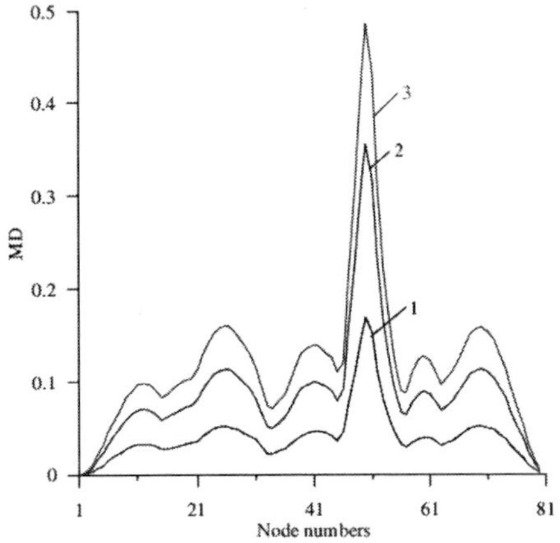

Fig. 1. Modal displacements based damage indexes for three different depth of the cuts. 1 – Cut1; 2 - Cut2; 3 – Cut3

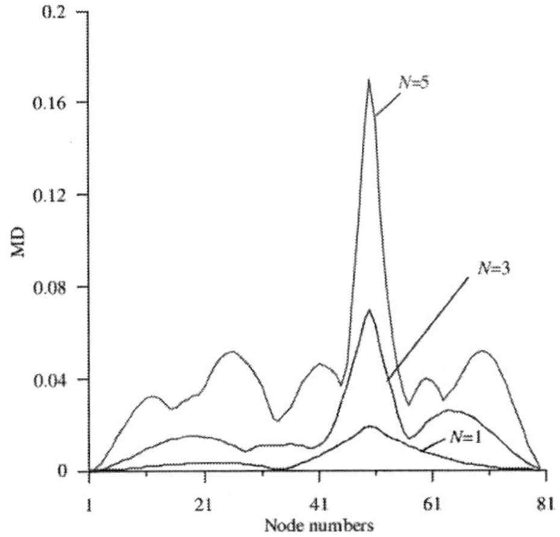

Fig. 2. Influence of the modes numbers taken to compute MD DI .

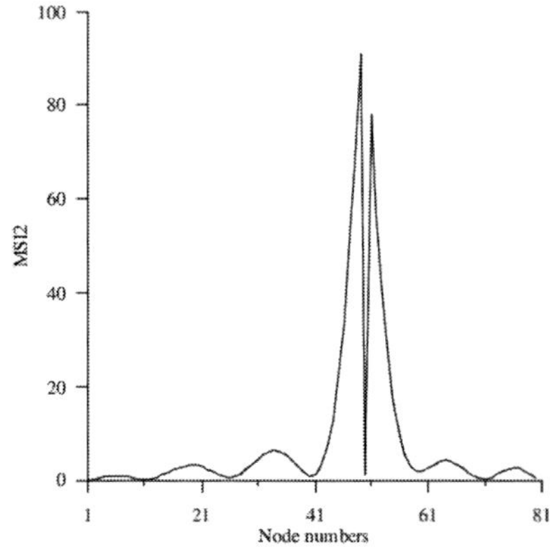

Fig.3 Mode shape slop criterion for the beam with CUT2. N=1.

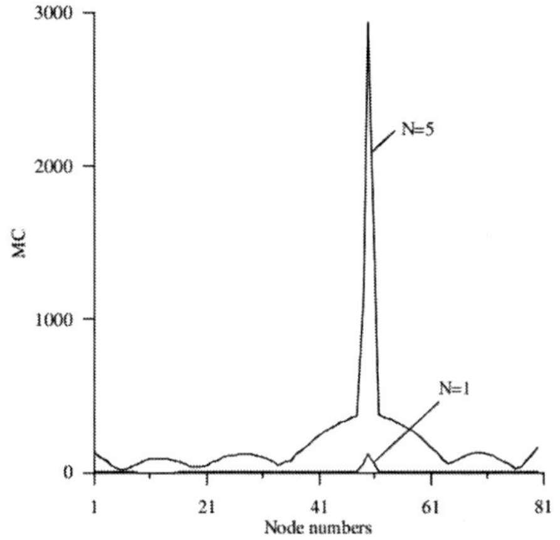

Fig. 4. Modal curvature criterion.

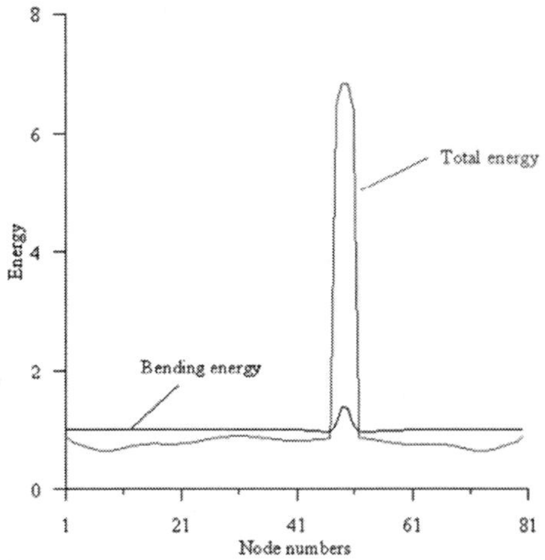

Fig.5. Modal strain energy criterion.

6. Experimental results

First step in the experimental study was to compare the natural frequency measured experimentally and numerically. The difference between the first 7 frequencies did not exceeded 5 %. This fact confirmed

DOI: 10.1051/matecconf/20168305007

that the experimental measurements are performed correctly.

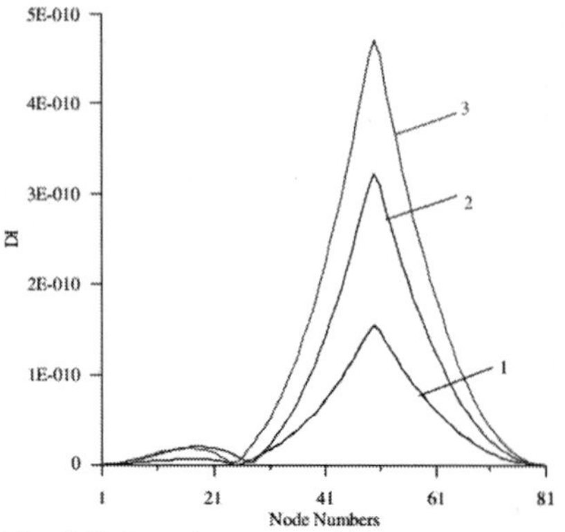

Fig. 6. Poincaré map based DI.

a)

b)

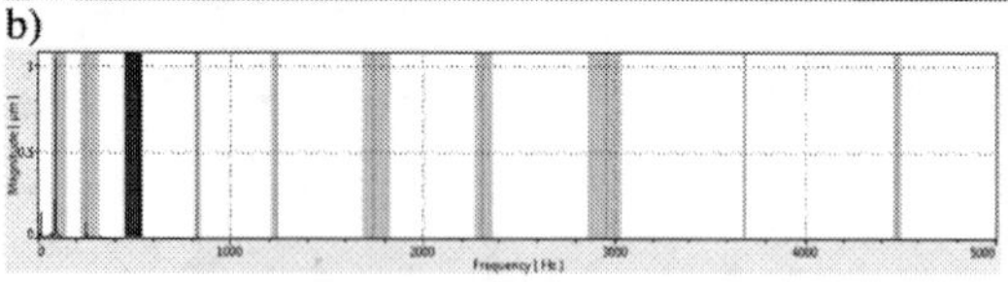

Fig. 7 Clamped-clamped beam with markers -(a) and amplitude-frequency response of the healthy beam- (b)

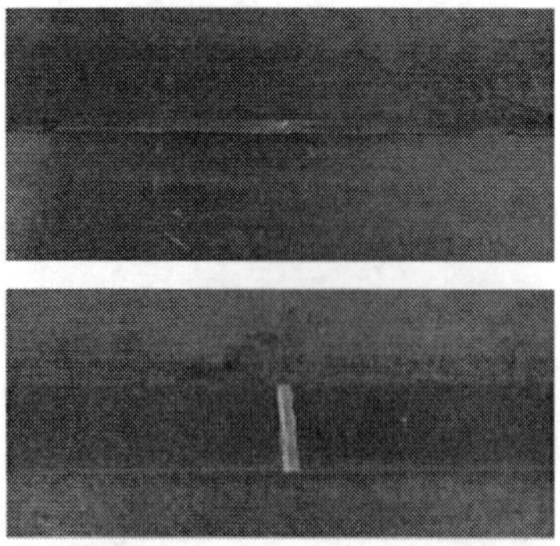

Fig. 8 Beam samples with a cut.

The DD methods listed above were checked on the base of the experimentally obtained data. The results obtained by the modal slope method did not predict the damage and its location and that is why they are not shown here. The modal displacement method was checked for N=1,3,5 and the obtained results were similar. The slope of the curve of MD index is not sharp but the method indicates exactly the damage location (figure 9). The similar are the results obtained by the Poincaré map based method (see figure 11). The beam was artificially considered to be subjected to harmonic loading with p= 10^3 MPa at three nodes at the beam center with frequency equal to the first natural frequency (ω_e=540 rad/s). The modal curvature method gives also good results , but smaller peaks appear along the beam length which could make a confusion in the case of multiple delaminations. The best results are obtained by the strain energy method (the bending energy is considered only) – figure 11. For the case of N=1, however this method give wrong results, i.e. the method requires more modes to be included in the analysis.

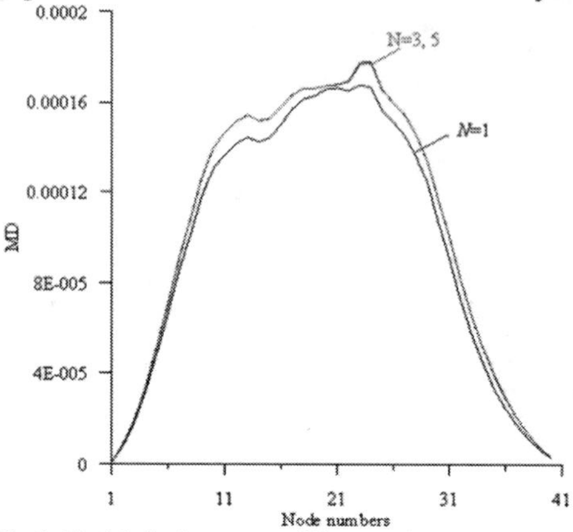

Fig.9. Modal displacements difference for beam with cut2.

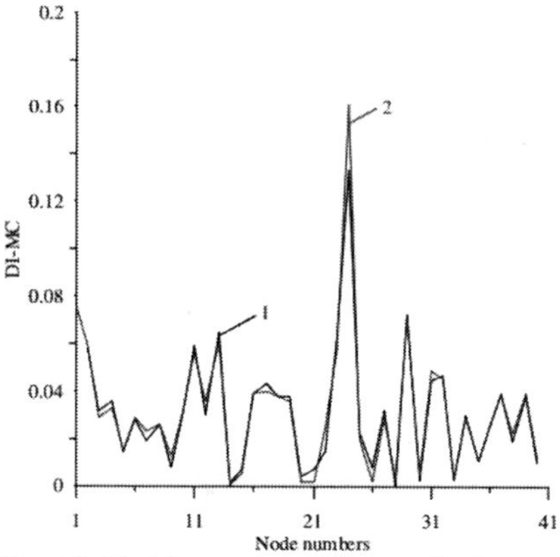

Fig. 10. Modal curvature difference for beam with Cut 2. 1-(black color) – N=1; 2 (-red color) N=7.

7. Conclusions

The numerical and experimental study shows that the modal analysis obtained by using SLDV can be used effectively for DD. Among the most popular modal based methods the one based on the modal strain energy gives the best results for experimentally obtained data (real data with noise). It requires, however, more modes to be considered. The modal properties of a structure could be successfully used for Poincaré map based forced response method and therefore for on line SHM. The improved strain energy method which includes the shear deformation could be much more effective when the shear stresses are essential.

DOI: 10.1051/matecconf/20168305007

Acknowledgements

The authors would like to acknowledge the financiall support by the Ministerial Research Project No. DEC-2013/11/B/ST8/04358 financed by the Polish National Science Centre. The support of this work through the Bulgarian NSF Grant DUNK-01/3 is gratefully acknowledged by the first author.

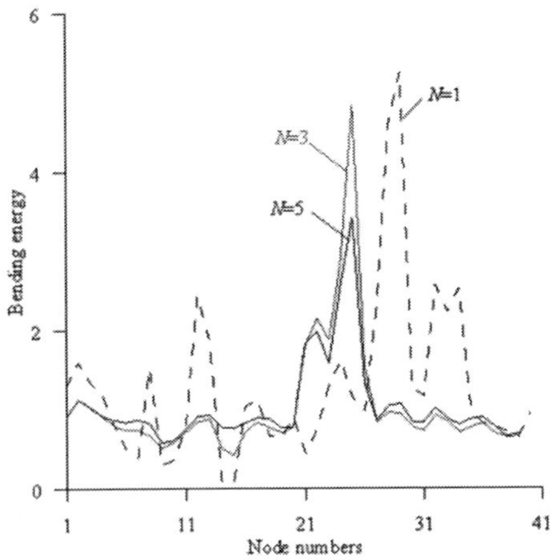

Figure 11 Strain energy based DI for beam with Cut3.

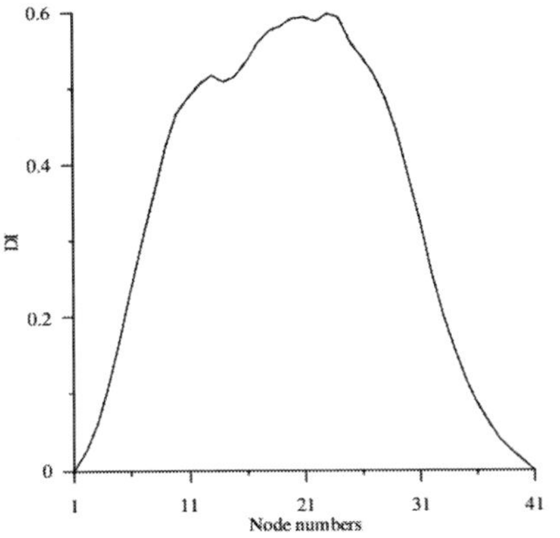

Fig. 12. Poincaré map based index for beam with Cut3.

References

1. P,Cawley, RD.Adams The Location of Defects in Structures from Measurements of Natural Frequencies. J. Strain Analysis, **14,** 49–57 (1979)
2. W.Ostachowicz, M.Krawczuk. Vibration Analysis of a Cracked Beam. Comp.&Struct;**36**, 245–250. (1990)
3. E. Manoach, S.Samborski, A. Mitura, J. Warminski, Vibration based damage detection in composite beams under temperature variations using Poincaré maps. Int J Mech Sci; **62**, 120–132, (2012).
4. S.W. Doebling, C.R. Farrar, M.B. Prime. Summary Review of Vibration-Based Damage Identification Methods, The Shock and Vibr. Digest, **30**, 91-105, (1998)
5. N.M.M. Maia, J.M M. Silva, E.A.M. Almas. Damage detection in structures: From mode shapes to frequency response function methods. Mechanical Systems and Signal Processing **17**, 489–498,(2003)
6. D. Montalvao, N. M. M. Maya, A. M. R. Ribeiro, A review of vibration based structural health monitoring with special emphasis on composite materials. The shock and vibr. Digest, **38**, 295-323 (2006)
7. J. V. Araujo dos Santos, N.M.M. Maia, C.M. Mota Soares. Structural damage identification: A Survey. *Trends in computational Structural Technologies*, Chapter 1 pp 1-24. Saxe-Coburg Publication, Stirlingshire, UK, 2008.
8. S. Gopalakrishnan, M. Ruzzene, S. Hanagud. *Computational Techniques for Structural Health Monitoring*, Spinger-Verlag London, Limited. (2011).
9. W. Fan, P. Qiao. Vibration-based damage identification methods: A Review and comparative study. Struct Health Monit **10** , 83-111, (2011)
10. A. Deraemaeker, E. Reynders, G. De Roeck, Vibration based SHM : comparison of the performance of modal features vs features extracted from spatial filters under changing environmental conditions, in: *ISMA*, pp. 849–86, (2006)
11. Y. K. Ho and D. J. Ewins, On the structural damage identification with mode shapes, in *European COST F3 Conference on System Identification and Structural Health Monitoring*, Madrid, Spain, 677–686, (2000)
12. AK Pandey, M Biswas, MM Sammam. Damage detection from changes in curvature mode shapes. J Sound Vib. **145** ,321-332 (1991)
13. A.K Pandey, M. Biswas, Damage Detection in Structures Using Changes in Flexibility. J. Sound Vib, **169**, 3-17 (1994.)
14. N. Stubbs, I.-T. Kim, C. R. Farrar, Field Verification of a Nondestructive Damage Localization and Sensitivity Estimator Algorithm, *13th Int. Modal Analysis Conf*, 210-218, (1995).
15. N.Stubbs, J.T. Kim, Damage localization in structures without baseline modal parameters, AIAA J., **34**, 1644–1649, (1996).
16. P. Cornwell, S. W. Doebling, C. R. Farrar . Application of the strain energy damage detection method to plate-like structures, J. Sound Vib **224**,. 359-374. (1999)
17. E.Manoach, I. Trendafilova, Large Amplitude Vibrations and damage detection of rectangular plates J. Sound Vib., 315, pp. 591-606. (2008)
18. I Trendafilova, E. Manoach, Vibration based damage detection in plates by using time series analysis, Mech. Syst. Signal Proc. **22**, 1092-1106. (2008)

Averaging in Parametrically Excited Systems – A State Space Formulation

Bastian Pfau[1,a], Thomas Breunung[1], Fadi Dohnal[2], and Richard Markert[1]

[1] Technische Universität Darmstadt, Institute for Applied Dynamics, 64287 Darmstadt, GERMANY
[2] Ansaldo Energia Switzerland Ltd, 5401 Baden, SWITZERLAND

Abstract. Parametric excitation can lead to instabilities as well as to an improved stability behavior, depending on whether a parametric resonance or anti-resonance is induced. In order to calculate the stability domains and boundaries, the method of averaging is applied. The problem is reformulated in state space representation, which allows a general handling of the averaging method especially for systems with non-symmetric system matrices. It is highlighted that this approach can enhance the first order approximation significantly. Two example systems are investigated: a generic mechanical system and a flexible rotor in journal bearings with adjustable geometry.

1 Introduction

Parametrically excited vibrations occur if one or more parameters of a mechanical system, for example stiffness or damping coefficients, are (mostly periodically) varying in time, independently of the system motion. A classic example is the pendulum with periodically moving support leading to the famous MATHIEU equation. A parametrically excited system may exhibit a destabilising parametric resonance, see for example [3,13]. In coupled systems of differential equations with time-periodic coefficients not only parametric resonances but also parametric anti-resonances may occur which induce an increased dissipation of vibration energy. The beneficial effect of a parametric anti-resonance on self-excited vibration was discovered by Aleš TONDL in his pioneering works [14,15]. This concept was transferred to general dynamic systems and interpret physically as an energy transfer between the vibration modes of the original system [6], followed by and experimental validation for simple systems including a flexible rotor. A recent summary on this topic can be found in [7]. The stability of such systems can be studied analytically by applying perturbation techniques following the methodologies in [2,12]. The present contribution revisits and translates the averaging method of first order, described by

$$\dot{\hat{\boldsymbol{x}}} = \varepsilon \sum_{k=1}^{K} \frac{1}{T_k} \int_0^{T_k} \boldsymbol{f}^k(\hat{\boldsymbol{x}}, t)\, dt = \varepsilon \langle \boldsymbol{f}(\hat{\boldsymbol{x}}, t) \rangle , \qquad (1)$$

to a state-space representation which enables the extension to systems with non-symmetrical matrices and strongly damped vibration modes and applies the methodology to the systems investigated in [9,11,8].

2 Procedure of Averaging

The equation of motion of a linear homogeneous mechanical system with N degrees of freedom can be written as

$$\boldsymbol{M}\ddot{\boldsymbol{q}} + \boldsymbol{B}(t)\,\dot{\boldsymbol{q}} + \boldsymbol{K}(t)\,\boldsymbol{q} = \boldsymbol{0} \qquad (2)$$

with an invertible mass matrix $\boldsymbol{M}$, a speed-proportional matrix $\boldsymbol{B}$ and a deflection-proportional matrix $\boldsymbol{K}$. In general, the matrices $\boldsymbol{B}$ and $\boldsymbol{K}$ are not symmetrical ($\boldsymbol{K} \neq \boldsymbol{K}^T$ and $\boldsymbol{B} \neq \boldsymbol{B}^T$), which can be caused by gyroscopic forces, inner damping or journal bearings. Furthermore, $\boldsymbol{B}$ and $\boldsymbol{K}$ alter with the parametric excitation frequency $\Omega_P = 2\pi/T_P$,

$$\boldsymbol{K}(t) = \boldsymbol{K}(t+T_P) \qquad \text{and} \qquad \boldsymbol{B}(t) = \boldsymbol{B}(t+T_P)\,. \qquad (3)$$

The periodic matrices can be expanded into a FOURIER series. Equation (2) then becomes

$$\begin{aligned} \boldsymbol{M}\ddot{\boldsymbol{q}} + \boldsymbol{B}_0\,\dot{\boldsymbol{q}} + \boldsymbol{K}_0\,\boldsymbol{q} = \\ -\varepsilon \sum_{k=1}^{K} \Big[\Big(\boldsymbol{K}^+ e^{ik\Omega_P t} + \boldsymbol{K}^- e^{-ik\Omega_P t} \Big) \boldsymbol{q} \\ + \Big(\boldsymbol{B}^+ e^{ik\Omega_P t} + \boldsymbol{B}^- e^{-ik\Omega_P t} \Big) \dot{\boldsymbol{q}} \Big] . \end{aligned} \qquad (4)$$

In this formulation, constant and varying parts are strictly separated on the left and right hand side. The factor ε is a small parameter. Formulating equation (4) in state space and transforming the system into modal space using left and right eigenvectors of the underlying system with constant coefficients,

$$\begin{bmatrix} \boldsymbol{q} \\ \dot{\boldsymbol{q}} \end{bmatrix} = \boldsymbol{Q} \begin{bmatrix} \boldsymbol{p} \\ \dot{\boldsymbol{p}} \end{bmatrix} = \boldsymbol{Q}\boldsymbol{z} , \qquad (5)$$

yields

$$\dot{\boldsymbol{z}} = \boldsymbol{\Lambda}\,\boldsymbol{z} - \varepsilon \sum_{k=1}^{K} \Big(\boldsymbol{P}^+ e^{ik\Omega_P t} + \boldsymbol{P}^- e^{-ik\Omega_P t} \Big) \boldsymbol{z} . \qquad (6)$$

Here, $\boldsymbol{\Lambda}$ is the diagonal matrix of the eigenvalues λ_n of the underlying system with constant coefficients, $\boldsymbol{z}^T = [\boldsymbol{p}^T, \dot{\boldsymbol{p}}^T]^T$ are the modal coordinates and the matrices $\boldsymbol{P}^+$ and $\boldsymbol{P}^-$ arise due to the parametric excitation and couple the modal degrees of freedom. For the n-th degree of freedom, equation (6) can be written as

$$\dot{z}_n = \lambda_n z_n - \varepsilon \sum_{m=1}^{2N} \sum_{k=1}^{K} \Big(P^+_{n,m} e^{ik\Omega_P t} + P^-_{n,m} e^{-ik\Omega_P t} \Big) z_m . \qquad (7)$$

[a] e-mail: pfau@ad.tu-darmstadt.de

Introducing the dimensionless time $\tau = \Omega_P t$ and allowing a small detuning $\Delta\Omega_P$ of the parametric excitation frequency,

$$\Omega_P = \Omega_{P0} + \varepsilon\Delta\Omega_P + O(\varepsilon^2), \tag{8}$$

the TAYLOR series expansion of equation (7) for small values of ε is given by

$$z_n' = \overline{\lambda}_n z_n - \frac{\varepsilon}{\Omega_{P0}}\left[\sum_{m=1}^{2N}\sum_{k=1}^{K}\left(P_{n,m}^{+}e^{ik\Omega_P t} + P_{n,m}^{-}e^{-ik\Omega_P t}\right)z_m + \Delta\Omega_P\overline{\lambda}_n z_n\right] + O(\varepsilon^2). \tag{9}$$

At this, $\overline{\lambda}_n = \lambda_n/\Omega_{P0}$ is a dimensionless eigenvalue and the mark $'$ denotes the derivative with respect to the dimensionless time τ. Making an ansatz of the type

$$z_n = u_n(\tau)\,e^{i\tau\,\Im\{\overline{\lambda}_n\}}, \tag{10}$$

which is equivalent to the approach of slowly changing phase and amplitude (see e. g. [2]), and rescale the real parts of the eigenvalues,

$$\Re\{\overline{\lambda}_n\} \mapsto \varepsilon\,\Re\{\overline{\lambda}_n\} \tag{11}$$

leads to the standard form and the method of averaging for quasiperiodic systems as outlined in [12] can be applied:

$$u_n' = \varepsilon f(u,t) + O(\varepsilon^2) = -\frac{\varepsilon}{\Omega_{P0}}\left[\sum_{m=1}^{2N}\sum_{k=1}^{K}\left(P_{n,m}^{+}e^{ik\tau} + P_{n,m}^{-}e^{-ik\tau}\right)e^{i\tau(\Im\{\overline{\lambda}_m\}-\Im\{\overline{\lambda}_n\})}u_m + (i\Delta\Omega_P\,\Im\{\overline{\lambda}_n\} - \Re\{\overline{\lambda}_n\}\Omega_{P0})u_n\right] + O(\varepsilon^2). \tag{12}$$

Averaging the harmonic functions yields always zeros, unless the term becomes resonant. The terms of equation (12) and their resonance condition are given in table 1. Summarizing the last two terms leads to the parametric resonance frequencies at which parametric resonances and anti-resonances can occur,

$$\Omega_{P0} = \frac{\mp\Im\{\lambda_n\} \pm \Im\{\lambda_m\}}{k}. \tag{13}$$

Compared to the standard literature, e. g. [3, 13], the imaginary parts of the eigenvalues decide on the parametric resonance frequencies, not the undamped eigenfrequencies.

summand	resonance condition
$i\Delta\Omega\,\Im\{\overline{\lambda}_n\}$	always
$\Re\{\overline{\lambda}_n\}\Omega_{P0}$	always
$P_{n,m}^{+}$	$-\Im\{\overline{\lambda}_n\} + \Im\{\overline{\lambda}_m\} + k = 0$
$P_{n,m}^{-}$	$-\Im\{\overline{\lambda}_n\} + \Im\{\overline{\lambda}_m\} - k = 0$

Table 1. Remaining terms (resonant terms) and their conditions after applying the method of averaging on equation (12)

From now on, higher orders of the FOURIER series are neglected, $K = 1$. A second order approximation in ε and a consideration of higher orders in k are discussed in [4].

General case
If the parametric excitation frequency is not equal to a parametric resonance frequency, averaging of equation (12) yields

$$\hat{u}_n' = -\frac{\varepsilon}{\Omega_{P0}}\sigma_n\hat{u}_n + O(\varepsilon^2) \tag{14}$$

with

$$\sigma_n = i\frac{\Im\{\lambda_n\}\Delta\Omega_P}{\Omega_{P0}} - \Re\{\lambda_n\}$$

for all $n = 1\ldots 2N$.

Parametric resonances and anti-resonances
If the parametric excitation frequency is equal to a parametric resonance frequency or in other words, equation (13) is fulfilled for certain values of $n = \bar{n}$ and $m = \bar{m}$, averaging of equation (12) yields

$$\begin{bmatrix}\hat{u}_{\bar{n}}'\\ \hat{u}_{\bar{m}}'\end{bmatrix} = -\frac{\varepsilon}{\Omega_{P0}}\begin{bmatrix}\sigma_{\bar{n}} & P_{\bar{n},\bar{m}}^{\pm}\\ P_{\bar{m},\bar{n}}^{\mp} & \sigma_{\bar{m}}\end{bmatrix}\begin{bmatrix}\hat{u}_{\bar{n}}\\ \hat{u}_{\bar{m}}\end{bmatrix} + O(\varepsilon^2). \tag{15}$$

This simplified equation covers the cases for fundamental parametric resonance frequencies (1st kind) and parametric combination resonance frequencies (2nd kind). Table 2 lists the indices for the substitution of $\bar{n}$ and $\bar{m}$ in equation (15). If the resonance condition with the upper sign (column 3 in table 2) is fulfilled, the upper signs in exponent of the counter diagonal elements are valid. Analogous holds for the lower signs.

resonance frequency kind	resonance frequency type	resonance condition	$\bar{n}$	$\bar{m}$
1.	-	$2\Im\{\overline{\lambda}_{\bar{n}}\} \mp 1 = 0$	$\bar{n}$	$\bar{n}+N$
2.	difference	$\Im\{\overline{\lambda}_{\bar{n}}\} - \Im\{\overline{\lambda}_{\bar{m}}\} \mp 1 = 0$	$\bar{n}$	$\bar{m}$
			$\bar{m}+N$	$\bar{n}+N$
	sum	$\Im\{\overline{\lambda}_{\bar{n}}\} + \Im\{\overline{\lambda}_{\bar{m}}\} \mp 1 = 0$	$\bar{n}$	$\bar{m}+N$
			$\bar{m}$	$\bar{n}+N$

Table 2. Indices for equation (15) for the modes $\bar{n}$ and $\bar{m}$ in dependence of kind and type of the parametric resonance frequency; $+N$ denotes the complex conjugated eigenvalue

Equation (15) is not valid in this simplified form if some eigenvalues have the same imaginary part as $\lambda_{\bar{n}}$ or $\lambda_{\bar{m}}$. This is for example the case if purely real eigenvalues are existing. In this case, averaging of equation (12) leads to a larger matrix which is shown in [4] in more detail.

3 Stability investigation

In this section, the calculation of the stability of the averaged System $\hat{\boldsymbol{u}}' = \boldsymbol{A}\,\hat{\boldsymbol{u}}$ (equation (14) respectively (15)) under the neglection of higher orders in ε is presented. If the parametric resonance frequencies are not close to each other, an interaction between them can be ruled out. In this case, it is sufficient to consider one parametric resonance frequency for the calculation of each stability domain, which is the common method in literature [12]. If two or more stability domains overlap then the isolated treatment of only one parametric resonance frequency is strictly no longer valid. The present investigation establishes an empiric approach which takes all involved parametric resonances into account.

3.1 No overlapping of parametric resonances

This subsection is, analogous to section 2, divided into a general case and into the case of parametric resonances and anti-resonances.

DOI: 10.1051/matecconf/20168305008

General case
Investigating the stability of equilibria of equation (14) yields one stability condition,

$$\Re\{\lambda_n\} < 0\,, \tag{16}$$

for all $n = 1 \ldots 2N$. This condition is also to find in every good book on structural dynamics [10] or systems theory.

Parametric resonances and anti-resonances
Investigating the stability of equilibria of equation (15) using the criteria from Routh and Hurwitz for complex matrices (see [1]) yields the stability conditions

$$\Re\{\lambda_{\bar{n}}\} + \Re\{\lambda_{\bar{m}}\} < 0 \tag{17}$$

and

$$\Delta\Omega_P^2 + 2a_1\Delta\Omega_P + a_0 > 0\,. \tag{18}$$

For the remaining modal degrees of freedom, $n \neq \bar{n}$, $\bar{m}$, $\bar{n}+N$, $\bar{m}+N$, equation (16) has to be fulfilled for stability. Setting equation (18) equal to zero, critical values for $\Delta\Omega_P$ are calculated to

$$\Delta\Omega_{P,c}^{\pm} = -a_1 \pm \sqrt{a_1^2 - a_0} = \Delta\Omega_{P,s} \pm \Delta\Omega_{P,w}\,. \tag{19}$$

Herein, $\Delta\Omega_{P,s}$ defines the distance between the parametric resonance frequency and the skeleton line of the stability domain, also denoted as frequency shift (see e. g. [5]), and $\Delta\Omega_{P,w}$ is the width of the stability domain. The parameters a_1 and a_0 are given by

$$2a_1 = C_1 C_2 \big(\Re\{\lambda_{\bar{m}}\} - \Re\{\lambda_{\bar{n}}\}\big) \tag{20}$$

and

$$a_0 = \frac{C_1\,\Omega_{P0}}{\Im\{\lambda_{\bar{m}}\} - \Im\{\lambda_{\bar{n}}\}}\Big[- C_2^2 + \big(\Re\{\lambda_{\bar{n}}\} + \Re\{\lambda_{\bar{m}}\}\big)^2 \cdot\Big(\Im\{P^{\pm}_{\bar{n}\bar{m}}\}\Im\{P^{\mp}_{\bar{m}\bar{n}}\} - \Re\{P^{\pm}_{\bar{n}\bar{m}}\}\Re\{P^{\mp}_{\bar{m}\bar{n}}\} + \Re\{\lambda_{\bar{n}}\}\Re\{\lambda_{\bar{m}}\}\Big)\Big] \tag{21}$$

with

$$\begin{aligned} C_1 &= \Omega_{P0}\Big[\Re\{\lambda_{\bar{n}}\}\Re\{\lambda_{\bar{m}}\}\big(\Im\{\lambda_{\bar{m}}\} - \Im\{\lambda_{\bar{n}}\}\big)\Big]^{-1}, \\ C_2 &= \Re\{P^{\mp}_{\bar{n}\bar{m}}\}\Im\{P^{\mp}_{\bar{m}\bar{n}}\} + \Re\{P^{\mp}_{\bar{m}\bar{n}}\}\Im\{P^{\pm}_{\bar{n}\bar{m}}\}\,. \end{aligned} \tag{22}$$

This defines a domain of stability (respectively instability) which is given by

$$\begin{aligned} &\Omega_{P0} + \varepsilon(\Delta\Omega_{P,s} - \Delta\Omega_{P,w}) < \Omega_P \\ &\qquad < \Omega_{P0} + \varepsilon(\Delta\Omega_{P,s} + \Delta\Omega_{P,w})\,. \end{aligned} \tag{23}$$

The given solution is only valid if the imaginary parts of all other eigenvalues λ_n ($n \neq \bar{n}, \bar{m}$) are not equal to $\Im\{\lambda_n\}$ or $\Im\{\lambda_m\}$. For these cases an analytical solution is not found yet and have to be treated using numerical calculations.

3.2 Overlapping of parametric resonances

If parametric resonances and anti-resonances are close to each other, the way described in section 3.1 may lead to contradictory results, due to the isolated treatment of only one parametric resonance frequency. An empirical approach is proposed here based on the knowledge that the eigenvalues of the averaged system $\hat{\boldsymbol{u}}' = \boldsymbol{A}\,\hat{\boldsymbol{u}}$ at parametric excitation (equations (15)) differ from the eigenvalues λ of the system with constant coefficients (see [6]). The interaction of R involved parametric resonance frequencies Ω^r_{P0} ($r = 1...R$) is investigated. The eigenvalue of the averaged system $\hat{\boldsymbol{u}}' = \boldsymbol{A}\,\hat{\boldsymbol{u}}$ for the n-th modal degree of freedom and the r-th parametric resonance frequency Ω^r_{P0} (see table 2) is calculated to

$$\rho^r_n = \begin{cases} \text{eig}(\boldsymbol{A}(\Omega^r_{P0})) & \text{if } n = \bar{n}, \bar{m} \\ \lambda_n & \text{if } n \neq \bar{n}, \bar{m} \end{cases}. \tag{24}$$

If the upper condition in equation (24) is fulfilled, the corresponding eigenvalue depends on the parametric excitation frequency, $\rho^r_n = \rho^r_n(\Omega_P)$. The real part of the parameter

$$\Delta\rho^r_n = \rho^r_n - \lambda_n \tag{25}$$

can be interpret as a measure of additional damping (negative or positive) which is caused by the r-th parametric resonance frequency Ω^r_{P0}. Summing these parameters over all r yields

$$\tilde{\rho}_n = \lambda_n + \sum_{r=R}^{R} \Delta\rho^r_n. \tag{26}$$

The real part of the parameter $\tilde{\rho}_n$ is used to evaluate the stability without a contradiction. For stability the real parts of $\tilde{\rho}_n$ need to be negative for all $n = 1 \ldots 2N$.

4 Examples

The presented equations are applied on two exemplary mechanical systems which have been studied in literature.

4.1 Generic mechanical system

First of all, a generic system with two degrees of freedoms from [9] is investigated, it is shown in figure 1. The system consists of two bodies that are connected by a spring with a time-periodic stiffness, $k_2(t) = \hat{k}_2(1+\varepsilon\cos\Omega_P t)$. The damping constant b_1 is negative which has an destabilizing effect on the system. The physical properties are listed in [9].

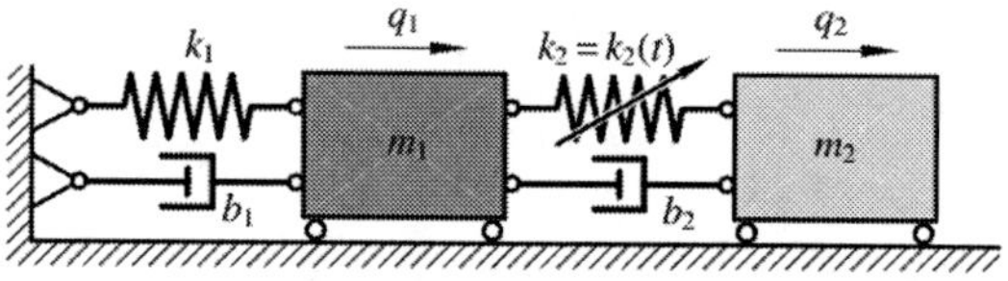

Fig. 1. Two degree of freedom system with stiffness excitation

Figure 2 shows a stability map in dependence of the parametric excitation frequency $\Omega_P / \sqrt{k_1/m_1}$ and the variable system parameter $Q^2 = k_2 m_1/(k_1 m_2)$. The gray and white areas depict stable and unstable areas, they are computed by numerical calculation using Floquet theory and are used for comparison purposes. The black and and gray lines are the stability boundaries calculated with the method of averaging of the first order presented in [9] and the state space representation in equation (12). The trivial equilibrium position of the system without parametric excitation is stable for $0.934 < Q < 0.964$. An additional stability domain occurs due to a parametric anti-resonance at the parametric resonance frequency $|\Im\{\lambda_2\} - \Im\{\lambda_1\}|$. Figure 2 visualizes that the black lines and the stability boundaries from Floquet theory match very well. Comparing the results from the different approaches of the averaging method show that the formulation in state space enhances the first order approximation significantly and shows the same accuracy as the higher order approximation in [9].

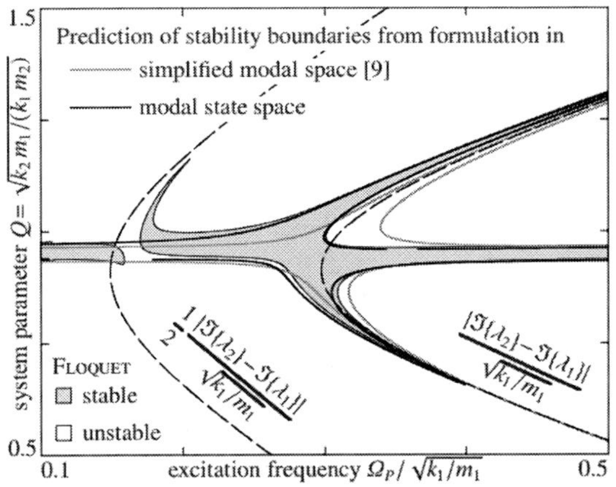

Fig. 2. Stability map for the generic system ($\varepsilon=0.3$)

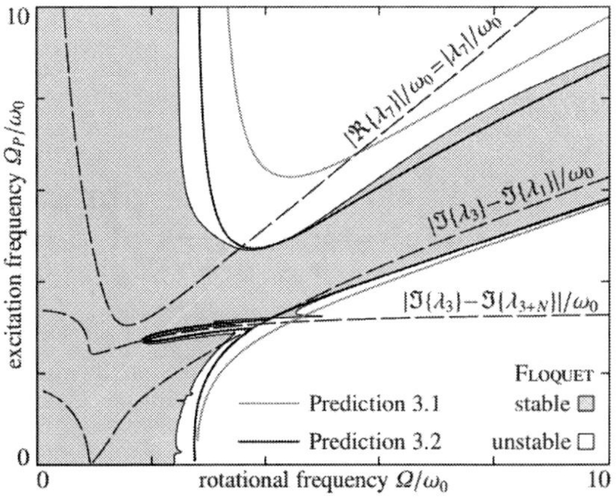

Fig. 3. Stability map for a rotor system in journal bearings with adjustable geometry (see also [11] and [8])

4.2 Flexible rotor in journal bearings

The second example is a flexible rotor in journal bearings with adjustable geometry, the problem is described in detail in [11,8]. The aim of the proposed journal bearing is a suppression of an instability induced by the bearing (oil whip) which has to be avoided in any case. Figure 3 shows an example of a stability map. The system without parametric excitation is stable up to a rotational speed of $\Omega/\omega_0 \approx 2.4$ which limits the permissible speed range. By exploiting a parametric anti-resonance ($|\Im\{\lambda_3\}-\Im\{\lambda_1\}|/\omega_0$), the operation speed range is increased significantly.

In this example, treating of one single parametric resonance frequency (chapter 3.1) causes a relatively large deviation in comparison to the results from FLOQUET theory. The reason for this is the interaction of several parametric resonance frequencies, especially a coupling with the strongly damped eigenvalue $|\lambda_7|$.Taking them into account and applying the empirical method described in 3.2 (equation (26)) yields a better approximation which allows for engineering design optimisations.

5 Conclusions

In this paper, the method of averaging was applied on linear parametrically excited systems. It was shown that the formulation in state space implicates several advantages: an improved approximation as well as an enhanced handling of the averaging method. An experimental validation for the suppression of bearing induced instabilities using parametric anti-resonances is in progress.

Acknowledgements

This work is dedicated to Aleš TONDL (*1925, †2016). He was an excellent scientist in the field of vibrations and rotor dynamics. The present work would not exist without his discovery of the beneficial effect of parametric anti-resonance.

References

1. H. Bilharz. Bemerkung zu einem Satze von Hurwitz. *Z. angew. Math. Mech.*, 24(2):77–82, 1944.
2. N. Boguliubov and Y. A. Mitropolskij. *Asymptotische Methoden in der Theorie der nichtlinearen Schwingungen.* Akademie Verlag Berlin, 1965.
3. V. Bolotin. *The Dynamic Stability of Elastic Systems.* Holden-Day, San Francisco, California, 1964.
4. T. Breunung. Semianalytische Berechnungen von Stabilitätskarten eines gleitgelagerten Rotorsystems mit Parametererregung. Masterthesis, Technische Universität Darmstadt, 2016.
5. F. Dohnal. Suppressing self-excited vibrations by synchronous and time-periodic stiffness and damping variation. *Journal of Sound and Vibration*, 306:136–152, 2007.
6. F. Dohnal. Damping by Parametric Stiffness Excitation: Resonance and Anti-Resonance. *Journal of Vibration and Control*, 14(5):669–688, 2008.
7. F. Dohnal. *A contribution to the mitigation of transient vibrations, Parametric anti-resonance: theory, experiment and interpretation.* habilitation thesis, Technische Universität Darmstadt, 2012.
8. F. Dohnal, B. Pfau, and A. Chasalevris. Analytical predictions of a flexible rotor in journal bearings with adjustable geometry to suppress bearing induced instabilities. In *Proceedings of the 13th Int. Conf. on Dynamical Systems - Theory and Applications*, Łódź, Poland, December 2015.
9. F. Dohnal and F. Verhulst. Averaging in vibration suppression by parametric stiffness excitation. *Nonlinear Dynamics*, 54:231–248, 2008.
10. R. Gasch, K. Knothe, and R. Liebich. *Strukturdynamik - Diskrete Systeme und Kontinua.* Springer, 2012.
11. B. Pfau, M. Rieken, and R. Markert. Numerische Untersuchungen eines verstellbaren Gleitlagers zur Unterdrückung von Instabilitäten mittels Parameter-Antiresonanzen. In *First IFToMM D-A-CH Conference*, Dortmund, Germany, 2015.
12. J. A. Sanders, F. Verhulst, and J. Murdock. *Averaging Methods in Nonlinear Dynamical Systems.* Spr., 2007.
13. G. Schmidt. *Parametererregte Schwingungen.* Deutscher Verlag der Wissenschaften, Berlin, 1975.
14. A. Tondl. *On the interaction between self-excited and parametric vibrations.* Monographs and Memoranda No. 25, National Research Institute for Machine Design, Běchovice, 1978.
15. A. Tondl. To the Problem of Quenching Self-Excited Vibrations. *Acta Techn. CSAV*, 43:109–116, 1998.

DOI: 10.1051/matecconf/20168305009

Nonlinear transverse vibrations of clamped beams carrying two or three concentrated masses at various locations

A.Rahmouni[1, 2] and R. Benamar[1]

[1]Laboratoire des Etudes en Simulation, Instrumentation et Mesures , EMI, Mohammed V University, Rabat, Maroc
[2]Laboratoire Mécanique Productique et Génie Industriel, EST, University Hassan II-Casablanca, Casablanca, Maroc

Abstract: In a recent work, a discrete model for geometrically nonlinear transverse free constrained vibrations of beams with various end conditions has been developed and validated via comparison with known results corresponding to nonlinear vibration of clamped beams carrying a concentrated mass. It is extended here to continuous beams carrying two or three concentrated masses at various locations and subjected to large vibration amplitudes. The discrete model used is an *N*-dof (*N*-Degrees of Freedom) system made of *N* masses placed at the ends of solid bars connected by springs, presenting the beam flexural rigidity. The large transverse displacements of the bar ends induce a variation in their lengths giving rise to axial forces modelled by longitudinal springs causing nonlinearity. The calculations made allowed application of the semi-analytical model developed previously for nonlinear structural vibration involving three tensors, namely the mass tensor m_{ij} , the linear rigidity tensor k_{ij} and the nonlinearity tensor b_{ijkl} presenting the effect of the change in the bar lengths. The addition of three concentrated masses studied here induces a change in the mass matrix. By application of Hamilton's principle and spectral analysis in the modal basis, the nonlinear vibration problem is reduced to a nonlinear algebraic system, using an explicit method, developed previously for non-linear structural vibration. This study shows that concentrated masses may be used for practical purposes to shift the resonant frequency; if the three masses locations are appropriately chosen.

1 Introduction

Application of the discrete model developed in the work [1, 2, 3] is made here to a Bernoulli beam carrying two or three concentrated masses at various locations and subject to geometrical nonlinear vibration due to large transverse displacements. This model focuses on the known physical phenomenon of the dynamic behavior: the stretching of the beam induces nonlinearity. To escape the occurrence of modal coupling, internal resonance, bifurcation points and chaos, the amplitudes considered here do not exceed the radius of gyration r of the cross section of the beam with respect to the neutral axis given by $r = \sqrt{I/S}$, where S is the beam section area and I is the quadratic moment of the beam with respect to the neutral fiber (in the case of a rectangular section, r is equal to the thickness of the beam divided by $\sqrt{12}$). This study shows that the developed model may be used to study successfully non-linear vibrations of beams carrying many concentrated masses simply by changing the mass matrix but without any change in the linear and nonlinear stiffness tensors, corresponding to a uniform beam defined in [1].

In addition, this discrete model may be very easily adapted to the study of beams with variable cross sections, with concentrated masses or stiffness, or with discontinuities in the section, the stiffness or the material properties.

2 Presentation and nomenclature

The studied model of a beam with three concentrate dmasses M_s, M_t and M_q is shown in Figure 1:

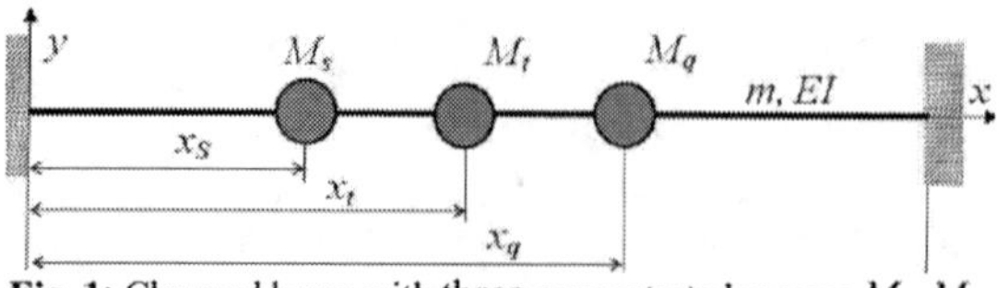

Fig. 1: Clamped beam with three concentrated masses M_s, M_t and M_q

Figure 2 shows the discrete system with *N*-dof considered in the present application, consisting on *N* masses m_1, ..., $M_s + m_{s-1}$, ..., $M_t + m_{t-1}$, ..., $M_q + m_{q-1}$, ..., m_N connected by N+2 torsional springs and N+2 longitudinal tension springs, considered in its neutral position.

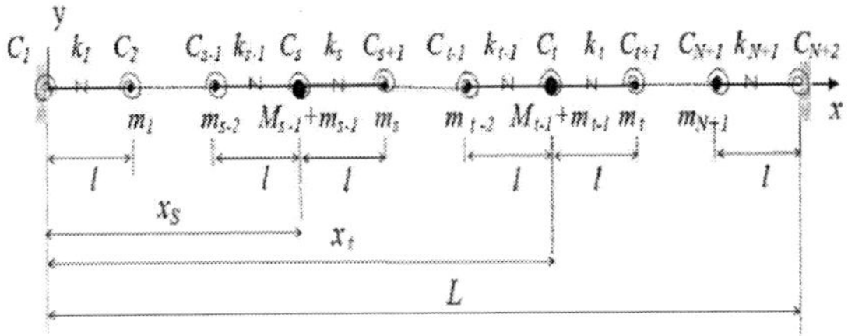

Fig. 2: Discrete system with several degrees of freedom (N - dof), modelling a clamped beam with two concentrated masses M_s and M_t

where the stiffness of the torsionals springs representing the flexural rigidity of the beam is $C_i = \frac{EI}{l}$. Equation (1) determines the location coordinates of the three masses:

$$i = \frac{x_i}{l} + 1 \quad \text{for } s, t \text{ and } q \tag{1}$$

N must be a common multiple of s, t and q.

3 Dimensionless formulations

The following equations link the dimensional values to the dimensionless ones (with an asterisk):

$$y^* = \frac{y}{r} \tag{2}$$

$$\omega^* = \frac{\omega L^2}{\sqrt{\frac{EI}{\rho S}}} \tag{3}$$

$$\eta_i = \frac{x_i}{L} \quad i = s \text{ ou } t \tag{4}$$

$$\alpha_i = \frac{\text{concentrated mass}}{\text{total mass of the beam}} = \frac{M_i}{m} = \frac{M_i}{\rho S L} \quad i = s \text{ ou } t \tag{5}$$

where ρ is the density of the beam density, E the Young's modulus, α_i le ratio of the concentrated mass i to the total mass of the beam, and η_i the non-dimensional location of the concentrated mass i.

4 General form of the masses matrix in the case of a beam with three concentrated masses

The presence of the three masses in the nodes s, t and q of the beam induces changes of the sth, t^{th} and q^{th} diagonal elements of the mass matrix. The matrix $[m_{ij}]$ of order N-2 has the form [2]:

$$[m_{ij}] = \frac{\rho S L}{(N+1)} \begin{pmatrix} 1 & 0 & 0 & \dots & & \dots & 0 & 0 \\ 0 & 1 & \dots & & 0 & & & 0 \\ 0 & \dots & & 0 & \dots & & & \\ \dots & & & (1+(N+1)\alpha_s) & \dots & & & \dots \\ 0 & & \cdots & 0 & 0 & & & 0 \\ \dots & & & \dots & (1+(N+1)\alpha_t) & & & 0 \\ & & & & \dots & 0 & \dots & 0 \\ 0 & \dots & & & & 1+(N+1)\alpha_q) & 0 & 0 \\ & & & & & 0 & 1 & 0 \\ 0 & 0 & \dots & & & \dots & 0 & 1 \end{pmatrix} \tag{6}$$

5 General form of the linear stiffness matrix $[k_{ij}]$

We consider the case where the masses do not change the bending stiffness. The general form of the linear stiffness matrix $[k_{ij}]$ to take into account (of order N-2) is unchanged, compared to that of a uniform beam. In the case of a beam carrying masses it can be written as follows [2]:

$$[\mathbf{K}_{cc(N-2)}] = \frac{(N+1)^3 EI}{L^3} \begin{pmatrix} 6 & -4 & 1 & 0 & . & . & . & 0 \\ -4 & 6 & -4 & 1 & 0 & . & . & . \\ 1 & -4 & 6 & -4 & 1 & 0 & . & . \\ 0 & . & . & . & . & . & . & . \\ . & . & . & . & . & . & . & . \\ . & . & . & . & . & . & . & . \\ . & . & . & 0 & 1 & -4 & 6 & -4 \\ . & . & . & . & 0 & 1 & -4 & 6 \end{pmatrix} \tag{7}$$

6 Results in the linear case for a beam with two concentrated masses

The results obtained for N = 49 are given in Table 1

Table 1: The first three natural frequencies of the discrete system (N=49), for $\alpha_i = 1$, $\eta_q = 0$ and different values of η_s, η_t

η_s	0	0.1	0.2	0.3	0.4	0.5
s	1	5	10	15	20	24
η_t	0	0.9	0.8	0.7	0.6	0.5
t	49	44	39	34	29	25
$\omega^{cc\,nl}_{disc\,1}$	22.37	22.21	17.02	12.35	9.95	9.16
$\omega^{cc\,nl}_{disc\,2}$	64.18	52.10	31.00	26.43	33.49	63.11
$\omega^{cc\,nl}_{disc\,3}$	125.65	78.65	64.81	99.52	115.53	94.97

It is noted that when the masses are placed in the middle of the beam, the first natural frequency decreases; this is due to the shape of the first vibration mode which has a maximum amplitude in the middle of the beam, while the second eigenmode remains unchanged at this location since the form of this mode has a node in the middle of the beam. Note also that when the masses are placed in 1/4 of the beam span, the second natural frequency decreases until the value of the first natural frequency (for $\eta_s = 0{,}3$; $\eta_t = 0{,}7$ and $\eta_q = 0$; this is due to the shape of the second vibration mode having three antinodes at these locations of the beam.

7 Cases of non-linear vibration of a beam carrying two or three concentrated masses

It should be noted that the coefficients of linear rigidity and non-linear tensors $\bar{k}_{ij}$ and $\bar{b}_{ijkl}$ calculated in the modal basis are different from those of a continuous beam, because the transition matrix $[\Phi]$ has undergone a change, due to addiction of concentrated masses, as illustrated in the following equations.

$$\bar{m}_{ij} = \Phi_{si}\Phi_{tj} m_{st} \qquad i,j = 1,...,N \tag{8}$$

$$\bar{k}_{ij} = \varphi_{si}\varphi_{tj} k_{st} \qquad i,j = 1,....,N \tag{9}$$

$$\bar{b}_{ijkl} = \varphi_{si}\varphi_{tj}\varphi_{pk}\varphi_{ql} b_{stpq} \qquad i,j,k,l = 1,....,N \tag{10}$$

In the displacement basis, the nonlinear system is written as:

$$[\mathbf{M}]\{\ddot{\mathbf{y}}\} + [\mathbf{K}]\{\mathbf{y}\} + \frac{3}{2}[\mathbf{B}(y)]\{\mathbf{y}\} = \{\mathbf{0}\} \tag{11}$$

where [M], [K], and [B (*y*)] are respectively the matrix mass, the linear matrix rigidity, and the nonlinear rigidity tensor. It should be noted here that Equation (11) becomes, if the non-linear term is neglected, identical to the classical equation of linear modal analysis:

$$[\mathbf{M}]\{\ddot{\mathbf{y}}\}+[\mathbf{K}]\{\mathbf{y}\}=\{\mathbf{0}\} \tag{12}$$

The tonsorial form of the non linear energy V_{nl} is:

$$V_{\text{nl}} = \frac{1}{2} y_i y_j y_k y_l b_{ijkl} \tag{13}$$

Using the Pythagorean Theorem, we express the nonlinear energy versus the characteristics of the beam [1], see Figure3:

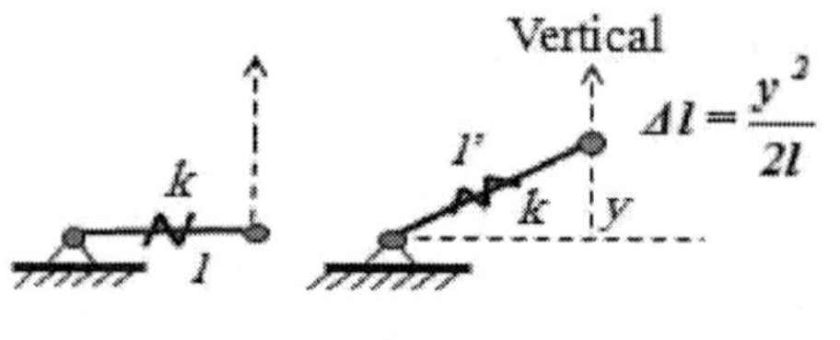

Fig. 3: modelling the tensile bars rigidity $k = ES/l$ and calculating the nonlinear energy with neglecting horizontal displacements

Using the known formula $V_{nl} = \frac{1}{2} k (\Delta l)^2$, the nonlinear potential energy is deduced as $V_{nl} \cong \frac{k}{8l^2} y^4$.

By the same approach, we calculate the nonlinear energy for all of the bars (see Figure 4), which we compare to the tonsorial expression to determine the coefficients of the nonlinear stiffness tensor [B (*y*)] [1] (see equations (14) to (17)).

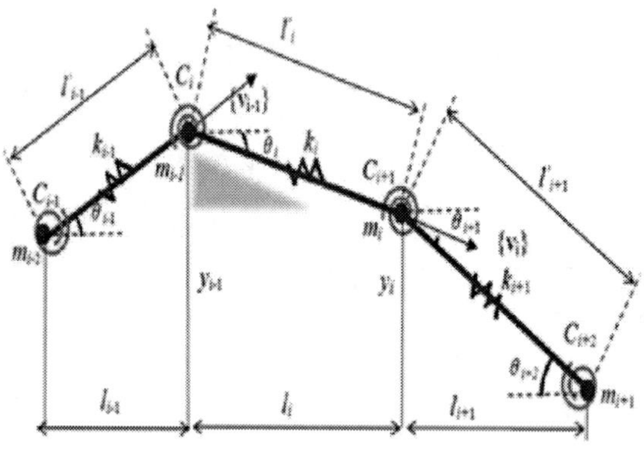

Fig. 4 Discrete system zoomed in the neighbourhood of bar *i*

$$b_{iiii} \cong 2\frac{ES}{8l^3} \cong 2(N+1)^3 \frac{ES}{8L^3} \qquad i = 1,....,N \tag{14}$$

$$b_{i(i-1)(i-1)(i-1)} = b_{(i-1)i(i-1)(i-1)} = b_{(i-1)(i-1)i(i-1)} = b_{(i-1)(i-1)(i-1)i} = \frac{ES}{8l^3} \qquad i = 2,....,N \tag{15}$$

$$b_{i(i-1)(i-1)i} = b_{(i-1)i(i-1)i} = b_{(i-1)(i-1)ii} = b_{(i-1)ii(i-1)} = b_{i(i-1)i(i-1)} = b_{ii(i-1)(i-1)} \cong \frac{ES}{8l^3} \qquad i = 2,....,N \tag{16}$$

$$b_{iii(i-1)} = b_{ii(i-1)i} = b_{i(i-1)ii} = b_{(i-1)iii} \cong -\frac{ES}{8l^3} \qquad i = 2,...,N \tag{17}$$

The other coefficients are equal to zero.

Our method consists on applying Hamilton's principle in the modal basis:

$$\delta \int_0^{2\pi/\omega^{\text{nl}}_{\text{disc}}} (V - T)\,\text{dt} = 0 \tag{18}$$

We obtain a system of nonlinear algebraic equations in which time is not involved:

$$3a_i a_j a_k \bar{b}_{ijkr} + 2a_i \bar{k}_{ir} - 2a_i (\omega^{\text{nl}}_{\text{disc}})^2 \bar{m}_{ir} = 0 \quad i,j,k,r = 1,N \tag{19}$$

We used the explicit method presented in [4] in the Modal basis to solve this equation: The explicit formulation is based on an approximation which consists on assuming, when the first nonlinear mode shape is under examination, that the contribution vector $\{a\}^T = \left[a_1 a_2 \ldots . a_N\right]$ can be written as $\{a\}^T = \left[a_1 \varepsilon_2 \ldots \varepsilon_i \ldots \varepsilon_N\right]$ with ε_i being small compared to a_1 ($a_1^2 \varepsilon_i = 0$ $\varepsilon_i \varepsilon_j \varepsilon_k \approx 0$ $a_1 \varepsilon_i \varepsilon_j \approx 0$). Neglecting in the expression $a_i a_j a_k \bar{b}_{ijkr}$ of Eq. (19), which involves summation for the repeated indices i, j, k the first, second and third-order terms so that the only remaining term $a_1^3 \bar{b}_{111r}$, leads to:

$$\left(\overline{k_{jr}} - (\omega_{disc}^{nl})^2 \overline{m_{jr}}\right)\varepsilon_j + \frac{3}{2} a_1^3 \bar{b}_{111r} = 0 \quad \text{for } r=2,\ldots,N \qquad (20)$$

The contributions $\left\{a_{1disc}\ \varepsilon_1\ \varepsilon_2 \ldots \varepsilon_N\right\}$ in the modal basis are calculated by [1]:

$$\varepsilon_r = \frac{\frac{3}{2} a_1^3 \overline{b_{111r}}}{((\omega_{disc}^{nl})^2 \overline{m_{rr}} - \overline{k_{rr}})} \qquad \text{for} \quad r=2,\ldots,N \qquad (21)$$

The non linear frequency is calculated by:

$$(\omega_{disc}^{nl})^2 = \frac{\overline{k_{11}}}{m_{11}} + \frac{3}{2}\frac{\overline{b_{1111}}}{m_{11}} a_1^2 \qquad (22)$$

After solving the equation in the modal basis, the amplitudes are calculated in the displacement basis:

$$\left[A_1\ A_2 \ldots\ldots A_N\right]^T = \left[\varphi\right]^{-1}\left\{a_1\ \varepsilon_1\ \varepsilon_2 \ldots \varepsilon_N\right\} \qquad (23)$$

The amplitude is the maximum of $(A_1 A_2 \ldots A_N)$, and we plot the dimensionless frequency curve as a function of the dimensionless amplitude

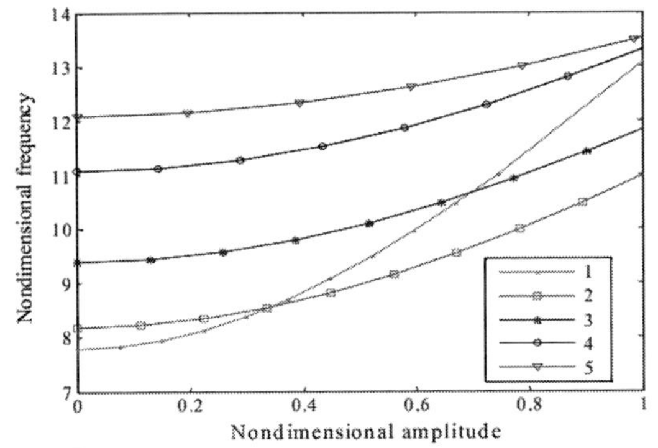

Fig. 5: Frequency curves according to the amplitude corresponding to the discrete systems with $N = 49$ dof, $\alpha_i = 1$, and (**1**:$\eta_s = \eta_t = \eta_q = 0.5$; **2**: $\eta_s = 0.4$, $\eta_t = 0.5$ and $\eta_q = 0.6$; **3**: $\eta_s = 0.3$, $\eta_t = 0.5$ and $\eta_q = 0.7$; **4**: $\eta_s = 0.2$, $\eta_t = 0.5$ and $\eta_q = 0.8$; **5**: $\eta_s = 0.1$, $\eta_t = 0.5$ and $\eta_q = 0.9$)

We note that when the three masses are in the middle of the beam, stretching are important which increases the nonlinearity.

8 Conclusions

The discrete model developed and validated in the case of a continuous beam presented in [1, 2, 3], was applied to beams with two or three concentrated masses. Linear and nonlinear vibrations were examined. This shows the effectiveness of this discrete model, its formulation and the associated program for the study of linear and nonlinear vibrations of a beam with discontinuities in the distribution of masses. The concentrated masses change significantly the beam dynamic response.

References

1. A. Rahmouni, Z. Beidouri and R. Benamar, *A discrete model for geometrically nonlinear transverse free constrained vibrations of beams with various end conditions*, Journal of Sound and Vibration **332** (2013) 5115–5134.
2. A. Rahmouni, Z. Beidouri and R. Benamar, *A discrete model for the natural frequency and mode shapes of beams with various boundary conditions* MATEC Web of Conferences **01**(2012); 1.DOI: 10.1051/matecconf/20120110015
3. A. Rahmouni and R. Benamar *A discrete model for geometrically nonlinear transverse free constrained vibrations of clamped beam carrying a concentrated mass at various locations,* EURODYN (2014) ISSN 2311-9020 ISBN 978-972-752-165-4 , 2093-2099
4. M. El. Kadiri, R. Benamar, R.G. White, *Improvement of the semi-analytical method, based on Hamilton's principle and spectral analysis, for determination of the geometrically non-linear free response of thin straight structures. Part I: application to C-C and SS-C beams,* Journal of Sound and Vibration, **249**(2002) 263-305.

Nonlinear vibration of rotating system near resonance

G. Sypniewska-Kamińska[1], R. Starosta[1], and J. Awrejcewicz[2]

[1]Poznan University of Technology, Institute of Applied Mechanics , Piotrowo 3, 60-965 Poznań, Poland
[2]Technical University of Lodz, Department of Automatics and Biomechanics Sciences, Stefanowskiego 90-924, Łódź, Poland

Abstract. The paper concerns analysis of nonlinear vibration of the rotating systems consisted of rigid disks mounted on the elastic massless shaft. The investigations are focused on their behaviour under the resonance conditions. The analytical method of multiple scales in time domain is applied to the analysis of dynamics of the system near main resonance. The transition phenomenon depending on the value of the nonlinearity parameter is discussed.

1 Introduction

Torsional vibration belongs to major problems in design of the power transmission systems. The dynamic stresses caused by torsional oscillations, especially near resonance when the amplitudes grow significantly, may be very large and can lead to failure of the whole system.

Both discrete and continuous models are commonly used in order to investigate the torsional vibration of the power transmission systems [1,4,6].

The problem of vibration in nonlinear systems has to be treated with due care. The numerical methods commonly used in solving nonlinear problems do not create the possibility of a synthetic regard on the diversity of the dynamic behaviour of the systems. Some surprising types of the vibrational motion may be not detected when only the numerical methods are used. The Multiple Scales Method (MSM) in time domain, belonging to the asymptotic methods, not only allows obtain the approximate solutions, but also gives the possibility to exact investigate the influence of the particular parameters on the motion. This method is applied in the paper to solve the problem of torsional vibration of the shaft on which the rigid disks are mounted. The relationships, obtained in the analytical form, between the amplitudes and phases of vibration at resonance have been examined. Associating the MSM with an interesting approach proposed by Manevich [3] known as Limiting Phase Trajectory (LPT) Method allows to predict the sudden change of the shape of the amplitude modulation which appears in resonance at a certain value of the nonlinearity parameter. A similar approach, but based completely on the proposed by Manevich formulation in the field of complex functions, was applied to investigate the torsional vibration in the system with one degree of freedom [5].

2 The description of the system

Two rotating and vibrating systems presented in figure 1 are considered in the paper. The systems consist of two or three disks fasten to the elastic shaft. The shaft is relatively light, so its mass may be neglected. Elastic properties of the shaft are nonlinear of the cubic type. Therefore, two coefficients of stiffness, marked by K_j and K_{jn}, are introduced for the j-th part of the shaft. Moreover, viscous damping is taken into account, and C_j for j=1,2 denotes the damping coefficient. The whole system is mounted on frictionless bearings which are also ideal in the geometric sense. One of the disks is under the action of the harmonically changing torque $M(t) = M_0 \cos(Pt)$.

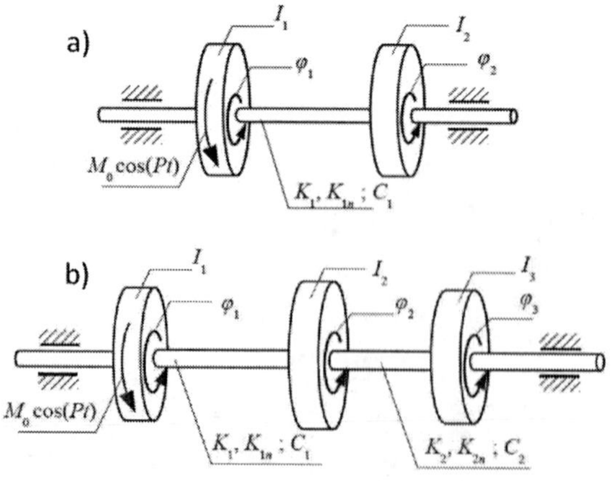

Fig. 1. Model of rotating systems with a) two and b) three disks.

3 Mathematical model

The systems have two or three degrees of freedom, respectively. The rotation angles φ_j of the *j*-th wheels are chosen as the generalized coordinates. The Lagrange formalism is used to derive the mathematical model describing the system.

The Lagrangian of the system in the most general form is

$$L = \frac{1}{2}\sum_{j=1}^{N} I_j \dot{\varphi}_j^2 - \frac{1}{2}\sum_{j=1}^{N-1} K_j(\varphi_j - \varphi_{j+1})^2 - \frac{1}{4}\sum_{j=1}^{N-1} K_{jn}(\varphi_j - \varphi_{j+1})^4 \quad (1)$$

where I_j is the moment of inertia of the *j*-th disk and *N* is the total number of them.

The external excitation and damping effects are included into mathematical model as the generalized forces. The equations of motion derived from the Lagrange equation of the second type are transformed into dimensionless form.

In order to investigate the vibration, it is very convenient to introduce the function describing the internal motion i.e. the screw angle of the *j*-th shaft

$$\Phi_j(t) = \varphi_j(t) - \varphi_{j+1}(t), \quad j = 1,\ldots,N-1. \quad (2)$$

In this way we exclude from calculation the rigid rotation of the shaft and further investigation concerns only the vibrational effects.

The governing equation for the system with two disks takes the form of the Duffing equation

$$\ddot{\phi}_1(\tau) + c_1\dot{\phi}_1(\tau) + \phi_1(\tau) + \alpha_1\phi_1(\tau)^3 = m_0 \cos p\tau. \quad (3)$$

For the case with three disks the governing equations are

$$\ddot{\phi}_1 + c_1\dot{\phi}_1 + \phi_1 + \alpha_1\phi_1^3 - c_2\gamma_3\dot{\phi}_2 - \gamma_3 w^2\phi_2 - \alpha_2\gamma_3\phi_2^3 = m_0\cos(p\tau), \quad (4)$$

$$\ddot{\phi}_2 + c_2\dot{\phi}_2 + w^2\phi_2 + \alpha_2\phi_2^3 - c_1\gamma_1\dot{\phi}_1 - \gamma_1\phi_1 - \alpha_1\gamma_1\phi_1^3 = 0, \quad (5)$$

where $m_0 = M_0/(I_1 k_1)$, $p = P/\sqrt{k_1}$, $c_j = \hat{c}_j/\sqrt{k_1}$, $\gamma_3 = I_3/(I_3 + I_2)$, $\gamma_1 = I_1/(I_1 + I_2)$, $\alpha_j = k_{jn}/k_1$, $w = k_2/k_1$, dimensionless time $\tau = \sqrt{k_1}\,t$, and $k_j = K_j(I_j + I_{j+1})/(I_j I_{j+1})$, $k_{jn} = K_{jn}(I_j + I_{j+1})/(I_j I_{j+1})$, $\hat{c}_j = C_j(I_j + I_{j+1})/(I_j I_{j+1})$ for *j*=1, …, *N*-1.

The equations (3) and (4) – (5) are supplemented by the initial conditions

$$\phi_j(0) = \phi_{0j}, \dot{\phi}_j(0) = \beta_{0j}, \quad j = 1,\ldots,N-1 \quad (6)$$

where ϕ_{0j} and β_{0j} are known values.

4 Asymptotic solution

The initial value problems are solved in the asymptotic way, using Multiple Scales Method (MSM) in time domain. Let us assume that the system is weakly nonlinear and the damping coefficient and the amplitude of the external torque are also of order of the small parameter ε

$$m_0 \to \tilde{m}_0\varepsilon, \alpha_j \to \tilde{\alpha}_j\varepsilon, c_j \to \tilde{c}_j\varepsilon. \quad (7)$$

The assumed form of the asymptotic solution is as follows

$$\phi_j(\tau) = \phi_{j0}(\tau_0,\tau_1) + \varepsilon\phi_{j1}(\tau_0,\tau_1), \quad (8)$$

where τ_0 and τ_1 are slow and fast time scales, respectively.

The differential operators relating to these time scales take the form

$$\frac{d}{d\tau} = \frac{\partial}{\partial\tau_0} + \varepsilon\frac{\partial}{\partial\tau_1}, \qquad \frac{d^2}{d\tau^2} = \frac{\partial^2}{\partial\tau_0^2} + 2\varepsilon\frac{\partial^2}{\partial\tau_0\partial\tau_1} + o(\varepsilon^2) \quad (9)$$

Then relations (7) – (9) are introduced into equations of motion (3) or (4) – (5). In this way, the small parameter appears in the model. After arranging the equation with respect to the power of the small parameter, the equations of the first and the second order approximation are obtained.

4.1 Case I – shaft with two disks

The analysis is concerned on the case of the main resonance, that occurs when $p \approx 1$. In order to consider this case, the small detuning parameter σ is introduced in the form $p = 1 + \sigma = 1 + \varepsilon\tilde{\sigma}$.

After substituting (7) – (9) into Eq. (3) and arranging it with respect to the powers of the small parameter ε we obtain

– the equation of order ε^0

$$\frac{\partial^2\phi_{10}}{\partial\tau_0^2} + \phi_{01} = 0, \quad (10)$$

– the equation of order ε^1

$$\frac{\partial^2\phi_{11}}{\partial\tau_0^2} + \phi_{11} = \tilde{m}_0\cos(1 + \tilde{\sigma}\tau_1) - \tilde{\alpha}_1\phi_{10}^3 - \tilde{c}_1\frac{\partial\phi_{10}}{\partial\tau_0} - 2\frac{\partial^2\phi_{10}}{\partial\tau_0\partial\tau_1}. \quad (11)$$

The solution of the first order equation (10) is introduced into Eq. (11).

We are interested in the bounded solutions, so all secular terms ought to vanish. Zeroing the secular terms leads to the solvability condition

$$\frac{dA(\tau_1)}{d\tau_1} + 3\tilde{\alpha}_1 A^2(\tau_1)\bar{A}(\tau_1) + i\tilde{c}_1 A(\tau_1) = \frac{1}{2}\tilde{m}_0 e^{i\tau_1\tilde{\sigma}}, \quad (12)$$

where $A(\tau_1)$ is the unknown complex function.

It is convenient to express the function $A(\tau_1)$ by the real-valued functions $a(\tau_1)$, $\psi(\tau_1)$ describing amplitude and phase respectively

$$A(\tau_1) = \frac{1}{2}a(\tau_1)e^{i\psi(\tau_1)}. \quad (13)$$

Taking advantage of relations (9) and (13), and returning to the original denotations of the system parameters, one can obtain

$$\frac{da}{d\tau}+ia\frac{d\psi}{d\tau}+\frac{1}{2}c_1 a-\frac{3}{8}i\alpha_1 a^3=\frac{m_0}{2}e^{i(\sigma\tau)}e^{-i\psi}. \qquad (14)$$

Separating in (14) real and imaginary parts, we can write

$$\begin{aligned}&\frac{da}{d\tau}+\frac{1}{2}c_1 a=\frac{m_0}{2}\cos\theta,\\&-a\frac{d\theta}{d\tau}+a\sigma-\frac{3}{8}\alpha_1 a^3=\frac{m_0}{2}\sin\theta,\end{aligned} \qquad (15)$$

where $\theta=\sigma\tau-\psi$ is the modified phase. Equations (15) describe the modulation of the amplitude a and modified phase θ.

Let us consider the non-damped vibrations ($c_1=0$). In that case, the set of Eqs (15) has the first integral

$$H=-a\frac{m_0}{2}\sin\theta+\sigma\frac{a^2}{2}-\frac{3}{32}\alpha_1 a^4=\text{const}, \qquad (16)$$

where the constant H depends on the initial conditions. Equation (16) represents one-parameter family of the curves on the plane (a, θ). We are especially interested in the case when maximal energy exchange between the system and the external loading appears. This situation takes place for $H=0$. In that case the first integral (16) has the form:

$$-16m_0\sin\theta+16\sigma a-3\alpha_1 a^3=0. \qquad (17)$$

It is easily to show that the curve given by Eq. (17) has extrema for $\theta=-\pi/2$ and $\theta=\pi/2$. The qualitative change in the behaviour of the system is observable for the critical value of the nonlinearity parameter

$$\alpha_c=\frac{64\sigma^3}{81m_0^2}. \qquad (18)$$

The trajectories on the plane (a, θ) for various values of α_1 are presented and discussed in details in paper [5].

When α_1 exceeds the critical value α_c, the shape of the amplitude modulation rapidly changes. The amplitude modulation in time τ, obtained from Eqs (15), is presented in figures 2 and 3 for α_1 a bit smaller, and a bit greater than α_c.

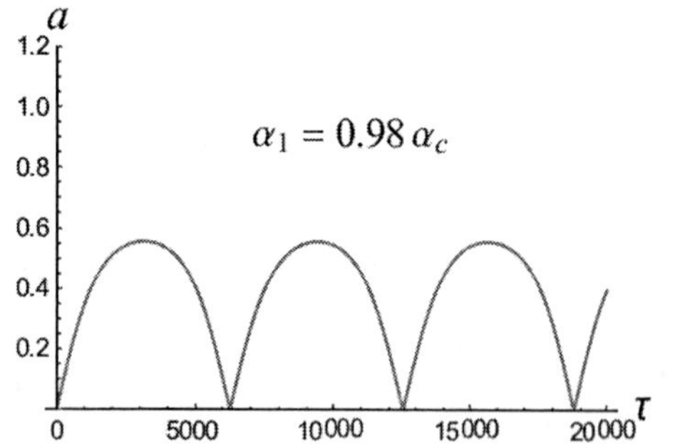

Fig. 2. Amplitude modulation for α_1 just below α_c ($\sigma=0.01$, $m_0=0.002$, $\phi_{01}=0$, $\beta_{01}=0$).

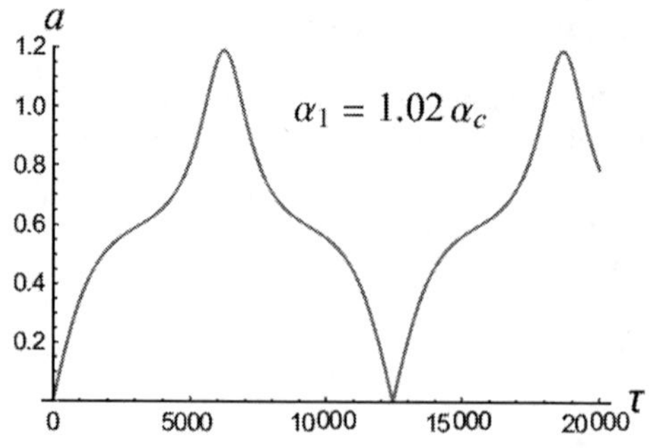

Fig. 3. Amplitude modulation for α_1 just above α_c ($\sigma=0.01$, $m_0=0.002$, $\phi_{01}=0$, $\beta_{01}=0$).

The transition which occurs at α_c indicates stepwise change in the intensity of energy exchange between the system and external loading.

4.2 Case II – shaft with three discs

The procedure applied to solve the problem is in general similar but more complicated to that one in the case I. Now the equations of order ε^0 reads

$$\begin{aligned}&\frac{\partial^2\phi_{10}}{\partial\tau_0^2}+\phi_{10}-\gamma_3 w^2\phi_{20}=0,\\&\frac{\partial^2\phi_{20}}{\partial\tau_0^2}+w^2\phi_{20}-\gamma_1\phi_{10}=0,\end{aligned} \qquad (19)$$

and the equations of order ε^1 are

$$\begin{aligned}&\frac{\partial^2\phi_{11}}{\partial\tau_0^2}+\phi_{11}-\gamma_3 w^2\phi_{21}=-\tilde{\alpha}_1\phi_{10}^{\ 3}+\tilde{\alpha}_2\gamma_3\phi_{20}^{\ 3}-\tilde{c}_1\frac{\partial\phi_{10}}{\partial\tau_0}+\\&\tilde{c}_2\gamma_3\frac{\partial\phi_{20}}{\partial\tau_0}-2\frac{\partial^2\phi_{10}}{\partial\tau_0\partial\tau_1}+\tilde{m}_0\cos(p\tau_0),\\&\frac{\partial^2\phi_{21}}{\partial\tau_0^2}+w^2\phi_{21}(\tau_0,\tau_1)-\gamma_1\phi_{11}(\tau_0,\tau_1)=\tilde{\alpha}_1\gamma_1\phi_{10}(\tau_0,\tau_1)^3-\\&\tilde{\alpha}_2\phi_{20}(\tau_0,\tau_1)^3+\tilde{c}_1\gamma_1\frac{\partial\phi_{10}}{\partial\tau_0}-\tilde{c}_2\frac{\partial\phi_{20}}{\partial\tau_0}-2\frac{\partial^2\phi_{20}}{\partial\tau_0\partial\tau_1}.\end{aligned} \qquad (20)$$

The linear differential operator in Eqs (19) – (20) has two different self-frequencies

$$\omega_1=\sqrt{\frac{1+w^2-\sqrt{1-2w^2+w^4+4w^2\gamma_1\gamma_3}}{2}}, \qquad (21)$$

$$\omega_2=\sqrt{\frac{1+w^2+\sqrt{1-2w^2+w^4+4w^2\gamma_1\gamma_3}}{2}}. \qquad (22)$$

The linear combination of the normal modes forms the solution of the first order equations. It can be written in the exponential form

$$\phi_{10}=A_{11}(\tau_1)e^{i\tau_0\omega_1}+A_{12}(\tau_1)e^{i\tau_0\omega_2}+CC, \qquad (23)$$

$$\phi_{20}=\frac{1-\omega_1^2}{\gamma_3 w^2}A_{11}(\tau_1)e^{i\tau_0\omega_1}+\frac{1-\omega_2^2}{\gamma_3 w^2}A_{12}(\tau_1)e^{i\tau_0\omega_2}+CC, \qquad (24)$$

where $A_{1j}(\tau_1)$ for j=1,2 are unknown complex functions and CC denotes the complex conjugate parts.

It is convenient to introduce the real-valued functions $a_{1j}(\tau_1)$, $\psi_{1j}(\tau_1), j=1,2$, similarly as in the previous section (see Eq. (13)).

Let us consider the case $p \approx \omega_1$ of main resonance. The detuning parameter σ is introduced in the form $p=\omega_1+\sigma=\omega_1+\varepsilon\tilde{\sigma}$.

The requirement of zeroing the secular terms in the equations (20) leads to the modulation equation

$$\frac{d}{d\tau}a_{1j}(\tau)=F_j(\alpha_1,\alpha_2,c_1,c_2,w,\gamma_1,\gamma_3,m_0,\sigma_1)\quad j=1,2 \qquad (25)$$

$$\frac{d}{d\tau}\psi_{1j}(\tau)=G_j(\alpha_1,\alpha_2,c_1,c_2,w,\gamma_1,\gamma_3,m_0,\sigma_1)\quad j=1,2 \qquad (26)$$

where functions F and G can be expressed explicitly.

The amplitude modulation obtained from Eqs (25) – (26) are presented in figures 4 and 5.

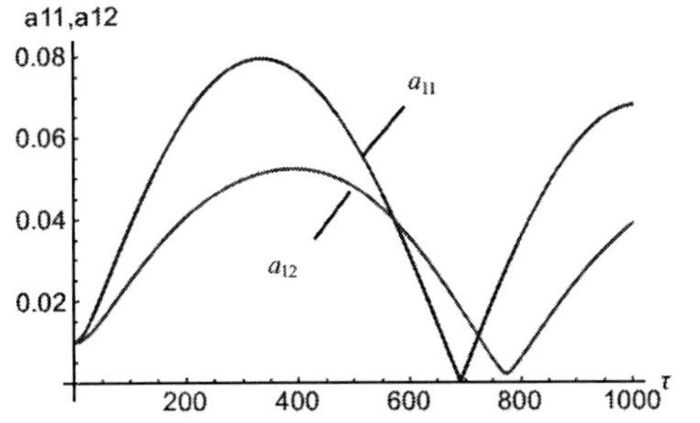

Fig. 4. The modulation of amplitude for $\phi_{10}(\tau)$ related to the first and second natural frequency ($\alpha_1=0.4$, $\alpha_2=0.15$, $c_1=c_2=0.001$ $\gamma_1=0.57$, $\gamma_3=2.5$, $\sigma=0.01$, $m_0=0.001$, $\phi_{01}=0$, $\beta_{01}=0$).

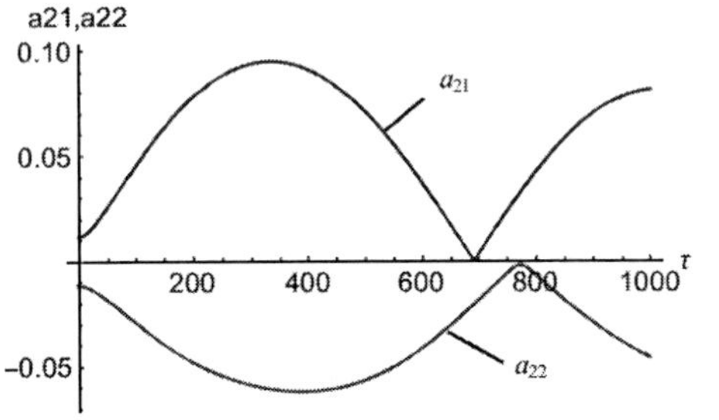

Fig. 5. The modulation of amplitude for $\phi_{20}(\tau)$ related to the first and second natural frequency($\alpha_1=0.4$, $\alpha_2=0.15$, $c_1=c_2=0.001$ $\gamma_1=0.57$, $\gamma_3=2.5$, $\sigma=0.01$, $m_0=0.001$, $\phi_{01}=0$, $\beta_{01}=0$).

The first mode vibration is connected with frequency ω_1 given in Eq. (21) and the second with frequency ω_2 given in Eq. (22).

5 Conclusions

Analysis of the nonlinear disks-shaft system has been done using the asymptotical Multiple Scales Method in time domain properly adopted to solve the problem.

The investigation has been focused mainly on the non-steady vibration of the forced system near the external resonance.

The main advantage of the asymptotic solutions consists in achieving qualitative information on the dynamics of the considered systems. The important non-linear dynamical transition has been detected for the critical value of the parameter connected with the spring nonlinearity in the case with two disks. The amplitudes modulation indicate the intensive energy exchange between the system and external excitation.

The second of considered problems which concerns the system reducible to the mathematical model of two degree of freedom is significantly more complicated and requires further studies. This model is strongly coupled what requires to introduce some modification into the classical formulation of the MSM. The eigenvalues of the differential operator i.e. the undamped natural frequencies are determined in the paper. The modulation equations for the amplitude and phase of the normal modes in case of the main resonance are also obtained.

Acknowledgements

This paper was financially supported by the National Science Centre of Poland under the grant MAESTRO 2, No. 2012/04/A/ST8/00738, for years 2013-2016.

References

1. Y.Huang Ying, S.Yang, F.Zhang, C.Zhao, Q.Ling, and H.Wang, Chinese J. Mech. Eng. **25**, 4 (2012)
2. J. Kožešnik, *Dynamics of machines (in Polish)*, (WNT, 1963)
3. L. I. Manevitch, A.I. Musienko, Nonlinear Dyn. **58** (2009)
4. B. Porter, J. Mech. Eng. Sci. **7**, 1 (1965)
5. R. Starosta, *Nonlinear dynamics of discrete systems – asymptotic approach (in Polish)*, Nr 457 Rozprawy, WPP, Poznań 2011
6. Z.Wenming, W.Bohua, Z.Shuangshuang and L.Shuang, Int. J. Com. Sci. **10**, 3 (2013)

 DOI: 10.1051/matecconf/20168305011

Electro-mechanical coupling of rotating 3D beams

S. Stoykov[1] and E. Manoach[2]

[1]Institute of Information and Communication Technologies, Bulgarian Academy of Sciences, Bulgaria
[2]Institute of Mechanics, Bulgarian Academy of Sciences, Bulgaria

Abstract. A rotating thin-walled beam with piezoelectric element is analysed. The beam is considered to vibrate in space, hence the longitudinal, transverse and torsional deformations are taken into account. The bending deformations of the beam are modelled by assuming Timoshenko's theory. Torsion is included by considering that the cross section rotates as a rigid body but can deform in longitudinal direction due to warping. The warping function is computed preliminary by the finite element method. The equation of motion is derived by the principle of virtual work and discretized in space by the Ritz method. Electro-mechanical coupling is included in the model by considering the internal electrical energy and the electric charge output. The piezo-electric constitutive relations are used in reduced form. The beam is assumed to rotate about a fixed axis with constant speed. The equation of motion is derived in rotating coordinate system, but the influence of the rotation of the coordinate system is taken into account through the inertia forces. Results in time domain are presented for different speeds of rotation and frequencies of vibration. The influence of the speed of rotation and of the frequency of vibration on the electrical output is presented and analysed.

1 Introduction

In the last decades, the need of independent energy sources for small devices has increased significantly. Small devices, such as sensors, have application in health monitoring and control of structures. The most popular devices that convert kinetic energy into electrical energy are the devices composed of piezoelectric elements [1].

In the current work, a rotating thin-walled beam with piezoelectric element is analysed. Three-dimensional displacement field is considered, i.e. the beam can vibrate in both transverse directions and it can perform longitudinal and torsional deformations. The beam equation of motion is derived by Timoshenko's beam assumption and it is considered that under torsion the cross section rotates as a rigid body but it can deform in longitudinal direction due to warping. The proposed model is suitable for beams with arbitrary open and closed thin-walled cross sections as well for solid cross sections. Clamped-free boundary conditions are assumed and rotation of the beam about fixed axis with constant speed is included in the model. The piezoelectric element is considered to be close to the clamped end. Geometrical nonlinearity is considered and electro-mechanical coupling is introduced into the model through the piezo-electric constitutive equations, used in reduced form.

The equation of motion is derived by the principle of virtual work and results in time domain are presented. Comparison between linear and nonlinear models is shown and the importance of the geometrical nonlinear terms on the displacements and on the electrical potential difference is outlined. The responses of the beam due to harmonic force and different speeds of rotation are compared.

2 Mathematical model

A thin-walled beam with piezoelectric element on the upper surface is considered (Fig. 1). The piezoelectric element does not cover the whole length of the beam. The displacements in longitudinal and transverse directions are expressed by the displacements and rotations on the reference line:

$$u(x,y,z,t) = u_0 - y\phi_z + z\phi_y + \psi\frac{\partial\theta_x}{\partial x}$$
$$v(x,y,z,t) = v_0 + y\cos(\theta_x) - y - z\sin(\theta_x) \quad (1)$$
$$w(x,y,z,t) = w_0 + y\sin(\theta_x) + z\cos(\theta_x) - z$$

where $u_0(x,t)$, $v_0(x,t)$ and $w_0(x,t)$ represent the displacements in longitudinal and transverse displacements in y and z axes, respectively. $\phi_y(x,t)$ and $\phi_z(x,t)$ represent the rotations of the cross section about y and z axes, $\theta_x(x,t)$ denotes the torsion and $\psi(y,z)$ is the warping function.

Fig. 1. Thin-walled beam with piezoelectric element.

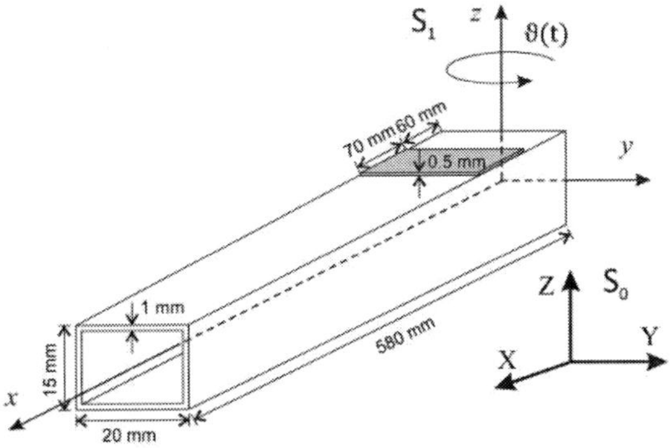

Analytical solution for the warping function does not exist for complex cross sections, thus it is computed numerically. The equation of equilibrium for the warping function is derived from the equilibrium equations of elasticity, where the stresses due to torsion are computed from the displacement field:

$$\tilde{C}_{66}^{k}\frac{\partial^2\psi}{\partial y^2}+\tilde{C}_{55}^{k}\frac{\partial^2\psi}{\partial z^2}=0 \quad \text{in } \Omega_k$$

$$\left(\tilde{C}_{66}^{k}-\tilde{C}_{66}^{k-1}\right)\left(\frac{\partial\psi}{\partial y}-z\right)n_y+ \qquad (2)$$

$$+\left(\tilde{C}_{55}^{k}-\tilde{C}_{55}^{k-1}\right)\left(\frac{\partial\psi}{\partial z}+y\right)n_z=0 \quad \text{on } \Gamma_k$$

The equation for the warping function is given for the case of beam with composite materials, because of the piezoelectric element. For the current beam, two warping functions are computed - one for the cross section without piezoelectric element and another one for the cross section with the piezoelectric element. In eq. (2) Ω_k is the area of the kth material and Γ_k is its contour. $\tilde{C}_{55}^{k}$ and $\tilde{C}_{66}^{k}$ are the shear moduli of the material from the reduced stiffness matrix. n_y and n_z are the components of the normal vector to the boundary. The normal vector points in the outer direction for the outer boundaries and it is considered to point in the lower numbered material for the boundaries between different layers. $\tilde{C}_{66}^{k-1}=\tilde{C}_{55}^{k-1}=0$ for the outer boundaries. Equation (2) is solved by the finite element method [2].

Geometrical nonlinearity is considered in the model. Hence the strains are expressed as:

$$\varepsilon_x=\frac{\partial u}{\partial x}+\frac{1}{2}\left(\frac{\partial v}{\partial x}\right)^2+\frac{1}{2}\left(\frac{\partial w}{\partial x}\right)^2$$

$$\gamma_{xz}=\frac{\partial w}{\partial x}+\frac{\partial u}{\partial z}+\frac{\partial v}{\partial z}\frac{\partial v}{\partial x}+\frac{\partial w}{\partial z}\frac{\partial w}{\partial x} \qquad (2)$$

$$\gamma_{xy}=\frac{\partial u}{\partial y}+\frac{\partial v}{\partial x}+\frac{\partial v}{\partial x}\frac{\partial v}{\partial y}+\frac{\partial w}{\partial x}\frac{\partial w}{\partial y}$$

The piezo-electric constitutive equations are used in their reduced form [1]:

$$\begin{Bmatrix}\sigma_x\\ \tau_{xz}\\ \tau_{xy}\\ D_3\end{Bmatrix}=\begin{bmatrix}E & 0 & 0 & -\bar{e}_{31}\\ 0 & G & 0 & 0\\ 0 & 0 & G & 0\\ \bar{e}_{31} & 0 & 0 & \bar{\varepsilon}_{33}^{S}\end{bmatrix}\begin{Bmatrix}\varepsilon_x\\ \gamma_{xz}\\ \gamma_{xy}\\ E_3\end{Bmatrix} \qquad (3)$$

where σ_x, τ_{xz} and τ_{xz} are bending and shear stresses, E is the Young's modulus of the material, G is the shear modulus, D_3 represents electric displacement component, $\bar{e}_{31}$ is an effective piezo-electric stress constant, $\bar{\varepsilon}_{33}^{S}$ is permittivity component and E_3 is electric field component. The electric field component can be presented as $E_3=-\frac{v(t)}{h_p}$ where $v(t)$ is an electric potential difference (voltage) and h_p is the thickness of the piezoelectric element.

The rotation of the beam is taken into account by considering two coordinate systems: one fixed, denoted by S_0, and another one rotating about the fixed coordinate system denoted by S_1. The displacement fields (1) are written in the rotating coordinate system S_1, it is considered that it rotates with constant speed $\dot{\vartheta}$ (rad/s). The rotation of the beam is introduced into the equation of motion through the inertia forces. For that purpose, the absolute acceleration of an arbitrary point P(x,y,z) from the beam is expressed with respect to the fixed coordinate system S_0 [3]:

$$\boldsymbol{a}_{\mathrm{P}S_0}=\boldsymbol{a}_{\mathrm{P}S_1}+\boldsymbol{\omega}\times(\boldsymbol{\omega}\times\boldsymbol{r})+2\,\boldsymbol{\omega}\times\boldsymbol{v}_{\mathrm{P}S_1} \qquad (4)$$

where $\boldsymbol{a}_{\mathrm{P}S_0}$ is the absolute acceleration of point P, i.e. the acceleration of point P with respect to S_0, $\boldsymbol{a}_{\mathrm{P}S_1}$ is the relative acceleration of point P, i.e. the acceleration of point P with respect to S_1, $\boldsymbol{v}_{\mathrm{P}S_1}$ is the relative velocity of point P, ω is the angular velocity vector of the rotating coordinate system, and $\boldsymbol{r}$ is the position vector of point P with respect to the origin of S_1. In eq. (4) $2\,\omega_{10}\times v_{\mathrm{P}S_1}$ represents the acceleration of Coriolis.

The equation of motion is derived by the principle of virtual work:

$$\delta W_V+\delta W_{in}+\delta W_E+\delta W_{ie}+\delta W_{nc}=0 \qquad (5)$$

where δW_V is the virtual work of internal forces, δW_{in} is the virtual work of inertia forces, δW_E is the virtual work of external forces, δW_{ie} is the virtual work of internal electrical energy and δW_{nc} is the non-conservative virtual work due to electric charge output.

The displacement components are expressed by shape functions and generalized coordinates. The shape functions used in [2] are used also here. They have to satisfy the essential boundary conditions. Applying the principle of virtual work, the following system of ordinary differential equations is obtained:

$$\mathbf{M}\ddot{\mathbf{q}}+\beta\mathbf{K}_{\mathrm{l}}\dot{\mathbf{q}}+\mathbf{CR}(\mathbf{q},\dot{\vartheta})\,\dot{\mathbf{q}}+\mathbf{K}_{\mathrm{l}}\mathbf{q}+\mathbf{K}_{\mathrm{nl}}(\mathbf{q})\mathbf{q}+$$

$$+\mathbf{T}(\mathbf{q},\dot{\vartheta})\,\mathbf{q}+\boldsymbol{\theta}_{\mathrm{l}}v+\boldsymbol{\theta}_{\mathrm{nl}}(\mathbf{q})v=\mathbf{f}-\mathbf{r}(\dot{\vartheta}) \qquad (6)$$

$$C_pv+Q-\boldsymbol{\theta}_{\mathrm{l}}^{\mathrm{T}}\mathbf{q}-\frac{1}{2}\boldsymbol{\theta}_{\mathrm{nl}}^{\mathrm{T}}(\mathbf{q})\mathbf{q}=0$$

where $\mathbf{M}$ is the mass matrix, $\mathbf{K}_{\mathrm{l}}$ is the stiffness matrix of constant terms, $\mathbf{K}_{\mathrm{nl}}(\mathbf{q})$ is the stiffness matrix that depends on the vector of generalized coordinates $\mathbf{q}$. $\mathbf{K}_{\mathrm{nl}}(\mathbf{q})$ results from the geometrical nonlinearity of the model and introduces quadratic and cubic terms at the equation of motion, $\boldsymbol{\theta}_{\mathrm{l}}$ is vector of constant terms which

DOI: 10.1051/matecconf/20168305011

results from the electro-mechanical coupling, $\mathbf{\theta}_{nl}(\mathbf{q})$ is vector which also results from the electro-mechanical coupling but it depends linearly on the vector of generalized coordinates $\mathbf{q}$. $\mathbf{f}$ represents the vector of generalised external forces, Q represents the electric charge output on the electrodes and C_p is the corresponding capacitance. $\mathbf{CR}(\dot{\vartheta})$ is matrix that depends on the speed of rotation, it results from the acceleration of Coriolis and $\mathbf{T}(\dot{\vartheta})$ and $\mathbf{r}(\dot{\vartheta})$ are matrix and vector that also arise from the rotation of the beam and depend on the speed of rotation. Stiffness proportional damping is introduced in the model. The equation of motion (6) is solved in time domain by the Newmark's method.

3 Numerical results

3.1. Validation

The derived equation of motion is validated by comparing the linear natural frequencies with results from the three-dimensional equation of elasticity applied to the same structure. The software Elmer [4] is used for that purpose and space discretization is achieved by quadratic tetrahedrons. The beam is considered to be composed of brass and the piezoelectric element is considered to be PZT-5A [1]. The material properties are given in Table 1.

Table 1. Material properties of the structure.

	Brass	PZT-5A
Young modulus (E)	105 GPa	60.6 GPa
Shear modulus (G)	40 GPa	23 GPa
Density (ρ)	9000 kg/m^3	7500 kg/m^3
Piezo-electric stress constant ($\bar{e}_{31}$)	-	-10.4 C/m^2
Permittivity constant ($\bar{\varepsilon}^S_{33}$)	-	13.3 x 10^{-9} F/m

In the comparison of the natural frequencies, only the elastic properties of the piezoelectric element are taken into account. Clamped-free boundary conditions are considered. The first six natural frequencies are presented in Table 2. The results confirm that very good approximation is obtained by the proposed beam model.

Table 2. Validation of natural frequencies (rad/s) of the beam with piezoelectric element.

Mode	Elmer 3D 300 000 DOF	Beam model 60 DOF	Difference %
1	215.5521	215.8158	0.12
2	268.1170	268.1142	0.01
3	1312.9806	1321.1389	0.62
4	1646.1054	1651.9888	0.36
5	3587.8169	3645.1821	1.60
6	4490.5568	4533.8394	0.96

3.2 Influence of nonlinear terms

A comparison of the response of the beam between the linear and the geometrically nonlinear models is presented in this sub-section. A transverse force in z direction is applied on the free end of the beam and the rotation is not considered. The transverse displacement w_0 is shown in Fig. 2 and the electric potential difference, which results from the deformation of the beam due to the external force, is presented in Fig. 3. The excitation frequency is chosen to be one third from the fundamental frequency. The difference between the linear and nonlinear models is obvious not only on the displacements, but also on the electrical potential difference, which confirms that geometrical nonlinear terms are essential for accurate results.

Fig. 2. Transverse displacement w_0, at the free end, due to harmonic force $F_z = 100\cos(71.94t)$ applied at the free end in z direction, β=10^{-3}.— nonlinear model,- - - - linear model.

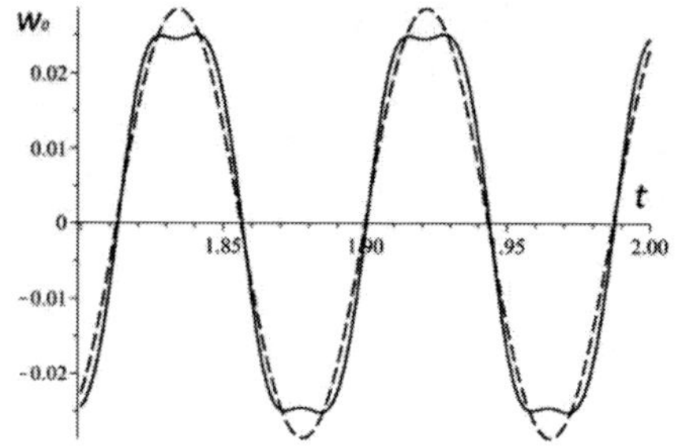

Fig. 3. Electric potential difference $v(t)$ due to harmonic force $F_z = 100\cos(71.94t)$,— nonlinear , - - - linear model.

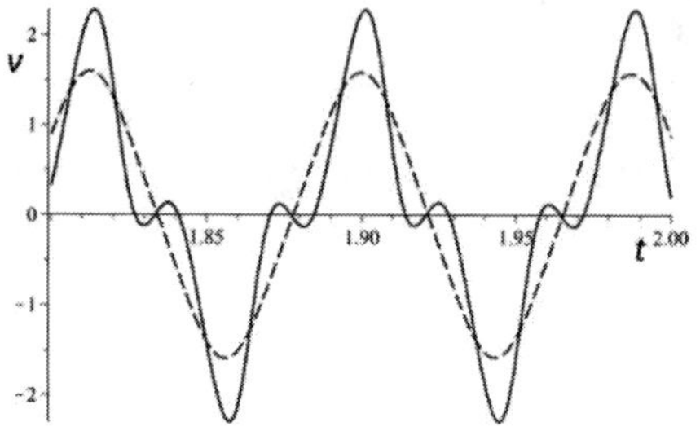

3.2 Influence of speed of rotation

The influence of the speed of rotation on the displacements and on the electrical potential difference is analysed in this sub-section. It is known that the speed of rotation makes the beam stiffer, due to the centrifugal forces, and the natural frequencies become higher. The first four natural frequencies for different speeds of rotation are presented in Table 3.

Table 3. Natural frequencies (rad/s) of rotating beam with different speeds of rotation (rad/s).

Mode\Speed	0	50	146	250
1	215.82	222.61	268.07	288.42
2	268.11	269.01	275.48	346.47
3	1321.14	1327.20	1371.95	1465.22
4	1651.99	1656.08	1686.55	1751.46

In order to excite both transverse displacements and the torsion and to have vibration in the three-dimensional space, harmonic forces are applied on the free end of the beam in directions of both transverse axes. An excitation frequency equal to the second natural frequency of the beam, when it does not rotate, is considered. This frequency is related with the first mode of vibration in *xy* plane, while the fundamental frequency is related with first mode in *xz* plane. The excitation frequency becomes fundamental frequency of the beam when the beam rotates with speed of 146 rad/s.

Fig. 4. Electric potential difference $v(t)$ due to external force $F_y = F_z = 100\cos(268\,t)$ and non-rotating beam.

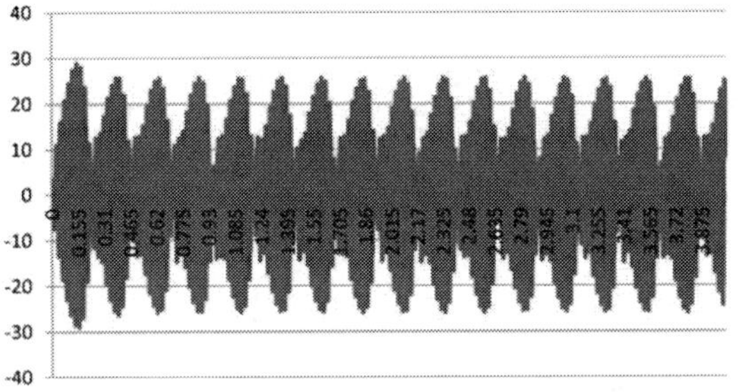

The resulting electric potential difference is presented in figures 4 to 7, for the cases of non-rotating beam, and for rotating beam with speeds of rotation respectively 50 rad/s, 146 rad/s and 250 rad/s.

Fig. 5. Electric potential difference $v(t)$ due to external force $F_y = F_z = 100\cos(268\,t)$ and rotation with speed 50 rad/s.

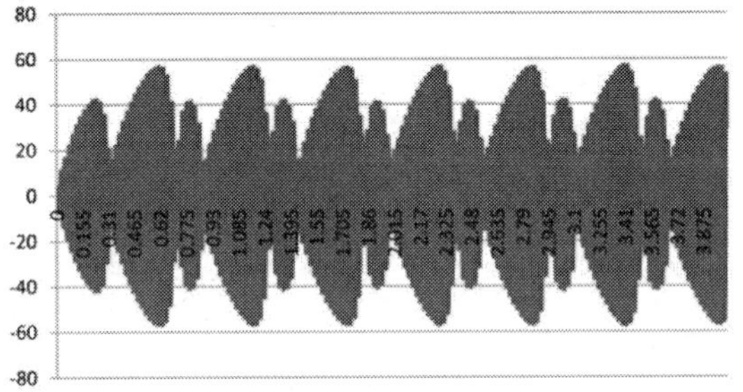

The results demonstrate that the speed of rotation influences significantly on the electrical potential difference and depending on the speed of rotation, different amounts of energy are generated.

Fig. 6. Electric potential difference $v(t)$ due to external force $F_y = F_z = 100\cos(268\,t)$ and rotation with speed 146 rad/s.

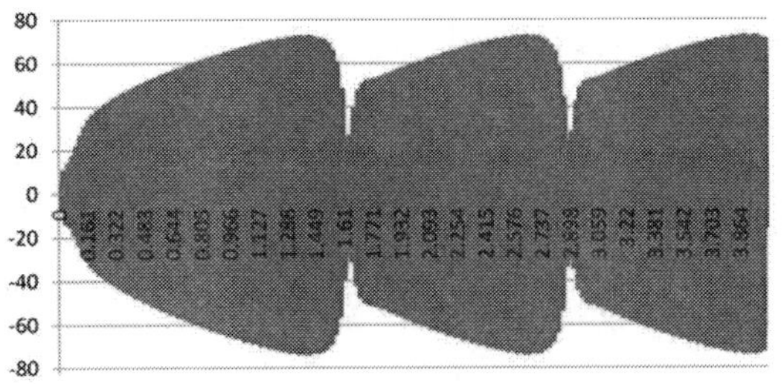

Fig.7. Electric potential difference $v(t)$ due to external force $F_y = F_z = 100\cos(268\,t)$ and rotation with speed 200 rad/s.

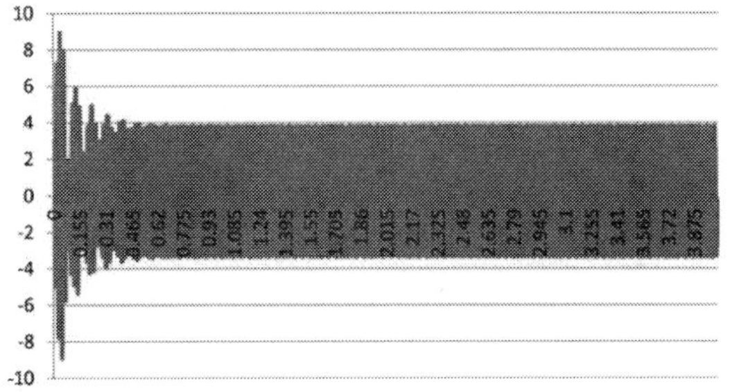

For the cases of non-rotating beam and beam rotating with speeds of 50 rad/s and 146 rad/s the excitation frequency of the external force is between first two natural frequencies. Because of the nonlinear terms and the interaction between first two modes the response of the beam becomes complex. Consequently the electrical potential difference is also complex. For the case of beam rotating with speed of 250 rad/s, the excitation frequency is smaller than the fundamental one, and the response of the beam is similar to harmonic with small amplitude. Thus, the resulting potential difference is also similar to harmonic and the amplitude is smaller than in the previous cases.

4 Conclusion

The equation of motion of rotating beam with piezoelectric element is derived considering geometrical type of nonlinearity and introducing electro-mechanical coupling. The importance of the nonlinear terms on the response of the beam is presented. The influence of the speed of rotation on the electric potential difference is outlined and it is shown that the speed of rotation can change significantly the response of the beam and the resulting electrical output.

Acknowledgements

The support of this work through the Bulgarian NSF Grant DUNK-01/3 is gratefully acknowledged. The second author acknowledges the support through H2020 Marie Sklodowska-Curie ITN grant (grant agreement No 643095).

References

1. A. Erturk, D. Inman, *Piezoelectric energy harvesting*, (John Wiley & Sons Ltd. 2011)
2. S. Stoykov, E. Manoach, S. Margenov, An efficient 3D numerical beam model based on cross sectional analysis and Ritz approximations, Z. Angew. Math. Mech. (2015), DOI 10.1002/zamm.201400139 (In Press)
3. A. Shabana, *Dynamics of multibody systems*, (Cambridge University Press, 2013)
4. Elmer web site: www.csc.fi/elmer (last time visited 02.06.2016)

MATEC Web of Conferences **83**, 05012 (2016) DOI: 10.1051/matecconf/20168305012
CSNDD 2016

On a Method for Solving of Multidimensional Equations of Mathematical Physics

M.D. Todorov[a]

Faculty Appl. Math. & Comp. Sci., Technical University of Sofia, Bulgaria

Abstract. We consider linear multidimensional evolutionary equations or the linear part of nonlinear ones. The complex structure and the presence of terms with different physical sense requires the coordinate splitting to be preceded by splitting by physical factors (processes). In contrast to the coordinate splitting this kind of splitting can be exact in some nodes and in the intervals it can be controlled. The method in question is developed in 80s of 20th century by G. Marchuk [1] and till now it is applied very successfully for solving various problems in ecology, air and water pollution, diffusion, etc. Here we show that this method is relevant for studying of propagating ultrashort localized pulses in nonlinear waveguides as well.

1 Introduction

In this paper we aim to demonstrate that the splitting by physical factors is applicable and can be efficient for solving multidimensional evolutionary problems in the optics. These equations possess very rich phenomenology, they contain nonlinear terms and do not admit analytic solutions being nonintegrable in the general case. Without loosing of generality we pay attention to an initial-value problem of (3+1)d equations of Schrödinger kind. We split the linear part of the differential operator and the initial conditions to two consequent Cauchy problems and by using a spectral analysis we proof that the original and the resulting splitting problems are equivalent in the ends of a given interval. For constant coefficients the above assertion holds in the whole interval. In opposite case for small enough step the approximation error varies in an acceptable range. The nonlinear terms are linearized by using of so-called inner iteration.

The method is applied successfully for numerical solving (3+1)d Schrödinger equation with sign-variable group velocity as well for (3+1)d amplitude equations (of envelope), when the propagation regimes of the light pulses may be ultra-short (femtosecond). The obtained results are reliable and give good predictions for the material quantities and dynamics of the light pulses.

2 (3+1)d Schrödinger Equation

Let us consider the Cauchy problem

$$\mathrm{i}\frac{\partial u}{\partial z} + \frac{1}{2k}\left(\frac{\partial^2 u}{\partial x^2} + \frac{\partial^2 u}{\partial y^2}\right) - \frac{\beta}{2}\frac{\partial^2 u}{\partial \tau^2} = 0, \qquad (1)$$

$$u(x, y, \tau, z = 0) = g(x, y, \tau), \qquad u|_{\Gamma} = 0$$

where Γ is the boundary of 3d spatial region, g is a given function, $\tau = t - \frac{z}{v_g}$ is the local time in a moving frame with group velocity of the pulse v_g, k is the wave number, β is the dispersion, and z is the evolutionary coordinate of the pulse. Eq. (1) describes two different physical process: the first one is the diffraction of the pulse represented by the equation

$$\mathrm{i}\frac{\partial u}{\partial z} + \frac{1}{2k}\left(\frac{\partial^2 u}{\partial x^2} + \frac{\partial^2 u}{\partial y^2}\right) = 0$$

$$u(x, y, \tau, z = 0) = g(x, y, \tau), \; u|_{\Gamma} = 0.$$

The second process relates to the dispersion and is governed by the equation

$$\mathrm{i}\frac{\partial u}{\partial z} - \frac{\beta}{2}\frac{\partial^2 u}{\partial \tau^2} = 0$$

$$u(x, y, \tau, z = 0) = g(x, y, \tau), \; u|_{\Gamma} = 0.$$

When the initial function and the sought solution approach fast zero uniformly in all arguments there exist Fourier integrals

$$g = \int_{-\infty}^{\infty}\int_{-\infty}^{\infty}\int_{-\infty}^{\infty} \hat{g}(m, n, p, z) \exp\left[\mathrm{i}(mx + ny + p\tau)\right] \mathrm{d}m\mathrm{d}n\mathrm{d}p \qquad (2)$$

$$u = \int_{-\infty}^{\infty}\int_{-\infty}^{\infty}\int_{-\infty}^{\infty} \hat{u}(m, n, p, z) \exp\left[(mx + ny + p\tau)\right] \mathrm{d}m\mathrm{d}n\mathrm{d}p \qquad (3)$$

where $\hat{g}$ and $\hat{u}$ are the Fourier images of the functions g and u.

On the other hand the triple Fourier transform of Eq. (1) with respect to x, y, and τ gives a Cauchy problem of the ordinary differential equation

$$\mathrm{i}\frac{\partial \hat{u}}{\partial z} - \frac{1}{2k}(m^2\hat{u} + n^2\hat{u}) + \frac{\beta}{2}p^2\hat{u} = 0 \qquad \hat{u}(z = 0) = \hat{g},$$

which admits the exact solution

$$\hat{u} = \hat{g}\exp\left[\mathrm{i}\left(-\frac{1}{2k}m^2 - \frac{1}{2k}n^2 + \frac{\beta}{2}p^2\right)\right]. \qquad (4)$$

[a] e-mail: mtod@tu-sofia.bg

DOI: 10.1051/matecconf/20168305012

Then the solution of the original problem, Eq. (1) has the following integral form

$$u = \int_{-\infty}^{\infty}\int_{-\infty}^{\infty}\int_{-\infty}^{\infty} \hat{g}(m,n,p)$$
$$\exp\left\{\mathrm{i}\left[mx + ny + p\tau - \left(\frac{m^2}{2k} + \frac{n^2}{2k} - \frac{\beta}{2}p^2\right)z\right]\right\}dmdndp. \tag{5}$$

2.1 Fourier Analysis, Splitting by Physical Processes

Let us introduce an interval $0 \le z \le \xi$ and consider the following initial-value problems

$$\mathrm{i}\frac{\partial u_1}{\partial z} + \frac{1}{2k}\left(\frac{\partial^2 u_1}{\partial x^2} + \frac{\partial^2 u_1}{\partial y^2}\right) = 0 \tag{6}$$
$$u_1(x,y,\tau,z=0) = g(x,y,\tau)$$

and

$$\mathrm{i}\frac{\partial u}{\partial z} - \frac{\beta}{2}\frac{\partial^2 u}{\partial \tau^2} = 0 \tag{7}$$
$$u(x,y,\tau,z=\xi) = u_1(x,y,\tau,\xi).$$

If function g decades on infinity then the Fourier-images of above problems exist. They admit exact solutions

$$\hat{u}_1(z) = \hat{g}\exp\left[-\frac{\mathrm{i}}{2k}(m^2+n^2)z\right]$$
$$\text{and}\quad \hat{u}(z) = \hat{u}_1(\xi)\exp\left(\mathrm{i}\frac{\beta_2}{2}p^2 z\right). \tag{8}$$

After obvious algebraic manipulations we obtain

$$u = \int_{-\infty}^{\infty}\int_{-\infty}^{\infty}\int_{-\infty}^{\infty} \hat{g}(m,n,p)$$
$$\times\exp\left\{\mathrm{i}\left[mx + ny + p\tau - \left(\frac{m^2}{2k} + \frac{n^2}{2k} - \frac{\beta}{2}p^2\right)\xi\right]\right\}dmdndp. \tag{9}$$

For $z = \xi$ the solutions (5) and (9) are equivalent for any ξ while for $0 \le z \le \xi$ the solution of the split problem is approximate. The latter means that in a discrete set of nodes z_j, $j = 0, 1, 2, \ldots$ one has to solve consequently the initial-value problems

$$\mathrm{i}\frac{\partial u_1}{\partial z} + \frac{1}{2k}\left(\frac{\partial^2 u_1}{\partial x^2} + \frac{\partial^2 u_1}{\partial y^2}\right) = 0 \tag{10}$$
$$u_1(x,y,\tau,z=z_j) = u^j$$

and

$$\mathrm{i}\frac{\partial u}{\partial z} - \frac{\beta}{2}\frac{\partial^2 u}{\partial \tau^2} = 0 \tag{11}$$
$$u(x,y,\tau,z=z_j) = u_1^{j+1}.$$

This is the gist of the method of splitting by physical processes applied for multidimensional problems.

By using Fourier analysis we showed that the original problem (1) is equivalent to solve consequently initial-value problems (10) and (11) at a given set of nodes z_j, $j = 0, 1, 2, \ldots$. In the concrete case (1) is (2+1)d and needs one more splitting – this time coordinate and after that it can be solved numerically by alternative directions implicit (ADI) method, e.g. by the Peaceman-Rachford scheme or by the scheme of stabilizing correction [2]. In opposite, Eq. (11) is (1+1)d and can be solved direct by Crank-Nicolson scheme in complex arithmetic. The resulting three equations possess sparse matrices and can be easily treated by Thomas algorithm and its complex-valued modifications [3,4].

Splitting by physical processes is applied for more complicate Eq. (1) with cubic nonlinearity and an initial condition – function $g(x,y,\tau) \equiv \exp\left(-\frac{x^2}{2} - \frac{y^2}{2} - \frac{\tau^2}{2}\right)$. To get an implicit scheme about the nonlinear part also we use so-called internal iteration, which is convergent for small enough steps in z-direction [5,6].

3 (3+1)d Envelope Equation

Let consider one more Schrödinger-like equation – the envelope equation, which is (3+1)d dispersion equation with cubic, quintic and rational nonlinearities. The linear part of the equation is

$$\mathrm{i}\frac{\partial u}{\partial z} + \frac{1}{2k}\Delta u - \mathrm{i}a_1\frac{\partial}{\partial \tau}\Delta u - a_2\frac{\partial^2 u}{\partial \tau^2} - \mathrm{i}a_3\frac{\partial^3 u}{\partial \tau^3} = 0, \tag{12}$$

where $\Delta = \frac{\partial^2}{\partial x^2} + \frac{\partial^2}{\partial y^2}$, a_1, a_2, a_3 are dimensionless real constants. For simplicity we will not comment their physical sense in this paper. The splitting by physical processes leads to following two initial-value problems

$$\mathrm{i}\frac{\partial u}{\partial z} + \frac{1}{2k}\Delta u - \mathrm{i}\frac{a_1}{2}\frac{\partial}{\partial \tau}\Delta u = 0 \tag{13}$$
$$u(x,y,\tau,z=0) = g,\quad u|_\Gamma = 0$$

and

$$\mathrm{i}\frac{\partial u}{\partial z} - \mathrm{i}\frac{a_1}{2}\frac{\partial}{\partial \tau}\Delta u - a_2\frac{\partial^2 u}{\partial \tau^2} - \mathrm{i}a_3\frac{\partial^3 u}{\partial \tau^3} = 0 \tag{14}$$
$$u(x,y,\tau,z=0) = g,\quad u|_\Gamma = 0.$$

The Fourier analysis results in consecutive solving in discrete set of nodes z_j, $j = 0, 1, 2, \ldots$ of initial-value problems

$$\mathrm{i}\frac{\partial u_1}{\partial z} + \frac{1}{2k}\Delta u_1 - \mathrm{i}a_1\frac{\partial}{\partial \tau}\Delta u_1 = 0 \tag{15}$$
$$u_1(x,y,\tau,z=z_j) = u_j$$

and

$$\mathrm{i}\frac{\partial u}{\partial z} - a_2\frac{\partial^2 u}{\partial \tau^2} - \mathrm{i}a_3\frac{\partial^3 u}{\partial \tau^3} = 0 \tag{16}$$
$$u(x,y,\tau,z=z_j) = u_1^{j+1}.$$

Eq. (15) requires a coordinate splitting, while Eq. (16) can be solved directly by complex-valued Crank-Nicolson scheme. Due to the higher dispersion the band matrix is four-diagonal. In order to use again Thomas algorithm modifications the mixed derivative in both problems should be put on the old z-stage. In this way the scheme of stabilizing correction for Eq. (15) yields

$$\mathrm{i}\frac{\tilde{u}-u^j}{\Delta z/2} = -\frac{1}{2k}(\Lambda_{xx}\tilde{u}+\Lambda_{yy}u^j) + \mathrm{i}a_1(\Lambda_{\tau xx}+\Lambda_{\tau yy})u^j, \tag{17}$$

$$\mathrm{i}\frac{u^{j+1}-\tilde{u}}{\Delta z/2} = -\Lambda_{yy}(u^{j+1}-u^j).$$

Here $(\tilde{\cdot})$ stands for the values of the grid function on the z-half-stage, $(\cdot)^j$ – old z-stage $(\cdot)^{j+1}$ – new z-stage, $\Lambda_{xx} \approx \frac{\partial^2}{\partial x^2}$, $\Lambda_{yy} \approx \frac{\partial^2}{\partial y^2}$, $\Lambda_{\tau xx} \approx \frac{\partial^3}{\partial\tau\partial x^2}$, $\Lambda_{\tau yy} \approx \frac{\partial^3}{\partial\tau\partial y^2}$. The first equation is implicit in x-direction, while the second one – in y-direction. The fully fledged (3+1)d nonlinear Envelope equation is quite more complicate compared to (3+1)d nonlinear Schrödinger equation

$$\mathrm{i}\frac{\partial u}{\partial z} + \frac{1}{2k}\Delta u - \mathrm{i}a_1\frac{\partial}{\partial\tau}\Delta u - a_2\frac{\partial^2 u}{\partial\tau^2} - \mathrm{i}a_3\frac{\partial^3 u}{\partial\tau^3}$$
$$+ b_1|u|^2u + \mathrm{i}b_2\frac{\partial}{\partial\tau}(|u|^2u) - b_3|u|^4u - \mathrm{i}\frac{\partial}{\partial\tau}(|u|^4u)$$
$$+ \mathrm{i}\rho u + \mathrm{i}b_6\frac{\partial}{\partial\tau}(\rho u) + \mathrm{i}b_7\frac{u}{|u|^2} = 0. \tag{18}$$

The last three terms describe the ionization processes, the cubic and quintic nonlinearities – the effects of self-steepening, the spatial self-focusing (self-defocusing), $b_1,...,b_7$ are real constant coefficients, ρ is the electron concentration, which is a dynamical quantity. The latter requires to solve Eq. (18) together with the kinetic equation

$$\frac{\partial\rho}{\partial\tau} = t_1(u)\rho + t_2,\ t_1\text{– functional coefficient},\ t_2\text{– real constant}. \tag{19}$$

The nonlinearities can be linearized again by an inner iteration convergent for small enough z-steps. Eq. (19) being 1d for given values of set function u is treat by Crank-Nicolson scheme. An important advantage of the linear parts of the above splitting schemes is their absolute stability and full approximation of the differential operators.

The above considered equations are femtosecond (1fs = 10^{-15}sec). In order to emphasize the significance of these equations we stress that they govern and are a tool for study of the spatio-temporal dynamics of the molecular and atomic processes.

4 Discussion and Conclusion

We implement the above splitting in finite differences for (3+1)d nonlinear Schrödinger equation and (3+1)d nonlinear envelope equation considering real regimes of pulse propagation. In this paper we do not intend to describe the physical results because they are considered thoroughly in our earlier papers [7], [8] and [9]. We aim only to motivate the splitting by physical processes combined with the inner iteration as a reliable method for solving of multidimensional wave and amplitude equations governing ultrashort regimes when exact solutions are impossible and the experimental observations and measurements are difficult and expensive.

References

1. G.I. Marchuk, *Mathematical Models in Environmental Problems*, in J.L. Lions et al (Eds), Studies in Mathematics and its Applications, vol. **16** (North-Holland, Amsterdam 1986) chapter 3, pages 82–89
2. N.N. Yanenko, *Method of Fractional Steps* (Gordon and Breach, New York 1971) pages 148–161
3. C.I. Christov, Gaussian Elimination with Pivoting for Multidiagonal Systems, Internal Report. University of Reading, UK **4**, (1994).
4. M.D. Todorov and C.I. Christov, Discrete and Continuous Dynamical Systems, Supplement **2007**, (2007) 982–992
5. C.I. Christov, S. Dost, G. Maugin, Physica Scripta **50**, (1994) 449–454
6. M.D. Todorov, Mathematics and Computers in Simulation **127**, (2016) 273–286
7. I.G. Koprinkov, M.D. Todorov, M.E. Todorova, T.P. Todorov, Journal of Physics B: Atomic, Molecular and Optical Physics **40**, (2007) F231–F236
8. T.P. Todorov, M.E. Todorova, M.D. Todorov, I.G. Koprinkov, Optics Communications **323**, (2014) 128–133.
9. M.E. Todorova, T.P. Todorov, M.D. Todorov, I.G. Koprinkov, *Spatiotemporal Dynamics of High-Intensity Ultrashort Laser Pulses in Strongly Nonlinar Regime*, in V.M. Mladenov and P.C. Petrov (Eds), 22nd International Conference "Nonlinear Dynamics of Electronic Systems", Communications in Computer and Information Science, vol. **438**, (Springer International Publishing Switzerland 2014) pages 165–172

Dynamics of a Circular Mindlin Plate under Mechanical Loading and Elevated Temperature

Anna Warminska[1,a], Emil Manoach[2,b], and Jerzy Warminski[3,c]

[1] Department of Thermodynamics, Fluid Mechanics and Aviation Propulsion Systems, Lublin University of Technology, Lublin, Poland
[2] Institute of Mechanics, Bulgarian Academy of Sciences, Sofia, Bulgaria
[3] Department of Applied Mechanics, Lublin University of Technology, Lublin, Poland

Abstract. Dynamics of a nonlinear circular Midlin plate is studied in the paper. The mathematical model represented by partial differential equations includes nonlinear geometrical terms resulted from large displacements. The plate is subjected to mechanical and thermal loadings. The dynamics of a coupled thermo-mechanical problem is reduced from partial to ordinary differential equations. Considering the first mode reduction and uniformly distributed temperature just a single nonlinear differential equation is obtained. The bifurcation analysis shows that elevated temperature shifts the rezonanse curve and new solutions arise. Depending on initial conditions this may lead to buckling phenomenon and then relatively small oscillations around this state, symmetric periodic oscillations of large amplitude, or irregular oscillations.

1 Introduction

Dynamics of structures influenced by thermal field plays a very important role in the advanced and critical structural elements in mechanical or aerospace engineering [1]. The problem of thermally induced oscillations or oscillations generated by coupled mechanical and thermal loadings is crucial if the studied element is nonlinear [2]. The important coupled field analysis in a case of beam-like structural elements has been presented in [3] and [4]. Authors evidently presented that elevated temperature or heat flux may drastically change the system response. They attacked the problem solving a set of partial differential equations (PDE) using an iterative procedure based on, so called, the pseudo-normal mode method. The method gave accurate results but the computation cost was quite high and therefore rather difficult to deal with bifurcation analysis. Therefore in papers [5], [6] the reduced thermo-mechanical model of a beam has been proposed. The model allowed to study a bifurcation scenario demonstrating how the elevated temperature may change the beam response. It has been shown that among periodic also chaotic oscillations may arise and a small temperature variation may transit the system to regular motion. The bifurcation analysis based on a reduced beam model but for a nonuniform temperature distribution has been presented in [7]. The analysis performed for beams has been extended in [7] for plates. The preliminary results for circular plate model for elevated temperature and two variants external excitation frequencies have been presented. In this paper we reduce the thermo-mechanical model of the circular plate described by PDEs presented in paper [7]. The preliminary results are obtained for clamped boundary conditions.

[a] e-mail: a.warminska@pollub.pl
[b] e-mail: e.manoach@imbm.bas.bg
[c] e-mail: j.warminsk@pollub.pl

2 Plate model

A nonlinear model of the plate is based on Mindlin plate theory. The model proposed in this paper is extended and takes into account nonlinear displacement field, thermal and mechanical loadings. The schematic plate with indicated coordinates and dimensions is presented in Fig. 1.

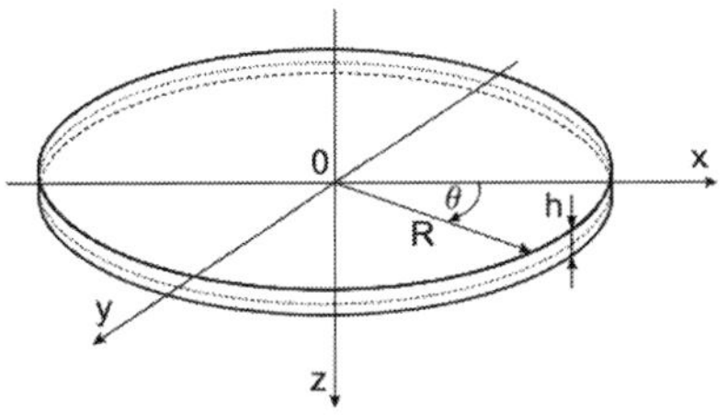

Fig. 1. Circular plate model with indicated coordinates and dimensions.

The partial differential equations of fully coupled, axis-symmetric vibration of a nonlinear thermo-elastic plate is presented in [7] and has the form

$$\begin{aligned}
&Ah\left[\frac{\partial^2 u}{\partial r^2} + \frac{\partial w}{\partial r}\frac{\partial^2 w}{\partial r^2} + \frac{1}{r}\frac{\partial u}{\partial r} - \frac{u}{r^2} + \frac{1}{2r}(1-\nu)\left(\frac{\partial w}{\partial r}\right)^2\right] \\
&= A(1+\nu)\alpha_T\frac{\partial \gamma_T}{\partial r} \\
&D\left[\frac{\partial^2 \psi}{\partial r^2} + \frac{1}{r}\frac{\partial \psi}{\partial r} - \frac{\psi}{r^2}\right] - k^2 Gh\left(\frac{\partial w}{\partial r} + \psi\right) - c_2\frac{\partial \psi}{\partial t} \\
&-\frac{\rho h^3}{12}\frac{\partial^2 \psi}{\partial t^2} = A(1+\nu)\alpha_T\frac{\partial \kappa^T}{\partial r} \\
&k^2 Gh\left(\frac{\partial^2 w}{\partial r^2} + \frac{1}{r}\frac{\partial w}{\partial r} + \frac{\partial \psi}{\partial r} + \frac{\psi}{r}\right) \\
&+Ah\left[\frac{\partial u}{\partial r} + \frac{1}{2}\left(\frac{\partial w}{\partial r}\right)^2 + \frac{\nu}{r}u - \frac{1+\nu}{h}\alpha_T\gamma_T\right]\left(\frac{\partial^2 w}{\partial r^2} + \frac{1}{r}\frac{\partial w}{\partial r}\right) \\
&+Ah\left(\frac{\partial^2 u}{\partial r^2} + \frac{\nu}{r}\frac{\partial u}{\partial r} - \frac{\nu}{r^2}u + \frac{\partial w}{\partial r}\frac{\partial^2 w}{\partial r^2} - \frac{1+\nu}{h}\alpha_T\frac{\partial \gamma_T}{\partial r}\right)\frac{\partial w}{\partial r} \\
&+c_1\frac{\partial w}{\partial t} - \rho h\frac{\partial^2 w}{\partial t^2} = -p(r,t)
\end{aligned} \tag{1}$$

where h and R are the plate thickness and radius, $u(r,t)$ is the in-plane displacement, $w(r,t)$ is the transverse displacement, $\psi(r,t)$ is the cross-section rotation angle, E, G are Young and shear modulus, ρ is material density, α_T is the coefficient of thermal expansion, and $A = \frac{E}{1-\nu^2}$, $D = \frac{Eh^3}{12(1-\nu^2)}$, $\gamma^T(r) = \int_{-h/2}^{h/2} \Delta T(r,z)dz$, $\Delta T = T - T_0$, T and T_0 are current and reference temperatures and $p(r,t)$ is external loading. We note that in Eq. 1(a) inertia term $\frac{\partial^2 u}{\partial t^2}$ is neglected.

3 Model reduction

Taking into account that temperature is uniformly distributed along the plate radius and thickness, Eq. (1) can be simplified. Furthermore, Eq. 1(a) can be solved analytically and coordinate u can be related to w. Thus, assuming solution in the form

$$w(r,t) = \sum_{n=1}^{N} q_n(t) W_n(r), \quad \psi(r,t) = \sum_{n=1}^{N} q_n(t) \Psi_n(r) \quad (2)$$

where $W_n(r)$, $\Psi_n(r)$ are linear eigenmodes and $q_n(t)$ generalised coordinates, the PDEs can be transformed to ordinary differential equations (ODE) applying the Galerkin method. In this paper we take just one mode reduction of the clamped around plate, considering excitation distributed according to the first mode and acting around the first natural frequency. The first vibration mode is defined by Bessel functions which satisfy boundary conditions for clamped beam.

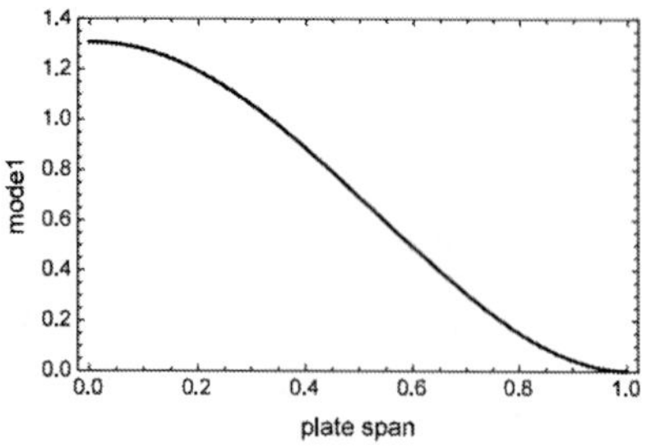

Fig. 2. First vibration mode of the clamped plate based on modified Bessel functions.

The mode shape against normalised radius is presented in Fig. 2. This mode applied in Galerkin procedure results in just one nonlinear differential equation

$$\ddot{q}_1 + 2\xi_1\omega_1\dot{q}_1 + \omega_1^2 q_1 + F_{NL_1} q_1^3 + F_{NT_1}\Delta T q_1 = F_{NP_1} \sin\omega t \, . \quad (3)$$

Above equation is written in the dimensionless form by introducing dimensionless time $\bar{t} = tc/R$ and displacements $\bar{u} = u/R$, $\bar{w} = w/R$, where $c = \sqrt{\frac{E}{\rho(1-\nu^2)}}$. The 'dot' denotes dimensionless time derivative and q_1 first dimensionless generalised coordinate. We note that in Eq. (3) ω_1 is the first natural frequency, F_{NL_1} coefficient occurring due to nonlinear geometrical terms, F_{NT_1} - coefficient related to temperature variation, F_{NP_1} - amplitude of mechanical loading and modal damping is introduced by coefficient ζ_1.

4 Numerical analysis

Numerical study of the plate dynamics under mechanical and thermal loadings is performed for a clamped plate with physical parameters: $h = 0.0025\,\mathrm{m}$, $R = 0.1\,\mathrm{m}$, $E = 0.7 \times 10^{11}\,\mathrm{Pa}$, $\nu = 0.34$, $\rho = 2778\,\mathrm{kg/m^3}$. This physical parameters after the modal reduction lead to dimensionless coefficients present in Eq. (3). Values of the coefficients are given in Table 1.

Table 1. Dimensionless parameters of the plate reduced model.

parameter	ω_1	c_{NL_1}	c_{T_1}	ζ_1
value	0.07359	10.3516	-2.0033×10^{-4}	0.01

At first a resonance curve is computed on the basis of Eq. (3) for $\Delta T = 0$. The amplitude of excitation is fixed $p_1 = 4.331 \times 10^{-6}$ while frequency is varied around the first natural frequency. The influence of the geometrical nonlinear terms is demonstrated in Fig. 3 by stiffening effect present on the resonance curve. The curve has typical Duffing type nature with stable and unstable branches.

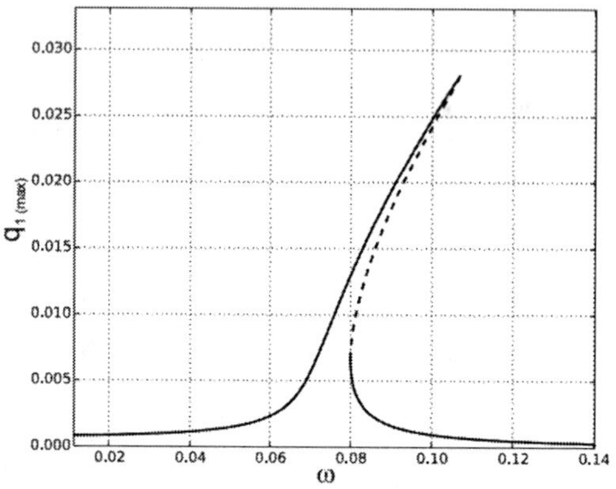

Fig. 3. Bifurcation diagram versus varied excitation frequency and fixed temperature, $\Delta T = 0$, $p_1 = 4.331 \times 10^{-6}$.

When the temperature is elevated then we observe a shift of the resonance zone into the direction of lower frequencies. However, above a certain temperature threshold apart from the shift also qualitative changes appear. The resonance curve obtained for elevated temperature $\Delta T = 28$ is plotted in Fig. 4. In fact, the resonance curve is shifted into lower frequency direction but furthermore, the left part of the branch becomes unstable (dashed line) and lower right branch becomes unstable too, with arising new stable solutions plotted in red. This red colour branch arises in bifurcation point leading to a buckling phenomenon. The buckling is observed by offset of the oscillation centre in a positive or negative direction.

The example of time histories is presented in Fig. 5(a) for fixed excitation frequency $\omega = 0.05$. In blue and red colours the oscillations with positive or negative offset are

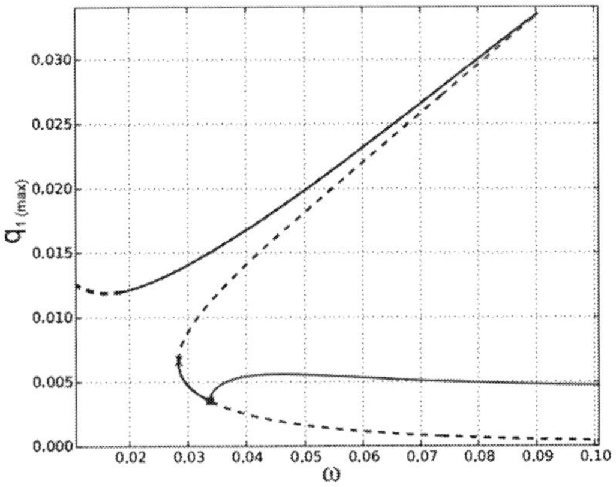

Fig. 4. Bifurcation diagram versus varied excitation frequency and fixed temperature, $\Delta T = 28$, $p_1 = 4.331 \times 10^{-6}$.

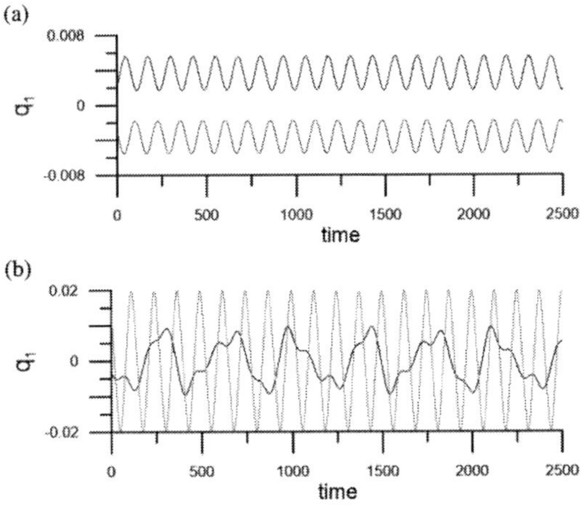

Fig. 5. Time histories for temperature $\Delta T = 28$, (a) solution with positive shift (blue) and negative shift (red); (b) periodic solution (green) and irregular motion (black); $\omega = 0.05$, $p_1 = 4.331\times10^{-6}$.

demonstrated, respectively. The plate can buckle in a positive or negative direction and then oscillates around that new dynamic steady state, performing relatively small oscillations.

For the same excitation frequency $\omega = 0.05$ but different initial conditions the plate does not buckle but oscillates symmetrically with large amplitude equal to about 0.02 which corresponds to upper stable branch of the resonance curve (black solid line in Fig. 4). The time history of this oscillation is presented in Fig. 5(b) in a green colour. We see that in fact this solution is symmetric around zero position. The performed detailed numerical simulations for the same parameters but different initial conditions exhibited also possible irregular motion. This irregular solution is plotted in Fig. 4(b) by black line.

5 Conclusions

The extended model of Mindlin plate which takes into account geometrical nonlinear terms due to large deformation and influence of thermal field are presented in the paper. It is shown that the complex model, represented by partial differential equations can be reduced to ordinary differential equations taking into account a selected number of modes. The computations based on the first mode reduction show the stiffening effect of the resonance curve. The elevated temperature shifts the resonance curve into lower frequencies direction. Moreover, due to the temperature increase above a certain threshold the resonance curve changes qualitatively. For example, for temperature increase up to $\Delta T = 28$, some parts of the curve become unstable and a new stable branch arises. The new solutions correspond to buckling phenomenon and periodic oscillations around this state. These small amplitude oscillations are characterised by a positive or negative offset. But for the same parameters and varied initial conditions the response can be periodic with no offset or irregular. The detailed bifurcation analysis and detection of the transition to irregular (chaotic) motion is planed to be studied in future investigations.

Acknowledgements

The first author would like to acknowledge the support received from the Polish NSF Grant S-38/M/2015, the second author the partial support from Bulgarian NSF Grant DUNK-01/3 and H2020 Marie Sklodowska-Curie ITN under grant agreement No 643095, and the third author the support from Polish NSF grant S-3/M/2015.

References

1. E.A. Thorton, *Thermal structures for aerospace applications* (AIAA Education Series, 1996) 1–479.
2. M. Amabili, S. Carra, J. Sound and Vibration, **321**, (2009) 936–954.
3. E. Manoach, P. Ribeiro, Int. J. Mech. Sci., **46**, (2004) 1589–1606.
4. P. Ribeiro, E. Manoach, J. Sound and Vibration, **285**, (2005) 1093–1107.
5. A. Warminska, E. Manoach, J. Warminski, Meccanica, **49**, (2014) 1775–1793.
6. A. Warminska, E. Manoach, J. Warminski, S. Samborski, Continuum Mechanics and Thermodynamics, (2014), DOI 10.1007/s00161-014-0381-6.
7. A. Warminska, E. Manoach, J. Warminski, Proceedings of ICoEV 2015 Int. Conf. on Engineering Vibration, Ljubljana, Slovenia, (2015) 1057–1066.

Statistical Analysis of Reinforced Concrete Bridge Defects for Optimum Maintenance Planning and Budgeting

A. Abdul-Ameer and K. Alhefeiti

The British University in Dubai, PO Box 345015, Dubai, UAE

Abstract. Concrete degradation is arguably the main restriction on longevity of service life of buildings and infrastructure, most developed economies spend more on rehabilitation of existing structures than on new one. Concrete can be damaged by a range of chemical , physical and biological cause that include; fire, sea water, aggressive environment, corrosion of reinforcement, impact, bacterial attach to name a few. There is a continuing argument about the main cause of the premature deterioration of structures containing reinforced concrete, these range from inadequate specification, poor design and detailing, poor workmanship and aggressive environment. By far the biggest cause of deterioration is the corrosion of the steel reinforcement, with its expansive byproducts that result in the cracking and spilling of concrete and can ultimately cause total service life failure of the structure if not checked and remedied. Maintenance engineers face an ever ending problem of when is the optimum time to intervene and rehabilitate a deteriorated structural element or whole structure. Budget restrictions and need to optimize the maintenance cycle. The paper will report on a large database of deterioration records of concrete bridges covering a period of over 60 years and include data from over 400 bridge structures. Statistical analysis were used to estimate the deterioration rates of various reinforced concrete defects in a rage of structural elements and bridge types and to arrive a mean service life of the structure, such service life data can then be used by the maintenance manager to budget and prioritize his maintenance work.

1 Introduction

Concrete is the most widely used construction material in the bridge structures because of its durability and versatility. The durability of concrete is often challenged by the deterioration, because of which most of the concrete bridges do not serve for the full service life as it is designed for. Deterioration of the concrete bridges has caused great loss in infrastructure maintenance. The increasing number of concrete bridge deterioration has increased the repair, rehabilitation or replacement of these structures in short term in the highway networks of most European and North American countries.

Reinforced concrete bridges are designed to perform satisfactorily for their service life. Depending upon the parameters taken in to account during design, service life of bridges varies 50 to 100 years or even more. Minimum requirements for reinforced concrete structures design can be found in EURO CODE 2. However most of the reinforced structures do not perform well for the period it is designed for, as they deteriorate with time.

There have several research been done [1], [2] & [3] and several studies are being carried out to identify the root cause of deterioration. Research shows that the main cause of deterioration is corrosion of reinforcing steel [4], [5] which make structure weak and leads to collapse of structure after certain period if it is left ignored.

1.1 Deterioration of reinforced concrete

There are four primary causes of concrete degradation, which include chemical, physical, mechanical and biological issues [6]. Chemical factors are primarily linked to salination as well as acidification in the atmosphere, which can lead to corrosion in the case of steel reinforcements, or hydrolysis of the concrete itself [6]. Physical factors may include any temperature or environmental causes that abrade or erode the structure. Mechanical factors may include the amount of load that is placed on the structure, which can induce cracking and the movement of external fluids or grit into these cracks, which can result in breaks. Finally, when there is a possibility of moss, ivy or fungi to have an impact on the structure, among other similar materials, this is a biological factor [6]. Many of these factors may overlap or have an effect on one another, as well [7]. This is due to the fact that concrete is extremely porous and therefore when there is a corrosive or physical effect on the material it is likely to face structural failure [7].

Fig. 1: Damage to a Concrete Wall Caused by Reinforcement Corrosion

Climate change can have a significant effect on all of these factors, which needs to be taken into account now more than ever [8]. As there are more frequent shifts in temperature, an increase in extreme weather events, and unpredictable changes in certain areas of the world and in microclimates, there is a need to recognise that this can have an effect on the way that concrete reacts, and also this can make an impact on the biological factors that may come into play, as these may be very different from in the past [8], [9].

The use of steel reinforcement also has an impact on the way that concrete degrades. Corrosion may be able to decrease the structural capacity of the concrete as a whole through decreasing the cross sectional area of steel reinforcement, which is at an increased risk when it comes to managing the greater loads that are expected in pre-stressed concrete structures [7].

In other words, when it comes to the evaluation of the factors that lead to concrete degradation, it is important to take into context the impact of different materials on the process. Steel is especially problematic, as well, in areas with shifting climate change effects because of its temperature effects relative to the concrete itself [8], [9]. Whereas concrete will erode during temperature changes, the expansion and contraction of steel may increase cracks as well as erosion, but could also lead to structural breaks [6].

1.2 Existing service life models

Service life model is the mathematical presentations that show the service life of a structure as a function of different design parameters. They are often derived from degradation or performance models when limit states of maximum degradation or minimum performance are known. Service life models are the very important tools in the bridge management system as well as an engineering infrastructures management. A service life models helps in following:

- Predicting the time it takes to reach relevant limit states for different design and maintenance options.
- Ensuring safety, serviceability & functionality.
- Evaluating life cycle costs–optimize design, inspection and maintenance.

Because of its importance in structures management many research has been under taken in the service life prediction models. Many approaches have been put forward to estimate the service life concrete structures. According to [10] up to date they are based on the concept of delay time, factor method or stochastic/ probabilistic approach. Redmond [11] has well explained about the delay time modelling of the growth of the defects in concrete bridges components. Likewise many researchers have explained about the factor method and probabilistic approach of service life prediction models.

Clifton [12] has given a good review of methods for predicting the service life of concrete. According to him all these developed methods of concrete service life estimation are subjected to concrete deterioration processes and are based on the following points, which sometimes they may be used in combination of each other.

I. Experience
II. Performance of similar materials
III. Accelerated testing
IV. Mathematical modelling based on chemistry and physics of degradation processes.
V. Application of reliability and stochastic concepts.

Lounis [13] has identified several existing service life models to date in his report "A Probabilistic cumulative damage model for service life prediction of concrete structures" they are: a) Empirical models (accelerated life aging) ; b) Heuristic models; c) Factorial models; d) Logistic models; e) Mechanistic models; f) Cumulative damage models; g) Shock models; h) Probabilistic models. Generally, however, service life prediction models developed for concrete bridges are based on either a deterministic or a probabilistic approach and the corresponding models are called as deterministic models & probabilistic/stochastic model.

1.3 Regression model

Most of the developed degradation models are complex and expensive. They predict deterioration rates by deriving a physical degradation process, which are developed by large amount of test data collected from expensive physical testing. Hence these are not feasible for the modelling deterioration of bridges. There are some alternative models which use historical inspection records. Hence are more feasible than the previous one. The most commonly used concrete bridge deterioration model by the Bridge Management System (BMS) based on historical inspection data are:

I. Regression Model
II. Markovian Model
III. Factor Method

DOI: 10.1051/matecconf/20168306001

Regression analysis is primarily used to find the best-fitted mathematical model, so that a dependent or response variable can be predicted from independent or predictor variable(s). Descriptions and controls of the cause-and-effect relationship between variables are also major purposes that regression analysis serves. However, conventional regression techniques are suitable for dealing with non-fuzzy data. A detail evaluation about this model has been done by [13]. A commonly used regression model is the regression function obtained by doing a regression analysis, which was used by [14].
According to Pan [15] among the existing regression model the model proposed by Tanka in 1982 is quite popular and useful, the only problem associated with this model is that it is restricted to symmetric triangular fuzzy numbers. To overcome those problems, a fuzzy least square regression model has developed. However this approach still require considerable computation as in that regression coefficients are derived from non- linear programming problem.
To overcome these problem of complex mathematical computations and asymmetric & symmetric triangular membership functions, Pan [15] have proposed matrix driven multiple fuzzy linear regression model. They use matrix algebra as the calculation tool which makes the model comparatively simpler to follow and easy to apply than other existing regression model. In the paper "Estimating bridge performance based on a matrix-driven fuzzy linear regression model " published in 2009 , Pan [15] have explained about this model and have used for the estimation of overall bridge condition in Taiwan by using bridge inspection data.

2 Extensive analysis of the database

The bridge database consists of massive information gathered from the bridge inspections. In order to determine the most useful data, an extensive analysis was under taken. During this process in order to determine the natural deterioration of the structures over their service life, the records of inspection which only relates to the first occurrence of the defect observed were used and the records related with the inspection carried out after remedial measures were not used.
At first the data was analysed to determine the total number of the structures and their corresponding members. The overall data were then analysed on the basis defect type and its frequency of occurrence.
In order to determine the change of the defects with time, the database was again analysed. The records were organised according to their age at inspection and exposure condition. Then they were grouped into ten – year intervals. Bridges under mild and moderate and severe exposure were investigated and then a bar chart was produced to show the proportion of defects in each time interval on the basis of the exposure condition.

3 deterioration of the structure

The main part of the analysis of the database is to investigate the degradation of various structures and their corresponding members over their service life. The main cause of the deterioration was found to be due to corrosion cause by chloride attack or carbonation. Chloride attack was found to be the severe cause of deterioration which arises due to use of de-icing salt .During the analysis following steps were carried out:

I. The inspection records were sorted into groups according to structure type and then to member type.
II. Each group was divided into sub-groups representing the five possible condition states described in Section 4.2.1.3.
III. The average age for each sub-group of records was calculated and used to chart the deterioration of the structure types over their service lives.

By organising, the main data was group according to the exposure type recorded and repeating steps (1) to (3) then the effects of exposure conditions on the deterioration of the structures were investigated.

4 Statistical analysis of bridge deterioration

The distribution of possible cause was studies in fig. 2. The cause was not known in 93.0% of the cases. Impact (4.2%) is the main cause of fault/defect. The reason for this was the fact that the inspectors are not trained engineers and only indicated cause where the reason was obvious, such as bomb damage, impact or clear indication of de-icing salts.

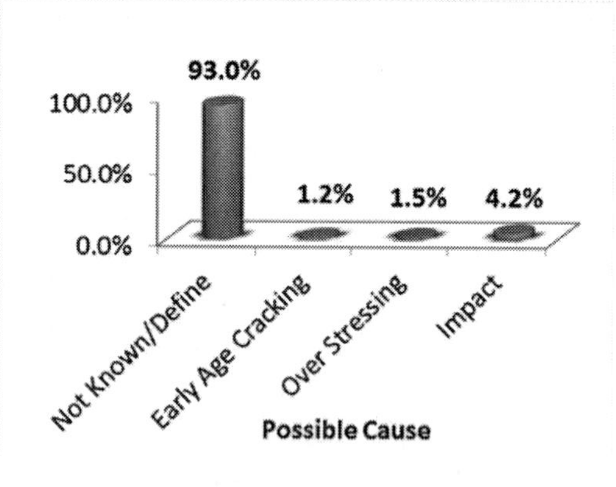

Fig. 2, Distribution of Possible Cause

The distribution of member type was studies in fig. 3. Flexural (47.6%) was the main type seen in in the study sample followed by compression (25.5%). This is to be expected as cracks mainly appear in flexural members, where cracking in the tension zone is the norm.
The exposure condition was studies in fig. 4. About 36.1% of samples were mild, 53.9% were moderate and 10.0% were severe. The severe exposure is normally associated with exposure to de-icing salts for deck

members of over bridges and pier members in underbridges.

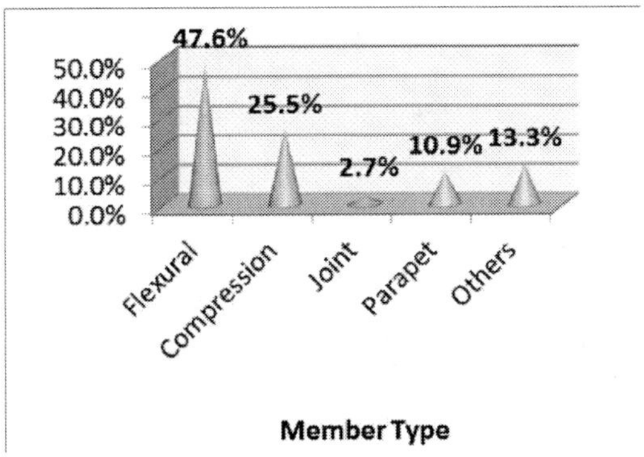

Fig. 3, Distribution of Member Type

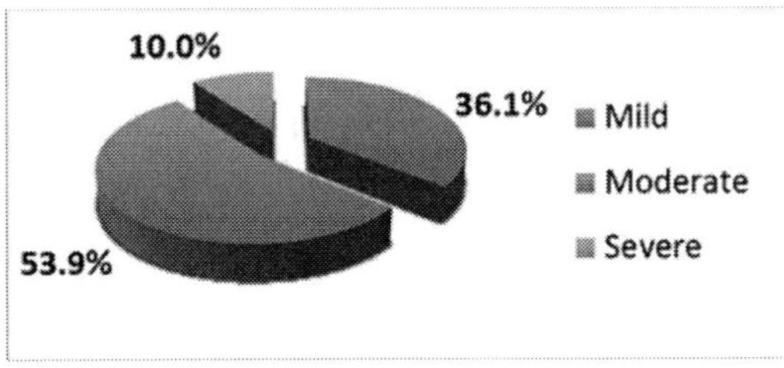

Fig. 4, Distribution of Exposure Condition

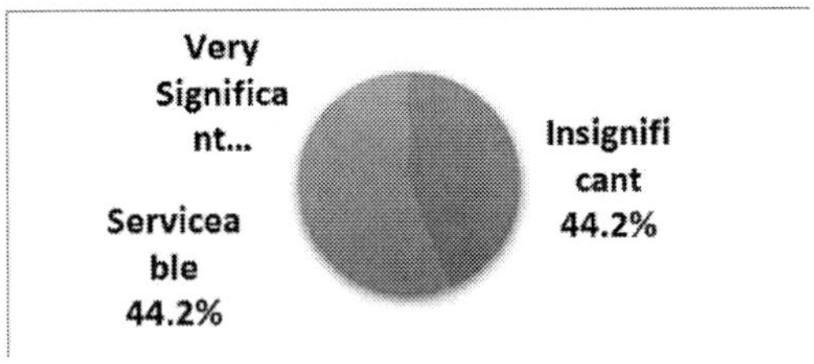

Fig. 5, Distribution of Repair/Urgency Ranking

Repair/urgency ranking was studies in fig. 5. About 44.2% of samples were insignificant, 44.2% were serviceable and 11.5% were very significant. The excellent annual inspection cycle of visual inspections and four yearly principle inspections, kept structures in good state with almost ninety percent in good service condition and only 11% in significant need for repair. However, this number is significant as the bridge owners have over 5000 bridges and that these are in temperate climate.

Table 1 shows that the average age of building is high in severe spalling (35.9 years with standard deviation 18.5) followed by minor spilling (34.7 years with standard deviation 16.2). The average age of building is 25.8 years with standard deviation 16.2 in multiple cracks 3.0mm, 23.4 years with standard deviation 16.3 in multiple cracks 1.5mm and 20.2 years with standard deviation 16.6 in hairline crack. It is clear that on average, deterioration is fairly rapid from minor spalling to sever spalling, this is to be expected since the removal of parts of the concrete in the minor spalling expose totally the steel to the elements, resulting in an accelerated deterioration.

Table 1, Descriptive Statistics of Building Age in Each Subset

Descriptive Statistics	Hairlin e Crack	Multiple Cracks 1.5mm	Multiple Cracks 3.0mm	Minor spalling	Severe spalling
Mean	20.2	23.4	25.8	34.7	35.9
Median	16.0	20.0	23.5	34.5	36.5
Std. Deviation	16.6	16.3	16.2	16.2	18.5
Variance	276	265	264	262	344
Range	62.0	71.0	62.0	62.0	80.0
Quartile - Q1	6.0	12.8	13.0	23.5	21.0
Quartile - Q2	16.0	20.0	23.5	34.5	36.5
Quartile - Q3	30.5	30.0	41.3	46.3	46.5
IQR	24.5	17.3	28.3	22.8	25.5

Table 2, 95% Confidence Interval of Age in Each Subset

Fault/Defect	**95% Confidence Interval**	
	Lower Bound	**Upper Bound**
Hairline Crack	16.23	24.14
Multiple Cracks 1.5mm	19.56	27.32
Multiple Cracks 3.0mm	21.19	30.41
Minor spilling	30.84	38.56
Severe Spilling	31.46	40.31
Total	26.20	30.08

Table 2 shows that the severe spilling has the highest lower and upper bounds (31.46-40.31) followed by minor spilling (30.84-38.56). Hairline crack has the 95% confidence interval 16.23-24.14, multiple cracks 1.5mm has 19.56-27.32 and multiple cracks 3.0mm has 21.19-30.41. The almost constant difference between the lower bound and upper bound is indicative of the normal distribution of the sample. Although the different defects has each, large standard deviation, the population of defects is of a normal distribution.

Table 3, Chi-square test to check the independency of subsets

Variables	**Chi-square**	**df**	**p – value**
Possible Cause	36.055	12	0.000
Member Type	33.424	16	0.006
Exposure Condition	12.766	8	0.120
Repair/Urgency Ranking	67.551	8	0.000

H_0: possible cause/member type/exposure condition/urgency ranking and fault/defect are independent
H_1: possible cause/member type/exposure condition/urgency ranking and fault/defect are not independent
Table 3 shows the results of chi-square test. Chi-square test was used to check whether possible cause/member type/exposure condition/urgency ranking and subset are independent.
The p-value corresponding to possible cause is less than the significance level 0.05; the null hypothesis rejected and accept the alternative hypothesis that the possible cause and fault/defect are not independent. Then can conclude that the possible cause is depends on fault/defect.
The p-value corresponding to member type is less than the significance level 0.05; the null hypothesis rejected and accept the alternative hypothesis that the member type and fault/defect are not independent. Then can conclude that the exposure condition is depends on fault/defect.
The p-value corresponding to exposure condition is greater than the significance level 0.05; the null hypothesis accept that the exposure condition and fault/defect are independent. Then can conclude that the exposure condition is not depending on fault/defect.
The p-value corresponding to repair/urgency ranking is less than the significance level 0.05; the null hypothesis rejected and accept the alternative hypothesis that the repair/urgency ranking and fault/defect are not independent. Then can conclude that the repair/urgency ranking is depends on fault/defect.

Table 4: Check the Difference in Age of Building in All Subsets Using ANOVA

	Sum of Squares	df	Mean Square	F - value	p - value
Between Groups	13459	4	3365	11.888	0.000
Within Groups	91990	325	283		
Total	105449	329			

H_0: the average age of building is same in all subsets
H_1: the average age of building is different in all subsets
Table 4 shows the results of analysis of variance (ANOVA). ANOVA was used to check the difference in average age of building in all subsets. Here the p-value is less than the significance level 0.05; the null hypothesis reject and accept the alternative hypothesis that the average age of building is different in all subsets. Then can conclude that the age of building is significantly different in each data subsets.
H_0: the average age of building is same between each pair of subset.
H_1: the average age of building is different between each pair of subset.
Table 5 shows the results of independent sample t-test. Independent sample t-test was used to check the difference in average age of building between each pair of subset. The table shows that age in hairline crack is significantly different from minor spilling and severe spilling as the corresponding p-values are less than the significance level 0.05. The age in multiple cracks 1.5mm is significantly different from minor spilling and severe spilling as the corresponding p-values are less than the significance level 0.05. The age in multiple cracks 3.0mm is significantly different from minor spilling and severe spilling as the corresponding p-values are less than the significance level 0.05.

Table 5, T-Test to Check the Pair-Wise Difference in Age between Subsets

Pair	Mean Difference	t - value	df	p - value
Hairline Crack -Multiple Cracks 1.5mm	3.257	1.172	138	0.243
Hairline Crack - Multiple Cracks 3.0mm	5.614	1.843	118	0.068
Hairline Crack - Minor Spalling	14.514	5.239	138	0.000
Hairline Crack - Severe Spalling	15.700	5.277	138	0.000
Multiple Cracks 1.5mm - Multiple Cracks 3.0mm	2.357	0.783	118	0.435
Multiple Cracks 1.5mm - Minor Spalling	11.257	4.105	138	0.000
Multiple Cracks 1.5mm - Severe Spalling	12.443	4.219	138	0.000
Multiple Cracks 3.0mm - Minor Spalling	8.900	2.967	118	0.004
Multiple Cracks 3.0mm - Severe Spalling	10.086	3.091	118	0.002
Minor Spalling –Severe Spalling	1.186	0.403	138	0.688

5. General discussion

The statistical analysis of the database has shown than deterioration of the concrete bridges although is varied and has several causes, it follows a normal distribution. This gives confidence in using the database to draw service life values that can be used in the asset and maintenance management of concrete bridges for an optimum repair and rehabilitation planning of such as important asset in the transport system in question. Although the exposure condition is not always explicitly stated, the statistical values given are a good indication for the exposure condition of these bridges in the temperate climate they exist in.
The value of the annual visual inspection, which is low cost and takes relatively little time is evident in the low percentage of structures in need of major rehabilitation. This inspection is supported by a major principle inspection, where more extensive inspection is conducted and tests carried out where necessary.
The analysis of the data to check the significance of the service life estimates of the various bridges shows that the differences are real and statistically significant. This helps the maintenance engineer in having the confidence in planning and budgeting of the repair and maintenance of his bridge stock and maintaining his assess serviceable

and functional.

6. Conclusion

The hairline crack was started to see when the average age of the building will be at almost 20.2 years. Multiple cracks 1.5mm and multiple cracks 3.0mm are started to build when the age gone past 23 and 25 years respectively. The spall cases are seen when the building got much older. Minor spilling and severe spilling are started to build when the age will be at 34.7 and 35.9 years respectively.
Age of building in is normally distributed in all faults/defects. Severe spilling has the highest confidence interval for age followed by minor spilling. Possible cause, member type and urgency ranking are dependent with fault/defect and exposure condition is independent. Building age and Building size have no relationship and they are independent. The difference in building age between defects/faults is highly significant.

ACKNOWLEDGMENT

1. The authors wish to acknowledge the support and encouragement for this research provided by the Vice Chancellor, The British University in Dubai-UAE.

2. The authors wish to thank Prof Abid Abu-Tair for providing the historical structures data used in this research.

References

1. P.T. Lees, *Deterioration Mechanism, In Durability of concrete structures. Investigation, repair, protection*. Edited by Mays Geoff, E & FN SPON, an imprint of chapman & Hall, London, (1992).
2. S. Rostam, *Durability of Concrete Structures- The State Of The Art*, In: Durability of Materials & Structures in Building & Civil Engineering, edited by C.W.Yu & John W. Bull, Whittles Publishing. CRC press, UK, (2006).
3. A. M. Neville, *"Properties of Concrete"*, In : Properties of concrete, Fourth edition, pp 482-483, London: Longman, (1995).
4. C. B. Mc Parland, A. I. Abu-Tair, A. Nadjai, *Modelling Deterioration of Concrete Bridges using a Large Database of Inspection Records*, 9th International Conference on Structural Faults and Repair, 'Extending the Life of Bridges, Civil & Building Structures', London, 4th-6th July, (2001).
5. J. Hu, S. J. Lee, Y. J. Kim and M. Mahgoub, Inspection and Non-destructive Evaluation of Concrete Bridge with Environmental Deterioration. *Climatic Effects on Pavement and Geotechnical Infrastructure*, p.193, (2014).
6. S. Loos, *Probabilistic Evaluation of Carbonation Induced Deterioration in Prestressed Concrete Structures*. Ohio State University Working Paper, (2015).
7. M. G. Stewart, X. Wang, and M. N. Nguyen, Climate change adaptation for corrosion control of concrete infrastructure. *Structural Safety*, *35*, pp.29-39, (2012).
8. X. Wang, M.G. Stewart, and M. N. Nguyen, Impact of climate change on corrosion and damage to concrete infrastructure in Australia. *Climatic Change*, *110*(3-4), pp.941-957, (2012).
9. A. I. Abu-Tair, C. McParland, J. F. Lyness, and A. Nadjai, Predictive Models of Deterioration Rates of Concrete Bridges Using The Factor Method Based on Historic Inspection Data, Paper 262, Proceedings of 9th International Conference on the Durability of Building Materials and Components, Brisbane(9DBMC), Brisbane, Australia, 17-20. March 2002. Paper 262, (2002).
10. D. F. Redmond, A. H. Christer, S. R. Rigden, E. A. Burley, A. Tajelli, and A. I. Abu-Tair, *O.R. modelling of the deterioration and maintenance of concrete Structures*, European Journal of Operational Research 99 (1997) 619-631, (1997).
11. J. R. Clifton, *Predicting the Service Life of Concrete*, ACI Materials Journal, Vol. 90, No. 6, 611-617 (1993).
12. Z. Linous, *A Probabilistic cumulative damage model for service life prediction of concrete structures*, National Research Council Canada, (2007).
13. G. Morcous, H. Rivard, and A. M. Hanna, *Predicting the Condition of Bridge Decks Using Case-Based Reasoning*, Proceedings of the CSCE 29th Annual Conference, Edited by Noakes, D., Victoria, BC, Canada, June (2001).
14. D. Veshosky, C. R. Beidleman, G. W. Buetow, and M. Demir, *Comparative Analysis of Bridge Superstructure Deterioration*, ASCE Journal of Structural Engineering, Vol. 120(7), pp. 2123-2136 (1994).
15. Pan, Nang – Fei, Lin, Tzu – Chieh and Pan, Nai – Hsin, *Estimating bridge performance based on a matrix – driven fuzzy linear regression model*, Automation in Construction 18(2009) 578 – 586, (2009).F. De Lillo, F. Cecconi, G. Lacorata, A. Vulpiani, EPL, **84** (2009)

Use of Just in Time Maintenance of Reinforced Concrete Bridge Structures based on Real Historical Data Deterioration Models

A. Abu-Tair and A. Al Hussain

The British University in Dubai, PO Box 345015, Dubai, UAE

Abstract: Concrete is the backbone of any developed economy. Concrete can suffer from a large number of deleterious effects including physical, chemical and biological causes. Large owning bridge structures organizations are facing very serious questions when asking for maintenance budgets. The questions range from needing to justify the need for the work, its urgency, to also have to predict or show the consequences of delayed rehabilitation of a particular structure. There is therefore a need for a probabilistic model that can estimate the range of service lives of bridge populations and also the likelihood of level of deteriorations it can reached for every incremental time interval. A model was developed for such estimation based on statistical data from actual inspection records of a large reinforced concrete bridge portfolio. The method used both deterministic and stochastic methods to predict the service life of a bridge, using these service lives in combination with the just in time (JIT) principle of management would enable maintenance managers to justify the need for action and the budgets needed, to intervene at the optimum time in the life of the structure and that of the deterioration. The paper will report on the model which is based on a large database of deterioration records of concrete bridges covering a period of over 60 years and include data from over 400 bridge structures. The paper will also illustrate how the service life model was developed and how these service lives combined with the JIT can be used to effectively allocate resources and use them to keep a major infrastructure asset moving with little disruption to the transport system and its users.

1 Introduction

Although concrete is considered a very durable and strong material, there are many factors that can lead to deterioration of RCC. Bridges specially can be borne to failure and deterioration due to the huge moving loads on their decks.

Corrosion is the most common cause of deterioration for bridges because of the severe exposure and excessive environment conditions. Whether the corrosion is due to chemical attack, physical change or internal reaction the corrosion of steel reduces the residual strength of reinforcement. This can happen when strength of reinforcement is decreased due to the loss of steel section, losing the internal bond and decreasing the concrete cross section by cracking and spalling [1]. Chemical reaction is one of the causes of bridges failure and deterioration. These chemical can range from Sulfuric, chlorides, carbonation and Alkali Silica Reaction (ASR). ASR is more common in high humid and high temperature, the reaction occurs internally and develops gel that oppose internal pressure that reduces the concrete stiffness and strength. It is difficult to predict the occurrence of ASR to concrete structures [2]. Other faults which can require repair include: insufficient cover to steel, honeycombing or voids in the concrete, blemishes such as conspicuous or excessive blow-holes which are not structurally significant but are aesthetically displeasing.

2.1 Just in time maintenance

JIT maintenance was first developed in Japan particularly by Toyota engineers who are why it is called Toyota Production System. Later it was adopted by United States and Europe to control the production. From concept of JIT in manufacturing construction industry has adapted the concept to improve the quality of construction and reduce money and time.

Construction and manufacturing has a lot of differences of the end product. The on-site production, uniqueness of projects and the complexity is what construction difficult to adapt to just in time production. Lean Project Delivery System into four interconnected phases: project definition, lean design, lean supply, and lean assembly [3]. For large infrastructure maintenance with good service records and inspection regimes, the use of JIT to plan maintenance priorities would be most useful, this would replace the crises management run by many organisation of leaving a structure to deteriorate until there is a serious danger to public or system before action is taken, such work is usually more costly, less effective and results in management playing catch up, with large backlog of maintenance work being delayed or cancelled.

2.2 Bridge model for service life

Highway bridges are subjected to deterioration due to the environmental attacks; therefore it is necessary to be able to predict the life-expectancy of the bridge. This is done by developing a model for bridge's life, the maintenance and repairs are easier to perform. The advancement in software and computational hardware has enabled a better modelling to the structure and better understanding of their lifetime behaviour.

Research in life-cycle performance prediction and service-life estimation of bridges has not fully caught up with the impressive advances in today's technology. The use of incremental nonlinear finite element analyses (INL-FEA) in the computation of system reliability of bridge structures has emerged in part due to the advances attained in the finite-element (FE). Quadratic response surface (RS) modelling using design of experiments concepts, and Latin hypercube sampling [4].

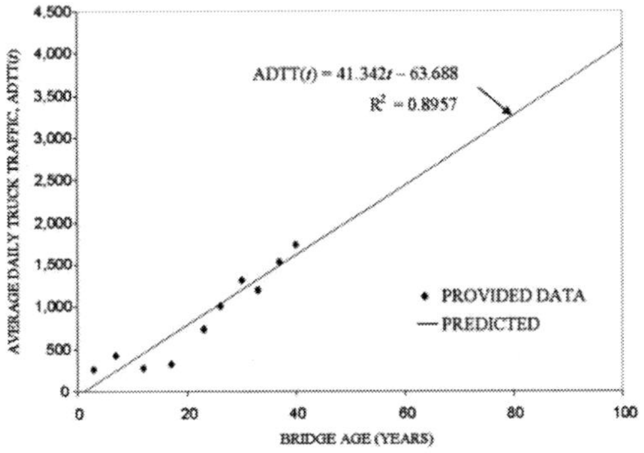

Figure 1: Forecasting ADTT (Average Daily Truck Traffic)

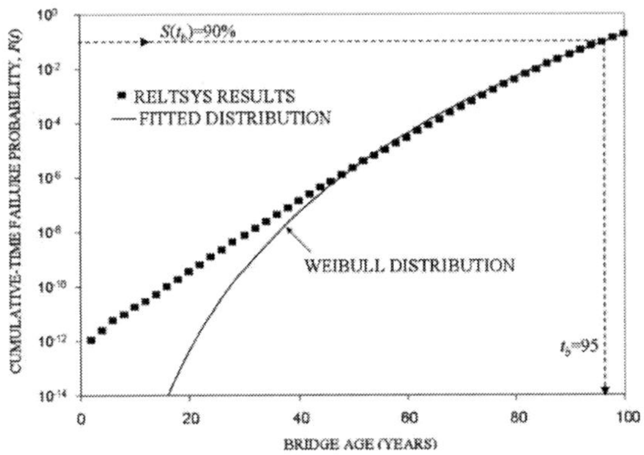

Figure 2: Cumulative-time failure probability of failure of the bridge

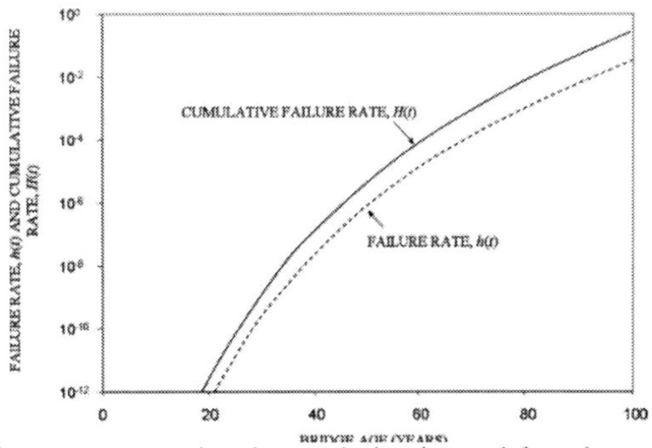

Figure 3: Hazard and cumulative hazard function of the bridge

New method used to predict the life expectancy of concrete bridges using time series variant reliability where load and resistance are considered time dependant. This methods is mainly used to predict the corrosion of steel via time [5].

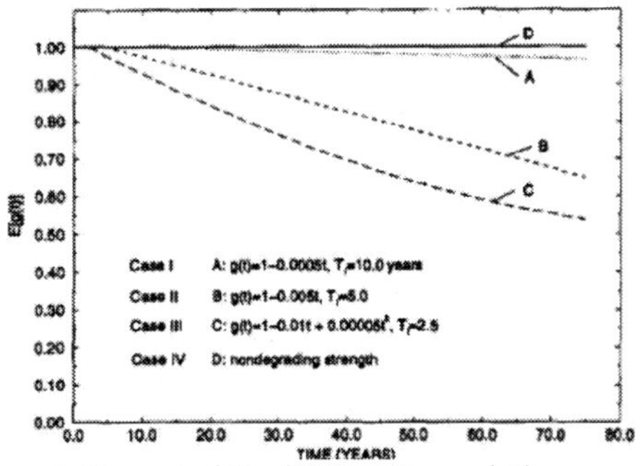

Figure 4: Expected Resistance Degradation versus Time for Cases I-IV

2.2.1 Stochastic modelling

The modelling of bridge deterioration has the problem of uncertainty due to the realistic physical process and limited data on the materials, environment and loading characteristics condition exposed. In principle, the methods of structural reliability enable efficient Bayesian updating of any stochastic model with any kind of information Dynamic Bayesian Networks (DBNs) can be applied for the modelling of deterioration [6].

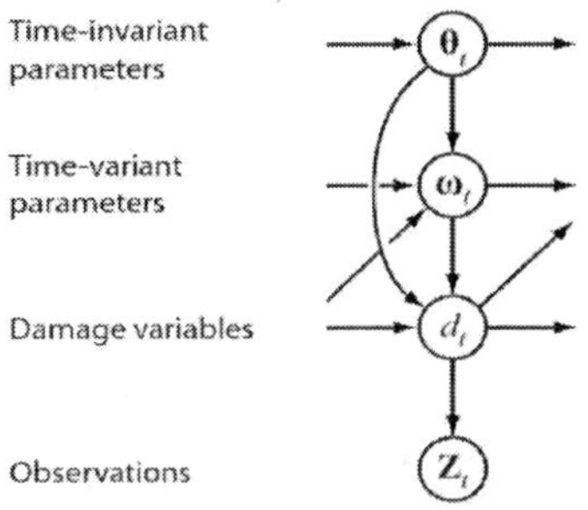

Figure 5 : Dynamic Bayesian Network (DBN) Observation

DOI: 10.1051/matecconf/20168306002

3 Analysis of defects in concrete bridges

The database contains the defect histories of over 430 concrete bridges in the London area. The life span of many of these bridges begins in the early 1930's with the associated inspection records covering most of this period. A number of these bridges were constructed as long ago as 1889 [7-9]. Figure 6 shows a deteriorated bridge affected by reinforcement corrosion. A sample of the data collected and the fields is given in Table 1. Figure 7 shows the defects recorded from the historical inspection records, over 420 bridge structures were investigated and analysed.

Figure 6: Footbridge suffering from severe spalling caused by corrosion.

Code	Yr built	Inyr	Type	Exp		Urg	Fault	Age
TC001	1937	1966	5	2		2	7	29
TC001	1937	1970	5	2		1	11	33
TC002	1937	1970	5	3		1	8	33
TC002	1937	1976	5	3		1	9	39
TC002	1937	1982	5	3		3	9	45

Table 1 : Example of the data contained in the query table.

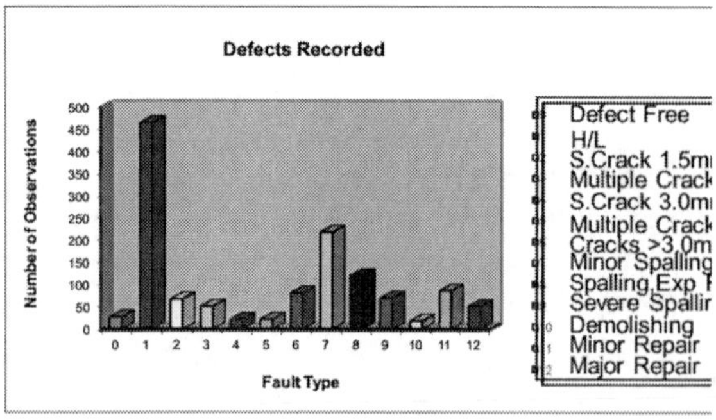

Figure 7 Defects recorded.

3.1 Extensive Analyses of the Database

To begin, the data was analysed to determine the types of structure and number of members inspected. A number of graphs were produced to portray this information. These are outlined below.

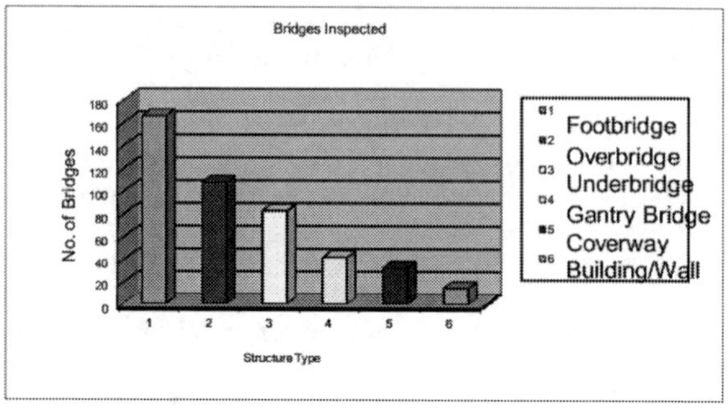

Figure8 Bridges inspected.

- Figure 8 illustrates the range of structures that have been inspected.
- Figure 9 highlights the type of member inspected on each structure and the frequency of that occurrence.

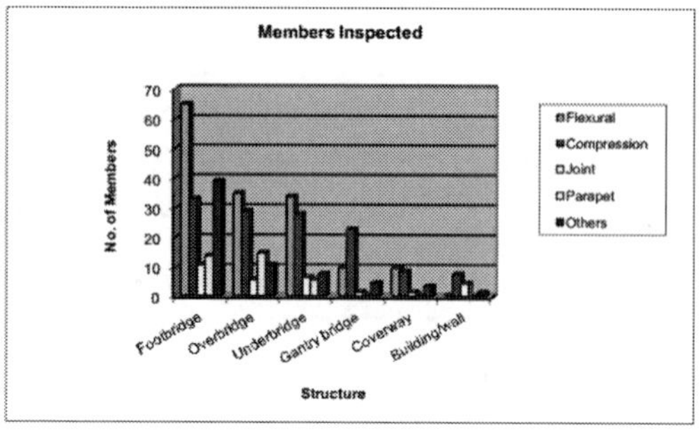

Figure 9: Members inspected.

Perhaps the most significant data recorded during an inspection is the defect of the particular component. During inspections, members are assigned a value according to the most serious defect observed. There are ten possible states, ranging from 0 (defect free) to 9 (severe spalling). Members are assigned the value 10 if they should be demolished, 11 if they require minor repair and 12 if major repair is necessary.

Figure 9 displays the number of each type of defect recorded during the numerous inspections. As this information is recorded for each component the degradation of a particular member can be charted throughout its service life. This information is important in the development of a model predicting the deterioration rate of various components.

 DOI: 10.1051/matecconf/20168306002

It is possible that some defects may have been incorrectly classified because of the large number of defect types, which increases the influence of subjectivity during inspections. To minimise this ambiguity, defects were sorted into five discrete states:

(i) Defect Free (Defects classified as 0 or 1),
(ii) Minor Cracking (Defects classified as 2, 3 or 4),
(iii) Major Cracking (Defects classified as 5, 6 or 7),
(iv) Spalling (Defects classified as 8 or 9),
(v) Failure (Defects classified as 10, 11 or 12).

These condition states are used throughout the analysis.

3.2 Example of results

The tables below gives examples of the service life estimates from the data base for both different structural types and different structural members.

	Mild Exposure (Years)	Moderate Exposure (Years)	Severe Exposure (Years)
Flexural	*38.0*	*N/A*	*39.0*
Compression	*N/A*	*41.3*	*44.0*

Table 2 Service failure of members.

	Flexural (Years)	Compression (Years)
Footbridge	36.7	43.3
Overbridge	38.8	N/A
Underbridge	36.0	46.0

Table 3 Service failure of structures.

4. Discussion

The most probable cause of deterioration is most likely to be due to corrosion caused by chloride attack or carbonation. Much of the most severe concentrations of chloride levels from de-icing salt occur where bridge deck joints have leaked, allowing chlorides in water to trickle over the substructure. Too many deck joints leak and, therefore, abutments and piers become contaminated. Severe corrosion will inevitably result. Spray from passing traffic is also causing chloride contamination of roadside piers and abutments, but the concentrations tend not to be as high as for leakage. Members were analysed according to the structure type they were part of, flexural and compression members followed similar trends of degradation with the former deteriorating faster [7-9].

From the results obtained by the interrogation of the database it became apparent that there was a definite pattern in the way the information changed under certain conditions [7]. It confirmed the belief that, for a given structure or member, the age and subsequent exposure conditions were key factors in determining the deterioration rate of the element. Using these factors it may then be possible to develop a model that could readily predict the deterioration rate of a given concrete structure.

A combination of Service Life Statistical values combined with good budgeting and JIT intervention would optimise the maintenance structure of any infrastructure authority, like highway and railway entities. This would eliminate the need for major unpredictable budget items to manage crises, which occurs when there is a sudden failure. The service life values where available as in Tables 2 and 3 can be used to plan budgets for the maintenance of a complete portfolio of structures such as rail bridges and for a structured mai8ntenance plan that would deliver an efficient use of resources, maintaining the system running and under full control of the managing authority.

References

1) Tapan, M., & Aboutaha, R. S. (2008) Strength evaluation of deteriorated RC bridge columns. *Journal of Bridge Engineering, 13*(3), 226-236. doi:10.1061/(ASCE)1084-0702(2008)13:3(226)
2) Alnaggar, M.,Cusatis, G., & Di Luzio, G. (2013).Lattice discrete particle modeling of alkali silica reaction deterioration of concrete structures.*Cement and Concrete Composites, 41*, 45. doi:10.1016/j.cemconcomp.2013.04.015
3) Ballard, G., & Howell, G. (2003). Lean project management. *Building Research & Information, 31*(2), 119-133. doi:10.1080/09613210301997
4) Okasha, N. M., & Frangopol, D. M. (2010). Advanced modeling for efficient computation of life-cycle performance prediction and service-life estimation of bridges. *Journal of Computing in Civil Engineering, 24*(6), 548-556. doi:10.1061/(ASCE)CP.1943-5487.0000060
5) Straub, D. (2009). Stochastic modeling of deterioration processes through dynamic Bayesian networks. *Journal of Engineering Mechanics, 135*(10), 1089-1099. doi:10.1061/(ASCE)EM.1943-7889.0000024
6) Enright, M. P & Frangopol, D. M. (1998) Service-life prediction of deteriorating concrete bridges. *Journal of Structural Engineering, 124*(3), 309-317. doi:10.1061/(ASCE)0733-9445(1998)124:3(309)
7) Rigden, S. R. et al, *Predicting future performance of concrete bridges using long term inspection records*, Proceedings of the 5th International Conference on Structural Faults & Repair, 1993, 43-46.
8) Christer, A. H. et al, *Modelling the deterioration and maintenance of concrete structures*, presented at the International Conference of Operation Research, 1993.

9) Rigden, S. R., et al., *Service life prediction of concrete bridges*, Concrete Repair, Rehabilitation and Protection, 1996, 705-714.

Multi-cracks identification based on the nonlinear vibration response of beams subjected to moving harmonic load

H. Chouiyakh[1], L. Azrar[1,2,3], O. Akourri[1], and K. Alnefaie[3]

[1] Mathematical Modeling and Control, FST of Tangier, Abdelmalek Essaâdi University; Tangier; Morocco
[2]LaMIPI, Higher School of Technical Education of Rabat (ENSET), Mohammed V University in Rabat, Morocco
[3] Mechanical Engineering Department, Faculty of Engineering, King Abdulaziz University, Jeddah, Saudi Arabia

Abstract. The aim of this work is to investigate the nonlinear forced vibration of beams containing an arbitrary number of cracks and to perform a multi-crack identification procedure based on the obtained signals. Cracks are assumed to be open and modelled trough rotational springs linking two adjacent sub-beams. Forced vibration analysis is performed by a developed time differential quadrature method. The obtained nonlinear vibration responses are analyzed by Huang Hilbert Transform. The instantaneous frequency is used as damage index tool for cracks detection.

1 Introduction

Vibration based techniques of damage identification aim to combine mathematical models with signal processing techniques. Relevant work has been published in this regard [1], but they almost assume that structures and damages behave linearly while in reality signals are nonlinear. Thus, the structural health monitoring researchers appeal to mathematical models of nonlinear dynamics [2, 3]. The problem of vibration of multi-cracked beams subjected to moving loads has also attracted many researchers [4] where free and forced vibrations of the beam are investigated. However, the linear case is usually considered and the forced response is computed following classical schemes of integration mainly the Runge-Kutta method [5, 6]. On the other hand, cracks identification is concerned by analysing the vibration signals using adapted techniques. The focus will be here on time frequency methods known for their local properties in both time and frequency domains [7].

In parallel with our previous work [8, 9], this paper focuses on the nonlinear behaviour of multi-cracked beams subjected to moving harmonic load. For the free and forced responses, a numerical method based on the differential quadrature method has been developed. Crack identification procedures are elaborated based on the numerically computed nonlinear responses.

2 Mathematical formulation

Consider a multi-cracked Euler-Bernoulli beam with length L, cross-section A, mass density ρ, moment of inertia I, and modulus of elasticity E that is subjected to a moving harmonic load of magnitude F_0, speed v and excitation frequency Ω, and R is the number of existing cracks as shown in figure 1. It is assumed that the cross-sectional area of the beam is rectangular and its material is homogenous. The 'R' cracks are assumed to be open and modelled through rotational springs which flexibilities are given by fracture mechanics [8]. The whole beam is sub-divided into (R+1) sub-beams.

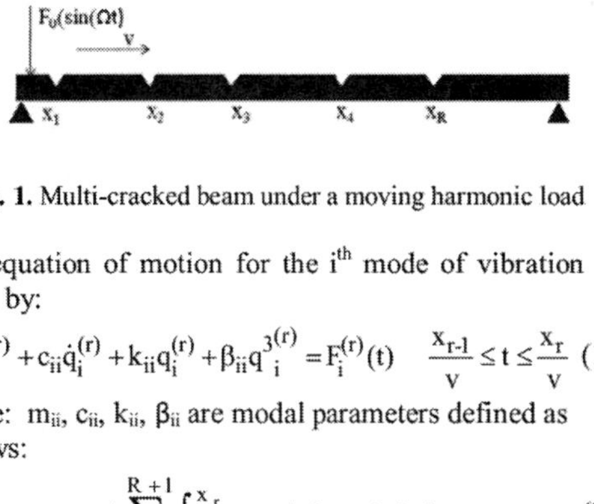

Fig. 1. Multi-cracked beam under a moving harmonic load

The equation of motion for the ith mode of vibration is given by:

$$m_{ii}\ddot{q}_i^{(r)} + c_{ii}\dot{q}_i^{(r)} + k_{ii}q_i^{(r)} + \beta_{ii}q_i^{3(r)} = F_i^{(r)}(t) \qquad \frac{x_{r-1}}{v} \le t \le \frac{x_r}{v} \quad (1)$$

where: m_{ii}, c_{ii}, k_{ii}, β_{ii} are modal parameters defined as follows:

$$m_{ii} = \rho A \sum_{r=1}^{R+1} \int_{x_{r-1}}^{x_r} w_i(x)\, w_i(x)\, dx \quad (2)$$

$$c_{ii} = \eta \sum_{r=1}^{R+1} \int_{x_{r-1}}^{x_r} w_i(x)\, w_i(x)\, dx \quad (3)$$

$$k_{ii} = EI \sum_{r=1}^{R+1} \int_{x_{r-1}}^{x_r} w_i''''(x)\, w_i(x)\, dx \quad (4)$$

$$\beta_{ii} = -\frac{EA}{2L}\sum_{r=1}^{R+1}\int_{x_{r-1}}^{x_r} w_i''(x)\left(\int_{x_{r-1}}^{x_r}(w_i'(x))^2\,dx\right)w(x)dx \qquad (5)$$

$$F_i^{(r)}(t) = \rho A\sum_{r=1}^{R+1}\int_{x_{r-1}}^{x_r} F_0\sin(\Omega t)\delta(x - vt)w_i(x)\,dx \qquad (6)$$

x_0 and x_{R+1} correspond respectively to the left and right boundaries of the beam (x=0 and x=L).
Note that the eignemode of a multi-cracked beam is written as [8]:

$$w_i(x) = w_1(0) + \sum_{r=1}^{n} w_{ri}(x)(H(x - x_{r-1}) - H(x - x_r)) + w_{R+1}(L) \qquad (7)$$

The present problem is a set of coupled differential equations which has been solved for the linear case (β=0) method [8].
Due to the fact that the excitation term depends on piecewise mode, the classical numerical methods cannot be used. In this work, a new numerical approach based differential quadrature method (DQM) has been developed in time domain for nonlinear analyses.

2.1 Differential quadrature method

The Differential quadrature method (DQM) was first introduced by Bellman and developed by many researchers [10]. The DQM requires the discretization of the problem into N points. The derivatives at any point are approximated by a weighted linear summation of all the functional values along the discretized domain, as follows [10]:

$$\begin{cases} \ddot{q}(t_i) = \sum_{i=1}^{N} b_{ij}^{(2)} q(t_j) \\ \dot{q}(t_i) = \sum_{i=1}^{N} b_{ij}^{(1)} q(t_j) \end{cases} \qquad (7)$$

N is the number of distretizing points, $b_{ij}^{(1)}$ and $b_{ij}^{(2)}$ are the first and second order weighting coefficients respectively. The weighting coefficients for the first order derivative to the functional values can be obtained as:

$$b_{kj}^{(1)} = \begin{cases} \dfrac{L(t)}{(t_k - t_j)L_1(t_j)} & k \neq j \\ -\sum_{j=1, j\neq k}^{N} b_{kj}^{(1)} & k = j \end{cases} \qquad (8)$$

$$L(t_k) = \prod_{i=1}^{N}(t_i - t_k) \qquad (9)$$

The second, third and higher derivatives can be calculated as:

$$b_{kj}^{(m)} = \sum_{j=1, j\neq k}^{N} b_{kl}^{(1)} b_{lj}^{(m-1)} \qquad (10)$$

The N discretizing points are calculated through:

$$t_j = \frac{1}{2}\left[1 - \cos\left(\frac{j-1}{N-1}\pi\right)\right] \quad j = 1,2,\dots, N \qquad (11)$$

2.2 Nonlinear forced response using the time differential quadrature method

In this work, a new approach for solving nonlinear differential equations by introducing a correction loop in for calculating nonlinear response is presented. Applying the DQ method in each sub-beam Eq.(1) can be discretized as:

$$\left([M] + [C] + [K] + [B]\{q^2\}^r\right)\{q\}^r = \{F\}^r \qquad (12)$$

where: $[M] = m_{ii}\, b_{kj}^{(2)}$; $[C] = c_{ii}\, b_{kj}^{(1)}$; $[K] = k_{ii}\,\mathrm{Id}(N,N)$; and $[B] = \beta_{ii}\,\mathrm{Id}(N,N)$.
We first set [B] = 0. The linear solution for the rth time interval is obtained by solving the algebraic problem.

$$\{q\}^r = ([M] + [C] + [K])^{-1}\{F\}^r \qquad (13)$$

In order to calculate non linear response, for the sub-beam 'r ', we propose an iterative process written as:

$$\{q\}_{i+1}^r = ([M] + [C] + [K])^{-1}\left(\{F\}^r - [B]\{q^2\}_i^r\right) \qquad (14)$$

The residue R_{i+1} is calculated by:

$$R_{i+1}^r = \left|[M]^r\{q\}_{i+1}^r + [C]^r\{q\}_{i+1}^r + [K]^r\{q\}_{i+1}^r + [B]^r\{q^3\}_{i+1}^r - \{F\}^r\right| \qquad (15)$$

The stopping criterion is taken as

$$\frac{R_{i+1}^r}{R_1^r} \leq \varepsilon \qquad (16)$$

3 Huang Hilbert transform: an overview

As the identification process, used in this paper, will be based on the Huang Hilbert transform, an overview on the empirical mode decomposition and Hilbert transform is given

3.1. Empirical mode decomposition

The Empirical mode decomposition (EMD) is a technique representing non linear and non-stationary signals as sum of simpler components called Intrinsic Mode Functions (IMFs). An IMF should satisfy the following conditions:

a) An IMF may only have one zero between successive extrema
b) An IMF must have zero local mean

The decomposition is performed through a repeated sifting procedure. At the end, the time signal x(t) can be expressed in terms of n number of IMFs:

$$s(t) = \sum_{i=1}^{n} \mathrm{IMF}_i(t) + \text{residue} \qquad (17)$$

The Hilbert transform is then applied to each of those components, in order to get instantaneous amplitude and frequency plots.

3.1. Hilbert transform

The Hilbert Transform (HT) of a signal s(t), is an integral transformation, from time domain to time domain, defined by [7] :

 DOI: 10.1051/matecconf/20168306003

$$H(s(t)) = \frac{1}{\pi}\int_{-\infty}^{+\infty}\frac{s(\tau)}{t-\tau}d\tau \tag{18}$$

The HT is the convolution of s(t) with 1/t and hence emphasizes the local properties of s(t). The real signal s(t) and its HT h (t), form an analytical complex signal $\tilde{S}(t)$ of the form :

$$\tilde{S}(t) = s(t) + i\,h(t) = A(t)e^{i\theta(t)} \tag{19}$$

The instantaneous $A(t)$ and phase $\theta(t)$ change with time. The instantaneous amplitude A(t) or envelope, is given by:

$$A(t) = \pm\sqrt{(s(t))^2 + (h(t))^2} \tag{20}$$

where the " ± " signs correspond to the upper positive and the lower negative envelops

$$\theta(t) = \text{Arc}\tan\left(\frac{h(t)}{s(t)}\right) \tag{21}$$

The instantaneous frequency (IF) is defined as the derivative of the phase:

$$\omega(t) = \frac{d\theta(t)}{dt} \tag{22}$$

It measures the rate and direction of a phase in the complex plane. It can be estimated by different algorithms [7].

4 Numerical results and discussion

3.1. Nonlinear forced response

In order to validate the previous developments, an Euler-Bernoulli beam with the following material properties is considered: Young's modulus E= 210 GPa, material mass density ρ= 7860 kg/m^3 and Poisson ratio ν= 0.3. The geometrical parameters of the beam are selected as: depth h = 0.01 mm, thickness b = 0.01mm. First, comparison is made with the reference [11] for the non cracked linear case. The obtained results are plotted in figure 2 and are the same as those presented in [11]

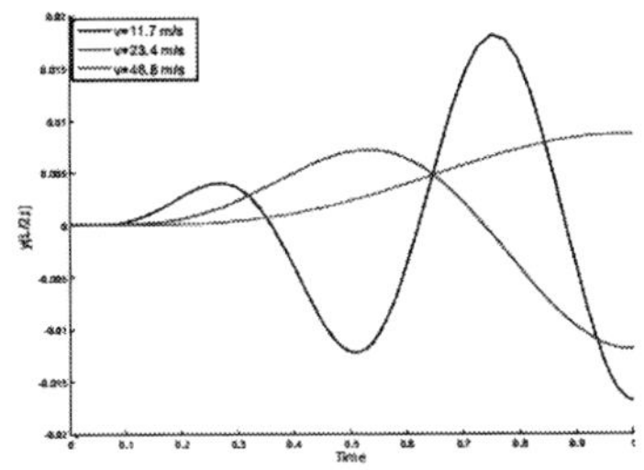

Fig. 2. Dynamic response of a pinned-pinned beam, varying the speed for Ω=$\omega_{\text{resonance}}$

For the nonlinear case, the numerically obtained forced responses of a multi-cracked beam are depicted for various speeds and different frequencies in figure 3. It is assumed that the beam contains four equally spaced cracks of equal depth (a/h=0.1) located at x/L=0.1; 0.3; 0.5 and 0.7.

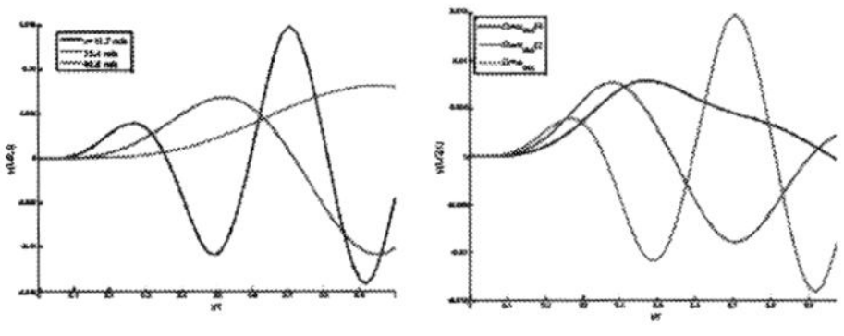

Fig. 3. Dynamic response of a pinned-pinned beam, varying the speed for different speeds (Ω=$\omega_{\text{resonance}}$) and different load frequencies (v=11.7 m/s)

3.2. Multi-cracks identification

As the main aim of this work is to investigate crack detection from nonlinear signals, we propose to combine Huang Hilbert transform to moving load properties for a better cracks identification. For that, the nonlinear signal, depicted in figure 4, is decomposed into simpler components (IMFs) using the EMD, then Hilbert spectral analysis is applied to each of those IMFs.

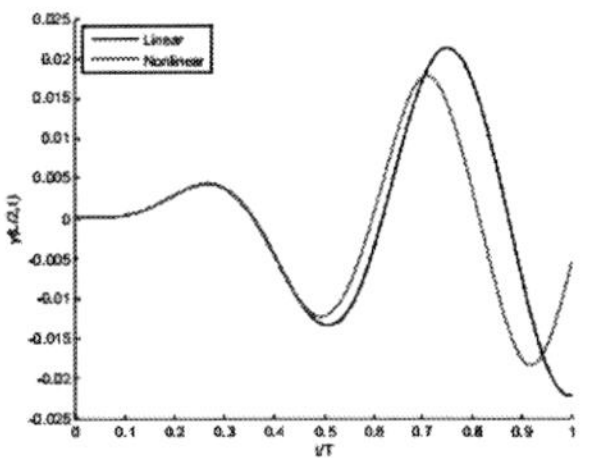

Fig. 4. Analyzed nonlinear signal

We notice that instantaneous frequency of the first IMF identifies positions of all cracks that are localized by sharp transitions in the curve. This is due to the presence of high frequency components in the signal at these locations as shown in figure 5. It should be noted that large peaks are obtained leading to clear crack position detection.

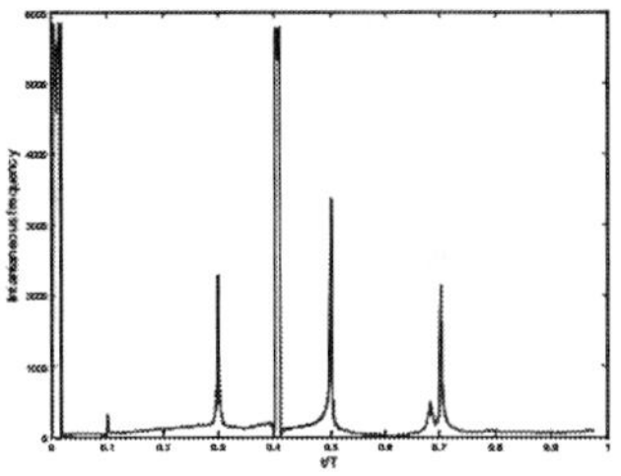

Fig. 5. Instantaneous frequency of the first IMF for v=11.7 and Ω=$\omega_{\text{resonance}}$

Moreover, a better detection is obtained for selected values of the speed v and excitation frequency Ω. In this work we obtain them by trial and found that the best detection is obtained for $\Omega=\omega_{resonance}$ and v= 11.7 m/s as shown in figure 5. In spite of that, there are other peaks in the curve which means that there are other higher frequency components in the analyzed signals. This is due to the fact that the nonlinear signal contains various other harmonics and these harmonics may contain other information about cracks positions. For efficient multi-cracks identification based on the nonlinear responses some filters and a deep analysis is required.

5 Conclusion

We developed a numerical algorithm based on the time differential quadrature method in order to get the forced responses of multi-cracked beams under moving harmonic load.

A large number of cracks can be easily considered for the direct problem and identified for the inverse problem. We used for cracks identification Huang Hilbert transform. Higher frequency components are first detected and for the nonlinear case, not only cracks produce sharp transitions in the curve of instantaneous frequency but also some nonlinear signal components. We cannot accurately define those components since we lack of explicit analytic solutions for the multi-cracked beam vibration problems.

The identification is performed for selected values of the speed and excitation frequency. Adjusted values lead to better multi-cracks detection. An optimization procedure can be elaborated to predict the best v and Ω parameters.

References

1. C. Boller, C. Fou-Kuo, and F. Yozo. Encyclopedia of structural health monitoring. John Wiley & Sons, (2009).
2. A.H.Nayfeh, F. Pai. Linear and Nonlinear Structural Mechanics. Wiley, May (2004)
3. Y.C Chu, H.H Shent. Analysis of Forced Bilinear Oscillators and the Application to cracked beam dynamics. AIAA journal, (1992): 2512-2519.
4. Bajer, Czesław I., and Bartłomiej Dyniewicz. Numerical analysis of vibrations of structures under moving inertial load. Vol. 65. Springer Science & Business Media, 2012.
5. N.Roveri, and A. Carcaterra. Damage detection in structures under traveling loads by Hilbert–Huang transform. Mechanical Systems and Signal Processing **28** (2012): 128-144.
6. A. Ariaei, S. Ziaei-Rad, and M. Ghayour. Repair of a cracked Timoshenko beam subjected to a moving mass using piezoelectric patches. International Journal of Mechanical Sciences **52**.8 (2010): 1074-1091.
7. M. Feldman, Hilbert transform in vibration analysis. Mechanical systems and signal processing **25**, 3 (2011): 735-802.
8. H. Chouiyakh, L. Azrar, K. Alnefaie and O. Akourri. Multicracks identification of beams based on moving harmonic excitation. Structural Engineering and Mechanics, **58**, 6 (2016)
9. H. Chouiyakh, L. Azrar, K. Alnefaie and O. Akourri. Vibration and multi-crack identification based on the free and forced responses of Timoshenko beams under moving mass using the differential quadrature method. Submitted to International Journal of Mechanical Sciences
10. Z. Zong and Z. Yingyan, Advanced differential quadrature methods. CRC press, 2009.
11. M. Abu-Hilal and M. Mohsen. Vibration of beams with general boundary conditions due to a moving harmonic load. Journal of Sound and Vibration **232**, 4 (2000): 703-717.

MATEC Web of Conferences **83**, 06004 (2016) DOI: 10.1051/matecconf/20168306004
CSNDD 2016

Early observation of modal parameter changes by an enhanced frequency evaluation algorithm

I.C. Mituletu, G.R. Gillich, M. Tufoi, N. Gillich

Universitatea "Eftimie Murgu" din Resita, Department of Mechanical Engineering, 320085 Resita, Romania

Abstract. By monitoring the vibration response of mechanical or civil engineering systems, the modal parameter changes due to materials or geometrical properties can be observed. Fine frequency shift identification is decisive to determine particularly small changes and thereby small cracks or damages. This paper firstly presents some existing methods to accurately find the frequency values. All these methods show non-linear, relevant errors in interpretation. Afterward a simple method is proposed, consisting in iterative decreasing of the number of samples from the original signal, performing a standard frequency analysis for each truncated signal and finally overlapping of achieved spectra. Thus, the resulted spectrum has a much fine resolution due to the important addition of spectral lines. In this way a more accurate identification of frequencies is possible. In fact, the relevant frequency is accomplished under a lobe shape. In the peak position of this lobe the best frequency value can be clearly identified. With the help of these more accurate frequency values, modal analysis becomes more reliable and efficient.

1 Introduction

Methods which based on the changes of the structural modal parameters are commonly involved to establish the global health of a mechanical or civil engineering structure and to assess damages. The methods based on frequency changes are easiest applied, but problems occur because small damage produces small changes in the natural frequencies. Thus, observe the changes in the frequency spectrum are hard to be found [1]. In standard frequency evaluation, the problem is that frequencies are indicated on spectral lines, whose location directly dependents on the analyzed signal length. As a consequence, for small damages, the frequency changes are not indicated in the spectrum until the peak-amplitude does not move to the inferior spectral line. For early damage observation, it becomes necessary to involve advanced algorithms, providing denser spectral lines, with direct consequence increasing the frequency evaluation accuracy. In standard frequency evaluation often false results are signalized. To understand why this happens, we review some notions of mathematics and signal analysis.

By applying the Fourier Transform (FT) to an infinite long signal in time, an infinitely small frequency resolution is achieved. Thus, the frequency is precisely identified. For a finite-time length signal, the resolution becomes relevant, depending on the time length. To calculate the frequency content for discrete signal x, the Discrete Fourier Transform (DFT) can be used. It provides, in a simple way, information about the frequency spectrum of the signal, by turning N samples of the signal from the time domain in N complex values in the frequency domain. Its mathematical expression is:

$$X(f_k)=\sum_{n=0}^{N-1}x(t_n)e^{-j2\pi f_k t_n}=\sum_{n=0}^{N-1}x(t_n)e^{-j2\pi\frac{k}{N-1}n} \qquad (1)$$

with $k = 0, 1, 2,\dots N\!-\!1$.

In Eq. (1) we denoted $x(t_n)$ the instantaneous input signal amplitude at time $t_n = nT$, T the sampling period, $X(f_k)$ the spectrum of x at frequency f_k and N the number of time samples. The signal time length results in t_{N-1}.

The Power Spectrum Density (PSD) is a positive real function of a frequency variable. While the Power Spectrum (PS) calculates the area under the signal plot using the discrete FT, the power spectrum density assigns units of power to each unit of frequency and so, enhances periodicities.

Appling the PSD to a sequence representing a sinusoidal signal, one can distinguish between two cases. First, if the sequence contains an integer number of periods, as depicted in figure 1a, the frequency is read precisely. It is exemplified in figure 1c for a signal whit a created frequency of 4 Hz, analyzed in the time interval of 1 s. The PSD, represented by a red line, indicates the correct frequency.

In the second case, the signal sequence does not contain an integer number of periods. This is illustrated in figure 1b. For this signal, also of 4 Hz, if the time interval is 1.1 s the virtual instrument allow no accurate frequency reading. The achieved spectrum is shown in figure 1c by blue lines. Obviously, correct frequency is not identified.

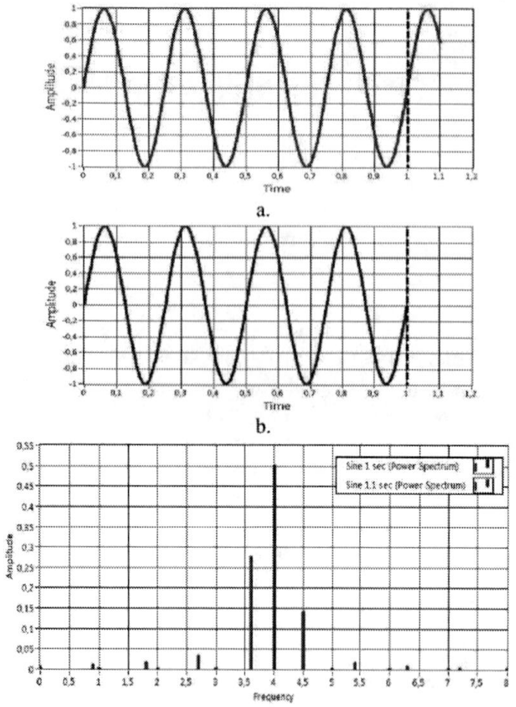

Fig. 1. Two signals, one containing an integer number of periods (a) and the second with non-integer number of periods (b) and their frequency spectra (c)

Having a look on figure 1c, we remark that although the spectral lines are denser, the frequency cannot be read accurately, while we found the peak amplitude at 3.63Hz instead of 4Hz. Even for an improved frequency resolution, the frequency value cannot be properly read because the spectral line does not coincide with the frequency value. This has as source the phenomenon known as "leakage", which is the energy dispersion from the true frequency of the signal into several false frequencies. Because this, leakage causes an apparent amplitude decrease, while the frequency is indicated at a spectral line that does not coincide with the real one [2]. The rest of the energy is found, erroneously, on the other spectral lines.

2 Actual methods to improve frequency identification

Several simple methods to improve the frequency readability are known. One of the most popular is zero-padding, which consists in adding a string of zeros to the signal before the PSD is calculated [3]. It allows a better approximation to the frequencies contained in a signal. By using this method, the frequency distance between the spectral lines is decreased; therefore, an improved frequency evaluation is possible [3]. A faster algorithm than zero-padded FFT is presented in [4]. The attained results are similar with those resulted by as zero-padding, but the computational effort is essentially reduced.

Windowing is another method often used to improve frequency evaluation. For a finite-length signal that consists of a non-integer number of cycles, the application of a window smoothes its ends, making the amplitude to vary gradually towards zero at the ends [5-6]. This minimizes the discontinuities of the signal edges, reducing spectral leakage. However, this method works well for identification of close-located frequencies but does not improve the readability.

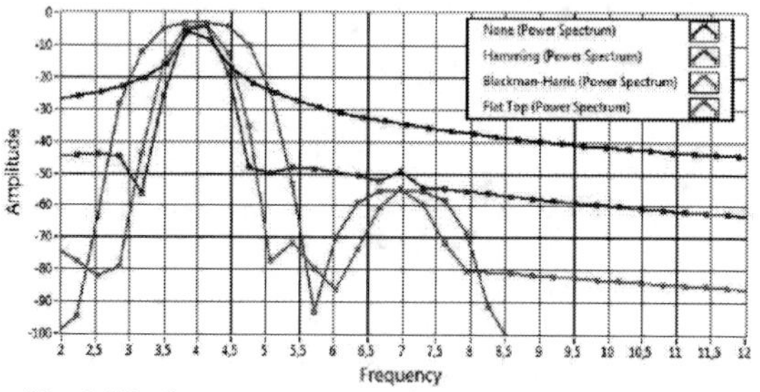

Fig. 2. The frequency spectrum for a two-tone signal in the absence and the presence of windowing - dB representation

It is also possible to evaluate the frequency by plotting a curve best fitting to three consecutive points, located around the estimated frequency, obtained by DFT. In figure 3, these points are $B(f_{j-1},A_{j-1})$, $C(f_j,A_j)$, $D(f_{j+1},A_{j+1})$, where k-1, k and k+1 are the spectral line numbers, f_{j-1}, f_j and f_{j+1} the frequencies indicated at that lines and A_{k-1}, A_k and A_{k+1} the frequency amplitudes associated with these lines. The presumed correct frequency f_{corr} and amplitude A_{max} are considered at the point in that the graph has a maximum. This peak is located on a virtual spectral line which does not exist in the calculated spectrum. It is found via a fractional correction term δ, as $f_{corr} = f_j + \delta$. There are a number of spectral peak location estimator methods for precise frequency estimation in the DFT domain [7]. In [8] parabola approximation for peak determination is used, so the correction term is given by

$$\delta = \frac{a_{k+1} - a_{k-1}}{4a_k - 2a_{k-1} - 2a_{k+1}} \qquad (2)$$

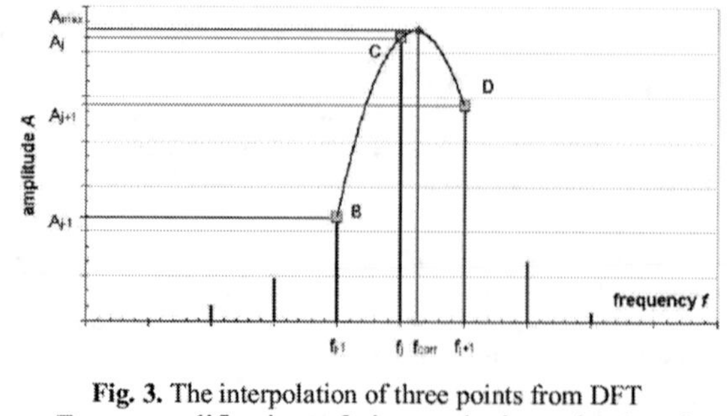

Fig. 3. The interpolation of three points from DFT

For exemplification of the method, again the sine signal of 4 Hz is used. The frequency resolution $\Delta f = 1/t$

is dependent on the signal time length t. Spectral analyses for different time lengths (t_{min} = 0.9 s and t_{max} = 1.12 s with a step of 0.02 s) are considered, resulting in different frequency resolutions. The results are indicated in table 1.

Having a look onto this table we find the amplitudes of the three points involved in the curve fitting, the fractional correction term, the values of read frequency and the values of corrected frequency obtained with the presented method. As can be seen in table 1, by using the so-called three-point method, the error between a read frequency and the corrected frequency is less than 7%. In figure 4 the actual frequency, that obtained by the three-point method and that by standard evaluation (direct reading the virtual instrument) are plotted. Observe that there are areas where the frequency values obtained by the two methods are similar (white background), while in other areas (gray background) large differences between the obtained frequencies values occur. The differences become smaller with the increase of the analysis time interval, as shown in Table 1.

From figure 4 it can be also deduced that the error is cyclical, depending on how the three considered points are placed around the real frequency.

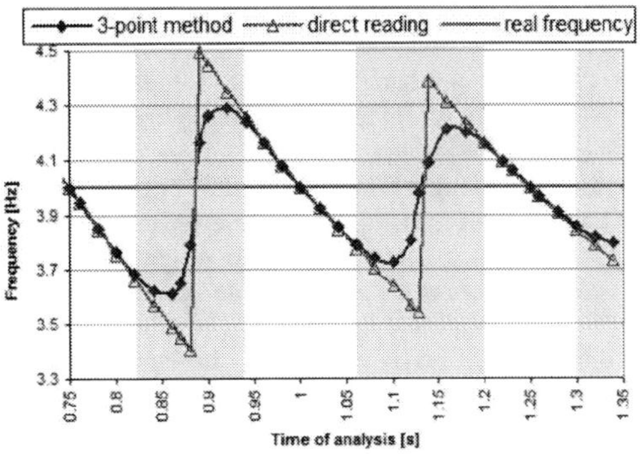

Fig. 4. Comparison between the real frequency, obtained by 3-points method and by direct reading (a) and the difference between the corrected and read frequency (b)

Table 1. The read frequencies and fractional correction term

t [s]	Δf [Hz]	f_{read} [Hz]	A_{k-1} [mm/s^2]	A_k [mm/s^2]	A_{k+1} [mm/s^2]	δ [Hz]	f_{corr} [mm/s^2]	Error [%]
1.12	0.892857	3.57143	0.03298	0.24430625	0.166872	0.23184	3.80327	-4.91829
1.1	0.909091	3.63636	0.031359	0.30906909	0.114683	0.088249	3.72461	-6.8847
1.08	0.925926	3.70370	0.024568	0.36470093	0.073694	0.038918	3.74262	-6.43445
1.06	0.943396	3.77358	0.015547	0.41093113	0.041821	0.017184	3.79077	-5.23079
1.04	0.961538	3.84615	0.007444	0.44977019	0.018447	0.006297	3.85245	-3.68873
1.02	0.980392	3.92157	0.002029	0.48110196	0.004398	0.001239	3.92281	-1.92981
1	1	4.00000	0	0.5	0	0	4.00000	0
0.98	1.020408	4.08163	0.002906	0.49826531	0.003295	0.000196	4.08183	2.045723
0.96	1.041667	4.16667	0.01473	0.46947917	0.010113	-0.00253	4.16414	4.103522
0.94	1.06383	4.25532	0.041063	0.41407979	0.016013	-0.01624	4.23908	5.976895
0.92	1.086957	4.34783	0.085721	0.34072065	0.018509	-0.05822	4.28960	7.240122
0.9	1.111111	4.44444	0.147096	0.26275333	0.017829	-0.17925	4.26520	6.629933

3 The proposed method

The method proposed in this paper requires an iterative time signal truncation, followed by a spectral analysis. The aim is to obtain numerous spectra with different frequency resolutions and superpose them in an "overlapped spectrum". This helps to increase the number of spectral lines in the final spectrum, as shown in figure 5, where three spectra with different frequency resolutions: Δf_1 = 1 Hz, Δf_2 = 1.05 Hz and Δf_3 = 1.11 Hz are involved.

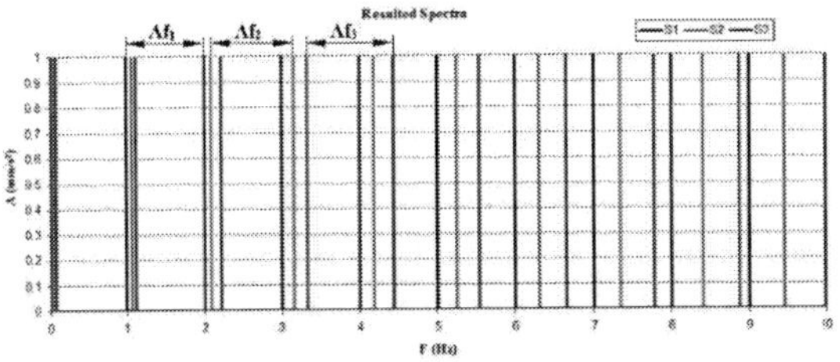

Fig. 5. Spectral lines distribution for the overlapped spectrum

From the overlapped spectrum the frequency associated with the amplitude presenting local maxima is extracted, indicating the most credible frequency value. Figure 6 exemplifies this for the frequency of 9.753 Hz. Note that 47 iterations are performed in order to ensure a dense spectral line net.

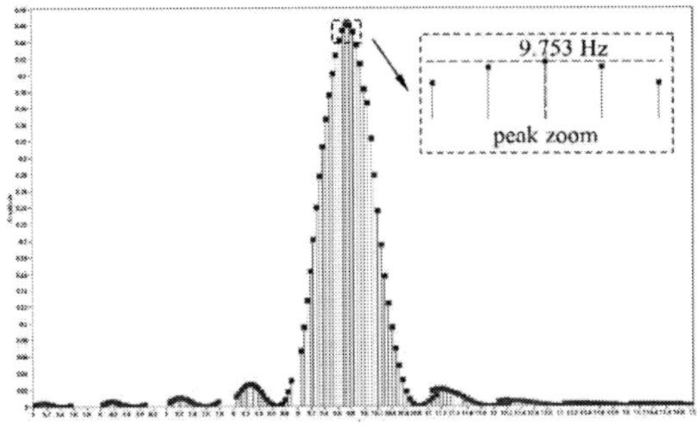

Fig. 6. The amplitude dependence on time intervals

DOI: 10.1051/matecconf/20168306004

The process is simple and contains several steps to be performed:

- a rough evaluation of the frequencies is made using PS, by direct reading;
- the frequencies of interest for the damage detection process are identified;
- the analysis time is set for in order to ensure a frequency resolution placing a spectral line close under the presumed frequency;
- the signal length is iteratively shortened by applying a rectangular window and the PS is derived;
- the observed spectral line will migrate to superior frequencies, because the analysis time decreases;
- first an amplitude increase will be observed until a maximum is achieved, afterward the amplitudes will decrease;
- the frequency associated with the peak amplitude is extracted as the real frequency.

The time length of the vibration signal acquired for damage detection should exceed five cycles of the investigated frequency. This time was to be truncated until one period is suppressed. In this way the achievement of the amplitude peak is certain.

It should be mentioned that for each frequency an individual analysis is necessary because the time periods are different for each of them. This constitutes no impediment to the damage detection process, while the analysis for several frequencies is performed in seconds.

Fine frequency evaluation ensures reliable data for vibration-based damage detection [9, 10], making possible the observation of small modal parameter changes. Figure 7 illustrates a lab stand designed to test the sensitivity of the proposed method.

Fig. 7. Stand for testing the sensitivity of the proposed frequency evaluation method

The laptop screen shows the virtual instrument designed in LabVIEW to automatically perform the seven steps described at the beginning of this section. The two graphs at the right present the original signal and signal resulted after truncation. The upper left graph shows the overlapped spectrum which has a clear peak. In addition, the frequency and amplitude values associated to this peak are numerically indicated.

4 Conclusion

In this paper, a simple method for an accurate frequency evaluation is proposed and compared with actual methods presented in the literature. Even if these methods improve the frequency readability, they are not enough accurate to evaluate the changes occurring in structures due to small damages. The proposed method bases on overlapping numerous spectra achieved from the same signal analysed for different time lengths. By overlapping, the number of spectral lines in the frequency domain of interest is dramatically increased, thus results the possibility to identify the peak amplitude and consequently the frequency with high precision. Test performed to reveal the frequency sensitivity shown that the frequency is identified with a precision of three digits after unit.

Acknowledgement

The work has been funded by the Sectoral Operational Programme Human Resources Development 2007-2013 of the Ministry of European Funds through the Financial Agreement POSDRU/159/1.5/S/132395.

References

1. Z. Praisach, G.R. Gillich, D.E. Birdeanu, *Considerations on Natural Frequency Changes in Damaged Cantilever Beams Using FEM*, Latest Trends on Engineering Mechanics, Structures, Engineering Geology (2010)
2. S.T. Abed, Z.E. Dallalbashi, F.A. Taha, *Studying the Effect of Window Type on Power Spectrum Based on MatLab*, Tikrit J. Eng. Sciences **19**(2) (2012)
3. S.A. Neild, *Using nonlinear vibration techniques to detect damage in concrete bridges*, PhD thesis, Univ. of Oxford, Oxford, U.K (2001)
4. C. Donciu, M. Temneanu, *An alternative method to zero-padded DFT*, Measurement **70** (2015)
5. G. Andria, M. Savino, A. Trotta, *Windows and interpolation algorithms to improve electrical measurement accuracy*, IEEE Transactions on Instrumentation and Measurement, 38(4) (1989)
6. S.T. Abed, Z.E. Dallalbashi, F.A. Taha, *Studying the Effect of Window Type on Power Spectrum Based on MatLab*, Tikrit J. Eng. Sciences 19(2) (2012)
7. Ç. Candan, *A Method For Fine Resolution Frequency Estimation From Three DFT Samples*, IEEE Signal Processing Letters, **18**(6) (2011)
8. P. Voglewede, *Parabola approximation for peak determination*, Global DSP Mag., **3**(5), (2004)
9. G.R. Gillich, Z.I. Praisach, M.A. Wahab, O. Vasile, *Localization of Transversal Cracks in Sandwich Beams and Evaluation of Their Severity*, Shock Vib. Article Number: 607125 (2014)
10. G.R. Gillich, Z.I. Praisach, *Modal identification and damage detection in beam-like structures using the power spectrum and time–frequency analysis*, Signal Process. **96** (2014)

Liquid-Liquid Polymerization Front under Quasiperiodic Gravitational Modulation

Saadia Assiyad[1,a], Karam Allali[1,b], and Mohamed Belhaq[2,c]

[1] University Hassan II-Casablanca, FSTM, Department of Mathematics, PO Box Mohammadia-Morocco
[2] University Hassan II-Casablanca, FS, Department of Physics, Casablanca-Morocco

Abstract. The influence of quasi-periodic gravitational modulation of polymerization front with liquid monomer and liquid polymer is studied in this paper. The model considered includes the heat equation, the equation for the concentration and the Navier-Stokes equations under the Boussinesq approximation. The linear stability analysis of the problem is conducted and by using numerical simulation, the convective instability boundary is carried out. Results obtained show that the convective instability threshold depends on the amplitudes and on the frequencies ratio of the quasi-periodic gravitational modulation. Effect of Prandtl number is also examined.

1 Introduction

Frontal polymerization is the process of converting monomer to polymer via reaction front [1]. In this paper, we study the influence of quasi-periodic (QP) gravitational modulation on the convective instability of liquid-liquid polymerization front.

The influence of convective instability on polymerization front, in the absence of vibrations, with liquid monomer and solid polymer was studied in [2] and the case of liquid polymer was examined in [3]. Other works were dedicated to analyze the effect of a periodic gravitational modulation on the convective instability with liquid-solid polymerization front in [4] and it was found that the propagation of the reaction is affected by the parameters of vibration, while the case of liquid-liquid polymerization was studied in [5] and it was concluded that, for small vibration amplitudes, the reaction front remains stable and it loses its stability for sufficiently large amplitude of vibrations. Recent works were devoted to the influence of a QP gravitational modulation on the convective instability of reaction front. For instance, the influence of the QP gravitational modulation on reaction front was examined in the case of porous media described by the Darcy equation [6]. A work was dedicated to the case of liquid-solid polymerization front in [7] using the Navier-Stokes equations instead of the Darcy equation and it was revealed that the amplitudes and the frequencies ratio influence the stability region of the reaction. Also, it was shown that for appropriate values of vibration amplitudes and increasing values of the frequencies ratio, a stabilizing effect is reached. The effect of the wave number on the reaction front was examined showing that increasing the wave number widens the stability domain.

Recently, the case of liquid-liquid polymerization front was studied in [8]. This case is different from the previous one [7] where the equation of motion is considered only after the reaction zone since the polymer is in the solid phase. Therefor, the equation of motion is considered before and after the reaction zone because, both the monomer and the polymer are liquids. In this latter work, the Prandtl number is taken as a constant parameter. However, many studies show a significant effect of low or high Prandtl number on convective instability (see for instance [9-11] and the references therein).

This paper is devoted to study the influence of QP gravitational modulation and Prandtl number on convective instability of liquid-liquid frontal polymerization. To this end, we consider that the acceleration b acting on the fluid is given by $g+b(t)$, where g is the gravity acceleration, $b(t) = \lambda_1 sin(\sigma_1 t) + \lambda_2 sin(\sigma_2 t)$ and λ_1, λ_2 and σ_1, σ_2 are, respectively, the amplitudes and the frequencies of the QP vibrations.

The paper is organized as follows. The next section introduces the frontal polymerization model, while the interface problem is given in Section 3. The numerical investigations are carried out in Section 4 and the last section summarizes the results.

2 Frontal polymerization model

The propagation of reaction polymerization front is modelled by the following system of equations:

$$\frac{\partial T}{\partial t} + (v.\nabla)T = \kappa \Delta T + qW, \tag{2.1}$$

$$\frac{\partial \alpha}{\partial t} + (v.\nabla)\alpha = W, \tag{2.2}$$

$$\frac{\partial v}{\partial t} + (v.\nabla)v = -\frac{1}{\rho}\nabla p + \nu \Delta v$$
$$+g(1 + \lambda_1 \sin(\sigma_1 t) + \lambda_2 \sin(\sigma_2 t))\beta(T - T_0)\gamma, \tag{2.3}$$

$$div(v) = 0, \tag{2.4}$$

with the following boundary conditions:

$$z \to +\infty: \quad T = T_i, \quad \alpha = 0, \quad and \quad v = 0, \tag{2.5}$$

$$z \to -\infty: \quad T = T_b, \quad \alpha = 1, \quad and \quad v = 0. \tag{2.6}$$

[a] e-mail: assiyad.saadia@gmail.com
[b] e-mail: allali@hotmail.com
[c] e-mail: belhaq@yahoo.fr

and the notation:

$$\Delta = \frac{\partial^2}{\partial x^2} + \frac{\partial^2}{\partial y^2} + \frac{\partial^2}{\partial z^2}, \quad \nabla = (\frac{\partial}{\partial x}, \frac{\partial}{\partial y}, \frac{\partial}{\partial z}).$$

Here (x, y, z) are the spatial coordinates, $-\infty < x, y, z < +\infty$, T is the temperature, α the concentration of the reaction product, v the mean velocity, p the pressure, κ the coefficient of thermal diffusivity, q the adiabatic temperature rise, ρ an average value of density, ν the coefficient of kinematic viscosity, γ the unit vector in the z-direction (upward), β the coefficient of thermal expansion, g the gravitational acceleration, T_0 a mean value of temperature, T_i the initial temperature and $T_b = T_i + q$ is the temperature of the burned mixture. We then suppose the chemical reaction is one-step of order zero. The reaction rate is defined as follows :

$$W = k(T)\phi(\alpha), \qquad \phi(\alpha) = \begin{cases} 1 \ if \ \alpha < 1 \\ 0 \ if \ \alpha = 1 \end{cases}.$$

The temperature dependence of the reaction rate is shown mathematically in the Arrhenius Law $k(T) = k_0 exp(-E/R_0T)$ [12], where k_0 is the pre-exponential factor, E the activation energy (we will assume that the activation energy is large) and R_0 the universal gas constant. We assume that the liquids are incompressible and we neglect the term of diffusivity in the concentration equation in the way that the diffusivity coefficient is very small comparing to the thermal diffusivity coefficient.

To find the dimensionless model, we introduce the dimensionless spatial variables undermentioned :

$$x_1 = \frac{xc_1}{\kappa}, \quad y_1 = \frac{yc_1}{\kappa}, \quad z_1 = \frac{zc_1}{\kappa}, \quad t_1 = \frac{tc_1^2}{\kappa},$$

$$p_1 = \frac{p}{c_1^2\rho}, \quad c_1 = \frac{c}{\sqrt{2}}, \quad v_1 = \frac{v}{c_1},$$

$$\theta = \frac{T - T_b}{q} \quad and \quad c^2 = \frac{2k_0\kappa R_0 T_b^2}{qE}\exp(-\frac{E}{R_0 T_b}),$$

Here c represents the stationary reaction front velocity and can be calculated asymptotically for large Zeldovich number [13]. To simplify, we keep the same notation for the other variables and pressure. We can rewrite the system (2.1)-(2.4) with the two boundary conditions (2.5)-(2.6) in the following form :

$$\frac{\partial\theta}{\partial t} + (v.\nabla)\theta = \Delta\theta + Z\exp\left(\frac{\theta}{Z^{-1} + \delta\theta}\right)\phi(\alpha), \tag{2.7}$$

$$\frac{\partial\alpha}{\partial t} + (v.\nabla)\alpha = Z\exp\left(\frac{\theta}{Z^{-1} + \delta\theta}\right)\phi(\alpha), \tag{2.8}$$

$$\frac{\partial v}{\partial t} + (v.\nabla)v = -\nabla p + P\Delta v + PR(1+\lambda_1\sin(\sigma_1 t)+\lambda_2\sin(\sigma_2 t))(\theta+\theta_0)\gamma, \tag{2.9}$$

$$div(v) = 0, \tag{2.10}$$

with the following conditions at infinity :

$$z \to +\infty : \qquad \theta = -1, \quad \alpha = 0, \quad and \quad v = 0,$$

$$z \to -\infty : \qquad \theta = 0, \quad \alpha = 1, \quad and \quad v = 0.$$

Here $P = \frac{\nu}{\kappa}$ is the Prandtl number, $R = g\beta q\kappa^2/(\nu c^3)$ is the Rayleigh number, $Z = qE/R_0T_b^2$ is the Zeldovich number, δ and θ_0 are given, respectively, by $\delta = R_0T_b/E$ and $\theta_0 = (T_b - T_0)/q$ and $\mu = 2\kappa\sigma/c^2$.

3 The interface problem

To study the propagation of frontal polymerization reaction with a liquid reactant and liquid product, we consider the equation of motion before and after the reaction zone. This changes the jump conditions and influences the stability conditions of the reaction process. It is worthy to notice that in the case of liquid-solid polymerization front, the equation of motion is considered after the reaction zone. The linear stability analysis leads to the following interface problem :

In the liquid monomer $z > \zeta$:

$$\frac{\partial\theta}{\partial t} + (v.\nabla)\theta = \Delta\theta, \tag{3.1}$$

$$\alpha = 0, \tag{3.2}$$

$$\frac{\partial v}{\partial t} + (v.\nabla)v = -\nabla p + P\Delta v + Q(1 + \lambda_1\sin\sigma_1 t + \lambda_2\sin\sigma_2 t)(\theta + \theta_0)\gamma, \tag{3.3}$$

$$div(v) = 0. \tag{3.4}$$

In the liquid polymer $z < \zeta$:

$$\frac{\partial\theta}{\partial t} + (v.\nabla)\theta = \Delta\theta, \tag{3.5}$$

$$\alpha = 1, \tag{3.6}$$

$$\frac{\partial v}{\partial t} + (v.\nabla)v = -\nabla p + P\Delta v + Q(1 + \lambda_1\sin\sigma_1 t + \lambda_2\sin\sigma_2 t)(\theta + \theta_0)\gamma, \tag{3.7}$$

$$div(v) = 0. \tag{3.8}$$

At the interface $z = \zeta$:

$$\theta\big|_{\zeta-0} = \theta\big|_{\zeta+0}, \tag{3.9}$$

$$\left.\frac{\partial\theta}{\partial z}\right|_{\zeta=-0} - \left.\frac{\partial\theta}{\partial z}\right|_{\zeta=+0} =$$

$$\left(1 + \left(\frac{\partial\zeta}{\partial x}\right)^2 + \left(\frac{\partial\zeta}{\partial y}\right)^2\right)^{-1} \times \left(\frac{\partial\zeta}{\partial t} + \left(v_x\frac{\partial\zeta}{\partial x} + v_y\frac{\partial\zeta}{\partial y} - v_z\right)\Big|_\zeta\right), \tag{3.10}$$

$$\left(\frac{\partial\theta}{\partial z}\right)^2\Big|_{\zeta-0} - \left(\frac{\partial\theta}{\partial z}\right)^2\Big|_{\zeta_1+0} =$$

$$-2Z\left(1 + \left(\frac{\partial\zeta}{\partial x}\right)^2 + \left(\frac{\partial\zeta}{\partial y}\right)^2\right)^{-1}\int_{-\infty}^{\theta|_\zeta}\exp\left(\frac{\tau}{Z^{-1} + \delta\tau}\right)d\tau, \tag{3.11}$$

$$v_z\big|_{\zeta=-0} = v_z\big|_{\zeta=+0}, \tag{3.12}$$

$$\left.\frac{\partial v_z}{\partial z}\right|_{\zeta=-0} = \left.\frac{\partial v_z}{\partial z}\right|_{\zeta=+0}, \tag{3.13}$$

$$\left.\frac{\partial^2 v_z}{\partial z^2}\right|_{\zeta=-0} = \left.\frac{\partial^2 v_z}{\partial z^2}\right|_{\zeta=+0}, \tag{3.14}$$

$$\left.\frac{\partial^3 v_z}{\partial z^3}\right|_{\zeta=-0} - \left.\frac{\partial^3 v_z}{\partial z^3}\right|_{\zeta=+0} = -R\left(1 + \left(\frac{\partial\zeta}{\partial x}\right)^2 + \left(\frac{\partial\zeta}{\partial y}\right)^2\right)^{-1} \times$$

$$\left(\left(1 + \left(\frac{\partial\zeta}{\partial x}\right)^2 + \left(\frac{\partial\zeta}{\partial y}\right)^2\right)^{-1} - 1\right)(1+\lambda_1 sin(\sigma_1 t)+\lambda_2 sin(\sigma_2 t))\left.\left(\frac{\partial\theta}{\partial z}\right)\right|_{\zeta=+0}. \tag{3.15}$$

with the conditions at infinity :

$$z = -\infty : \qquad \theta = 0, \quad v = 0,$$

$$z = +\infty : \qquad \theta = -1, \quad v = 0. \tag{3.16}$$

DOI: 10.1051/matecconf/20168307001

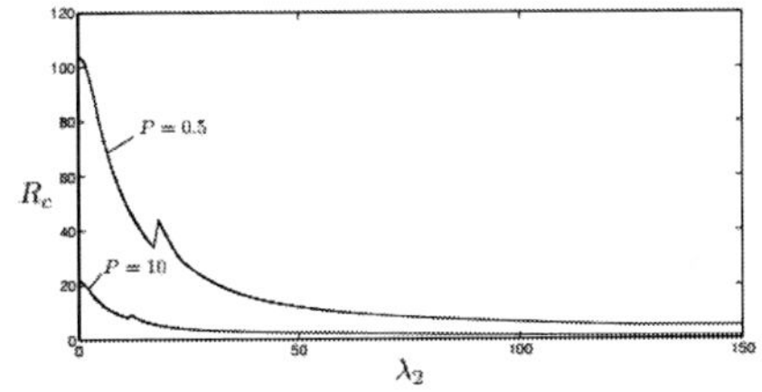

Fig. 1. The critical Rayleigh number versus the amplitude λ_2 for two different values of Prandtl number.

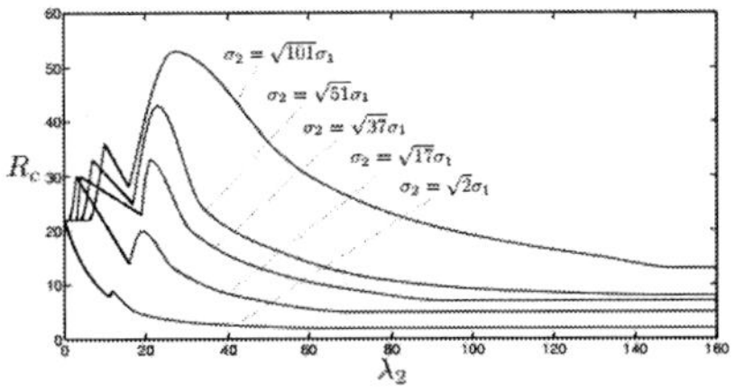

Fig. 2. The critical Rayleigh number versus the amplitude λ_2 for $P = 10$ and different frequencies ratio.

4 Numerical results

To study the stability of reaction front, we seek the solution of the problem as follows :

$$\theta = \theta_s + \tilde{\theta}, \qquad p = p_s + \tilde{p}, \qquad v = v_s + \tilde{v},$$

where $\tilde{\theta}$, $\tilde{p}$ and $\tilde{v}$ are respectively small perturbation of temperature, pressure and velocity.

We consider the perturbation as follows :

$$\hat{\theta}_i = \theta_i(z_2, t) \exp j(k_1 x + k_2 y),$$

$$\hat{v}_{zi} = v_{zi}(z_2, t) \exp j(k_1 x + k_2 y),$$

$$\xi = \varepsilon_1(t) \exp j(k_1 x + k_2 y),$$

where k_i, $i = 1, 2$ and ε_1 denote, respectively, the wave numbers (in x and y directions) and the amplitude of the perturbation and $j^2 = -1$.

The propagating reaction front problem can be reduced to the following system:

$\underline{z_2 < \xi}$:

$$\begin{cases} \dfrac{\partial w}{\partial t} - uw' - P(w'' - k^2 w) = -Qk^2(1 + \lambda_1 \sin(\sigma_1 t) + \lambda_2 \sin(\sigma_2 t))\theta, \\ w = v'' - k^2 v, \\ \dfrac{\partial \theta}{\partial t} - \theta'' - u\theta' + k^2\theta = 0. \end{cases}$$

$\underline{z_2 > \xi}$:

$$\begin{cases} \dfrac{\partial w}{\partial t} - uw' - P(w'' - k^2 w) = -Qk^2(1 + \lambda_1 \sin(\sigma_1 t) + \lambda_2 \sin(\sigma_2 t))\theta, \\ w = v'' - k^2 v, \\ \dfrac{\partial \theta}{\partial t} - \theta'' - u\theta' + k^2\theta = u \exp(-uz_2)v. \end{cases}$$

With the following boundary conditions :

$$\theta_2(0, t) - \theta_1(0, t) = u\varepsilon_1(t),$$

$$\theta_2'(0, t) - \theta_1'(0, t) = -u^2\varepsilon_1(t) - \varepsilon_1'(t) + v_1(0, t),$$

$$\theta_2'(0, t) + \frac{Z}{u}\theta_1'(0, t) = -u^2\varepsilon_1(t),$$

$$v_1^{(i)}(0, t) = v_2^{(i)}(0, t), \quad i = 0, 1, 2, 3.$$

By combining the boundary conditions, we get :

$$\theta_1'(0, t) + \frac{Z}{u}\theta_1(0, t) = \frac{1}{u}\left(\theta_2(0, t) - \theta_1(0, t)\right)_t' - v_0.$$

Fig. 1 shows the critical Rayleigh number as a function of the amplitude λ_2 of QP vibrations for $k = 0.7$, $\lambda_1 = 5$, $\mu_1 = 5$, $\frac{\sigma_2}{\sigma_1} = \sqrt{2}$, $Z = 8$ and for two different values of Prandtl number. It can be observed that for increased value of Prandtl number, the reaction front loses its stability.

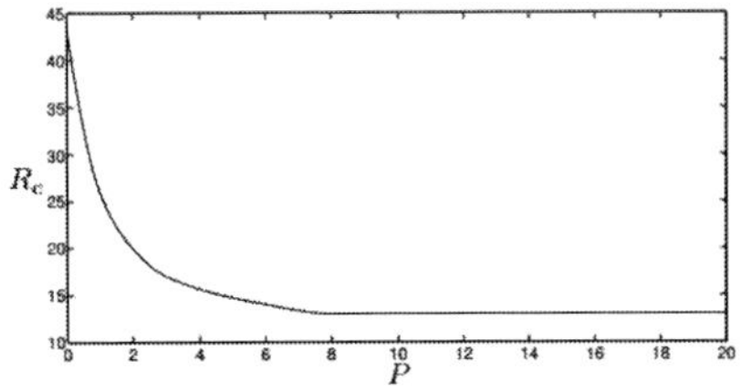

Fig. 3. The critical Rayleigh number versus the Prandtl number P for $\sigma_2 = \sqrt{2}\sigma_1$ and $\sigma_1 = 5$.

Fig. 2 illustrates the critical Rayleigh number versus the amplitude λ_2 for different frequencies ratio, $P = 10$, $Z = 8$, $k = 0.7$, $\lambda_1 = 5$ and $\mu_1 = 5$. From this figure we can conclude that by increasing the frequency ratio, the stability domain becomes large in certain interval of λ_2. By comparing this figure with the results found in [8] where the Prandtl number, we can observe that by increasing the Prandtl number ($P = 10$), the frequency ratio has a destabilizing effect for small values of the amplitude λ_2 ($0 \leq \lambda_2 \leq 10$).

The critical Rayleigh number as a function of the Prandtl number P is given in Fig. 3 for $\sigma_2 = \sqrt{2}\sigma_1$, $\sigma_1 = 5$, $\lambda_1 = \lambda_2 = 5$, $Z = 8$ and $k = 0.7$. It can be concluded that, for a fixed frequency ratio, small values of the Prandtl number has a stabilizing influence on the reaction front and by increasing P, we can observe that the stability region is decreased. When the Prandtl number exceeds certain value ($P \geq 7$), the effect of this parameter on the stability of the polymerization front is negligable.

5 Conclusion

In this paper, we have studied the influence of the QP gravitational modulation on the convective instability of frontal polymerization with liquid monomer and liquid polymer. To achieve this, we have considered the model which includes the heat equation, the concentration equation and the Navier-Stokes ones under boussinesq approximation.

Assuming the polymerization reaction occurs in a small zone when the activation energy is high, we have used the Zeldovich Frank-Kamenetskii method, called also narrow zone approximation. We have performed an asymptotic analysis to establish the interface problem and the solution has been chosen as a perturbed stationary solution. The numerical results show that the convective instability strongly depends on the parameters of the QP vibrations (frequencies ratio and amplitudes) and the Prandtl number.

References

1. P.M. Goldfeder, V.A. Volpert. V.M. Ilyashenko, A.M. Khan, J.A. Pojman, S.E. Solovyov. Mathematical modeling of free-radical polymerization fronts. The Journal of Physical Chemistry B, 101 (1997) 3474-3482.
2. G. Bowden, M. Garbey, V. M. Ilyashenko. Effect of convection on a propagating front with a solid product: comparison of theory and experiments. The Journal of Physical Chemistry B, 101 (1997) 678-686.
3. M. Garbey, A. Taik, V. Volpert. Linear stability analysis of reaction fronts in liquids. Quarterly of Applied Mathematics, 54 (1996) 225-247.
4. K. Allali, J. A. Pojman, V. Volpert. Influence of vibrations on convective instability of polymerization fronts. Journal of Engineering Mathematics, 41 (2001) 13-31.
5. K. Allali, F. Bikany, A. Taik, V. Volpert. Influence of vibrations on convective instability of reaction fronts in liquids. Mathematical Modelling of Natural Phenomena, 5 (2010) 35-41.
6. K. Allali, M. Belhaq, K. El Karouni. Influence of quasiperiodic gravitational modulation on convective instability of reaction fronts in porous media. Communications in Nonlinear Science and Numerical Simulation, 17 (2012) 1588-1596.
7. K. Allali, S.Assiyad, M. Belhaq. Convection of polymerization front with solid product under quasi-periodic gravitational modulation. Nonlinear Dynamics and Systems Theory, 14 (2014) 323-334.
8. S. Assiyad, K. Allali, M. Belhaq. Influence of quasiperiodic gravitational modulation on convective instability of liquid-liquid polymerization front. Mathematical Problems in Engineering, (2015) ID 484562.
9. F. H. Busse, J. A. Whitehead. Instabilities of convection rolls in a high Prandtl number fluid. Journal of Fluid Mechanics, 47 (1971) 305-320.
10. F. H. Busse. The oscillatory instability of convection rolls in a low Prandtl number fluid. Journal of Fluid Mechanics, 52 (1972) 97-112.
11. H. Salehipour, W. R. Peltier, A. Mashayed. Turbulent diapycnal mixing in stratified shear flows: The influence of Prandtl number on mixing effeciency and transition at high Reynolds number. Journal of Fluid Mechnaics, 773 (2015) 178-223.
12. M. Menzinger, R. Wolfgang. The meaning and use of the Arrhenius activation energy. Angewandte Chemie International Edition in English 8 (1969) 438-444.
13. B.V. Novozhilov. The rate of propagation of the front of an exothermic reaction in a condensed phase. J. Dokl. Phys. Chem., 141 (1961) 836-838.

Heat Explosion In Porous Media Using Radial Basis Functions

Karam Allali[1,a], Youssef Joundy[1,b], Ahmed Taik[1,c], and Vitaly Volpert[2,d]

[1] FSTM, Laboratory of Mathematics & Applications , University Hassan II-Casablanca, Mohammedia, Morocco,
[2] Institut Camille Jordan, University Lyon 1, UMR 5208 CNRS, 69622 Villeurbanne, France

Abstract. The paper is devoted to the numerical investigation of the interaction between natural convection and heat explosion in a fluid-saturated porous media in a rectangular domain. The model consists of Darcy equations for an incompressible fluid in a porous medium coupled with the nonlinear heat equation. Numerical simulations are performed using the radial basis functions method (RBFs). We study the bifurcation of the periodic oscillation of the response born by Hopf bifurcation. First, a symmetry-breaking bifurcations observed; then is followed by successive period-doubling bifurcations leading to chaos.

Keywords: convective heat explosion, porous media, period-doubling, radials basis functions method.

1 Introduction

The study of the interaction between natural convection and heat explosion is important for the understanding and predicting thermal explosion. The theory of thermal explosion was started with the works by Semenov and Frank-Kamenetskii [1,2]. In Semenov's theory, the temperature is considered to be uniform in the entire domain, while in Frank-Kamenetskii's model spatial temperature distribution is taken into account. The influence of natural convection on thermal explosion was first tackled in [3,4]. It was shown that vigorous convection occurs when the value of the Frank-Kamenetskii parameter or of the Rayleigh number exceed some critical values. Dynamics of thermal explosion with natural convection described by the Navier-Stokes equations was studied in [5] for a square domain. The same issue was studied in [7] but for a rectangular domain. The authors of these two works showed how complex regimes appeared through successive bifurcations. In this paper, we continue the investigation of thermal explosion in the case of porous media. To this end, we consider nonlinear heat equation coupled with Darcy equations for fluid motion in porous media:

$$\frac{\partial T}{\partial t} + \mathbf{v}.\nabla T = \kappa \Delta T + qK(T), \qquad (1)$$

$$\frac{\partial \mathbf{v}}{\partial t} + \frac{\mu}{K_p}\mathbf{v} + \nabla p = g\beta\rho(T - T_0)\gamma, \qquad (2)$$

$$\nabla.\mathbf{v} = 0, \qquad (3)$$

where T denotes the temperature, $\mathbf{v}$ is the velocity, p is the pressure, κ the coefficient of thermal diffusivity, μ the kinematic viscosity, ρ the density, q the heat release, g is the acceleration due to gravity, γ a unit vector in the vertical direction, T_0 the characteristic value of the temperature and K_p the permeability. Let us note that we consider unstationary form of the Darcy equation since conditions of heat explosion can be sensitive to the quasi-stationary approximation where the time derivative is cancelled.

In addition, the temperature dependence of the reaction rate $K(T)$ is given by the Arrhenius law:

$$K(T) = k_0 exp\left(-\frac{E}{RT}\right), \qquad (4)$$

where E is the activation energy, R the universal gas constant and k_0 the pre-exponential factor.

This system is considered in a $2D$ rectangular domain, $0 \le x \le n\ell$, $0 \le y \le \ell$. The boundary conditions will be specified below.

The paper is organized as follows. We first give the model in the stream function-vorticity formulation in the next section. The numerical method is described in Section 3. Section 4 deals with numerical simulation and Section 5 concludes the work.

2 The stream function-vorticity formulation

In order to write the dimensionless model, we introduce new spatial variables $x' = x/\ell$, $y' = y/\ell$, time $t' = \frac{\kappa}{\ell^2}t$, velocity $\frac{\ell}{\kappa}\mathbf{v}$ and pressure $\frac{K_p}{\mu\kappa}p$. Denoting $\theta = \frac{E(T-T_0)}{RT_0^2}$ and keeping for convenience the same notation for the other variables, we obtain:

$$\frac{\partial\theta}{\partial t} + u\frac{\partial\theta}{\partial x} + v\frac{\partial\theta}{\partial y} = \frac{\partial^2\theta}{\partial x^2} + \frac{\partial^2\theta}{\partial y^2} + F_K e^{\theta}, \qquad (5)$$

$$\frac{\partial u}{\partial t} + u + \frac{\partial p}{\partial x} = 0 \qquad (6)$$

$$\frac{\partial v}{\partial t} + v + \frac{\partial p}{\partial y} = R_p\theta \qquad (7)$$

[a] e-mail: `allali@hotmail.com`
[b] e-mail: `joundy.youssef@yahoo.fr`
[c] e-mail: `taik_ahmed@yahoo.fr`
[d] e-mail: `volpert@math.univ-lyon1.fr`

$$\frac{\partial u}{\partial x} + \frac{\partial v}{\partial y} = 0. \tag{8}$$

We introduce the stream function ψ

$$\begin{pmatrix} u \\ v \end{pmatrix} = \begin{pmatrix} \dfrac{\partial \psi}{\partial y} \\ -\dfrac{\partial \psi}{\partial x} \end{pmatrix}, \tag{9}$$

using incompressibility of the fluid, and the vorticity

$$\omega = curl\, \mathbf{v}. \tag{10}$$

We apply the rotational operator to equations (6)-(7) in order to eliminate the pressure. We get the equations

$$\frac{\partial \theta}{\partial t} + \frac{\partial \psi}{\partial y}\frac{\partial \theta}{\partial x} - \frac{\partial \psi}{\partial x}\frac{\partial \theta}{\partial y} = \frac{\partial^2 \theta}{\partial x^2} + \frac{\partial^2 \theta}{\partial y^2} + F_K e^{\theta}, \tag{11}$$

$$\sigma \frac{\partial \omega}{\partial t} + \omega = R_p \frac{\partial \theta}{\partial x}, \tag{12}$$

$$\omega = -\Delta \psi \tag{13}$$

with the following boundary condition:

$$x = 0, n: \ \frac{\partial \theta}{\partial x} = \frac{\partial \psi}{\partial y} = \psi = 0, \quad y = 0, 1: \ \theta = \frac{\partial \psi}{\partial x} = \psi = 0, \tag{14}$$

and some appropriate initial conditions on θ, ω, ψ.

Here $F_K = \dfrac{Eqk_0\alpha e^{-\frac{E}{RT_0}} \ell^2}{RT_0^2 \kappa}$ is the Frank-Kamenetskii parameter, $R_p = \dfrac{K_p \rho R_a}{\ell^2}$, $R_a = \dfrac{g\beta R T_0^2 \ell^3}{E\kappa\mu}$ is the Rayleigh number. Under the assumptions of large activation energy, $RT_0/E << 1$, we can perform the Frank-Kamenetskii transform, so that the nonlinear reaction rate in the equation (11) is taken to be $F_K exp(\theta)$ [2]. Next, $\sigma = \dfrac{1}{V_a}$ stands for the inverse of Vadasz number, $V_a = \dfrac{P_r}{D_a}$, $P_r = \dfrac{\mu}{\kappa}$ is the Prandtl number and $D_a = \dfrac{K_p}{\ell^2}$ is the Darcy number.

3 Numerical method

Let us assume that $(\theta^m, \omega^m, \psi^m)$ are known at time $t = m\Delta t$. They can be approximated as follows:

$$\hat{\theta}^m \simeq \sum_{j=1}^{N} \alpha_j^m \varphi(\|\mathbf{X_i} - \mathbf{X_j^c}\|_2, \varepsilon), \tag{15a}$$

$$\hat{\omega}^m \simeq \sum_{j=1}^{N} \beta_j^m \varphi(\|\mathbf{X_i} - \mathbf{X_j^c}\|_2, \varepsilon), \tag{15b}$$

$$\hat{\psi}^m \simeq \sum_{j=1}^{N} \gamma_j^m \varphi(\|\mathbf{X_i} - \mathbf{X_j^c}\|_2, \varepsilon). \tag{15c}$$

We choose the Hardy's *multiquadric* or the MQ-RBF [8].

Let $X_1^c, X_2^c, \ldots, X_N^c \in \Omega \subset R^2$ be a given set of centres where $X_i^c = (x_i^c, y_i^c)$. For each collocation points $X = (x, y) \in R^2$, the MQ-RBF is defined by

$$\varphi(\|\mathbf{X} - \mathbf{X}_j^c\|, \varepsilon) = \sqrt{(x - x_j^c)^2 + (y - y_j^c)^2 + \varepsilon^2}. \tag{16}$$

ε is called a *shape parameter*.

Using the implicit scheme for the time derivative, the equations (11)-(13) become:

$$\theta^{m+1} - \Delta t\left(-\psi_y^m \otimes \theta_x^{m+1} + \psi_x^m \otimes \theta_y^{m+1} + \theta_{xx}^{m+1} + \theta_{yy}^{m+1}\right) = \theta^m + \Delta t F_K e^{\theta^m}, \tag{17}$$

$$\omega^{m+1} = \left(\frac{\sigma}{\sigma + \Delta t}\right)\omega^m + \left(\frac{\Delta t}{\sigma + \Delta t}\right) R_p \theta_x^m, \tag{18}$$

$$\Delta \psi^{m+1} = -\omega^{m+1}, \tag{19}$$

where $\otimes$ is the element-by-element matrix product.

4 Numerical simulations

4.1 Periodic oscillations, symmetry-breaking and period-doubling

In our simulations, we fix the value of the Rayleigh number, $R_p = 10000$, the geometry of the domain, $n = 8$, and $\sigma = 0.01$. We will vary the Frank-Kamenetskii parameter F_K.

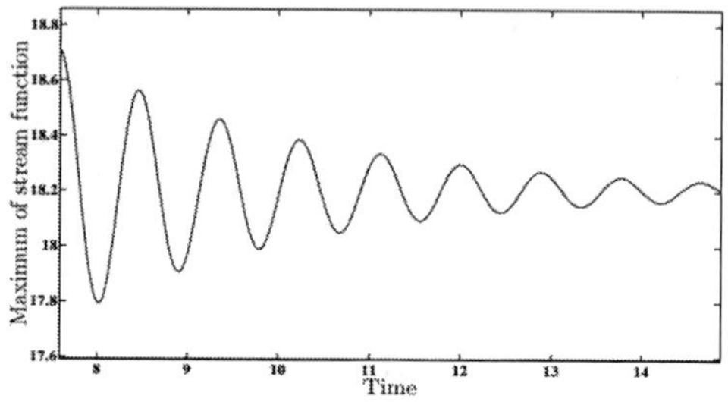

Fig. 1. Maximum of the stream function in time for $R_p = 10000$, $\sigma = 0.01$, $n = 8$ and $F_K = 1.87$.

Fig.1 shows the convergence of the solution to a stationary state for the value of $F_K = 1.87$. A small increase of this parameter leads to a Hopf bifurcation, after which the system oscillates. The corresponding limit cycle in the (ψ, θ) plane is an ellipse (Fig. 2(a)). When F_K increased slightly, the system loses its symmetry (Fig. 2(b)) [9].

Fig. 3(a) and 3(b) illustrate the first period-doubling where the system needs a $2T$ to browse the entire cycle. Further increase of this parameter to $F_K = 3.78$ and to $F_K = 3.812$ results in sequential bifurcations to the solutions with the periods $4T$ and $8T$ shown in Fig. 3(c) and 3(d), respectively.

4.2 Thermal explosion chaos

After successive period-doubling bifurcations, chaotic behavior of solutions is observed (Fig.4(a)). Continuous frequency spectrum (Fig.4(b)) is specific for chaotic oscillations.

DOI: 10.1051/matecconf/20168307002

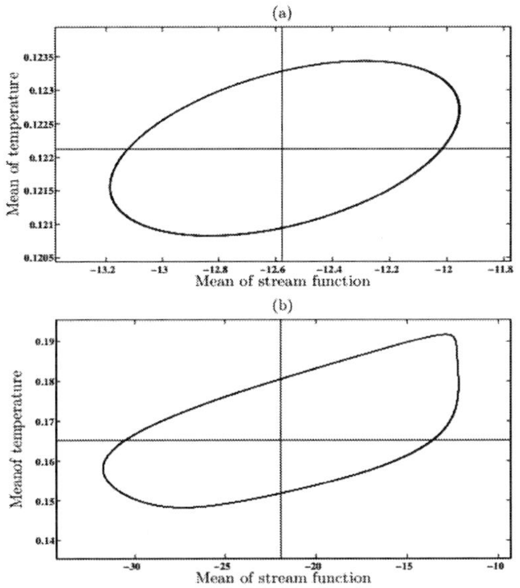

Fig. 2. Symmetric (a) and asymmetric (b) oscillations of the temperature and stream function; *(a)* $F_K = 1.907$ and *(b)* $F_K = 2.600$.

5 Conclusion

Thermal explosion in porous media is modelled by a system coupling a non-linear heat equation with a system describing the fluid motion in rectangular domain with unsteady terms in the Darcy quation. The model is studied numerically using multi-quadratic radial basis functions. The Frank-Kamenetskii number F_K was chosen as a bifurcation parameter. Increasing F_K, we found transitions between periodic and chaotic oscillations through a sequence of period-doubling bifurcations.

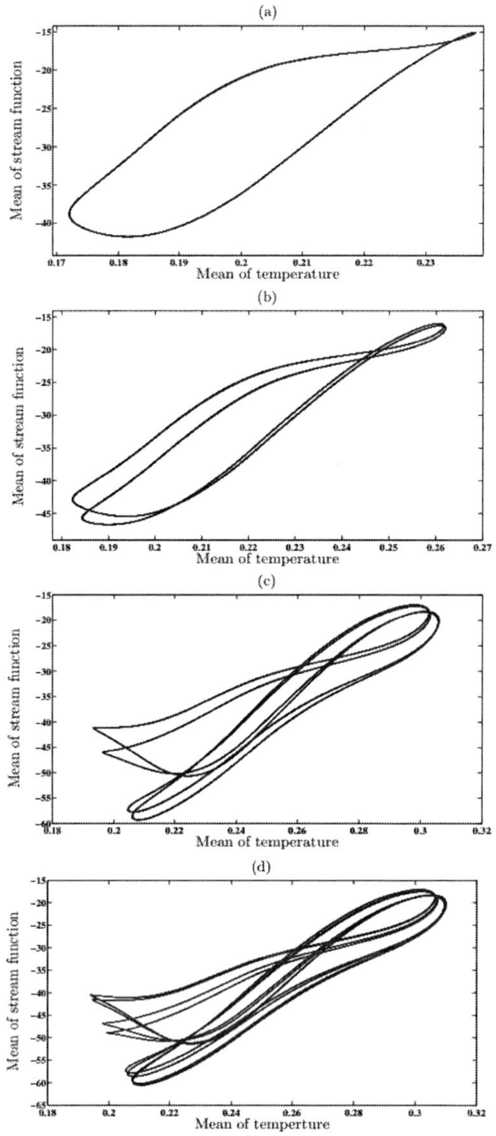

Fig. 3. Cascade of period-doubling: *(a)* period-1, *(b)* period-2, *(c)* period-4 and *(d)* period-8 respectively for $F_K = 3.10$, $F_K = 3.35$, $F_K = 3.78$ and $F_K = 3.82$.

References

1. N.N. Semenov, *Chemical Kinetics and Chain Reactions*. Clarendon Press, Oxford, 1935.
2. D.A. Frank-Kamenetskii, *Diffusion and Heat Transfer in Chemical Kinetics*. Plenum Press, New York 1969.
3. S.I. Khudyaev, E.A. Shtessel and K.V. Pribytkova. 1971. Numerical solution of the heat explosion problem with convection. Combust., Explosion Shock Waves, 2, 167.
4. A.S. Merzhanov and E.A. Shtessel. 1973. Free convection and thermal explosion in reactive systems. Acta Astronaut., 18, 191.
5. A. Lazarovici, V. Volpert and J.H. Merkin. 2005. Steady states, oscillations and heat explosion in a combustion problem with convection. Euro. J. Mech. B. Fluids, 24, 189.
6. K. Allali, F. Bikany, A. Taik and V. Volpert, *Numerical Simulations of Heat Explosion With Convection In Porous Media*. Combustion Sciences and Technology. DOI:10.1080/00102202.2014.938543. 2014.
7. M. Belk, V. Volpert, *Modeling of heat explosion with convection*, Chaos 14:263–273, 2004.
8. R.L. Hardy, *Multiquadric equations of topography and other irregular surfaces*. Journal of Geophysical Research, 76(8) :1905-1915, 1971.
9. M. Belhaq and M. Houssni, *Symmetry-Breaking and First Period-Doubling Following a Hopf Bifurcation in a Three Dimensional System*, Mechanics Research Communications, Vol. 22, No. 3, pp. 221-231, 1995.

DOI: 10.1051/matecconf/20168307002

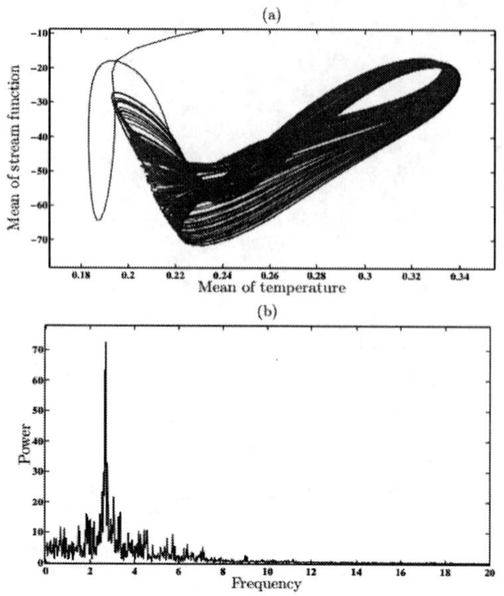

Fig. 4. (*a*) Chaotic behaviour for $F_K = 4.10$ and (*b*) the corresponding frequency spectrum.

The onset of Darcy-Bénard instability in a horizontal porous channel with a free surface using a thermal nonequilibrium model

Hajar Lagziri[1,2,a], Antonio Barletta[1], Michele Celli[1], and Mohammed Bezzazi[2]

[1] Department of Industrial Engineering, Alma Mater Studiorum University of Bologna, Viale Risorgimento 2, Bologna 40136, Italy.
[2] Department of Physics, University of Abdelmalek Essaadi, Tanger, Morocco.

Abstract The intent beyond this present work is to investigate the onset of instability in homogeneous horizontal porous channel saturated with Newtonian fluid bounded by free surface from the top and impermeable wall from the bottom. The upper free surface is cooled by a third kind boundary condition, while the lower one is heated by two different ways. The wall is modelled with uniform temperature in the first case and with a perfect heat flux (Model A) in the second case. Relaxing the assumption of local thermal equilibrium model can provide in general two different Biot numbers B_s and B_f as well as two temperature profiles at the basic state, one for each phase. The linear stability of the basic motionless state is studied by using the normal modes analysis. The neutral stability curves are obtained numerically by applying a sixth order Runge-Kutta method in combination with a shooting method. The critical values of R at which the instability initiates in the porous medium with isothermal wall or isoflux wall are calculated.

1 Introduction

The Darcy-Bénard problem is the equivalent of Rayleigh-Bénard problem with a porous material. The classical version of Darcy-Bénard configuration is defined as a plane porous layer saturated with Newtonian fluid and confined between two parallel isothermal walls heated from below [1]. The existence of the porous material in thermoconvective region implies a new expression of the critical Rayleigh number, which is reformulated as a product between the traditional Rayleigh number for the pure fluid and Darcy number. It is well-known as Darcy-Rayleigh number, R. By presence of the local thermal equilibrium condition, one energy balance equation is considered. In this case, the threshold value of the Darcy-Bénard instability is $R = 4\pi^2$ [4]. The thermal boundary conditions can change to various combinations between Dirichlet, Neumann and Robin boundary conditions. The same can also happen for the case of the hydrodynamic boundary conditions. Following Nield and Bejan [4], the critical value $R = 0$, can be observed while the boundary conditions are subjected to a uniform pressure and heat flux. Moreover, changing the boundary conditions into isoflux or free surface forces the basic flow to be more unstable. On the other hand, the local thermal equilibrium can be relaxed when the temperature of the solid and fluid phases are not the same. In this case the energy equation for the porous medium is replaced by two equations, one for the solid phase and one for the fluid phase. Under this model, the criterion for the onset of convection in porous system depends on the effect of two nondimensional parameters: the inter-phase heat transfer parameter H, the thermal conductivity ratio of the fluid and solid phase γ. Broadly speaking, the local thermal equilibrium model can appear inside a porous medium either by letting $H \to \infty$ with $\gamma \approx O(1)$ or by letting $\gamma \to \infty$ with $H \approx O(1)$ [2]. Thus, to study the effects of isoflux boundary instead of isothermal with local thermal non equilibrium condition on the onset of instability, Yang and Vafai [3] modelled a standard formula of the heat flux by two mathematical approaches, called model A and model B. The first one defines the total heat flux as sum between Fourier heat flux of the solid and fluid phase, weighted by the volumetric fraction of each phase, while the second one equals the Fourier heat flux considered in the solid phase and in the fluid phase. Usually, model A is used for the case of impermeable wall with higher thermal conductivity [3].

The aim of this current work is to study the comparative behaviour between the effect of the two Biot numbers B_s and B_f provided by Robin boundary condition, on the threshold values of the onset of instability in the porous channel bounded by isothermal wall from the bottom as first case, then by isoflux wall as second case when the condition of the local thermal equilibrium is abolished.

2 Mathematical model

The setup of the problem studied here is sketched in Fig 1, defined as an horizontal saturated porous layer confined between a free surface with uniform pressure at the upper boundary and rigid impermeable wall at the lower one. The free surface is subjected to Robin boundary condition, in meantime the rigid wall is modeled as isothermal T_w in the first case then as isoflux q_w (Model A) in the second case. With the existance of the local thermal non equilibrium model, the difference between the thermal conductivity coefficient of the solid and fluid phase, and the coefficient of the heat transfer between the solid phase h_s or the fluid phase h_f with an external environment yield two

[a] e-mail: lagziri-hajar7@hotmail.fr

DOI: 10.1051/matecconf/20168307003

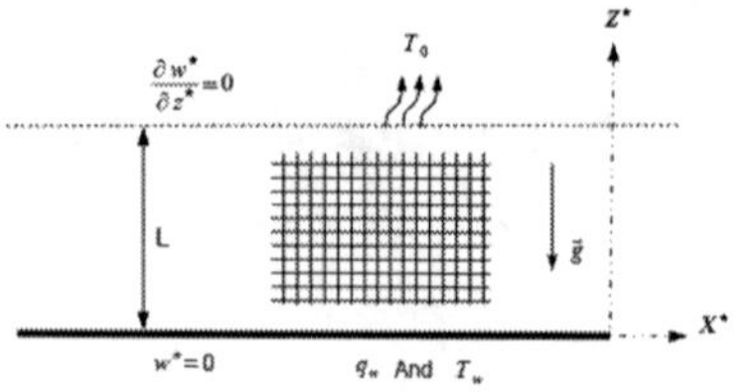

Figure 1. Schematic of the configuration.

Biot numbers, B_s and B_f as well as two energy balance equations coupled by inter-phase heat transfer coefficient h. The saturated porous layer is gouverned by Darcy's law and Boussinesq approximation. The curl operator is used in Darcy's law equation in order to neglect the pressure field. On the other hand, the governing equations are invariant by rotation around the z axes. Thus, the velocity field can be defined in plan (x, z) with the streamfunctions $w = -\psi_{,x}$ and $u = \psi_{,z}$. Hence, the system is reformulated as

$$\nabla^2\psi = R(\frac{\gamma+1}{\gamma})\nabla_x T_f, \quad (1a)$$

$$\lambda\, T_{s,t} = \nabla^2 T_s + \gamma H(T_f - T_s), \quad (1b)$$

$$T_{s,t} + \psi_{,z}T_{f,x} - \psi_{,x}T_{f,z} = \nabla^2 T_f - H(T_f - T_s). \quad (1c)$$

The subscript "s" and "f" denote a solid and fluid phases while ", z" and ", t" refer to first derivative with respect to z axes and time. The dimensionless boundary conditions are

$$\begin{aligned} z = 0, \quad \psi = 0, \quad T_f = T_s = 0 \quad &(\text{1st case}),\\ T_f = T_s, \quad T_{s,z} + \gamma T_{f,z} = -(1+\gamma) \quad &(\text{2nd case}).\\ z = 1, \quad \psi_{,z} = 0, \quad T_{s,z} = -B_s T_s, \quad &T_{f,z} = -B_f T_f. \end{aligned} \quad (2)$$

The Eqs. (1) and (2) are thus non-dimensionalized by the following transformations

$$\begin{aligned} &(x, y, w) \rightarrow (x^*, y^*, w^*)d, \quad t \rightarrow t^*\frac{\alpha_f}{d^2}, \quad \alpha_m \rightarrow \frac{k_m}{(\rho C)_f}.\\ &(u, v, w) \rightarrow (u^*, v^*, w^*)\frac{d}{\phi\alpha_f}, \quad T_{s,f} \rightarrow \frac{T^*{}_{s,f} - T_0}{\Delta T},\\ &\alpha_f \rightarrow \frac{k_f}{(\rho C)_f}, \quad \Delta T \rightarrow T_0 - T_w \quad (\text{1st case}),\\ &\Delta T \rightarrow \frac{q_w L}{k_m} \quad (\text{2nd case}). \end{aligned} \quad (3)$$

where the new dimensionless variables are

$$\begin{aligned} &H = \frac{hL^2}{\phi k_f}, \quad B_s = \frac{h_s L}{k_s}, \quad B_f = \frac{h_f L}{k_f}, \quad \gamma = \frac{\phi k_f}{(1-\phi)k_s},\\ &R = \frac{\beta\Delta T g K L}{\nu\alpha_m}, \quad \lambda = \frac{\alpha_f}{\alpha_s}, \quad k_m = (1-\phi)k_s + \phi k_f. \end{aligned} \quad (4)$$

Here, α_m is the effective thermal diffusivity, k_m is the effective conductivity in porous medium, K is the permeability, L is the thickness of the porous layer, ϕ is the porosity of the medium, α_s and α_f are the thermal diffusivities of the solid and fluid phases.

3 A basic non equilibrium state

Broadly speaking, when the local thermal non equilibrium is applied, the distribution of the temperature profile of the fluid phase and solid skeleton in both cases are different everywhere in the porous medium, see for instance [5]. Thus, to prescribe the temperature and velocity fields at the basic state, we solve Eqs. (1) by using Eqs. (2) then we report that

$$\psi_b = 0 \quad (5)$$

The temperature field for the isothermal case

$$\begin{aligned} T_{fb} = &\{(B_s - B_f)\sinh(\Omega z) + \sinh\Omega[B_f - (\gamma+1)\\ &(z-1)B_f B_s + \gamma B_s] + \Omega\cosh\Omega(\gamma B_f - z\gamma B_f\\ &-zB_s + B_s + \gamma + 1)\} \times \{\sinh\Omega[B_f(\gamma B_s + B_s + 1)\\ &+\gamma B_s] + \Omega\cosh\Omega(\gamma B_f + B_s + \gamma + 1)\}^{-1}, \end{aligned} \quad (6)$$

$$\begin{aligned} T_{sb} = &\{\gamma(B_f - B_s)\sinh(\Omega z) + \sinh\Omega[B_f - (\gamma+1)\\ &(z-1)B_f B_s + \gamma B_s] + \Omega\cosh\Omega(\gamma B_f - z\gamma B_f\\ &-zB_s + B_s + \gamma + 1)\} \times \{\sinh\Omega[B_f(\gamma B_s + B_s + 1)\\ &+\gamma B_s] + \Omega\cosh\Omega(\gamma B_f + B_s + \gamma + 1)\}^{-1}. \end{aligned} \quad (7)$$

The temperature fields for the isoflux case

$$\begin{aligned} T_{sb} = &\{\sinh\Omega\{B_f[1 - B_s(\gamma+1)(z-1)] + B_s\gamma\}\\ &+\gamma(B_f - B_s)\sinh(\Omega z)\} \times \{B_f B_s(\gamma+1)\sinh\Omega\}^{-1}\\ &-\{\Omega\cosh\Omega(B_f\gamma + B_s) - \Omega\cosh\Omega\{\gamma[B_f(z-1) - 1]\\ &+B_s(z-1) - 1\}\} \times \{B_f B_s(\gamma+1)\sinh\Omega + \Omega\cosh\Omega(B_f\gamma + B_s)\}^{-1}, \end{aligned} \quad (8)$$

$$\begin{aligned} T_{fb} = &-\{\sinh\Omega\{B_f[B_s(\gamma+1)(z-1) - 1] - B_s\gamma\} + (B_f - B_s)\\ &\sinh(\Omega z)\} \times \{B_f B_s(\gamma+1)\sinh\Omega + \Omega\cosh\Omega(B_f\gamma + B_s)\}^{-1}\\ &\times\{\Omega\cosh\Omega\{\gamma[B_f(z-1) - 1] + B_s(z-1) - 1\}\}\\ &\times\{B_f B_s(\gamma+1)\sinh\Omega + \Omega\cosh\Omega(B_f\gamma + B_s)\}^{-1}. \end{aligned} \quad (9)$$

The subscript "b" denotes the basic flow and $\Omega = \sqrt{H(1+\gamma)}$. The dimensionless parameters H and γ presente the inter-phase heat transfer coefficient, and thermal conductivity ratio of the fluid and solid phases, while R is the Darcy-Rayleigh number.

4 Stability analysis

The basic state is perturbed with an arbitrarily small disturbance reformulated in the following way

$$T_f = T_{fb} + \epsilon\tilde{\theta_f}, \quad T_s = T_{sb} + \epsilon\tilde{\theta_s}, \quad \psi = \psi_b + \epsilon\tilde{\psi}. \quad (10)$$

The parameter ϵ is a small perturbation, while $\tilde{\theta_f}, \tilde{\theta_s}$ and $\tilde{\psi}$ are the dimensionless fluid phase temperature disturbance, the dimensionless solid phase temperature disturbance and the dimensionless streamfuction disturbance, respectively. In order to obtain linear equations we introduce the decomposition of Eqs. (10) into Eqs. (1) and (2). Then, by taking into account $\epsilon << 1$, we remove all the additional terms of $O(\epsilon^2)$

$$\begin{aligned} \nabla^2\tilde{\psi} &= -R\frac{1+\gamma}{\gamma}\tilde{\theta}_{f,x}, \\ \lambda\tilde{\theta}_{s,t} &= \nabla^2\tilde{\theta}_s + \gamma H(\tilde{\theta_f} - \tilde{\theta_s}), \\ \tilde{\theta}_{f,t} - \tilde{\psi}_{,x}T_{fb,z} &= \nabla^2\tilde{\theta}_f + H(\tilde{\theta_s} - \tilde{\theta_f}). \end{aligned} \tag{11}$$

Furthermore, according to the normal modes method, the solution of the linearized equations are expressed as superposition of the plane waves, namly

$$\tilde{\psi}(x,z,t) = i\,\psi(z)e^{i(ax-\omega t)}, \quad \tilde{\theta}_s(x,z,t) = \theta_s(z)e^{i(ax-\omega t)}, \\ \tilde{\theta}_f(x,z,t) = \theta_f(z)e^{i(ax-\omega t)}. \tag{12}$$

Such that $\omega = \omega_r + i\omega_i$. The complex parameter of ω is the growth rate while the real one is the angular frequency. The parameter a is the dimensionless wave number and ψ, θ_s, θ_f are the dimensionless disturbance amplitudes. Thus, by substituting Eqs. (12) into Eqs. (11), the ordinary differential equations can be obtained as

$$\psi_{,zz} - a^2\psi + aR(\frac{\gamma+1}{\gamma})\theta_f = 0, \tag{13a}$$

$$\theta_{s,zz} - a^2\theta_s + \gamma H(\theta_f - \theta_s) + \omega\lambda\theta_s = 0, \tag{13b}$$

$$\theta_{f,zz} - a^2\theta_f + H(\theta_s - \theta_f) - a\psi T_{fb,z} + \omega\theta_f = 0, \tag{13c}$$

with the relative boundary conditions

$$\begin{aligned} z = 0: &\quad \psi(0) = 0, \quad \theta_s(0) = \theta_f(0) = 0 \quad \text{(1st case)} \\ &\theta_f(0) - \theta_s(0) = 0, \quad \theta_{s,z}(0) + \gamma\theta_{z,f}(0) = 0 \quad \text{(2nd case)} \\ z = 1: &\quad \psi_{,z}(1) = 0, \quad \theta_{s,z}(1) + \theta_s(1)B_s = 0, \\ &\quad \theta_{f,z}(1) + \theta_f(1)B_f = 0. \end{aligned} \tag{14}$$

Analytically, the variable coefficients defined in the temperature field of the fluid phase for both cases, Eqs. (6) and (9), prevent us to emphasize the validity of the principle of exchange of stabilities.

In the general case, specifically when the two Biot numbers are different, the principle of exchange of stabilities is proved only numerically by the absence of the oscillating branch in the thermoconvective instability which means $\omega_r = 0$. On the other hand, we may assume $\omega_i = 0$ since the system involves modes that describe the threshold condition of neutral stability, see [5]. Thus, the ordinary differential equations with $\omega = 0$ become

$$\psi_{,zz} - a^2\psi + aR(\frac{\gamma+1}{\gamma})\theta_f = 0, \tag{15a}$$

$$\theta_{s,zz} - a^2\theta_s + \gamma H(\theta_f - \theta_s) = 0, \tag{15b}$$

$$\theta_{f,zz} - a^2\theta_f + H(\theta_s - \theta_f) - a\psi T_{fb,z} = 0, \tag{15c}$$

where the subscript ", zz" refers to second derivative with respect to z axes.

5 Numerical solutions

The eigenvalue problem, Eqs. (15), described for the full case can be solved by using two numerical procedures based on sixth order Runge Kutta solver coupled with shooting method. However, to apply the solver function employed in the setup of the Runge Kutta method, we must add three addtional conditions at boundary $z = 0$ in order to complete the initial boundary conditions defined in Eqs. (14)

$$\begin{aligned} &\psi'(0) = 1, \quad \theta_f'(0) = c_1, \quad \theta_s'(0) = c_2 \quad \text{(1st case)}, \\ &\text{or} \quad \theta_s(0) = c_1, \quad \theta_s'(0) = c_2 \quad \text{(2nd case)}. \end{aligned} \tag{16}$$

The homogeneity displayed in the eigenvalue problem yields the condition $\psi_{,z}(0) = 1$. The unknown parameters c_1 and c_2 are assessed in the setup used by a shooting method, with the boundary condition defined at free surface. The numerical procedures employed here are implemented in Mathematica software (© Wolfram Research). The three last conditions prescribed at the target $z = 1$ allowing us to compute the parameters c_1, c_2 and the eigenvalue R, for any assigned input values of (a, γ, H, B_f, B_s). Moreover, the curves $R(a)$ drawn for each value given to the parameters (a, γ, H, B_f, B_s) yield the neutral stability curves.

6 Discussion

The numerical results displayed in Figs 2-4 are reported for special general case based on some specific conditions suggested by the authors, in order to optimize the number of the governing parameters data (γ, H, B_f, B_s) used in the study of the thermoconvective instability for both cases. On the other hand, giving more accurate details about the behaviour of each parameter data inside a porous layer. Thus, the porosity of the porous layer in this work is assumed as 1/2 while both external heat transfer coefficients at the upper free surface are supposed to coincide. These assumptions provide a convenient relationship between the two different Biot numbers and γ in the form of $B_s = \gamma B_f$. Figure 2 displays a neutral stability curves for fixed value of $H = 10$, with different values of γ: The graphic on the left hand side is for the case of isothermal wall and the graphic on the right hand side is for the case of isoflux wall. It is worth to mention that, in the case of isoflux wall the neutral stability curves are more sensitive to the effect of the parameter γ than the isothermal wall. The Figure 3 exhibits the trends of R_c versus B_f drawn for various values of γ with $H = 0.1$, while Figure 4 presents the plots of R_c versus B_s, drawn for different values of γ and with $H = 0.1$. The dashed horizontal lines describe the asymptotic values defined for $B_s \to \infty$ and $B_f \to \infty$.The values of R_c in the case of isoflux wall increase consequently as one of the Biot number increases and γ decreases, on the other hand, the values of R_c for isothermal wall display a non monotonic decrease with the Biot numbers and γ. In general, we can infer that, in the configurations of B_s or $B_f \to 0$, the system becomes more stable with the case of isothermal wall and unstable with the case of isoflux wall.

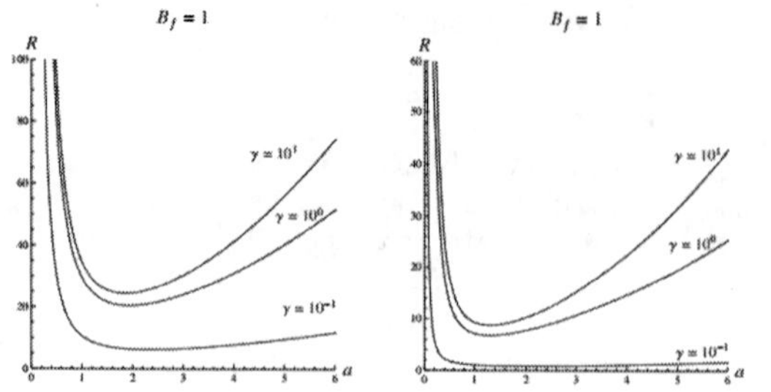

Figure 2. General case $B_s = \gamma B_f$: The neutral stability curves for different values of γ, with fixed $H = 10$ for the cases of isothermal wall (left frame) and isoflux wall (right frame)

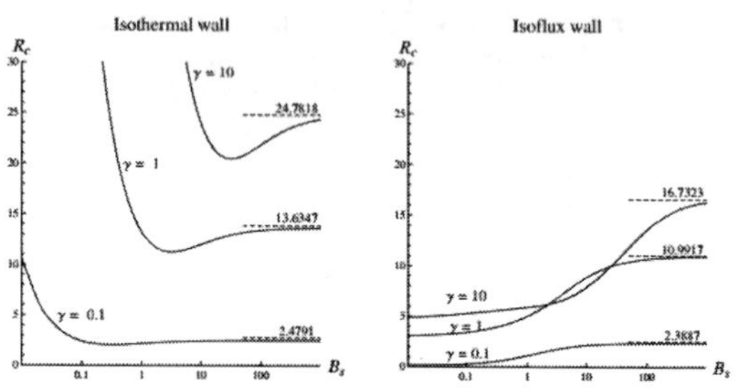

Figure 3. General case $B_s = \gamma B_f$: Plots of R_c versus B_s for assigned values of γ, with fixed $H = 0.1$ and $B_f = \frac{B_s}{\gamma}$.

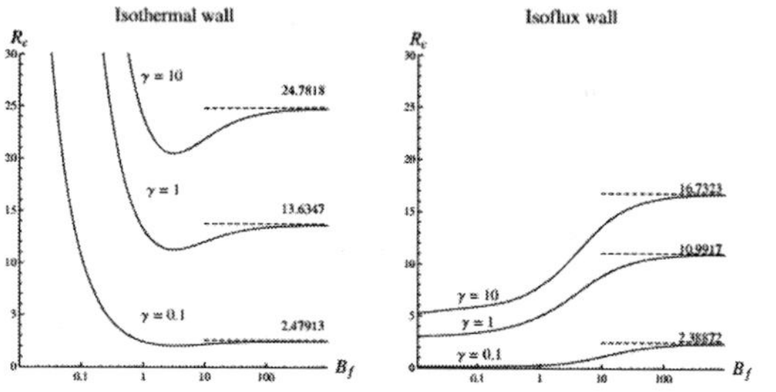

Figure 4. General case $B_s = \gamma B_f$: Plots of R_c versus B_f for assigned values of γ, with fixed $H = 0.1$.

7 Conclusion

The setup of Darcy-Rayleigh problem modified with free surface from the top, cooled by third boundary conditions and heated by uniform temperature from the bottom as first case, then by uniform heat flux (model *A*) as second case are numerically investigated. The effect of the local thermal non equilibrium model in the porous layer yields two different dimensionless Biot numbers, one for the fluid phase and the other for the solid skeleton. The analysis of the thermoconvective instability is carried out for special general case defined as $B_s = \gamma B_f$ in both cases. The numerical results obtained by employing a shooting method in conjunction with a sixth order Runge-Kutta method are characterized by the following behaviour:

- The dimensionless parameters B_f and B_s affect signficantly the temperature gradient of the medium in the two cases.
- The basic flow is more stable in the first case than in the second one.

References

1. E.R. Lapwood, *Convection of a fluid in a porous medium*, Proc. Cambridge. Phil. Soc. **44**,508-521, 1948.
2. N. Banu, D. A. S. Rees, *The onset of Darcy-Bénard convection using a thermal nonequilibrium model*, Int. J. Heat Mass Trans. **45**, 2221-2228, 2002.
3. K. Yang, K. Vafai, *Analysis of temperature gradient bifurcation in porous media-An exact solution*, Int. J. Heat Mass Trans. **53**, 4316-4325, 2010.
4. D. A. Nield, A. Bejan, *Convection in Porous Media*, 4th ed., Springer, New York, 2013.
5. A. Barletta, M. Celli, H. Lagziri, *Instability of a horizontal porous layer with local thermal non-equilibrium: Effects of free surface and convective boundary conditions*, Int. J. Heat Mass Trans. **89**, 75-89, 2015.

DOI: 10.1051/matecconf/20168307004

Heat removing under hypersonic conditions

Mikhail E. Semenov[1,2,3,a], Victor A. Nesterov[4], Ilya E. Kuznetsov[1], Andrey M. Solovyov[2], and Peter A. Meleshenko[1,2,b]

[1] Zhukovsky–Gagarin Air Force Academy, Starykh Bolshevikov st. 54 "A", 394064, Voronezh, Russia
[2] Digital Technologies Department, Voronezh State University, Universitetskaya sq. 1, 394006, Voronezh, Russia
[3] Voronezh State University of Architecture and Civil Engineering, XX-letiya Oktyabrya st. 84, 394006 Voronezh, Russia
[4] Moscow Aviation Institute (National Research University), Volokolamskoe shosse 4, 125993 Moscow, Russia

Abstract. In this paper we consider the heat transfer properties of the axially symmetric body with parabolic shape at hypersonic speeds (with a Mach number $M > 5$). We use the numerical methods based on the implicit difference scheme (Fedorenko method) with direct method based on LU-decomposition and iterative method based on the Gauss-Seigel method. Our numerical results show that the heat removing process should be performed in accordance with the nonlinear law of heat distribution over the surface taking into account the hypersonic conditions of motion.

1 Introduction

Rapid development of the aviation and space technology in the second half of XX century leads to a wide and exciting interest to the problems of hypersonic flow around bodies. In order to study the processes of heat exchange and heat removing in the design of aircraft as well as to obtain the stationary aerodynamic characteristics together with the traditional theoretical and experimental methods the methods of numerical simulation are widely used. The efficient numerical algorithms provide a possibility to obtain the data on the structure of flows around streamlined surfaces due to its comparative cheapness as well as to possibility to reproduce the natural conditions that are unattainable in the laboratory .

The thermal protection of the construction is a one of the most difficult problems in the interaction of high-enthalpy gas flows with aircraft. In order to realize the thermal protection in practice the passive and active methods are used, as well as the combined method: blowing the cooling gas in high-enthalpy gas flow from the porous structural elements combined with a flow of heat at the surface due to the choice of high thermal conductivity material for the composite shell. Study of the heat transfer (and heat removing) at these conditions is performed by solving the direct and inverse problems. In present time it is especially important to use the methods for solving the inverse problems due to the lack of the information on the processes under consideration as well as due to increasing requirements for the accuracy of determining of heat transfer characteristics of the streamlined body's shell taking into account the nonlinearity, multi-dimensionality and many-parametric character of the heat and mass transfer processes.

There are many works that are dedicated to the hypersonic aerodynamic regimes. However the problems of heat transfer and heat removing at hypersonic speeds remain not well studied. For example, an important problem is the high-temperature thermomechanical behavior of composite materials based on thermostable matrices and fillers [1–3]. At the same time the problem of numerical modeling of heat transfer on the surface of hypersonic aircraft [4–6] is also very actual and promisingly.

[a] e-mail: mkl150@mail.ru
[b] e-mail: melechp@yandex.ru

2 Heat transfer on parabolic surface

In this paper we consider a numerical simulation of the heat exchange on the surface of the axially symmetric body with parabolic shape at hypersonic speeds, which is a part of hypersonic aircraft (Fig. 1).

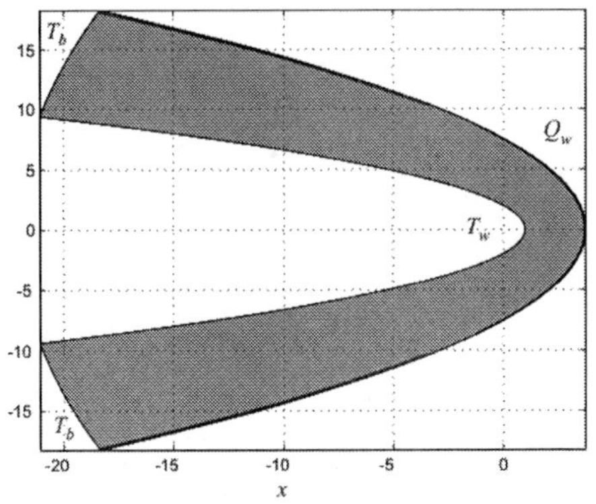

Fig. 1. Parabolic section of a body which is a part of hypersonic aircraft

In order to describe the distribution of heat on a flat surface we use the two-dimensional heat equation (equation of heat diffusion):

$$\frac{\partial T(x,y,t)}{\partial t} = D\left(\frac{\partial^2 T(x,y,t)}{\partial x^2} + \frac{\partial^2 T(x,y,t)}{\partial y^2}\right) + \varphi(x,y,t,T), \quad (1)$$

where D is a value which determines the rate of heat flow from the heated region to a less heated and may be a function of the coordinates and temperature. The function $\varphi(x,y,t,T)$ describes the influx of heat from outside, namely the heat sources that may also depend on the spatial coordinates (localization of sources), the time and temperature.

To simplify the modeling problem we will consider the special case where the stationary source of heat works for a long time and is determined by the initial conditions, and the transition processes caused by its appearance are ended.

In this case the heat conduction equation (1) is converted to the standard Laplace equation:

$$\frac{\partial^2 T(x,y,t)}{\partial x^2} + \frac{\partial^2 T(x,y,t)}{\partial y^2} = 0. \quad (2)$$

Taking into account the fact that the studied surface has a parabolic shape it is more appropriate to search the solutions of the equation (2) in the parabolic coordinates (Fig. 2).

2.1 Some notations on parabolic coordinates

Here we would like to recall some basic properties of the parabolic coordinates. The two-dimensional parabolic co-

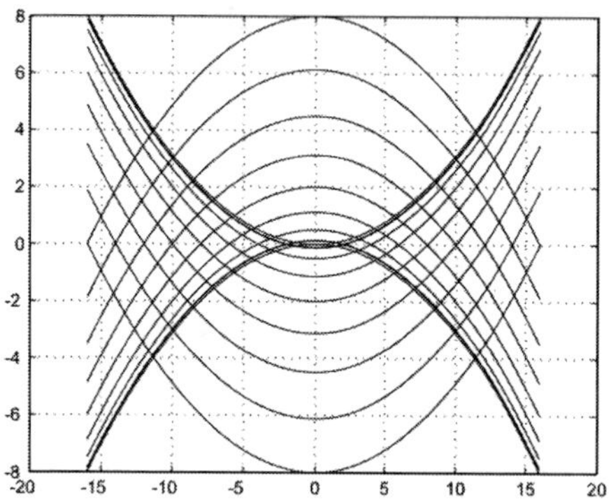

Fig. 2. Parabolic coordinates

ordinates are determined by the expression:

$$\begin{cases} x = \sigma\tau, \\ y = \frac{1}{2}(\tau^2 - \sigma^2), \end{cases} \quad (3)$$

while the surfaces of constant σ are confocal parabolas

$$2y = \frac{x^2}{\sigma^2} - \sigma^2, \quad (4)$$

expanding upward (along the ray $+y$), and the surfaces of constant τ are confocal parabolas

$$2y = -\frac{x^2}{\tau^2} - \tau^2, \quad (5)$$

expanding downward (along the ray $-y$). Focuses of all the parabolas are located at the origin.

It is obvious that the surface shown in Fig. 1 is a flat plate in parabolic coordinates, and the solution of the temperature distribution problem becomes simpler.

Let us consider the Laplace operator Δ in parabolic coordinates. Lame coefficients for parabolic coordinates are

$$H_\sigma = H_\tau = \sqrt{\sigma^2 + \tau^2}, \quad (6)$$

the area element is

$$dS = (\sigma^2 + \tau^2)d\sigma d\tau, \quad (7)$$

and the action of the Laplace operator has the form:

$$\Delta\Phi = \frac{1}{\sigma^2 + \tau^2}\left(\frac{\partial^2 \Phi}{\partial \sigma^2} + \frac{\partial^2 \Phi}{\partial \tau^2}\right). \quad (8)$$

Thus, the Laplace equation (2) in the parabolic coordinates can be written as:

$$\frac{\partial^2 T(\sigma,\tau)}{\partial \sigma^2} + \frac{\partial^2 T(\sigma,\tau)}{\partial \tau^2} = 0,\ \sigma \in [0,G],\ \tau \in [0,H]. \quad (9)$$

We assume that on the inner surface of the body there is a parabolic temperature distribution defined by the law $T_w(\sigma)$. In the field of T_b the heat exchange with the environment is absent and the temperature on the outer surface of the body depends on a number of aerodynamic parameters and describes by the function $Q_w(\sigma)$. Then the boundary conditions for the equation (9) can be written as follows:

$$\begin{cases} T(\sigma,0) = T_w(\sigma), \\ T(\sigma,H) = Q_w(\sigma), \\ \frac{\partial T}{\partial \sigma} = 0 \text{ at } \sigma = 0, \\ \frac{\partial T}{\partial \sigma} = 0 \text{ at } \sigma = G. \end{cases} \quad (10)$$

2.2 Heat transfer under hypersonic condition

As was shown in [7], in the case of hypersonic speeds, the heat distribution law $Q_w(\sigma)$ on the surface of the axially symmetric body will be determined by the following expression:

$$Q_w(\sigma) = F(M_\infty,\sigma)\left[\frac{1 + 0.096\sqrt{\varepsilon}}{1.068}\right]\frac{1 - g_w}{1 - g_{ws}}, \quad (11)$$

where ε is a coefficient depending on the speed of the external environment, g_w is an enthalpy of the outer surface of the body, g_{ws} is a critical enthalpy;

$$F(M_\infty,\sigma) = 2\sigma\sin\sigma\left[\left(1 - \frac{1}{\gamma_\infty M_\infty^2}\right)\cos^2\sigma + \right.$$
$$\left. + \frac{1}{\gamma_\infty M_\infty^2}\right]\frac{1}{\sqrt{D(M_\infty,\sigma)}}, \quad (12)$$

where γ_∞ is a heating rate, M_∞ is a Mach number, and

$$D(M_\infty,\sigma)=\left(1-\frac{1}{\gamma_\infty M_\infty^2}\right)\left[\sigma^2-\frac{\sigma\sin4\sigma}{2}+\right.$$
$$\left.+\frac{1-\cos4\sigma}{8}\right]+\frac{4}{\gamma_\infty M_\infty^2}\left[\sigma^2-\sigma\sin2\sigma+\frac{1-\cos2\sigma}{2}\right]. \quad (13)$$

Thus, the solution of equation (9) with the boundary conditions (10) using (11)-(13) determines the temperature distribution over the entire surface of the parabolic body which is a part of the hypersonic aircraft.

3 Numerical simulation

3.1 Difference scheme

Let us make the numerical simulation of the heat transfer on the surface of parabolic body by means of numerical solution of the equation (9). Let us approximate the equation (9) and the boundary conditions (10) on the five-point pattern of the "cross"-type. In this case we get the implicit difference scheme in the form

$$\begin{cases}\dfrac{T_{i+1}^j-2T_i^j+T_{i-1}^j}{h_i^2}+\dfrac{T_i^{j+1}-2T_i^j+T_i^{j-1}}{h_j^2}=0,\\ T(ih_i,0)=T_w(ih_i),\\ T(ih_i,Mh_j)=Q_w(ih_i),\\ \dfrac{T_1^j-T_0^j}{h_i}=0,\\ \dfrac{T_N^j-T_{N-1}^j}{h_i}=0,\\ i=\overline{0,N},\ j=\overline{0,M},\end{cases} \quad (14)$$

where h_i is a step along the σ-axis and h_j is a step along the τ-axis.

Let us prove the stability of the presented difference scheme with respect to perturbations of the boundary conditions using the spectral method [8]. We represent the solution of the system (14) in the form of harmonics

$$T_k^p=\lambda^p\exp(i\alpha k). \quad (15)$$

Substitution of (15) in to the first equation of the system (14) and simplification of the obtained expression gives

$$\frac{e^{i\alpha}+e^{-i\alpha}-2}{h_i^2}+\frac{(\lambda-1)^2}{\lambda h_j^2}=0. \quad (16)$$

The difference scheme is stable in the case of performing the required spectral Neumann conditions:

$$|\lambda(\alpha)|\leqslant 1. \quad (17)$$

From equation (16) it follows that

$$|\lambda(\alpha)|=1,\ \forall\alpha. \quad (18)$$

Thus, the spectral Neumann condition (17) is performed and the used difference scheme is absolutely stable.

Of course the system of equations (14) can be solved using the direct methods. However, for small values of h_i and h_j in this case we have a problem with the conditionality matrix describing the system, and this fact leads to increasing of the calculation errors. The iterative methods are connected with the searching of algorithms and parameters to ensure the optimum convergence. The most effective and used method which takes into account the above mentioned drawbacks is a Fedorenko method [9]. This method is a combination of direct and iterative methods for solving the grid problems. The essence of the method is as follows: at the first stage the required solution is obtained by direct methods such as Gaussian elimination, Cramer or LU-decomposition on the grid with a major step. Further approximation of the solutions is carried out on a grid with a fine step and made recalculation using the iterative methods such as Gauss-Seidel method with the control of the convergence process.

3.2 Numerical results

We solve the system (14) using the Fedorenko method [9]. As a direct method we use the LU-decomposition. As an iterative method we use the Gauss-Seidel method.

We use the following parameters of the system. Aerodynamic characteristics are: $g_w = 0.01$, $g_{ws} = 0.98$, $\varepsilon = 600$, $M_\infty > 5$, $T_w(\sigma) = 0$. For the direct method: $h_i = 1$, $h_j = 0.1$, $N = 10$, $M = 10$. For the iterative method: $h_i = 0.2$, $h_j = 0.02$, $N = 50$, $M = 50$.

The simulation results are presented in Fig. 3, 4, 5.

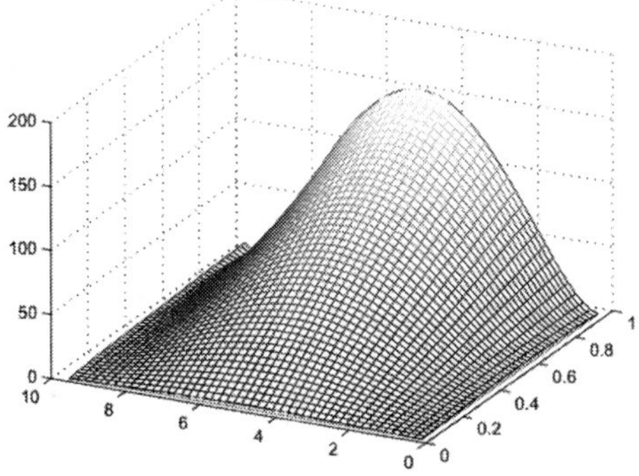

Fig. 3. Temperature distribution on the surface of parabolic body (parabolic coordinates).

As it can be seen from these figures, on the surface of the object there are regions with different degrees of heat that are nonlinearly distributed over the entire area. Therefore, if the object is a part of the hypersonic aircraft, the heat removing should be performed in accordance with the received law of distribution over the surface.

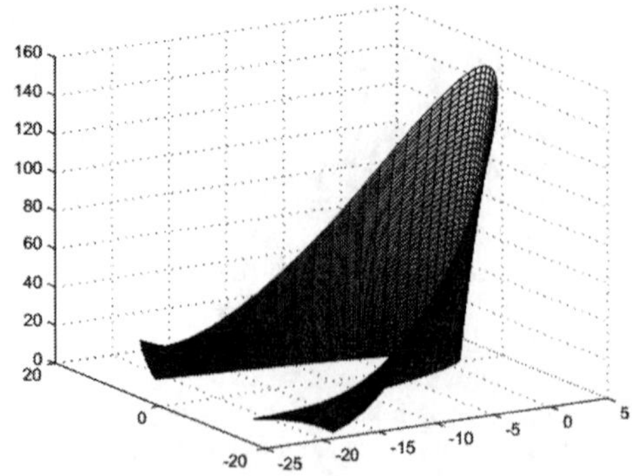

Fig. 4. Temperature distribution on the surface of parabolic body (Cartesian coordinates).

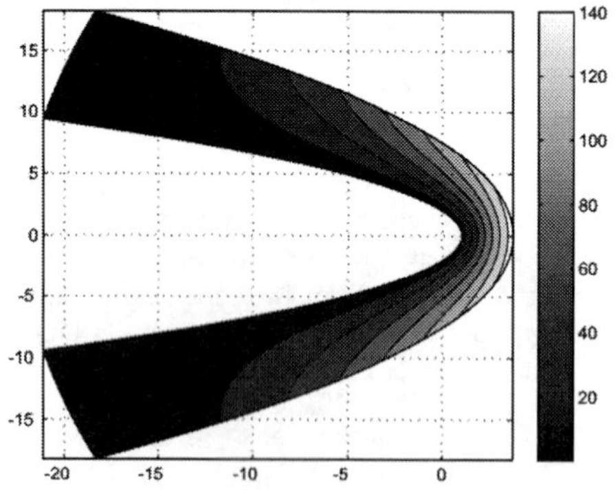

Fig. 5. Temperature distribution on the surface of parabolic body (contour plot).

4 Conclusions

In this paper we have considered the heat transfer properties of the axially symmetric body with parabolic shape at hypersonic speeds (with a Mach number $M > 5$). We have used the numerical methods for solution of the Laplace equation in parabolic coordinates based on the implicit difference scheme (we use the Fedorenko method) with direct method based on LU-decomposition and iterative method based on the Gauss-Seigel method. The fact that the considered parabolic body moves with a hypersonic velocity has taken into account by the special form of the heat distribution law $Q_w(\sigma)$, i.e. the heat distribution of the air flow on the outer surface at hypersonic velocity. Our numerical results show that the heat removing process should be performed in accordance with the nonlinear law of heat distribution over the surface (which is solution of the heat equation) taking into account the hypersonic conditions of motion.

This work is supported by the RFBR grant 16-08-00312-a.

References

1. Y.I. Dimitrienko, *Thermomechanics of Composites under High Temperatures* (Springer, 1999) 350pp
2. Y.I. Dimitrienko, Int. J. Heat Mass Transf. **38**, (1995) 139
3. Y.I. Dimitrienko, Int. J. Eng. Sci. **35**, (1997) 15
4. V.I. Zinchenko, A.Ya. Kuzin, J. Appl. Mech. Tech. Phys. (in Russian) **43**, (2002) 144
5. D.R. Ismagilov, R.V. Sidelnikov, "Almaz-Antey" Proc. (in Russian) **2**, (2015) 49
6. G. Han, Z. Jiang, Int. Commun. Heat Mass Transf. **43**, (2013) 46
7. W. Peng, Yu. He, X. Wang, J. Zhu, J. Han, Chin. J. Aeronautics, **28**, (2015) 121
8. S.K. Godunov, V.S. Ryabenkii *Difference schema* (Elsevier Science Ltd, 1987) 508 pp.
9. R.P. Fedorenko, USSR Comput. Math. Math. Phys., **1**, (1962) 1092

Simulation of simultaneous photonic and phononic band gaps in Sapphire

H. Bentarki[1,a], M. Tlidi[2], and A. Makhoute[1]

[1] Physique du Rayonnement et des Interactions Laser-Matière, Faculté des Sciences. Moulay Ismail University. B.P. 11201, Zitoune, Meknès, Morocco.
[2] Faculté des Sciences, Université Libre de Bruxelles (U.L.B.), CP. 231, Campus Plaine, B-1050 Bruxelles, Belgium.

Abstract The simultaneous existence of photonic and phononic band gaps in a simple square periodic array of holes drilled in sapphire substrate is investigated theoretically. The investigations are focused in a first part on a photonic gap obtaining with a phononic gap in the same time making use of some specific parameters. In the second part, using a cavity defect in which the strong confinement of acoustic and optic waves enhances the interaction, we obtain photonic and phononic modes classified according to their symmetry (even or odd), also allowed bands in gaps, which has led to the propagation of waves appropriate. The frequencies of the localized modes in the first band gap are computed with a finite element method.

1 Introduction

The photonic band gap materials [1-2] on one hand and phononic [3-4] on the other hand generate a significant interest in the scientific community in recent years. They are structures having a periodicity of one or more spatial directions. This periodic variation leads to the appearance of photonic and phononic bands which are defined as the frequency ranges for which any kind of wave cannot propagate in the structure, whatever its polarization and direction of propagation. These properties make these interesting crystals for many applications in integrated optics for the realization of new components such micro cavities with high quality factor, micro guides with low losses, micro sources lasers and spatial filters in wavelength (super prisms, filters,...). A new path is open in the field of acousto-optic devices [5-6 and 7] through the simultaneous existence of photonic and phononic structures, seeking to consolidate the principle of photonic and phononic crystal in one structure named phoXonic crystal. The interest of such artificial crystal is simultaneously confining of optical and acoustic energy. Such a structure is interesting for the study of interactions between photons and phonons but also for the development and improvement of devices of integrated acousto-optic, such modulators or multiplexers. The theme of 2D phoxonic crystals is very recent since the first theoretical papers published are from 2006 [8-9], when Maldovan and Thomas [10-11] showed theoretically that the phononic and photonic bandgap can be obtained in 2D crystals for a square or hexagonal network of air holes in a silicon matrix. Thereafter, the method of supercells [12] demonstrated the existence of localized modes, for confining optical and an acoustic wave in a cavity. Sadat-Saleh et al. [13] demonstrated the possibility of opening the phononic and photonic band gaps in complex matrices such as multiple cylinders (more than one atom per unit cell) into 2D structures in Lithium Niobate. The studies of periodic structures of pillars which allow the existence of forbidden bands were conducted, various structures containing only air inclusions [14-15]. The acousto-optical interaction started since the 60's, thanks to the invention of the laser and Leon Brillouin in 1914, which made the first theory of the coupling of a light wave with a hyper-sonic wave [16]. Leon Brillouin theory, published in 1921, provides that a liquid traversed by a hypersonic wave causes periodic variations of constraints that create, by photo-elastic effect, changes in the refractive index, then behaving as a network of diffraction for light. [16] This has resulted in the following years, a series of studies both theoretical and experimental involving photon-phonon interaction. Other studies, this time based on other than a 2D square lattice structures, [17-18] have emerged. For example using plates, [19] or the use of structured nano wave guides [20-21 and 22].

The main goal of this paper is to conduct a theoretical study of simultaneous phononic and photonic band gaps, in 2D square lattice constituted by a periodic array of holes deposited on a layer. We concentrate our calculations on a structure where the holes and the supporting plate are respectively made of air and sapphire. Taking sapphire as a matrix component allows us to take advantage of the anisotropy of the dielectric tensor in the microwave regime to show the possibility of confining optical energy in both polarizations (TE and TM) and energy acoustic simultaneously. In general, the phononic and photonic band structures are calculated by finite element (FE). However, finite difference time domain (FDTD) method has also been used in 2010 for the same structure, [23] just that we chose a different lattice parameter, to have localized modes at the introduction of defects in the perfect structure.

The paper is structured as follows. In section 2, we describe the geometrie considered in this paper as well as the method of calculation. This section presents also analysis of obtained results, and section 3 concludes the paper.

[a] e-mail: bentarki.houda@gmail.com

 DOI: 10.1051/matecconf/20168308001

2 Numerical method, results and discussion

2.1 Geometry and methodology

From the first work on photonic and phononic crystals, and after the publication of Yablonivitch in 1987, [24] the complexity of experimental studies and their cost have emerged. Then, the need for development of optical modeling methods for the study of these structures is imposed. It is then necessary to solve Maxwell's equations [25] for the study of photonic crystals and the elasticity equations for phononic crystals. To solve these equations numerically, some methods have been developed, such as the finite element [FE] method, as we have chosen to achieve our goal. The structure we used for this study was elaborated in a sapphire plate with a square lattice of cylindrical air holes drilled through the plate (see Fig. 1(a)). The lattice constant ($a = 70(nm)$) was set to this value in order to obtain a photonic and phononic band gap as large as possible in this structure. For the same reason, the filling ratio was fixed to about its maximum value ($f = 0.66$), leaving only a few nanometers between two adjacent holes. The unit-cell used for computation of the dispersion relation and the modes of a phoxonic crystal slab is shown in Fig. 1(b). By replacing one hole in the middle of the structure, we created a substitution-type defect forming a cavity in the phoxonic crystal, Fig. 1 (c) shows this structure.
The table 1 summarizes the acoustic and optical parameters of the material used in simulations.

Table 1. Please write your table caption here.

Table 2. Elastic and optical constants of Sapphire (Al_2O_3) cubic considered and used in numerical simulations. The refractive index (n), the density (ρ), the longitudinal (c_l) and transverse (c_t) velocities of sound along the [100] direction of propagation [26-27].

	n	$\rho(Kg/m^3)$	ε	$c_t(m/s)$	$c_l(m/s)$
Sapphire	1.77	3950	11.5	6040	11090

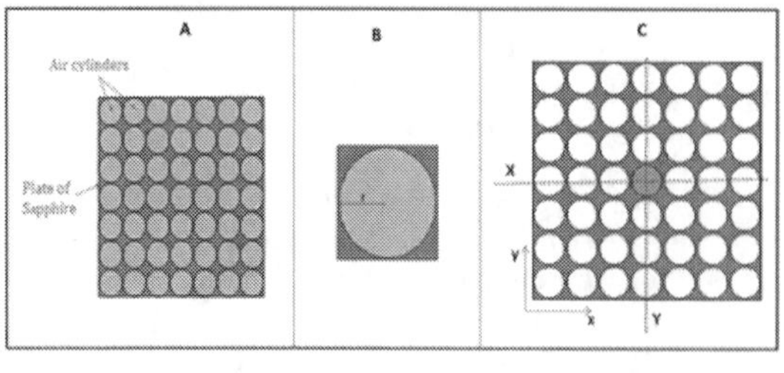

Figure 1. a) Two-dimensional photonic crystal composed of cylinders arranged periodically in a square arrangement on a plate of Sapphire. (b) Elementary cell. (c) Plans of symmetry.

3 Dual photonic and phononic band gaps

In this systematic study, the calculation of the band structure has been done to the main directions of the first Brillouin zone of the square lattice for Sapphire, that is to say along the closed path $\Gamma - X - M - \Gamma$, where ΓX and ΓM the directions of the first Brillouin zone. In the band structures presented in the paper, the frequencies are given in the dimensionless unit $\Omega = \omega a/2\pi c$ where c is the velocity of light in vacuum for electromagnetic waves and the transverse velocity of sound in sapphire for elastic waves. Photonic simulation is done in the range of microwave frequencies where the dielectric tensor of sapphire is anisotropic and the refractive indices are high $\varepsilon_{\parallel} = 11.5$ and $\varepsilon_{\perp} = 9.3$. Obviously spoken, obtaining a photonic band gap in an infrared frequency range is more difficult because of the low refractive index of the sapphire in this range ($n_{Sapphire} = 1.75$). Figure 2 shows the results obtained for the TE and TM polarizations, meaning the electric field and the magnetic field are parallel to the axis of the cylinders.

The dispersion curves in Figure 2 show the existence of a common gap between the TE mode and TM mode of the Sapphire matrix for a fill factor is 0.66. So these results are of great importance in view of completion of the optical and acoustic band gaps: this is the phoxonic crystal that crossed the world of telecommunications and radar.

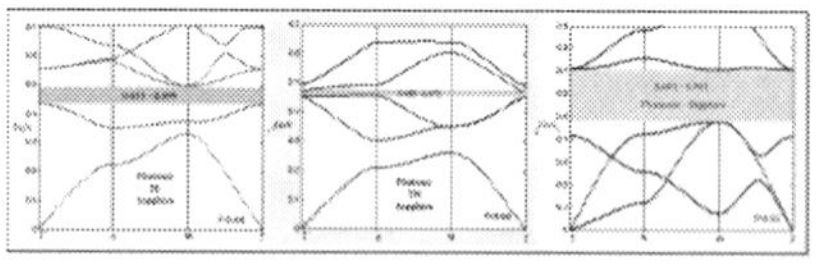

Figure 2. Band structure of a two-dimensional square lattice of air holes immersed in a matrix of Sapphire in TM and TE mode.

3.1 Photonic and phononic structure with defect

The objective in this section is the confinement of energy in a cavity. Punctuel defects completely surrounded by the photonic and phononic crystal, allowed to frequencies within the band gap to penetrate and confine, the photonic and phononic crystal with their defects behave like almost perfect mirrors. When we speak about optical cavities, the application that immediately comes to mind is the development of lasers. Just insert an active material within the cavity to obtain a microlaser.
The periodic structure is composed of a square array of air cylinders in a matrix of Sapphire (Al_2O_3). The width of the band gap of the defect-free crystal is proportional to the confinement within the cavity. This condition is fulfilled when the filling fraction is large. In this calculation, the radius of the rolls of air has been chosen equal to $0.4584a$, where a is the lattice parameter, which corresponds to a filling factor of $f = \pi r^2/a^2 = 0.66$, the defect created in the structure is an alternative default; the air hole in the center of the super cell is replaced by a silicon hole of same radius.
In this part, we will search the default mode when we break the periodicity of the material. then we use the concept of supercell. A supercell comprises a plurality of primitive cells. In this case, there is a supercell consisted of 7×7 primitive cells, meaning the cavity is repeated along the x and y directions with a super-period equal to $L = 7a$, where a is the lattice parameter, wherein a cavity is created by modifying a hole.

Figure 3 shows the mapping of fields of the 3 TM modes, 2 TE optical modes and 6 phononic modes localized in their respective photonic and phononic band.

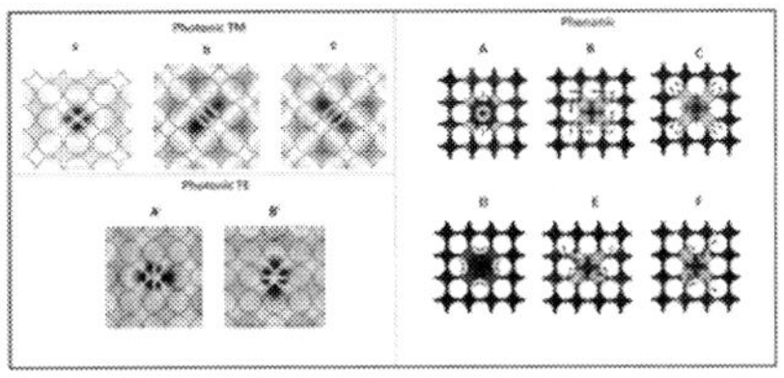

Figure 3. Mapping of the elastic and electromagnetic fields in the plan (x, y) for the photonic and phononic modes located in their respective band gap. For each mode of the cavity, the blue arrows indicate the magnitude and direction of the displacement field vector.

Photonic and phononic modes are classified according to their symmetry, or symmetrical (even (e)) antisymmetric (odd (o)), relative to the two planes of symmetry X and Y of the cavity (Figure 1.C). An analysis of the symmetry of the modes has been realized from the distributions of the fields in Fig. 3. As a result, the optical modes (b) and (c) are of symmetry (oo), (a) is of symmetry (ee), (A ') and (B') are of symmetry (oe) and (eo) respectively.
The phononic modes (A) and (D) are of symmetry (ee), (B) and (C) are of symmetry (oo). The other two modes (E) and (F) are degenerate and from fields of presentations, (oe) and (eo) are the respective symmetries of (E) and (F). The description of each phonon cavity mode during a vibration period is deducted from the direction of blue arrows, animations and modes of field maps. For the mode (A), the cavity is alternately stretched along one side of the square lattice, while the other side contracts. Mode (B) is a torsional mode perpendicular to the axis of the cavity with a rotation in the counterclockwise direction in the outer part of the cavity. The evolution in time of the mode (C) is such that the cavity is stretched along a diagonal and contracted along the other for a half acoustic period. The temporal evolution of the mode (D) substantially shows a breathing motion with homothetic deformation of the shape of the cavity. Finally, the field distribution of the modes (E) and (F) is orthogonal.

Punctuel defects completely surrounded by the photonic and phononic crystal, allowed to frequencies within the band gap to penetrate and confine, the photonic and phononic crystal with their defects behave like almost perfect mirrors. When we speak about optical cavities, the application that immediately comes to mind is the development of lasers. Just insert an active material within the cavity to obtain a micro laser.

4 Conclusion

In this work, we have simulated the propagation of electromagnetic and elastic waves in phononic and photonic crystals, and we have demonstrated that a periodic array of air holes deposited on a sapphire matrix exhibits simultaneous phononic and photonic complete band gaps in the square lattice. The confinement of electromagnetic and elastic energy is obtained by inserting a substitution defect in a silicon slab presenting simultaneously a phononic and a photonic band gap. The design of phoxonic structures depends on the geometrical parameters crystal, which were chosen, based on the finite element analysis of a perfect phoxonic crystal of circular holes. This does not stop only at the choice of the type of network but also the choice of materials, their respective cuts, or inclusions rays.

References

1. E. Yablonovitch, J. Opt. Soc. Am. B **10, 283**, (1993).
2. J. D. Joannopoulos, S. G. Johnson, J. N. Winn, and R. D. Meade, *Photonic Crystals: Molding the Flow of Light Book title* (Princeton University Press, 2008).
3. M. S. Kushwaha, P. Halevi, L. Dobrzynski, and B. Djafari-Rouhani, Phys. Rev. Lett. **71, 2022** (1993).
4. Y. Pennec, J. Vasseur, B. Djafari-Rouhani, L. Dobrzynski, and P. A. Deymier, Surf. Sci. Rep. **65, 229**, (2010).
5. T. Gorishny, M. Maldovan, C. Ullal, and E. L. Thomas, Phys. World, **18, 24** (2005).
6. M. Eichenfield, J. Chan, R. M. Camacho, K. J. Vahala, and O. Painter, Nature (London) **462, 78** (2009).
7. M. Eichenfield, J. Chan, R. M. Camacho, K. J. Vahala, and O. Painter, Nature (London) **459, 550** (2009).
8. M. Maldovan et E.L. Thomas: Simultaneous complete elastic and electromagnetic band gaps in periodic structures. Applied Physics B, **83(4): 595- 600**, (2006).
9. Martin Maldovan et Edwin L. Thomas: Simultaneous localization of photons and phonons in twodimensional periodic structures. Applied Physics Letters, **88 (25): 251907** (2006).
10. Said Sadat-Saleh, Sarah Benchabane, Fadi Issam Baida, Maria-Pilar Bernal et Vincent Laude: Tailoring simultaneous photonic and phononic band gaps. Journal of Applied Physics, **106 (7): 074912** (2009).
11. S. Sadat-Saleh, S. Benchabane, F.I. Baida, Maria-Pilar Bernal et S. Laude: Simultaneous photonic and phononic band gaps in a two-dimensional lithium niobate crystal. In Ultrasonics Symposium (IUS), 2009 IEEE International, **pages 1118- 1121** (2009).
12. B. Djafari Rouhani, Y. Pennec, E.H. Boudouti, J.O. Vasseur, Y. Hassouani, C. Li, A. Akjouj et D. Bria: Band gap engineering in simultaneous phononic and photonic crystal slabs. Applied Physics A, **103(3): 735- 739** (2011).
13. Y. El Hassouani, C. Li, Y. Pennec, E. H. El Boudouti, H. Larabi, A. Akjouj, O. Bou Matar, V. Laude, N.Papanikolaou, A. Martinez et B. Djafari Rouhani : Dual phononic and photonic band gaps in a periodic array of pillars deposited on a thin plate. Phys. Rev. B, **82: 155405** (Oct 2010).
14. Y. Pennec, B. Djafari Rouhani, E. H. El Boudouti, C. Li, Y. El Hassouani, J. O. Vasseur, N. Papanikolaou, S. Benchabane, V. Laude, et A. Martinez: Band gaps and waveguiding in phoxonic silicon crystal slabs. CHINESE JOURNAL OF PHYSICS, **49: 100** (Feb 2011).
15. Saeed Mohammadi, Ali A. Eftekhar, Abdelkrim Khelif et Ali Adibi: Simultaneous two dimensional

phononic and photonic band gaps in opto-mechanical crystal slabs. Opt. Express, **18(9): 9164- 9172** (Apr 2010).
16. L. Brillouin, Diffusion de la lumière et des rayons x par un corps transparent homogène, influence de l'agitation thermique. Ann. Phys., **17, 88-122** (1922). University Press (1995).
17. Q. Rolland, M. Oudich, S. El-Jallal, S. Dupont, Y. Pennec, J. Gazalet, J. C. Kastelik, G. Leveque, and B. Djafari-Rouhani. Appl. Phys. Lett. **101, 061109** (2012).
18. S. El-jallal, M. Oudich, Y. Pennec, B. Djafari-Rouhani, A. Makhoute, Q. Rolland, S. Dupont, and J. Gazalet. J. Phys.: Condens. Matter **26, 015005** (2014).
19. S. El-Jallal, M. Oudich, Y. Pennec, B. Djafari-Rouhani, V. Laude, J.-C. Beugnot, A. Martinez, J. M. Escalante, and A. Makhoute. Phys. Rev. B **88, 205410** (2013).
20. M. Oudich, S. El-Jallal, Y. Pennec, B. D. Rouhani, J. Gomis-Bresco, D. Navarro-Urrios, C. M. S. Torres, A. Martinez, and A. Makhoute, Phys. Rev. B **89, 245122** (2014).
21. J. Gomis-Bresco, D. Navarro-Urrios, M. Oudich, S. El-Jallal, A. Griol, D. Puerto, E. Chavez, Y. Pennec, B. D. Rouhani, F. Alzina, A. Martinez and C.M. Sotomayor Torres. Nature communications **5, 4452** (2014).
22. D. Navarro-Urrios, J. Gomis-Bresco, S. El-Jallal, M. Oudich, A. Griol, A. Pitanti, N. Capuj, A. Tredicucci, F. Alzina, A. Griol, Y. Pennec, B. Djafari-Rouhani, A. Martinez, and C. M. Sotomayor Torres. "Dynamical back-action at 5.5 GHz in a corrugated optomechanical beam". AIP Advances 4 , **124306** (2014).
23. D. Bria, M. B. Assouar, M. Oudich, Y. Pennec, J. Vasseur, and B. Djafari-Rouhani, J. Appl. Phys. **109, 014507** (2011).
24. E. Yablonovitch, Phys. Rev. Lett. **58, 2059** (1987).
25. A. Taflove, *the Finite-Difference Time-Domain Method*. (Artech, Boston (1998)).
26. Yariv. A and Yeh. P *Optical Waves in Crystals* (New York: Wiley 1984).
27. Renosi P, Sapriel J and Djafari Rouhani B 1993 Proc. 5th Int. Conf. on Indium Phosphide and Related Material (Paris) pp 592-5 doi: 10. 1109/ICIPRM. 1993. 380551 see also Renosi P and Sapriel J 1994. Appl. Phys. Lett. **64 2794**.

Bragg Fibers with Soliton-like Grating Profiles

S. Bugaychuk

Institute of Physics, NASU, 46 Prospect Nauki, 03680 Kyiv, Ukraine

Abstract. Nonlinear dynamical system corresponding to the optical holography in a nonlocal nonlinear medium with dissipation contains stable localized spatio-temporal states, namely the grid dissipative solitons. These solitons display a non-uniform profile of the grating amplitude, which has the form of the dark soliton in the reflection geometry. The transformation of the grating amplitude gives rise many new atypical effects for the beams diffracted on such grating, and they are very suitable for the fiber Brass gratings. The damped nonlinear Schrodinger equation is derived that describes the properties of the grid dissipative soliton.

1 Introduction

Fiber Bragg gratings (FBG) are very popular in the contemporary applications. FBG is an extended Bragg reflection grating formed in a fiber core. FBG can be used as inline optical filters to block certain wavelengths, or as a wavelength-specific reflector. The spectral selectivity of light beams reflected in FBG is due to periodic refractive index grating in the fiber core and beams' diffraction on this grating. FRG have narrow reflection spectrum and they are used in fiber lasers, fiber sensors and detectors, for stabilization and changes of the laser wavelength in laser diodes, and others [1]. But always the amplitude of the grating is constant, and only the changes of the grating period may occur due to different external influences such as mechanical stretching or compression of the fiber, changes the temperature, etc. In our work we consider the case, what would happened if the amplitude of the grating is changed and what changes must be. We show that the answers to these questions may be found in consideration of the dynamic holography. The grid dissipative soliton can be formed on the Bragg reflection grating during the wave-interaction in a dynamic dissipative medium. These solitons are very suitable for the FBG. We show that the proposed dynamical system is an effective tool for modeling many effects of wave interaction in FBG.

The dynamic holography covers the following processes: formation of a light grating (the interference pattern) due to interaction of coherent optical waves in a nonlinear medium, creation of a refractive index grating by this light grating, and transformation of the interacting wave owing to their self-diffraction on this grating. In the dynamical regime both the light grating and the material grating mutually influence to each others and both are changed in time and space.

If the nonlinear medium possesses a nonlocal response, i.e. the matter grating is shifted relative to the light grating in a space, the localized stated are emerging in such system. They display the non-uniform profile of the dynamical grating amplitude. We have found previously that they are described by the equation similar to the complex Ginzburg-Landau equation (CGLE) [2, 3]. The CGLE include dissipative soliton solutions that arise in non-equilibrium systems due to the balance between gain and loss [4]. The photoinduced refractive index would form the soliton-like patterns. The first experimental observation of localization effect for the photoinduced refractive index has been done in wave-mixing experiments in bulk $LiNbO_3$ crystals [5]. The same effects should be observed for the case, when the nonlinear response has time-delay regarding the impact of a light-pulse.

In present paper we derive the nonlinear Schrodinger (NLS) equation in a specific case, namely when the space shift between the light grating and the refractive index grating is equal to the quarter of the grating period and the refractive index grating is formed in the reflection geometry. The obtained damped NLS equation describes the properties of the grid dissipative soliton.

2 Grid dissipative soliton

The condition of the formation of the grid dissipative soliton in the reflection geometry is shown in Fig. 1. It appears, when two coherent waves interfere in an extended nonlinear medium. The straight lines show the maximums of the intensity pattern. But, the nonlinear response is nonlocal, and the dynamical grating formed by this intensity pattern is shifted in the space on quarter of the grating period, or on $\pi/2$ by phase. The maximums of the grating are shown by the dashed lines. The waves

DOI: 10.1051/matecconf/20168308002

experience the self-diffraction on the dynamical phase grating, which they are created themselves. As the result, both the light grating and the refractive index grating are changing in space and time. In this case the grating becomes with non-uniform amplitude. The amplitude is described by the dark-soliton solution. Then the grating will look like it is shown in Fig. 2, where $i\chi^{(3)}$ indicates the nonlocal Kerr-like nonlinear susceptibility, and the arrow shows the spatial direction of the nonlocal shift.

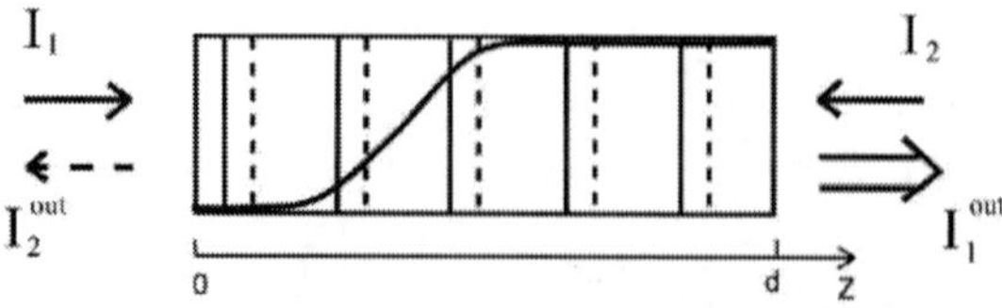

Fig. 1. The self-diffraction of two coherent waves in a dynamical dissipative medium in the reflection geometry. The curve illustrates the grating amplitude distribution.

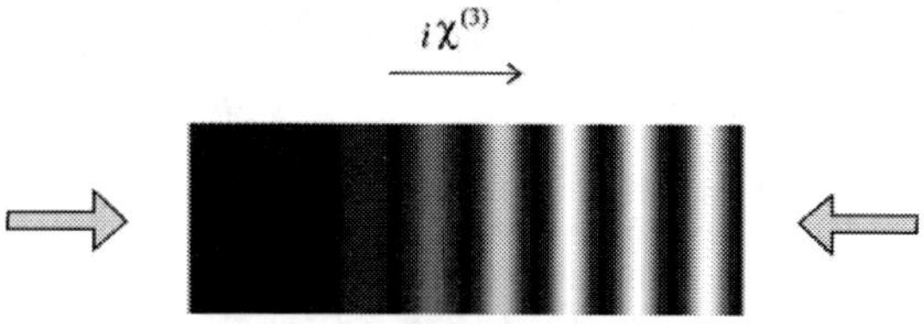

Fig. 2. The grating with soliton-like profile.

The basic test of the creation of such a grating is the amplification of one a wave, or, to say by other words, the energy transfer between waves. Since the dynamical grating is shifted in a determined direction relative to the intensity pattern, the two interacting waves become not equivalent. The amplitude of the grating is determined by the intensity ratio of two interacting waves. Such wave is amplified, which has the wave-vector directed to the grating shift. So, in our pictures, Figs.1,2, the wave 1 is a signal, and the wave 2 is the pump. The best case is when the pump is much higher than the signal. Then the pump will transfer its energy to the signal, which will be amplified. In the case of equal input intensities, or when the signal is higher than pump, the grating amplitude is decreased and the profile is moved, as the result there is almost no energy transfer.

The grid dissipative soliton has steady state solution in a form of the dark soliton in reflection geometry:

$$Q(z) = \{1+ tanh[\gamma_N \tau z - p + ln(4/I^2{}_d)/2]\}^{1/2} \cdot (0.5)^{1/2} \quad (1)$$

Where Q is the amplitude of the refractive index grating; γ_N is the coefficient of the nonlocal nonlinear gain; τ is the time constant; p is the constant, which depends on the input intensity ratio.

Interesting to note, to reach the optimal condition for the energy transfer, the medium should not have a strong nonlinear response. The calculations in Fig. 3 show that. These exists the optimal value of the nonlinear response, which is not very strong, when the created dynamical grating reaches its maximum.

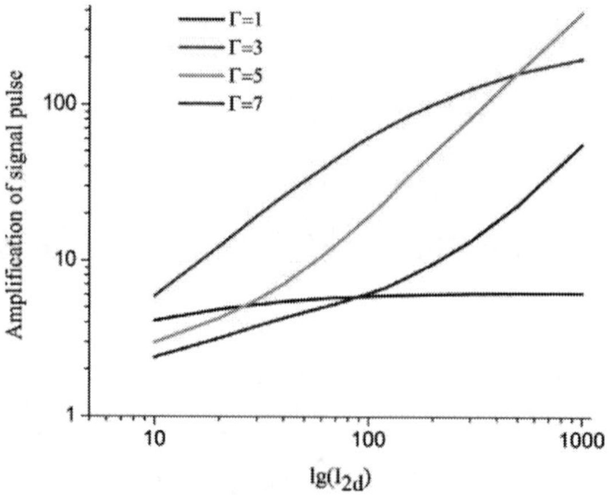

Fig. 3. Amplification of a signal pulse for various gain coefficients. ($\Gamma=\gamma_N\tau d$; $I_{10}=1$; $I_{2d}=100$).

3 Derivation of damped NLS equation for grid dissipative soliton

To derive the damped NLS equation, which describes the grid soliton, we consider the dynamical problem (see Fig. 4). It includes the interaction of two coherent plane waves from one laser source in an extended nonlinear medium.

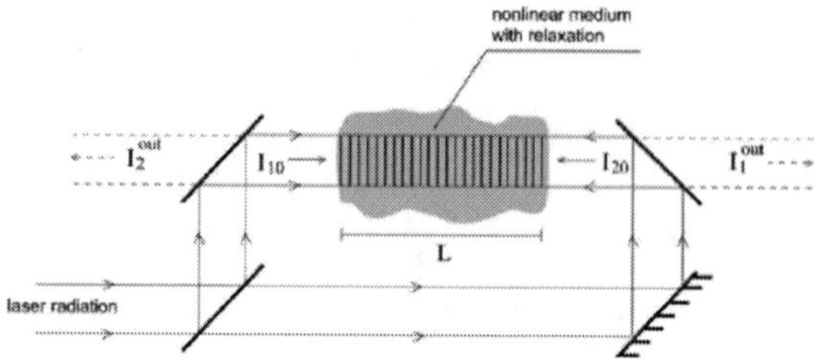

Fig. 4. Wave interaction in dynamical dissipative medium.

The physical mechanism of the nonlinearity is characterized by a physical value Q, for which we can write the evolution equation. The evolution equation for the amplitude of the nonlinearity in the simplest form, which includes the nonlinear gain proportional to the light intensity and the relaxation of the nonlinearity with the time constant τ, is the following:

$$\partial_t Q = \gamma_N (I_m/I_0) - Q/\tau \quad (2)$$

Also there are coupled-wave equations in the nonlinear medium, where the coupling coefficient is described by the function $Q(t,z)$:

$$\partial_z \boldsymbol{E}_1 = Q\,\boldsymbol{E}_2;\ \partial_z \boldsymbol{E}^*{}_2 = Q\,\boldsymbol{E}^*{}_1 \quad (3)$$

In the system (2)-(3) $\boldsymbol{E}_1$ and $\boldsymbol{E}_2$ are slow variable amplitudes (where $E_i(t,z)=E_i(t,z)\ exp[i\ \varphi_i(t,z)]$); ∂_t and ∂_z denote the partial derivative on t and z respectively; $I_0(t,z) = |\boldsymbol{E}_1|^2+|\boldsymbol{E}_2|^2$ is the total intensity; the sign * indicate the complex conjugate. The intensity of the interference pattern $I_m(t,z)$ is :

$$I_m = (1/2)\ [\,\boldsymbol{E}_1\boldsymbol{E}^*{}_2 + \boldsymbol{E}^*{}_1\boldsymbol{E}^*{}_2\,] = E_1E_2\ cos(\varphi_1-\varphi_2) \quad (4)$$

The coupling coefficient may be function, which describes changes of dielectric permittivity. Also the system (2)-(3) implies that there is the spatial shift between the matter grating and the light grating on the quarter of the period in the direction of z-axis [3].

The system (2)-(3) of three equations can be rewrote in the form of one nonlinear equation of the second order for the function Q:

$$\partial_t \partial_z Q + \partial_z Q/\tau - \gamma_N Q + (4/\gamma_N)\, Q\, [\partial_t Q + Q/\tau]^2 = 0 \qquad (5)$$

Further procedure permits to transform this nonlinear equation to the NLS equation. It would be convenient to use the presentation of the function Q in the following form: $Q=(\boldsymbol{Q}+\boldsymbol{Q}^*)/2$, where we introduce an auxiliary function $\boldsymbol{Q}$ which is a complex-valued function, and $\boldsymbol{Q}^*$ is its complex conjugate. Then we apply the reductive perturbation technique, in which we expand the function in the curved coordinates on a small parameter ε [6]:

$$\boldsymbol{Q}\,(t,z) = \varepsilon \boldsymbol{F}_0 + \varepsilon^2 \boldsymbol{F}_1 + \varepsilon^3 \boldsymbol{F}_2 + \dots \qquad (6)$$

$$T=\varepsilon T_0 + \varepsilon^2 T_1 + \varepsilon^3 T_2 + \dots;\; Z=\varepsilon Z_0 + \varepsilon^2 Z_1 + \varepsilon^3 Z_2 + \dots \qquad (7)$$

$$\boldsymbol{F}_j = \boldsymbol{F}_j(T_0, T_1, T_2, \dots, Z_0, Z_1, Z_2, \dots),\; j=0,1,2,3 \dots$$

We consider the nonlinear equations that appears at different orders of ε. Note, the reductive perturbation technique is applied to transform the sine-Gordon equation into the Nonlinear Schrodinger equation [6].

For the first order of ε we obtain the wave equation for function $\boldsymbol{F}_0$, for which we can write solution in a form of a plane wave relative to the coordinates T_0 and Z_0, but with the amplitude $\boldsymbol{A}$ depending on all others coordinates:

$$\varepsilon:\; L(\boldsymbol{F}_0)=0;\; L = \partial_{T0}\, \partial_{Z0} + (1/\tau)\, \partial_{Z0} - \gamma_N \qquad (8)$$

$$\boldsymbol{F}_0 = \boldsymbol{A}(T_1, T_2, \dots, Z_1, Z_2, \dots)\, exp\{i[(\gamma_N/q)T_0 - qZ_0] - T_0/\tau\} \qquad (9)$$

where L is the operator, and q is the wave-vector of the grating. For the second order of ε we obtain the following equation with two operators L and M:

$$\varepsilon^2:\; L(\boldsymbol{F}_1) + M(\boldsymbol{F}_0)=0;\; M=\partial_{T0}\, \partial_{Z1} + \partial_{T1}\, \partial_{Z0} + (1/\tau)\, \partial_{Z1} \qquad (10)$$

from which we can assume the function $\boldsymbol{F}_1$ is equaled to zero ($\boldsymbol{F}_1=0$) to avoid the appearance of secular terms, and for the function of $\boldsymbol{A}$ we obtain the following PDE:

$$[\partial_{T1} - (\gamma_N/q^2)\, \partial_{Z1}]\, \boldsymbol{A} = 0 \qquad (11)$$

Finally, for the case of the third order of ε, we obtain the equation with two new operators N and R:

$$\varepsilon^3:\; L(\boldsymbol{F}_2) + M(\boldsymbol{F}_1) + N(\boldsymbol{F}_0) + (4/\gamma_N)\, R(\boldsymbol{F}_0)=0 \qquad (12)$$

$$N=\partial_{T0}\, \partial_{Z2} + \partial_{T2}\, \partial_{Z0} + (1/\tau)\, \partial_{Z2} + \partial_{T1}\, \partial_{Z1} \qquad (13)$$

$$R(\boldsymbol{F}_0)=(1/4)[\, \boldsymbol{F}_0 + \boldsymbol{F}^*_0][\partial_{T0}(\boldsymbol{F}_0 + \boldsymbol{F}^*_0) + (\boldsymbol{F}_0 + \boldsymbol{F}^*_0)/\tau] \qquad (14)$$

from which we can assume the function $\boldsymbol{F}_2$ is equaled to be zero ($\boldsymbol{F}_2=0$) for the same reason as we did for the function $\boldsymbol{F}_1$: to avoid the secular terms.

We can apply new variables ζ and η determined by the following way:

$$\zeta = T_1 - \upsilon_g Z_1;\; \eta = T_2 - \upsilon_g Z_2;\; \upsilon_g = \gamma_N / q_2 \qquad (14)$$

We obtain the NLS equation with exponential damping on the time in the nonlinear term:

$$i\, \partial_\eta A + (\gamma_N/q^3)\, \partial_\zeta A^2 - (\gamma_N/q^3)\, |A|^2 A\, exp(-\, 2\, T_0/\tau) \qquad (15)$$

This equation represents the interest to further investigation in dependence on their parameters, as well as its connection with the experimental conditions.

4 Conclusion

We have shown that the nonlinear problem of the dynamic holography, which consists of the self-diffraction of waves in a dynamical nonlinear medium, essentially includes dissipative soliton solution, which we called the grid dissipative soliton. This soliton describes a localized profile of the grating amplitude. The distribution of the light intensity in the interference pattern will have the same profile. In this case the deal is with the interference pattern, which can be either standing or running. The same are the matter wave in a form of a sinusoidal phase grating.

We have derived the damped NLS equation for the case of the self-diffraction of coherent waves in the reflection geometry in a dynamic nonlinear medium with time relaxation. The obtained NLS equation has the steady state solution, which represents a dark soliton.

The grid dissipative soliton is defined by the intensity ratio of input waves, and in its own turn determines the intensities of output waves. The dynamical case predicts many effects, which may be implemented in Bragg fibers in the condition when to crate the grating with non-homogeneous amplitude, but with a certain determined profile calculated from the dynamical task. Among such effects are a giant amplification of a seed pulse, amplification of individual pulses in a series, and the pulse retardation. We show that for modelling and prediction of new effects the considered dynamical problem is an effective tool. There are many direction to expand the dynamical task, in particularly, to consider the dynamic holography with not coherent waves.

References

1. J. Canning, Laser and Photonics Reviews **2** (4), 275 (2008)
2. S. Bugaychuk, R. Conte, Phys. Rev. E **86**, 026603 (2012)
3. S. Bugaychuk, R. Conte, Phys. Rev. E **80**, 066603 (2009)
4. N. Akhmediev, A. Ankiew icz (eds.), *Dissipative solitons*. Lecture notes in physics **661** (Springer, Berlin, 2005)
5. S. Bugaychuk, L. Kovacs, G. Mandula, K. Poglar, R.A. Rupp, Phys. Rev. E **67**, 046603 (2003).
6. T. Dauxois, M. Peyrard, *Physics of solitons* (Cambridge University Press, Cambridge, 2006)

MATEC Web of Conferences **83**, 08003 (2016) DOI: 10.1051/matecconf/20168308003
CSNDD 2016

Supercontinuum generation in optimized photonic crystal fiber at 1.3 μm for optical coherence tomography

M. L.Ferhat[1] and L. Cherbi[1]

[1]Instrumentation Laboratory (LINS), University of Sciences and Technology Houari Boumediene, Algiers, Algeria

Abstract. In this paper, we have designed a high nonlinear photonic crystal fiber (HN-PCF) based on square-lattice geometry with the zero dispersion wavelength (ZDW) around 1300 nm. The exploitation of different nonlinear mechanisms in the pulse propagation allows supercontinuum generation, which is used to enhance the axial resolution of the optical coherence tomography (OCT) systems. First mechanism demonstrated is the soliton self-compression, we came up to realize pulse compression of 28.4 fs around 1300 nm by the generation of solitons of different orders to obtain ultrashort pulses of about 4 fs pulses in a PCF length of 66cm, then, we improved the pulse compression until 1.2 fs in a PCF length of 26 cm.The exploitation of the interplay between many nonlinear effects as self-phase modulation, intrapulse Raman scattering and self-steepening as second mechanism allows a generation of supercontinuum with a spectral bandwith of SBW=260 nm. The obtained spectral bandwidth could contribute to enhance the OCT imaging axial resolution which can be evaluated to 2.8 μm in air, working at 1.3 μm center wavelength which is widely used in several fields.

1 introduction

Optical coherence tomography (OCT) is a new technology for noninvasive cross-sectional imaging of tissue structure in biological system. Since OCT is based on low-coherence interferometry, the longitudinal image resolution, Δz, is inversely proportional to the bandwidth, $\Delta\lambda$, and proportional to the square of the center wavelength λ_0 [1] :

$$\Delta z = \frac{2\ln(2)}{\pi}\left(\frac{\lambda_0^2}{\Delta\lambda}\right) \quad (1)$$

The spectral region from 1.2 to 1.8 μm is of particular interest for OCT since it provides a high penetration deph in biological tissue. High resolution in this spectral region requires extremely broad bandwidths because of the $\lambda^2/\Delta\lambda$ dependence of the longitudinal resolution. Standard OCT use longitudinal resolution of 10-15 μm. An axial resolution of less than 5 μm at 1.3 μm can be achieved by using a broad bandwidth light source. High nonlinearity air-silica microstructure fibers can generate an extremely broad supercontinuum by the use of femtosecond laser sources.

The supercontinuum (SC) light using photonic crystal fibers (PCF) can be generated by different methods as pulse compression using solitons effects[2], the self phase modulation along with other nonlinear effects such as stimulated Raman scattering and four wave mixing when occurring simultaneously inside optical fibers [3].

In this work, a new design of high nonlinear photonic crystal fiber (HN-PCF) based on square-lattice geometry has been presented for supercontinuum generation, we demonstrate the supercontinuum generation using several mechanisms in anomalous dispersion regime. First, a pulse compression using soliton effects around 1300 nm is achieved [4], and 1.2 fs pulses are generated in only 26 cm of fiber, a supercontinuum of spectral bandwidth SBW $\approx$ 100 nm is generated. The interplay between several effects such as self phase modulation, self-steepening and intrapulse Raman scattering is used as a second mechanism for supercontinuum generation, it allowed a spectral bandwidth SBW=260 nm in only 1.89 mm length of PCF.

2 design and optimization procedures of hn-pcf

The proposed geometry shown in Fig. 1 is composed of six rings of air holes with silica as background of the core. d_i is the air hole diameter of the ring i(1-6) of the PCF and $d_{1\text{-}6}$=.637 μm, pitch (Λ=0.65 μm) is the distance between two successive air holes. Full-vector Finite-Element Method (FEM) with a Perfectly Matched Layer (PML) as bounding conditions has been used to investigate the optical characteristics of the proposed PCF.

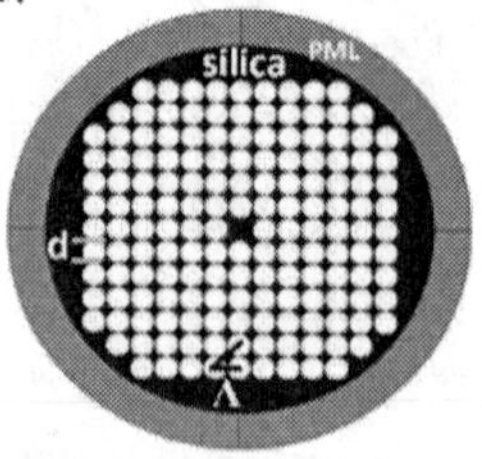

Fig. 1. The proposed PCF

The PML technique allows to strongly absorb outgoing waves from the interior of a computational region without reflecting them back into the interior [6].
The distribution of field and the effective index of the fundamental mode HE_{11} are calculated, then, the optical parameters such as the dispersion D and the group velocity dispersion (GVD) can be calculated by [3]:

$$D = -\frac{\lambda}{c}\frac{d^2 Re[n_{eff}]}{d\lambda^2} = \frac{2\pi c}{\lambda^2}\beta_2 \quad (2)$$

$Re[n_{eff}]$ is the real part of the effective index, c is the light speed and β_2 is the GVD. The bleu curve of the Fig. 2 illustrates the dispersion values of the regular PCF. By reducing the hole diameter of the first ring to d_1=0.34 µm and keeping the holes diameters of the other rings equal to 0.637 µm, we minimize the dispersion value to 4.67 [ps/nm.km] (red curve).

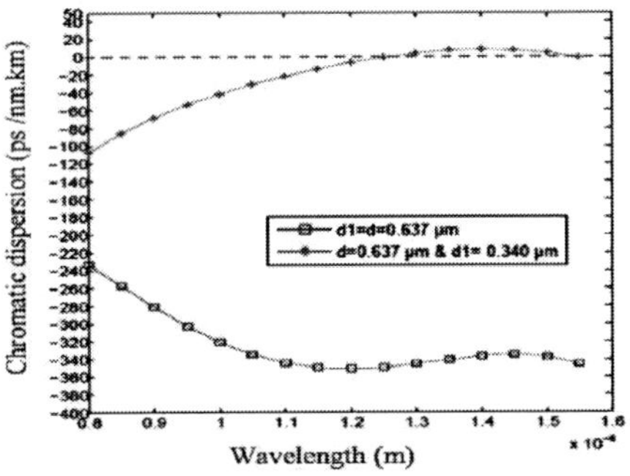

Fig. 2. Dispersion curves of regular PCF (bleu curve), and optimized air hole diameter of the first ring -PCF (red curve).

In order to acquire ultra-broadband SC in the proposed PCF, the dispersion curve and the geometry of the fiber are optimized through several steps. The first step illustrated the influence of the variation of the air hole diameters in the first ring. In the second step, we kept the optimized air holes diameters of the first ring and we optimized the distance between the air holes of the different rings (pitch).

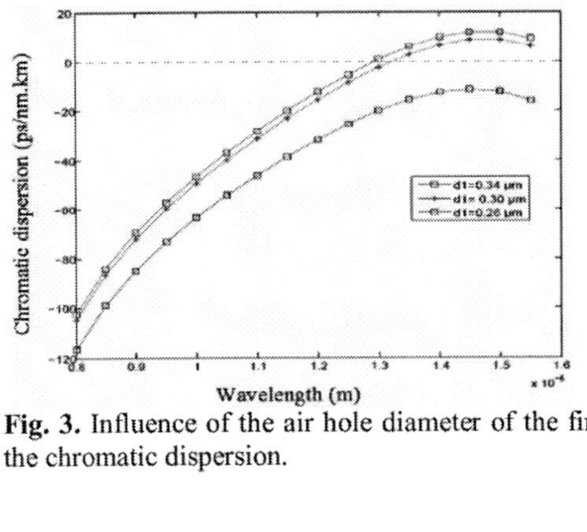

Fig. 3. Influence of the air hole diameter of the first ring on the chromatic dispersion.

In the last step, we modified the air holes diameters of the other rings by keeping the optimized d_1 and pitch. Fig. 3 illustrates the effect of varying the air hole diameter of the first ring, the dispersion value is minimizing by decreasing the value of (d_1). For d_1=0.26 µm, $d_{(2\text{-}6)}$=0.637 µm and Λ=0.65 µm the dispersion value is decreasing to be 2.34 [ps/nm.km].

Figure 4 shows the variation of the dispersion with the pitch value by maintaining the optimized value of d_1 and by keeping the other holes diameters at $d_{(2\text{-}6)}$=0.637 µm, the variation indicates that by decreasing the value of the pitch, the Zero Dispersion Wavelength (ZDW) is shifted to the blue side of the wavelength and the dispersion around ZDW=1300 nm in our case is minimized.

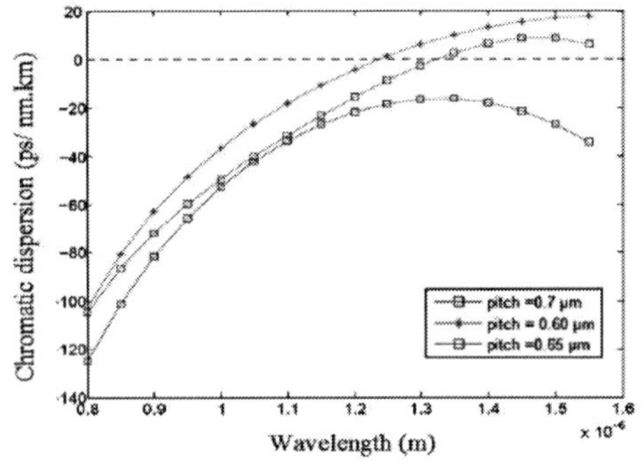

Fig. 4. Influence of the distance between two successive air holes on the chromatic dispersion

The last rings are used to locate the dispersion to a desirable position and there are kept at a higher air-filling fraction d/Λ from 0.4 to 0.98 to confine light better within the core and to reduce confinement loss as much as possible.

With the optimized parameters Λ= 0.65 µm, d_1/Λ= 0.4, d_2/Λ=0.9, d_3/Λ=0.89, d_4/Λ=d_5/Λ= d_6/Λ= 0.98, we performed the anomalous dispersion and the GVD to be 1.0677 [ps/nm.km] and - 0.9572 [ps^2/km] respectively (Fig. 5).

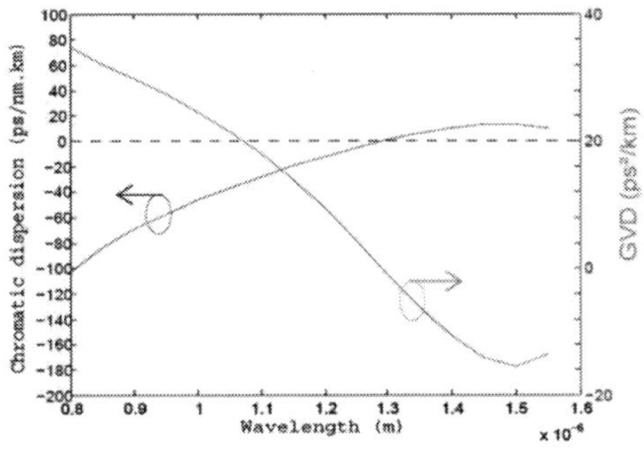

Fig. 5. Chromatic dispersion and group velocity dispersion of the optimized HNL-PCF.

The nonlinearity coefficient is primarily governed by two parameters: the nonlinear index of the material (n_2) and the effective area (Aeff). The nonlineaty coefficient is calculated using the relation [3] :

$$\gamma = (\frac{2\pi}{\lambda})(\frac{n_2}{A_{eff}}) \tag{3}$$

where n_2 is the nonlinear refractive index of material, λ is the pump wavelength and A_{eff} is the effective area of the fundamental mode, and it's calculated using the following equation:

$$A_{eff} = \frac{(\iint_{-\infty}^{\infty} |E(x,y)|^2 dx\, dy)^2}{\iint_{-\infty}^{\infty} |E(x,y)|^4 dx\, dy} \tag{4}$$

The numerical value of the nonlinear refractive index n_2 of pure silica is reported to be 2.2×10^{-20} m^2/W. The proposed design of HNL-PCF permits us to reduce the effective area; therefore, the nonlinear coefficient γ increases significantly to reach 55.45 $[W^{-1}.km^{-1}]$ as mentioned in Fig. 6.

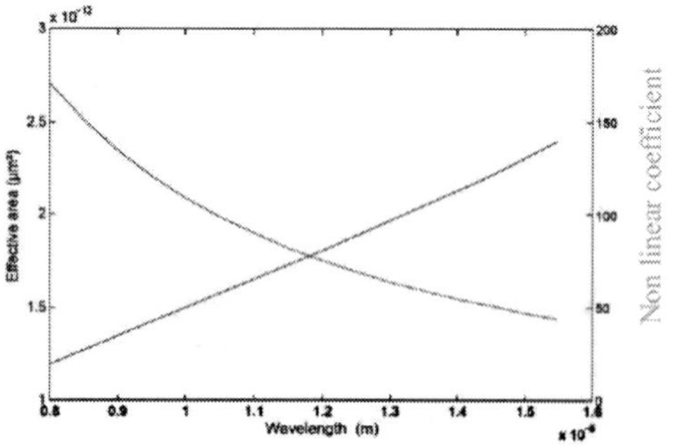

Fig. 6. Effective area and nonlinear coefficient of the optimized HNL-PCF

Figure 6 shows that effective area increases and the nonlinear coefficient decreases with wavelength, this can be explained by the dependence of the mode field diameter (MFD) on wavelength i.e. the more MFD is important the more mode profile is extended outside the core and the nonlinear effects are less important.

3 Supercontinuum generation and mechanisms

The numerical simulation of SC spectrum has been performed by solving the generalized nonlinear Schrodinger equation (GNLSE) using split-step Fourier method (SSFM) [3]. The first mechanism used for generating SC is based on soliton compression where the GNLSE takes into account the contributions of the linear effects (i.e. attenuation, chromatic dispersion and high order dispersions), and the nonlinear effects as kerr effect while in the second mechanism of SC generation, self-phase modulation (SPM), self-steepening and intrapulse Raman scattering are used. The GNLSE is given in following relation [3]:

$$\frac{\partial A}{\partial z} + \frac{\alpha}{2}A - \left(\sum_{n\geq 2} \beta_n \frac{i^{n+1}}{n!}\frac{\partial^n A}{\partial T^n}\right) = i\gamma\left(1 + \frac{1}{w_0}\frac{\partial}{\partial T}\right).[A(z,T)\int_{-\infty}^{\infty} R(T')|A(z,T-T')|^2 dT'] \tag{5}$$

where T is the time, A(z,t) is the complex amplitude of the optical field, α is the attenuation coefficient of the fiber, β_n(n=2 to 4) is the n-th order of propagation constant, γ is the nonlinear coefficient.
The nonlinear response function, R(T), which includes instantaneous electronic as well as delayed Raman conributions, is given by :

$$R(T) = (1 - f_r)\delta(T) + f_r h_r(T) \tag{6}$$

where $f_r = 0.18$, and the Raman response function is given by [3]:

$$h_R(t) = (\tau_1^{-2} + \tau_2^{-2})\tau_1 \exp(-t/\tau_2)\sin(\frac{t}{\tau_1}) \tag{7}$$

In our study, we consider the injected pulses having secant hyperbolic field profile emerging from a femtosecond laser:

$$U(0,T) = \sqrt{P_0}\, sech\left(\frac{T}{T0}\right) \tag{8}$$

where P_0 is the peak power, T_0 the temporal width of the pulse and $T_0 = T_{FWHM}/1.763$.

3.1 Soliton self-compression

In the anomalous dispersion region, the balance between the group velocity dispersion (GVD) and the nonlinear effect Self Phase Modulation (SPM) is responsible for the formation of solitons, whose order N increases with the pulse intensity.

$$N = (\sqrt{\frac{L_D}{L_{NL}}}) \tag{9}$$

with $L_D = \frac{T_0^2}{\beta_2}$ and $L_{NL} = \frac{1}{\gamma P_0}$.

In this work, a pulse with a full width at half maximum (FWHM) of 28.4 fs is compressed around 1300 nm, by generation of solitons of different orders (N= 2, N=3 , N=4) and varying the input power (table I).
The obtained pulse compression is characterized by the compression factor Fc defined by:

DOI: 10.1051/matecconf/20168308003

$$F_c = \frac{T_0}{T_{comp}} \tag{10}$$

where $T_{comp} = T_0 - T_{diff}$ is the width of the compressed pulse and T_{diff} is the difference between the width of the initial pulse T_0 and that of the compressed pulse.
The quality of compression is evaluated by the quality factor Qc that is given by the following relationship:

$$Q_c = \frac{P_{comp}}{F_c} \tag{11}$$

Pcomp is the peak power of the compressed pulse normalized to the input pulse.

Table 1. parameters of compression of N order-soliton

P_0(W)	N	L=18cm	L=26cm	L=66cm
86	2	T_{comp}=26fs F_c=1.09 Q_c=0.87 T_{diff}=2.4 fs	T_{comp}=17fs F_c=1.67 Q_c=0.55 T_{diff}=11.4fs	T_{comp}=4fs F_c=7.1 Q_c=0.29 T_{diff}=24.4fs
193	3	T_{comp}=9.1fs F_c=3.12 Q_c=0.34 T_{diff}=19.3fs	T_{comp}=1.2fs F_c=23.6 Q_c=0.19 T_{diff}=2.4 fs	Soliton fission region
342	4	T_{comp}=6.2fs F_c=4.58 Q_c=0.3 T_{diff}=22.2fs	Soliton fission region	Soliton fission region

First, we generated the second order-soliton in the proposed HN-PCF to obtain 4 fs pulses from a compression of 28.4 fs in a PCF length of 66 cm Fig. 7.

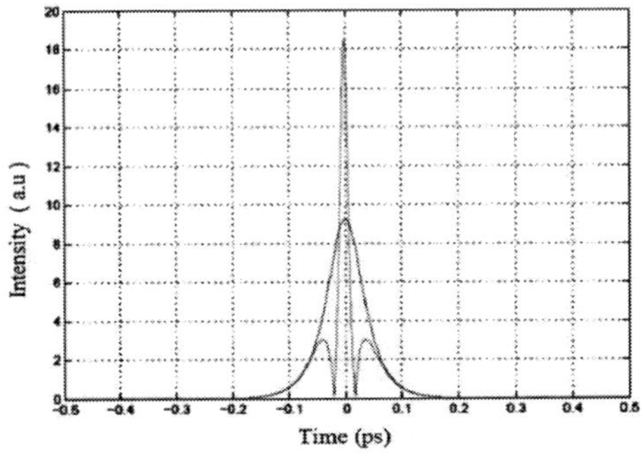

Fig. 7. Soliton self compression in 66 cm of HN-PCF: Input pulse of 28.4 fs (blue), and a compressed pulse of 4 fs (red).

We improved the pulse compression, by the generation of the third order soliton until 1.2 fs in a PCF length of 26 cm (Fig. 8).

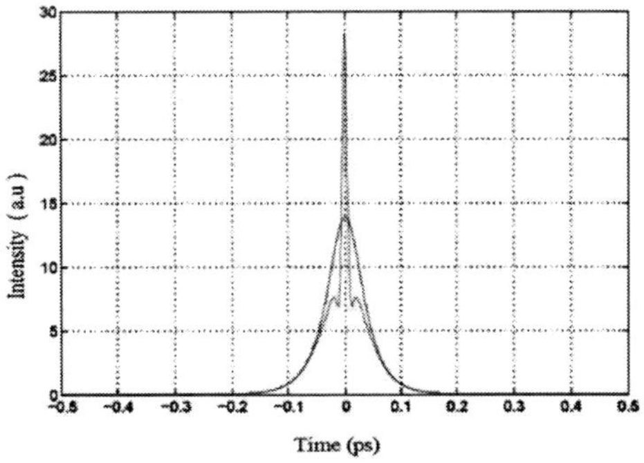

Fig. 8. Soliton self compression: input pulse of 28.4fs (blue), and 1.2 fs pulse compressed (red).

Figure 9 depicts the spectral evolution and approximately 50 nm supercontinuum is generated in 66 cm of PCF at 1300 nm pump wavelength.

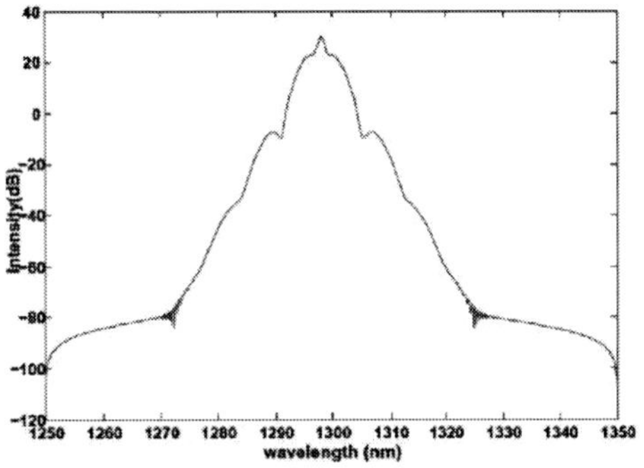

Fig. 9. Spectral evolution of the input pulse in 66 cm of PCF at 1300 nm.

3.2 Self-phase modulation, self-steepening and intrapulse Raman scattering

We consider the propagation of optical pulses as short as ≈ 50 fs. Figure 10 illustrates the effect of varying the input pulse power on the output spectra in 1.89 mm of HNL-PCF; we can see that output spectra are broadened as the incident pulse power increases. In fact, by increasing the input pulse power, nonlinear effects dominate the dispersion effects. This leads to spectral broadening of the propagated pulse; induced by the interplay between many nonlinear effects such self phase modulation, self-steepening and intrapulse Raman scattering [3]. Besides of spectral broadening due to Self-Phase Modulation (SPM), Self-Steepening leads to an asymmetry in the SPM-broadened spectra of ultrashort pulses, it creates an optical shock on the trailing edge of the pulse. As a result, the trailing edge becomes steeper along the distance of propagation (Fig. 11).

DOI: 10.1051/matecconf/20168308003

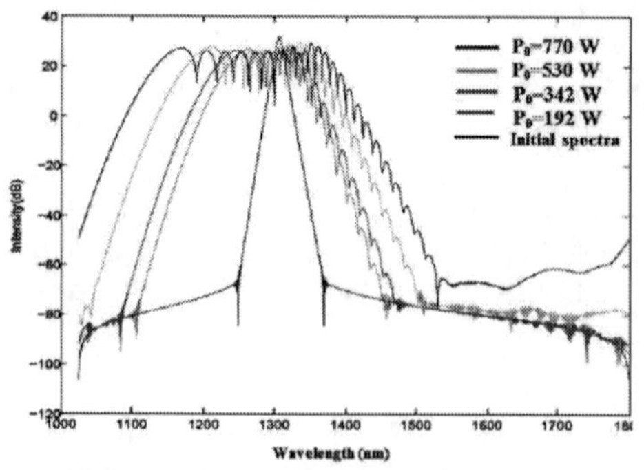

Fig. 10. Supercontinuum generation in 1.89 mm of HNL-PCF for different input pulse power.

This phenomenon is due to the intensity dependence of the group velocity that results in the peak of the pulse moving slower than the wings. A steeper trailing edge of the pulse implies larger spectral broadening on the blue sides as SPM generates blue components near the trailing edge (Fig. 11).

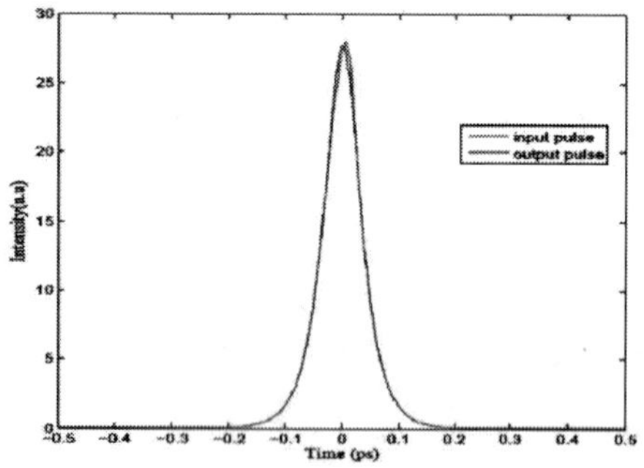

Fig. 11. Temporal evolution of the input pulse propagation inside the optimized HNL-PCF.

By pumping with pulses of T_{FWHM} = 50 fs, the Raman-gain amplify the low frequency components of the pulse, as a result, the pulse spectrum shifts toward the low frequency side (Near Infrared) as the pulse propagates inside the fiber. The introduction of this different effects in our work allows to generate a SC of Spectral Bandwidth SBW=260 nm at -20dB from the peak point at 1,3 μm wavelength using only 1.89 mm length of PCF and by pumping with 770 W (Fig. 12).

The obtained spectral bandwidth could contribute to enhance the OCT imaging axial resolution which can be evaluated to 2.8 μm in air, working at 1.3 μm center wavelength which is widely used in several fields.

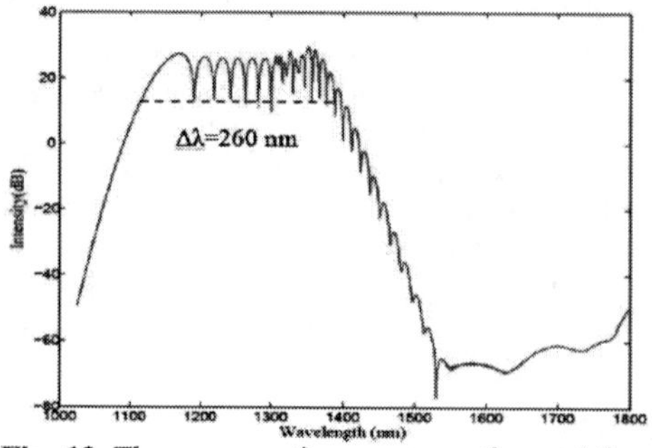

Fig. 12. The supercontinuum generation at 1.3 μm using 1.89 mm of HNL-PCF.

4 Conclusion

A highly nonlinear photonic crystal fiber with optimized dispersion around the pump wavelength is designed. It is demonstrated that the desired curve of dispersion can be performed by adjusting the air hole filling ratio of the composed fiber rings. Using two kinds of nonlinear mechanisms, we have generated two supercontinuums, first, using the soliton self-compression which allows a compression of the input pulses, a second solution is the use of the interplay between many nonlinear effets such as self-phase modulation, intrapulse Raman scattering and self-steepening to generate a second supercontinuum of SBW=260 nm, that can enhance the axial resolution of an optical coherence tomography system.

References

[1] M. A. Hossain, Y. Namihira, S. M. A. Razzak , M. A. Islam, J. Liu, S. F.Kaijage, and Y. Hirako," Design of all normal dispersion highly nonlinear photonic crystal fibers for supercontinuum light sources: Applications to optical coherence tomography systems", Opt. Optics and Laser Technology, Volum e 44, Issue 4, p. 976-980 (2012).

[2] L. Cherbi, N. Lamhene, F. Boukhelkhal, Anjan Biswas," Ultra-short pulse compression at 1065 nm in nonlinear photonic crystal fiber", International Journal for Light and Electron Optics 125 (2014).

[3] Agrawal, G. P. "Nonlinear Fiber Optics", 5th ed. Academic Press, 2013M.

[4] Ferhat, L. Cherbi Soliton self- compression in photonic crystal fiber at 1.3 μm, International Conference on Automatic Control, Telecommunications and Signals » ICATS 2015.

[5] J. M. Dudley, G. Genty, and S. Coen, "Supercontinuum generation in photonic crystal fiber," Rev. Mod. Phys. 78(4), 1135–1184 (2006).

[6] J.P. Berenger "A perfectly matched layer for the absorption of electromagnetic waves", *J. Comput. Phys*, **114**, pp 185-200 (1994).

Functional asynchronous networks: Factorization of dynamics and function

Christian Bick[1,a] and Michael Field[2,b]

[1] Department of Mathematics, University of Exeter, Exeter EX4 4QF, UK
[2] Department of Mathematics, Imperial College, SW7 2AZ, UK

Abstract. In this note we describe the theory of functional asynchronous networks and one of the main results, the Modularization of Dynamics Theorem, which for a large class of functional asynchronous networks gives a factorization of dynamics in terms of constituent subnetworks. For these networks we can give a complete description of the network function in terms of the function of the events comprising the network and thereby answer a question originally raised by Alon in the context of biological networks.

1 Introduction

Kastan & Alon [9] identify and describe the configurations of relatively simple and small subnetworks that occur more frequently in biological networks than would be the case if the network were random. They refer to these subnetworks as *network motifs*. Later, in his 2007 book on systems biology [1], Alon makes the following comment

> *"Ideally, we would like to understand the dynamics of the entire network based on the dynamics of the individual building blocks."* (Alon [1, page 27].)

The underlying premise behind this comment is that a modular, or engineering, approach to network dynamics is feasible. Identify building blocks, connect together to form networks and then describe dynamical properties of the resulting network in terms of the dynamics of its components. This approach works well in linear systems theory, where a superposition principle holds, or in, for example, the study of synchronization in weakly coupled systems of nonlinear approximately identical oscillators where network dynamics can be closely related to the dynamics of individual nodes (oscillators).

However, from the perspective of a mathematician or physicist, Alon's comment seems unhelpful for the study of dynamics of heterogenous networks modelled by systems of nonlinear ODEs. This is a well-known issue in complex systems [10]. Yet engineers do couple gadgets together to make more complex systems and so it is natural to ask if it is possible to reconcile these viewpoints.

In this note, we outline some of the basic ideas involved in an approach to network dynamics based on what we call *asynchronous networks*. We allow for features seen in networks from technology, engineering, and biology (for example, neuroscience or gene transcription networks). Network dynamics can involve a mix of distributed and decentralized control, adaptivity, event driven dynamics, switching, varying network topology and hybrid dynamics. Typically network dynamics will be piecewise smooth, time-irreversible, nodes may stop and later restart, and there may be no intrinsic global time. Significantly, many of these networks have a function: transportation networks bring people and goods from one point to another, neural networks may perform pattern recognition or computation, etc. Our goal is to address Alon's comment in the context of functional asynchronous networks. Specifically, we describe a factorization of dynamics theorem where it is possible to describe the function of a large class of functional asynchronous networks in terms of the function of constituent subnetworks. As this article is only intended to be an introduction, we work always with the simplest examples and models and omit technical details. We refer the reader to [3,4] for greater detail, generality and proofs.

2 Examples and properties of asynchronous networks

We briefly describe some characteristic examples of asynchronous networks (we refer to [3, §3] for more detail).

2.1 Threaded & parallel computation

In threaded or parallel computation, computation is broken into blocks or 'threads' which are then computed *independently* of each other at a speed that depends on the clock rates of the individual processors. As the computation proceeds, threads may need to exchange information with other threads. This process involves stopping and synchronizing (updating) the thread states. Each thread may have to stop and wait for other threads to complete their computations before it can continue with its own computation. Each thread has its own clock (determined by its associated processor). If threads run on separate processors, threads will be unaware of the clock times of other threads except during the stopping and synchronization events.

2.2 Production and transport networks

We give a simple detailed example of a transport network in section 4. In production networks, parts, compounds,

[a] e-mail: `c.bick@exeter.ac.uk`
[b] e-mail: `mikefield@gmail.com`

etc. are repeatedly built and combined as part of a production process leading to the desired item (for example, a car or protein). In particular, there is variation in both connection structure, typically intermittent, and in the set of nodes (nodes may be combined or decomposed).

2.3 Power grid models

A power grid consists of a network of various types of generators and loads connected by transmission lines. A microgrid is a local network, capable of existing independently of the main power grid, and typically powered by renewable energy sources (for example, solar or wind power). Critical questions involve the stability of the power grid in case of loss of a transmission line or generator (variation in network structure), and control and stability issues related to combining and separation (islanding) of a large set of microgrids from the main power grid.

2.4 Thresholds, spiking models and adaptation

Many mathematical models from engineering and biology incorporate thresholds. For networks, when a node attains a threshold, there are often changes (addition, deletion, weights) in connections to another nodes. For networks of neurons, reaching a threshold can result in a neuron firing (spiking) and short term connections to other neurons (for transmission of the spike). For learning mechanisms, such as Spike-Timing Dependent Plasticity (STDP) [7] relative timings (the order of firing) are crucial [8,5,11] and so each neuron, or connection between a pair of neurons, comes with a 'local clock' that governs the adaptation in STDP. In general, networks with thresholds, spiking and adaptation provide characteristic examples of asynchronous networks where dynamics is piecewise smooth and hybrid – a mix of continuous and discrete dynamics.

2.5 Properties of asynchronous networks

We summarize some of the characteristic features of asynchronous networks as revealed in the examples above.

1. Variable network structure and dependencies between nodes. Changes depend on the state of the system or are given by a stochastic process.
2. Synchronization events associated with stopping or waiting states of nodes.
3. Order of events may depend on the initialization of the system or stochastic effects.
4. Dynamics is only piecewise smooth and there may be a mix of continuous and discrete dynamics.
5. Aspects involving function, adaptation and control.
6. Evolution only defined for forward time – systems are not time reversible.

3 Asynchronous networks: formalism

We use the notational conventions that $\mathbf{k} = \{1, \ldots, k\}$ and $\mathbf{k}^\bullet = \mathbf{k} \cup \{0\}$, $k \in \mathbb{N}$. Let $\mathbb{R}_+ = \{x \in \mathbb{R} \mid x > 0\}$.

Assume given a network with k nodes, $N_1, \ldots, N_k$. Suppose that N_i has associated phase space M_i, $i \in \mathbf{k}$. Set $\mathbf{M} = \prod_{i \in \mathbf{k}} M_i$ – the network phase space. A vector field $\mathbf{f}$ on $\mathbf{M}$ is a *network vector field*.

Stopping, waiting, and synchronization are a feature of asynchronous networks. If nodes are stopped or partially stopped, node dynamics will be constrained to subsets of node phase space. We codify this situation by introducing a *constraining node* N_0 that, when connected to N_i, implies that dynamics on N_i is constrained. Set $\mathcal{N} = \{N_0, \ldots, N_k\}$.

3.1 Connection structures; admissible vector fields

Interactions between distinct nodes in the network are given by the network graph. Connections $N_j \to N_i$ encode *dependencies* if $i, j \in \mathbf{k}$, and *constraints* if $j = 0$, $i \in \mathbf{k}$.

A *connection structure* α is a directed network graph on the nodes $\mathcal{N}$ such that for all $i \in \mathbf{k}$, $j \in \mathbf{k}^\bullet$, $i \neq j$, there is at most one directed connection $N_j \to N_i$. We do not allow self-loops or connections to the constraining node N_0.

An *α-admissible* vector field $\mathbf{f}^\alpha$ is a network vector field with dependencies given by α. If $N_j \to N_i \notin \alpha$, $i \neq j$, then $\mathbf{f}^\alpha$ does not depend on $x_j \in M_j$ (and conversely, see [3, §2]).

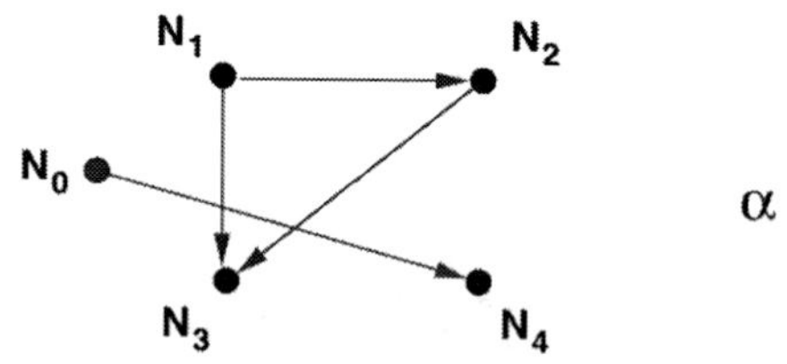

Fig. 1. Construction structure on $\{N_0, N_1, N_2, N_3, N_4\}$.

Referring to figure 1, suppose $\mathbf{f}^\alpha = (f_1^\alpha, \ldots, f_4^\alpha)$ is α-admissible. For $\mathbf{X} = (\mathbf{x}_1, \mathbf{x}_2, \mathbf{x}_3, \mathbf{x}_4) \in \mathbf{M}$, we have

$$f_1^\alpha(\mathbf{X}) = f_1(\mathbf{x}_1),\ f_2^\alpha(\mathbf{X}) = f_2(\mathbf{x}_2; \mathbf{x}_1)$$

$$f_3^\alpha(\mathbf{X}) = f_3(\mathbf{x}_3; \mathbf{x}_1, \mathbf{x}_2),\ f_4^\alpha(\mathbf{X}) = 0,$$

where here we have assumed that the constraint $N_0 \to N_4$ results in N_4 being stopped.

A *generalized connection structure* $\mathcal{A}$ is a (nonempty) set of connection structures on $\mathcal{N}$.

An $\mathcal{A}$-structure $\mathcal{F}$ is a set $\mathcal{F} = \{\mathbf{f}^\alpha \mid \alpha \in \mathcal{A}\}$ of network vector fields such that each $\mathbf{f}^\alpha \in \mathcal{F}$ is α-admissible.

3.2 The event map and asynchronous networks

Suppose that $\mathcal{A}$ is a generalized connection structure and $\mathcal{F}$ is an $\mathcal{A}$-structure.

Interactions between nodes in asynchronous networks may be state dependent or change over time (stochastically). Here we only consider state dependence.

We handle interactions and constraints using an *event map* $\mathcal{E} : \mathbf{M} \to \mathcal{A}$.

The quadruple $\mathfrak{N} = (\mathcal{N}, \mathcal{A}, \mathcal{F}, \mathcal{E})$ defines an *asynchronous network*. Dynamics on $\mathfrak{N}$ is given by the state dependent network vector field $\mathcal{F}$ defined by

$$\mathbf{F}(\mathbf{X}) = \mathbf{f}^{\mathcal{E}(\mathbf{X})}(\mathbf{X}),\ \mathbf{X} \in \mathbf{M}. \tag{3.1}$$

Remark 1 We refer to [3, §4.7] for the definition of an integral curve for (3.1) and note that the definition we use is different from that used in Filippov systems [6,2]. Under simple conditions on the event map [3, §§4.7, 4.8], it can be shown that if $\mathbf{M}$ is compact then for each $\mathbf{X} \in \mathbf{M}$ there is a unique piecewise smooth integral curve $\Phi_{\mathbf{X}} : [0,\infty) \to \mathbf{M}$ with initial condition $\mathbf{X}$ and corresponding semiflow $\Phi : \mathbf{M} \times \mathbb{R}_+ \to \mathbf{M}$. In general, Φ will not be continuous as a function of $\mathbf{X} \in \mathbf{M}$.

4 A simple transport example

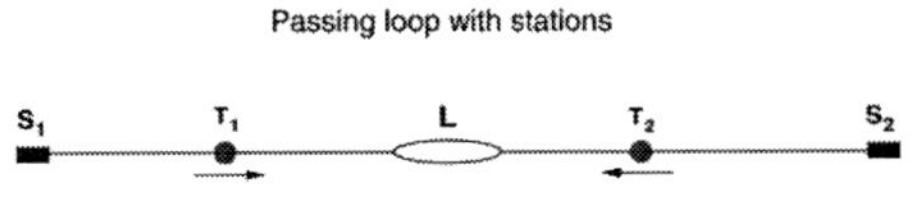

Fig. 2. A single track railway line with a passing loop.

We consider a single track railway line joining stations marked S_1, S_2 in figure 2. Suppose there is a passing loop at L. Trains, marked T_1 and T_2 in figure 2, start from stations S_1 and S_2 respectively and proceed towards the opposite station. There is no central control or communication between the train drivers except when both trains are in the passing loop. We further assume that both drivers are running a nonlinear oscillator. When a train enters the passing loop it stops. When both trains are in the passing loop, the drivers cross couple their nonlinear oscillators. In order for a train to leave the passing loop, two conditions must be satisfied.

1. Both trains must be in the passing loop.
2. The two nonlinear oscillators must be phase synchronized to within ε where $0 < \varepsilon \ll \pi$.

We model this setup using an asynchronous network with two nodes – corresponding to the two trains. The phase space for train T_i will be $M_i = [-a, b] \times \mathbb{T}$, $i \in \mathbf{2}$ ($\mathbb{T} = \mathbb{R}/2\pi\mathbb{Z}$), where the interval $[-a, b]$ models the line joining S_1 to S_2, S_1 has coordinate $-a < 0$, S_2 has coordinate $b > 0$, The passing loop L will be at $x = 0$, and $\mathbb{T}$ will be the phase space for the nonlinear oscillator.

Assume train motion given by V_1, V_2 where $V_1(x) > 0 > V_2(x)$ $x \in [-a, b]$. Note that neither V_1 or V_2 can be zero anywhere on $[-a, b]$ otherwise the trains will never both reach their destination stations in finite time.

We define four connection structures.

$$\alpha_i = N_0 \to N_i,\ i \in \mathbf{2}\ (T_i \text{ stopped})$$
$$\beta = N_0 \to N_1 \leftrightarrow N_2 \leftarrow N_0 (\text{stopped \& cross coupled})$$
$$\emptyset = \text{Empty connection structure}$$

Let $\mathcal{A} = \{\alpha_1, \alpha_2, \beta, \emptyset\}$ be the associated generalized connection structure.

Model the uncoupled oscillator dynamics for train T_i by $\dot{\theta}_i = \omega$, where $\omega > 0$, and the coupled dynamics by the Kuramoto phase oscillator system

$$\dot{\theta}_1 = \omega + \sin(\theta_2 - \theta_1)$$
$$\dot{\theta}_2 = \omega + \sin(\theta_1 - \theta_2)$$

It remains to define the admissible vector fields and event map that give the required dynamics for this example. As admissible vector fields (on $([-a, b] \times \mathbb{T})^2$) we take

$$\mathbf{f}^{\emptyset} = ((V_1, g), (V_2, g)),$$
$$\mathbf{f}^{\alpha_1} = ((0, g), (V_2, g)),\ (T_1 \text{ stopped},\ T_2 \text{ running})$$
$$\mathbf{f}^{\alpha_2} = ((V_1, g), (0, g)),\ (T_2 \text{ stopped},\ T_1 \text{ running})$$
$$\mathbf{f}^{\beta} = ((0, G_1), (0, G_2)),\ (T_1, T_2 \text{ stopped \& cross coupled})$$

We define the event map by

$$\begin{aligned} \mathcal{E}(\mathbf{X}, \boldsymbol{\theta}) &= \alpha_1,\ x_1 = 0, x_2 > 0 \\ &= \alpha_2,\ x_1 < 0, x_2 = 0 \\ &= \beta,\ x_1 = x_2 = 0,\ |\theta_1 - \theta_2| > \varepsilon \\ &= \emptyset,\ \text{otherwise} \end{aligned}$$

Finally, dynamics are given by the network vector field

$$F(\mathbf{X}, \boldsymbol{\theta}) = \mathbf{f}^{\mathcal{E}(\mathbf{X},\boldsymbol{\theta})}(\mathbf{X}, \boldsymbol{\theta}). \tag{4.2}$$

We leave it as an easy exercise for the reader to check that if T_i is initialized at $(x_i(0), \theta_i(0)) \in [-a, b] \times \mathbb{T}$, $i \in \mathbf{2}$, where $x_1(0) \le 0 \le x_2(0)$, then (4.2) has a well defined integral curve $\varphi : \mathbb{R}_+ \to ([-a, b] \times \mathbb{T})^2$, with specified initial condition, that gives the correct dynamics for the passing loop problem.

Remark 2 The passing loop gives an example of a simple *functional* asynchronous network. The function is for the trains to go from one station to the opposite station in finite time. Observe that for this example there is the possibility of a *dynamical deadlock*: if the trains start at the same time and if $\theta_1(0) = \theta_2(0) + \pi$, then the coupled phase oscillators will never phase synchronize – $\theta_1(t) = \theta_2(t) + \pi$ for all $t \in \mathbb{R}_+$ – and so the trains will never exit the passing loop. We refer to [4, §§2,3] for more details on deadlocks in functional asynchronous networks.

5 Functional asynchronous networks

We follow the notational conventions of section 3 and let $\mathfrak{N} = (\mathcal{N}, \mathcal{A}, \mathcal{F}, \mathcal{E})$ denote an asynchronous network. We assume that $\mathfrak{N}$ has associated semiflow

$$\Phi = (\Phi_1, \dots, \Phi_k) : \mathbf{M} \times \mathbb{R}_+ \to \mathbf{M}.$$

Suppose that we are given *initialization* and *termination* sets $\mathbb{I}, \mathbb{F} \subset \mathbf{M}$ where

$$\mathbb{I} = \prod_{i \in \mathbf{k}} \mathbb{I}_i, \quad \mathbb{F} = \prod_{i \in \mathbf{k}} \mathbb{F}_i,$$

Typically, $\mathbb{I}_i, \mathbb{F}_i \subset M_i$ will be closed disjoint hypersurfaces that separate M_i into three connected components, $i \in \mathbf{k}$. That is, $M_i = M_i^- \cup M_i^0 \cup M_i^+$ where $M_i^- \cap M_i^+ = \emptyset$ and

$$M_i^- \cap M_i^0 = \partial M_i^- = \mathbb{I}_i, \quad M_i^0 \cap M_i^+ = \partial M_i^+ = \mathbb{F}_i.$$

We call $\mathbf{N} = (\mathfrak{N}, \mathbb{I}, \mathbb{F})$ a *functional asynchronous network*. The network function is getting from $\mathbb{I}$ to $\mathbb{F}$ and is expressed by the transition and timing functions

$$G_0 : \mathcal{D} \subset \mathbb{I} \to \mathbb{F}, \quad \mathbf{S} : \mathcal{D} \subset \mathbb{I} \to \mathbb{R}_+^k.$$

That is, if $\mathbf{X} \in \mathcal{D}$, then for all $i \in \mathbf{k}$ there exists $S_i \in \mathbb{R}_+^k$ such that

$$\Phi_i(\mathbf{X}, S_i) \in \mathbb{F}_i, \quad \Phi_i(\mathbf{X}, t) \notin \mathbb{F}_i, t < S_i,$$
$$\mathbf{S}(\mathbf{X}) = (S_1, \dots, S_k),$$

Example 1 For the passing loop example discussed in the previous section, we take $\mathbb{I}_1 = M_1^- = M_2^+ = \mathbb{F}_2 = \{-a\} \times \mathbb{T}$, $\mathbb{F}_1 = M_1^+ = M_2^+ = \mathbb{I}_2 = \{b\} \times \mathbb{T}$. In this case, there is the implicit assumption that trains stop when they reach their termination set. Equally well, we could take $M_i = \mathbb{R} \times \mathbb{T}$ so that $M_1^- = (-\infty, a] \times \mathbb{T}$ etc. (see also [4, §3]). Finally, observe that $\mathcal{D} = \{((-a, \theta_1), (b, \theta_2)) \mid |\theta_1 - \theta_2| \neq \pi\}$.

More generally, we allow for general initialization times and define generalized transition and timing functions

$$G : \widehat{\mathcal{D}} \subset \mathbb{I} \times \mathbb{R}_+^k \to \mathbb{F}, \quad \widehat{\mathbf{S}} : \widehat{\mathcal{D}} \subset \mathbb{I} \times \mathbb{R}_+^k \to \mathbb{R}_+^k.$$

We refer to [4, §3.4] for details. For our main result, it is required that the network has a generalized transition and timing functions with $\widehat{\mathcal{D}} = \mathbb{I} \times \mathbb{R}_+^k$.

5.1 Functional networks built from events

In figure 3 we show a nine node functional asynchronous network that is built from the eight "events" $\mathbf{P}^a, \dots, \mathbf{P}^h$.

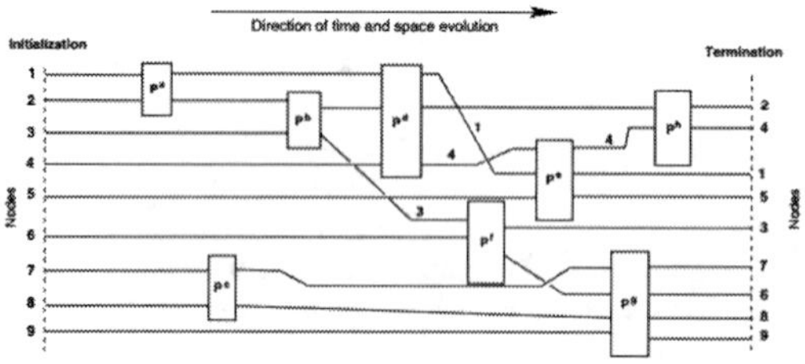

Fig. 3. A spatiotemporal decomposition of a functional asynchronous network

The initialization and termination sets are indicated on the left and right sides of the figure respectively. The events signify regions of phase space where there can be (state dependent) interaction between nodes. For example, the event labelled $\mathbf{P}^g$ involves interaction between nodes N_6, N_7, N_8, and N_9. Observe that there is only a partially ordered temporal structure on the events. Thus, the event $\mathbf{P}^g$ must occur after $\mathbf{P}^f$ but can occur before or after event $\mathbf{P}^h$.

5.2 Building blocks

In figure 4 we represent a basic building block with the same number of inputs and outputs.

The initialization sets are represented by the symbols ∘, termination sets by •. Interaction between nodes occurs only in the event region denoted by the rectangle. Outside of the event region, nodes evolve independently. More generally, we can allow for different number of inputs and outputs: nodes may merge or split.

Our immediate aim is describe some basic operations that we can define on functional asynchronous networks that allow us enable us to find a (maximal) decomposition of a functional asynchronous network into the form shown in figure 3.

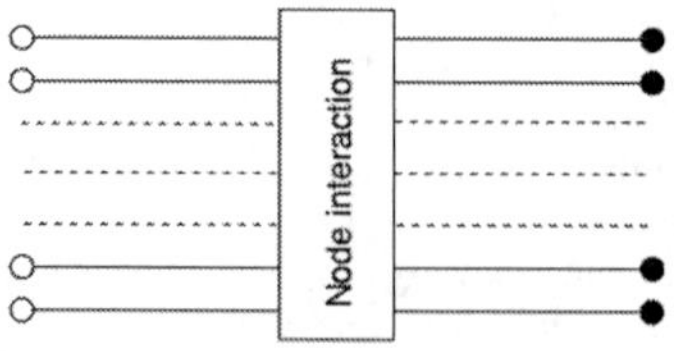

Fig. 4. Dynamical/functional module

5.3 Operations on functional asynchronous networks

If $\mathbf{N}^a = (\mathfrak{N}^a, \mathbb{I}^a, \mathbb{F}^a)$, $a \in \mathbf{q}$, are functional asynchronous networks with distinct node sets ($\mathcal{N}^a \cap \mathcal{N}^b \subset \{N_0\}$, $a \neq b \in \mathbf{q}$), define the *product* $\prod_{a\in\mathbf{q}} \mathbf{N}^a$ to be the functional asynchronous network $\mathbf{N} = (\mathfrak{N}, \mathbb{I}, \mathbb{F})$, where

$$\mathbb{I} = \prod_{a\in\mathbf{q}} \mathbb{I}^a, \quad \mathbb{F} = \prod_{a\in\mathbf{q}} \mathbb{F}^a$$

and $\mathfrak{N} = \prod_{a\in\mathbf{q}} \mathfrak{N}^a$ is defined in the obvious way to be the asynchronous network with node set $\mathcal{N} = \cup_{a\in\mathbf{q}} \mathcal{N}^a$ (we refer to [3, §6] for details).

We say the k-node functional asynchronous network $\mathbf{N} = (\mathfrak{N}, \mathbb{I}, \mathbb{F})$ is *trivial* if $\mathbf{N} = \prod_{a\in\mathbf{k}} \mathbf{N}^a$ where each $\mathbf{N}^a$ has exactly one node N_a. In particular, if $\mathbf{N}$ is trivial there are no interactions between nodes and no constraints.

Next let $\mathfrak{N}^a = (\mathfrak{N}^a, \mathbb{I}, \mathbb{F})$, $a \in \mathbf{q}$, be a family of functional asynchronous networks with common initialization set, termination set and node set $\mathcal{N} = \{N_0, N_1, \dots, N_k\}$. Suppose that for each $a \in \mathbf{q}$, there exists $\Sigma(a) \subset \mathbf{k}$ such that

1. $\mathfrak{N}^a = \mathfrak{N}_1^a \times \mathfrak{N}_2^a$ where $\mathfrak{N}_1^a$ has node set $\Sigma(a)$ and $\mathfrak{N}_2^a$ is trivial.
2. If $a \neq b$, $\Sigma(a) \cap \Sigma(b) = \emptyset$.

We define the *amalgamation* $\mathfrak{N} = \sqcup_{a\in\mathbf{q}} \mathfrak{N}^a$ to be the functional asynchronous network $(\prod_{a\in\mathbf{q}} \mathfrak{N}_1^a) \times \mathfrak{N}_2$, where $\mathfrak{N}_2$ is the trivial network defined as the product of the common trivial factors in $\mathfrak{N}_2^a$, $a \in \mathbf{q}$. Thus the node set of $\mathfrak{N}_2$ will be $\mathbf{k} \setminus \cup_{a\in\mathbf{q}} \Sigma(a)$.

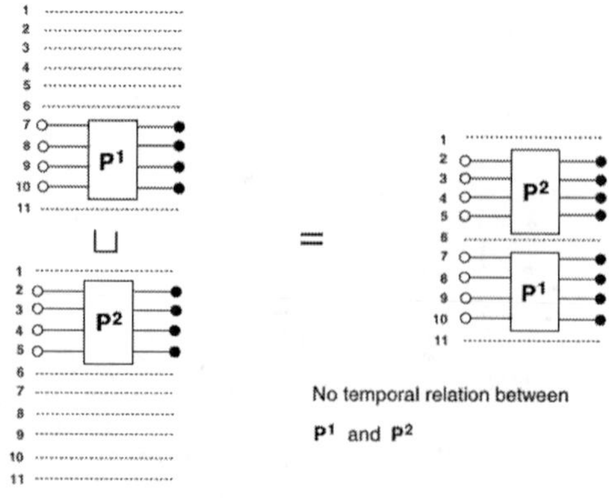

Fig. 5. Amalgamating two functional asynchronous networks.

Referring to figure 5, we have $\Sigma(1) = \{7, 8, 9, 10\}$ and $\Sigma(2) = \{2, 3, 4, 5\}$. The amalgamation $\mathbf{P} = \mathbf{P}^1 \sqcup \mathbf{P}^2$ is trivial when restricted to nodes $\{N_1, N_6, N_{11}\}$.

Finally, we outline the operation of concatenation, referring the reader to [4, §4] for the details (most) we omit. Suppose that $\mathfrak{N}^a = (\mathfrak{N}^a, \mathbb{I}^a, \mathbb{F}^a)$, $a \in \mathbf{2}$, are functional asynchronous networks with common node set. Assume that $\mathbb{F}^1 = \mathbb{I}^2$. The *concatenation* $\mathbf{N} = (\mathfrak{N}, \mathbb{I}, \mathbb{F}) = \mathbf{N}^2 \diamond \mathbf{N}^1$ will be a temporal merging $\mathbf{N}^1, \mathbf{N}^2$. We define

1. $\mathbb{I} = \mathbb{I}^1, \mathbb{F} = \mathbb{F}^2$.
2. $\mathcal{A} = \{\alpha_1 \vee \alpha_2 \mid \exists \mathbf{X} \in \mathbf{M}, \alpha_1 = \mathcal{E}^1(\mathbf{X}), \alpha_2 = \mathcal{E}^2(\mathbf{X})\}$,

where $\vee$ denote the join of the graphs. The definition of the set of admissible vector fields $\mathcal{F}$ for $\mathfrak{N}$ is trickier and requires additional conditions on $\mathbf{N}^1, \mathbf{N}^2$ – we refer to [4] for details. We define the event map by $\mathcal{E}(\mathbf{X}) = \mathcal{E}^1(\mathbf{X}) \vee \mathcal{E}^2(\mathbf{X}), \mathbf{X} \in \mathbf{M}$. We refer to figure 6 for the operation of concatenation.

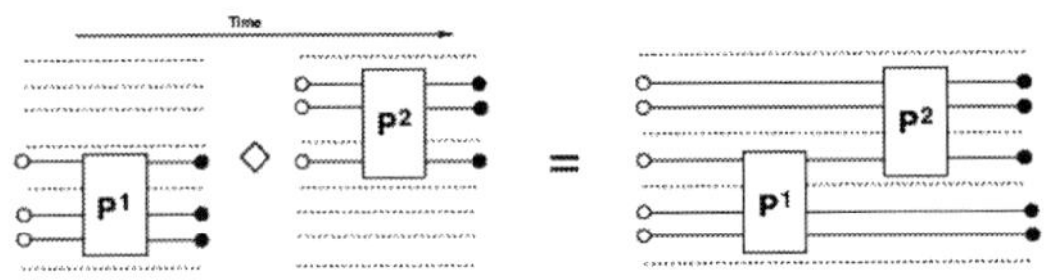

Fig. 6. Concatenating two functional asynchronous networks.

The concatenation $\mathbf{N}^2 \diamond \mathbf{N}^1$ has the important property that if $\mathbf{N}^a$ has generalized transition and timing functions $G^a : \mathbb{I}^a \times \mathbb{R}^k_+ \to \mathbb{F}^a, \widehat{\mathbf{S}}^a : \mathbb{I}^a \times \mathbb{R}^k_+ \to \mathbb{R}^k_+$, $a \in \mathbf{2}$, then $\mathbf{N}^2 \diamond \mathbf{N}^1$ has generalized transition function G given by $G(\mathbf{X}, \mathbf{T}) = G^2(G^1(\mathbf{X}, \mathbf{T}), \widehat{\mathbf{S}}^1(\mathbf{X}, \mathbf{T}))$ [4, Corollary 4.15].

Remark 3 We have deliberately avoided listing the detailed properties required of functional asynchronous networks in order to define amalgamations and concatenations. Briefly, apart from requiring the existence of generalized transition and timing functions, we require (1) the uncoupled vector vectors defining intrinsic dynamics of a node N_i to be transverse to $\mathbb{I}_i$, $\mathbb{F}_i$ and (2) a local product structure on the network. We refer to [4, §3] for the details.

6 Modularization of dynamics and function

A functional asynchronous network is *primitive* if it cannot be written as a nontrivial amalgamation or concatenation.

Theorem 1 *Under general conditions, a functional asynchronous network* $\mathbf{N}$ *has a unique (up to rearrangements) decomposition*

$$\mathbf{N} = \mathbf{N}^q \diamond \ldots \diamond \mathbf{N}^1,$$

where $\mathbf{N}^j = \mathbf{N}^{j,1} \sqcup \ldots \sqcup \mathbf{N}^{j,q(j)}$, $j \in \mathbf{q}$, *and and each* $\mathbf{N}^{j,\ell}$ *is primitive.*

The generalized transition function G *for* $\mathbf{N}$ *can be expressed in terms of the generalized transition and timing functions* $G^j, \widehat{\mathbf{S}}_j$ *of* $\mathbf{N}^j$ *(or* $G^{j,\ell}$ *for* $\mathbf{N}^{j,\ell}$*) by:*

$$G(\mathbf{X}, \mathbf{T}) = G^q(\ldots G^2(G^1(\mathbf{X}, \mathbf{T}), \widehat{\mathbf{S}}^1(\mathbf{X}, \mathbf{T}))\ldots),$$
$$\widehat{\mathbf{S}}(\mathbf{X}, \mathbf{T}) = \widehat{\mathbf{S}}^q(\ldots \widehat{\mathbf{S}}^2(G^1(\mathbf{X}, \mathbf{T}), \widehat{\mathbf{S}}^1(\mathbf{X}, \mathbf{T}))\ldots),$$
$$G^j = G^{j,1} \times \ldots \times G^{j,q(j)}, \quad j \in \mathbf{q}.$$

Example 2 Consider the network shown in figure 3 and assume that each event $\mathbf{P}^j$, $j \in \{a, \ldots, h\}$ is primitive. A decomposition satisfying the requirements of theorem 1 is indicated in figure 7 – the dashed lines indicate the initialization and termination sets for the subnetworks. The

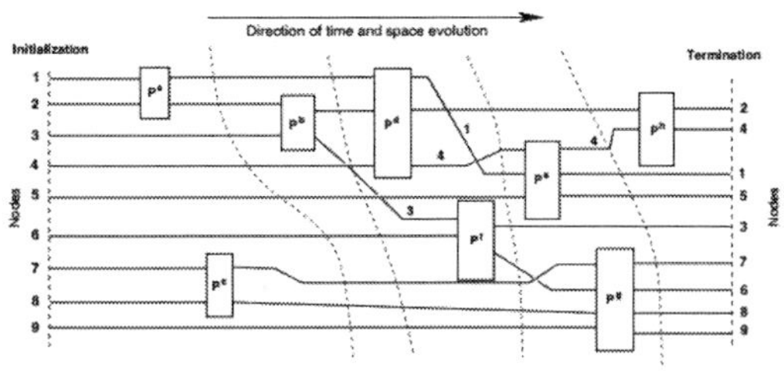

Fig. 7. Factorization of network of figure 3.

factorization for the network is

$$\mathbf{N} = \mathbf{P}^h \diamond (\mathbf{P}^e \sqcup \mathbf{P}^g) \diamond (\mathbf{P}^d \sqcup \mathbf{P}^f) \diamond \mathbf{P}^b \diamond (\mathbf{P}^a \sqcup \mathbf{P}^c).$$

This factorization corresponds to maximizing from the left hand side. However, if we maximize from the right we obtain the factorization

$$\mathbf{N} = (\mathbf{P}^h \sqcup \mathbf{P}^g) \diamond (\mathbf{P}^c \sqcup \mathbf{P}^e \sqcup \mathbf{P}^f) \diamond \mathbf{P}^d \diamond \mathbf{P}^b \diamond \mathbf{P}^a.$$

In either case there is a concatenation of five networks – that is the minimum number possible.

Theorem 1 allows us to write the function of a network explicitly in terms of the transition functions of the constituent subnetworks.

Results of this type depend crucially on intermittent connection structure and nonsmooth dynamics. For example, no such result is possible for a classical coupled network of phase oscillators.

The approach works because we have adopted an engineer's viewpoint: we emphasise function rather than dynamics. Indeed, we are indifferent to the specific dynamics occurring *between* the initialization and termination sets. Of course, both the timing and transition functions provide the key information about network function.

7 Concluding comments

1. Theorem 1 is a prototypical theorem providing proof of concept. The conditions for the theorem can be significantly weakened from those required in [4].
2. The theorem yields maximal feedforward structure on a functional asynchronous network (note that individual events may have feedback loops).
3. The result suggests the utility of starting with a small functional asynchronous network; understanding the structure in depth and then then evolving to optimize function (for example by adding feedback).
4. There are many as yet unexplored issues such as bifurcation, hidden deadlocks, and the effects of noise.
5. There is the problem of how far one can determine internal structure on the basis of input/output time series data.

References

1. U Alon. *An Introduction to Systems Biology. Design Principles of Biological Circuits* (Chapman & Hall/CRC, Boca Raton, 2007).
2. M di Bernardo, C J Budd, A R Champneys, & P Kowalczyk. *Piecewise-smooth Dynamical Systems* (Springer, Applied Mathematical Science 163, London, 2008).
3. C Bick & M J Field. 'Asynchronous networks and event driven dynamics', (2016), preprint.
4. C Bick & M J Field. 'Asynchronous networks: Modularization of Dynamics Theorem', (2016), preprint.
5. N Caporale & Y Dan. 'Spike timing dependent plasticity: a Hebbian learning rule' *Annu. Rev. Neurosci.* **31** (2008), 25–36.
6. A F Filippov. *Differential Equations with Discontinuous Righthand Sides* (Kluwer Academic Publishers, 1988).
7. W Gerstner, R Kempter, L J van Hemmen & H Wagner. 'A neuronal learning rule for sub-millisecond temporal coding', *Nature* **383** (1996), 76–78.
8. W Gerstner & W Kistler. *Spiking Neuron Models* (Cambridge University Press, 2002).
9. N Kashtan & U Alon. 'Spontaneous evolution of modularity and network motifs', *PNAS* **102** (39) (2005), 13773–13778.
10. J Ladyman, J Lambert, and K Wiesner. 'What is a complex system?', *Eur. J. for Phil. of Sc.* **3**(1) (2013), 33–67.
11. A Morrison, M Diesmann & W Gerstner. 'Phenomenological models of synaptic plasticity based on spike timing', *Biol. Cyber.* **98** (2008), 459–478.

Biomechanics of Natural Fiber Green Composites as Internal Bone Plate rafted

Ahmed M. Hashim[1;a], E. K. Tanner[2], and Jawad K. Oleiwi[3;2]

[1]Department of Materials Engineering, College of Engineering, Al-Qadisiyah University, Diwaniyah, Iraq.

[2]Biomedical Engineering Division, School of Engineering, University of Glasgow, Glasgow, G12 8QQ, UK.

[3] Department of Materials Engineering, University of Technology, Baghdad, Iraq.

Abstract. The main purpose of this paper was to introduce a new series of green composites as a plate for bone fracture applications. These green composites offer many advantages over traditional composites and biocomposites based on synthetic polymers. In current work, the new green composites bone plate based on two types of biopolymers involves biopolymethylmethacrylate thermoplastic material and bioepoxy thermosetting material reinforced with randomly oriented corn natural fibers at different weight fractions (5, 10, 15, and 20%) were fabricated by hand lay–up technique. Four forms of green composites ; treated and untreated corn biopolymethylmethacrylate groups A and B respectively, treated and untreated corn bioepoxy groups C and D respectively were investigated. The mechanical properties were tested (tensile, compression, and compact tension) under a flow of phosphate buffered saline PBS at 37 °C has been reported. The experimental results showed that the green composites group A have the best mechanical properties than other groups under PBS condition while the green composites group D have the weak mechanical properties due to plastization by immersion in PBS solution. Also, the analysis of femur bone fracture plates carried out by applying finite element method FEM using the ANSYS 16 software package.

1 Introduction

Internal fixation is a surgical procedure in orthopedics that stabilizes and joints the ends of fractured (broken) bone by mechanical devices such as plates, pins, wires, and screws positioned entirely inside the skin. Plates for long bone fractures are a major element of internal fixation [1] and must be bear external loads for 1.5 to 2 years [2]. Various materials can be used as bone plates including metals, ceramics, and polymer based composites [3]. Metallic materials such as stainless steel, cobalt alloys, pure titanium, and titanium alloys are most widely used as internal fixation for long bone fractures because of their outstanding mechanical properties [4]. However, these devices are not suitable for use as internal fixation plates inside the skin for long time due to numerous limitations. The large difference in elasticity between metals and bone causes stress shielding results in osteopenia, osteoporosis, plate loosening, and delay the healing period [5] as well as metallic bone plates are incompatibility with living tissue, bioactive, magnates effect [6], lower fabricability and high cost [7]. Although ceramic materials such as Al_2O_3 and ZrO_2 have been considered bioinert, but the higher degree of brittleness, hardness, and high elastic modulus lead to stress shield effect and thus making its unsuitable for use as bone plates [3]. Composites based on polymers derived from petrochemical materials results in production of harmful by degradation products when used in orthopedic surgery [8]. To eliminate these serious limitations, friendly material consist of natural fiber and biodegradable polymers are becoming highly attractive for alternative material use as bone plates which should satisfy the conditions of fully biodegradability and biocompatibility.

Biocomposites made from natural fibers as the matrix material and biodegradable polymers as the reinforcing material are called "green composites" [9]. The word "green" affixed with materials which are "biodegradable" and "renewable" in same time [8]. Frequently, they are referred as biomedical materials [10].

The current work investigates using two types of biopolymers: bio based epoxy thermosetting resin and bio based

[a] e-mail: ahmedmudhafarhashim@yahoo.co.uk

DOI: 10.1051/matecconf/20168309002

polymethylmethacrylate thermoplastic resin reinforced with corn natural fiber at various weight fractions in order to manufacture green composite bone plates. The main purpose for using green composite as plate fixation for long bones fractures due to their mechanical characteristics are expected to be closer to those of long bones.

2 Materials and Methods

2.1 Matrix, Reinforcing and Composite Materials

Two matrix materials were investigated in this study. Polymethylmethacrylate (PMMA) wasproducedusing 617H55 resin and 617P37 hardener (Otto Bock Healthcare plc, Egham, Surrey, UK). SUPER SAP® 100/1000 high bio-epoxy system composed of SUPER SAP® 100 resin and SUPER SAP® 1000 hardener (Professional Epoxy Coating, Kendal, Cumbria, UK).

Corn cobs were obtained from local shops (Glasgow), the fibers were extracted from the cobs and then dried at 37 °C for 240 hours in an incubator (Sanyo MIR262, Japan) before surface treatment. All the natural fiber types were cut in length of 10 mm. The corn fibers were soaked in 1% NaOH solution at room temperature for 90 minutes. The fibers were then dried at 37 °C for 240 hours in an incubator.

A PTFE mold of dimensions 240 × 110 × 4 mm was used to cast the composite sheets. A hand lay-up technique was used to prepare the samples. Before reaching hardening, pressurized to 2 tons for 30 minutes, giving a pressure of 27.58 MPa. Non-filled PMMA and epoxy were manufactured in the same manner to provide 0% filler samples.

2.2 Single-Fiber Tensile Test

Individual fibers of the corn fibers were tensile tested to BS ISO 11566 [26]. Weibull distribution is a statistical tool used to characterize the failure mode of brittle fibers. In this study, Weibull two parameter analysis was performed. For each individual corn fiber the failure stresses were ranked from minimum to maximum, according to following equation [11]:

$$P\left(\sigma_f\right) = 1 - exp\left\{-\left(\frac{\sigma_f}{\sigma_\circ}\right)\right\}^{\beta} \quad ...(2)$$

Where P is the probability of failure of a fiber length L at a stress less than or equal to σ. The constants $\sigma_\circ$ and β are the scale parameter (characteristic strength of life) and the shape parameter, respectively. The scale parameter corresponding to the fracture stress of the fiber. m is the shape parameter, also called the Weibull modulus. $L_\circ$ is the reference length (the minimum length to find a flaw); usually it is considered as unity or as the same length as the scale parameter.

The probability of failure, P_i, the probability of failure of sample i of the n samples in the group, was calculated using Bernard's correction [12]:

$$P_i = \frac{i-0.3}{n+0.4} \quad ...(3)$$

From P_i the Weibull number, Y, was obtained using equation (4)

$$Y = ln\, ln\left[\frac{1}{1-P_i}\right] \quad ...(4)$$

and σ_y was plotted against the Weibull number and the gradient and intercept found.

2.3 Composite Testing

Before testing all samples were soaked in phosphate buffered saline (PBS) solution (Sigma-Aldrich, United Kingdom) at 37 °C for between 1 and 3 weeks. The composite systems were prepared are divided into 4 groups as follows: treated and untrated corn/PMMA composites (groups A and B), and treated and untrated corn/epoxy composites (groups C and D). The mecahnical testing (tensile, comppresion) were performed according to ASTM D638-02 ans to ASTM D695 – 02a standards respectively, using an MTS 850 Mini Bionix® II universal testing machine (MTS, Minneapolis, USA).

3 Results and Discussion

3.1 Strength of Individual Fibers and Wiebull Analysis

The typical load – displacement curves selected corn natural fiber are shown in Figure 1. The curve shows that the tensile load increased proportionally with increasing strain until the point of ultimate load, which is the maximum load on the stress – strain curve. At this point the corn fiber broke and exhibited brittle behavior with no apparent yielding for corn fiber.
The modulus of elasticity is the slop of the stress – strain curve was determined in the elastic region of the curve. The values of the Young's modulus varies from 2.38 to 4.50 GPa, the tensile strength from 83.12 to 124.47 MPa, and the strain to failure from 4.10 to 7.58 %.

Figure 1. Load – displacement curves of selected corn naturals fibers.

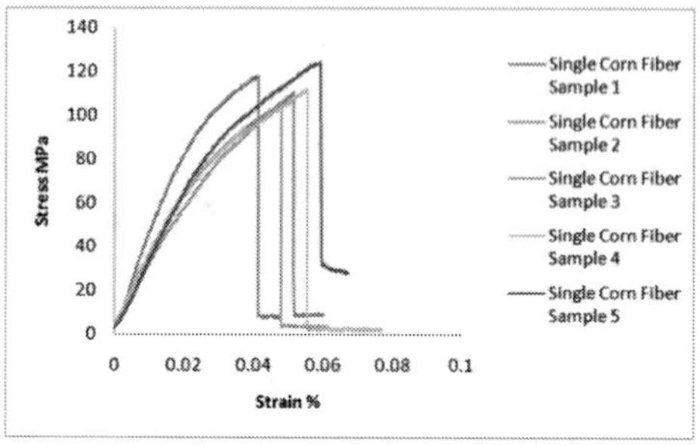

The Wiebull plot of tensile strength of corn natural fibers of 25 mm gauge length is shown in Figure 2. By applying the least squares method, the linear relationship between ln (σ_f) and ln (ln (1/(1-P))) is determined. From this figure, the R^2 coefficient of corn was 0.989. The values indicates a high degree of linearity.

For synthetic fibers, the Weibull modulus between 1-15, whereas in most common natural fibers types which intrinsically have more variation in properties lie in the range of 1-6 [13]. The Weibull modulus of corn fibers is β = 8.839, gives small variability in tensile strength. The little scatter in tensile strength and the lower variation (more regularity) in fiber diameter of randomly selected corn fiber (62 - 74 μm) as well as the distribution of flaws or defects on the corn fiber surface is less severe (more smoothness) results in higher value of Weibull modulus for corn fiber and exceed the normal range of most natural fibers.

DOI: 10.1051/matecconf/20168309002

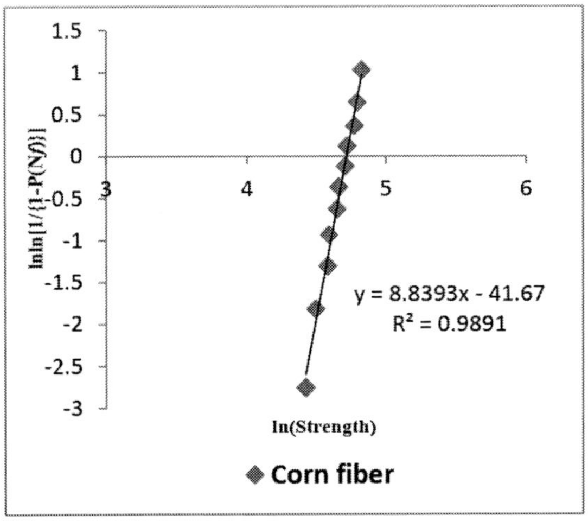

Figure 2. Weibull distribution of corn natural fibers tensile strength of 25 mm gauge length.

3.2 Tensile Test

The 1% NaOH alkali corn fiber treated have the major effect on both tensile strength and tensile modulus at degradation medium (groups A and C) as shown in Figures 3 and 4 respectively.

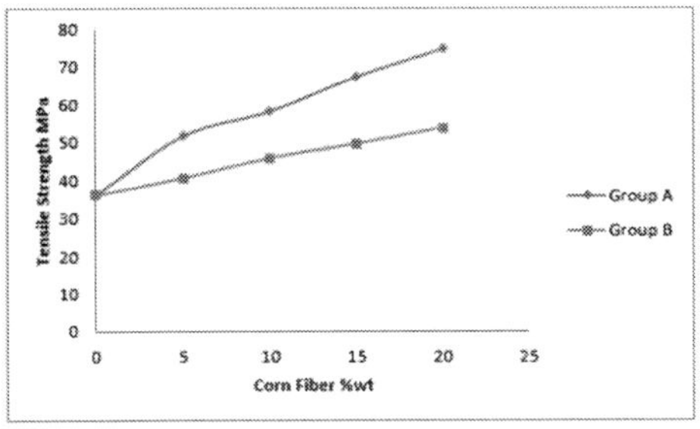

Figure 3. Effect of immersion in PBS at 37°*C* for 7 days on tensile strength of treated and untreated corn/PMMA green composites (Groups A and B).

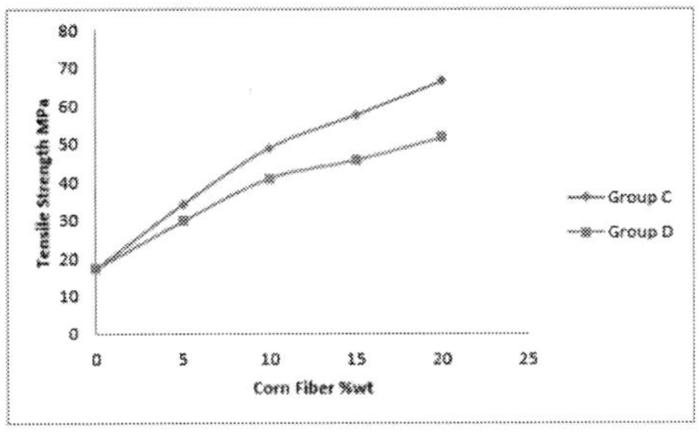

Figure 4. Effect of immersion in PBS at 37°*C* for 7 days on tensile strength of treated and untreated corn/ epoxy green composites (Groups C and D).

It was found that both tensile strength and tensile modulus increased as fiber concentrations increased. The tensile strength and tensile modulus were enhanced by 31%, 17.8%, 39%, and 18.6% of 20corn/80PMMA green composites in each groups from 53.68 MPa (group B) to 74.7 MPa (group A), and from 3.39 GPa (group B) to 4.02 GPa (group A) respectively. Immersio of treated and untreated fiber green composites (groups C and D) in PBS at 37°*C* for 7 days decreased the values of tensile strength and tensile modulus due to higher degree of plasticity for bioepoxy based resin. The largest results were found in group C of 66.23 MPa and 3.68 GPa respectively, while the lowest results in group D of 29.64 MPa and 2.79 GPa respectively.

3.3 Compression Test

The influence of both alkali treatment and immersion in PBS solution at 37 °*C* on compressive characteristics of corn/PMMA and corn/epoxy green composites are shown in Figure 5 and 6 respectively.

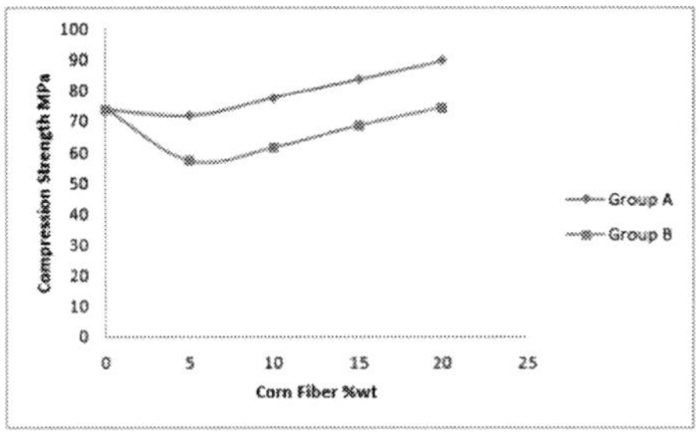

Figure 5. Effect on immersion in PBS at 37°*C* for 7 days on compression strength of treated and untreated corn/PMMA green composites (Groups A and B).

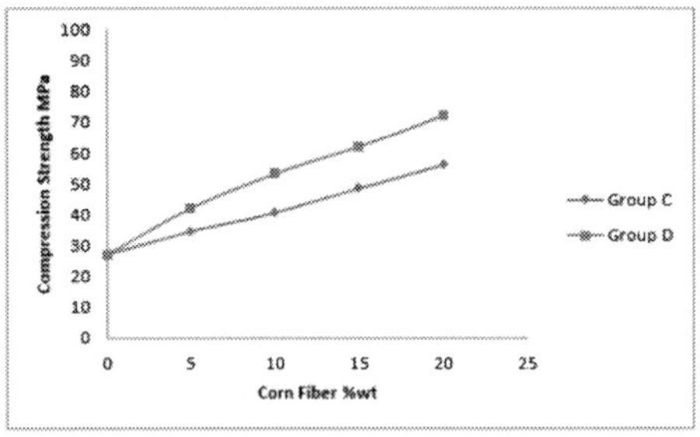

Figure 6. Effect on immersion in PBS at 37°*C* for 7 days on compression strength of treated and untreated corn/epoxy green composites (Groups C and D).

In both groups A and B, observed that there is a drop in flexural modulus of 5% wt corn fibers and then increased with increasing in fiber content. The drops in flexural modulus of both groups were not significantly. The largest increased in compressive strength of 89.7 MPa were occurs in PMMA based green composites containing 20% wt treated corn fiber (group A) compared to those of untreated fiber composites of same composition (group B) of 74.2 MPa.

The immersion in PBS solution at 37 °*C* led to higher drops in compressive strength for all samples of both treated and untreated corn epoxy composites (groups C and D). The results showed that the maximum compressive strengths were 56.12 MPa and 72.2 MPa of 20 % weight fraction of untreated and treated corn epoxy biocomposites respectively, which are lower than the value of the dry pure epoxy of MPa.

4 Numerical Results

The von Mises stress of femur-bone plate system has been calculated. The results are shown in Figure 7 for 20corn/80bioPMMA green composites bone plate, and Figure 8 for 20corn/80bioepoxy green composites bone plate, green composites bone plate that are prepared in this study. Also, the von Mises stress of femur-bone plate system has been calculated for stainless steel type 316 L and Ti6Al4V titanium alloy as shown in Figure 9 and 10 respectively. It is having been observed from this figure that maximum equivalent stresses at femur bone plate decreased significantly when using any group of the natural fiber green composites groups that are prepared in current study instead of traditional bone plates made from stainless steel type 316L or Ti6Al4V titanium alloy. This is related to the flexibility of natural fiber green composite plates compared to other metallic bone plates

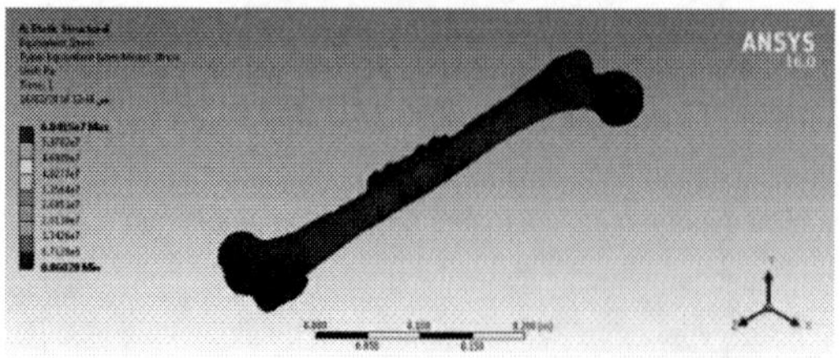

Figure 7 Contour plot for the maximum von Mises stresses for 20corn/80bioPMMA green composites.

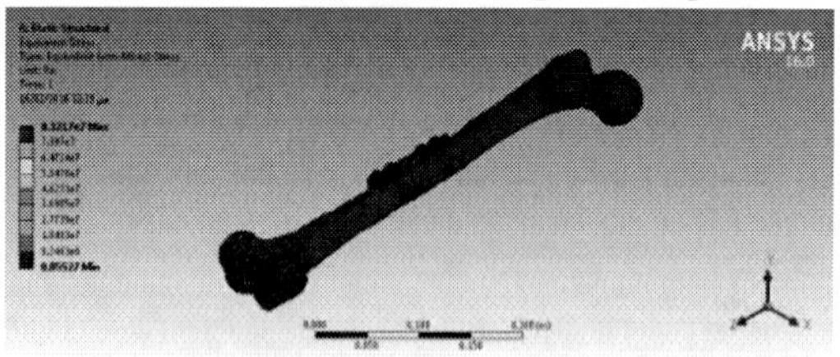

Figure 8 Contour plot for the maximum von Mises stresses of 20corn/80bioepoxy green composites.

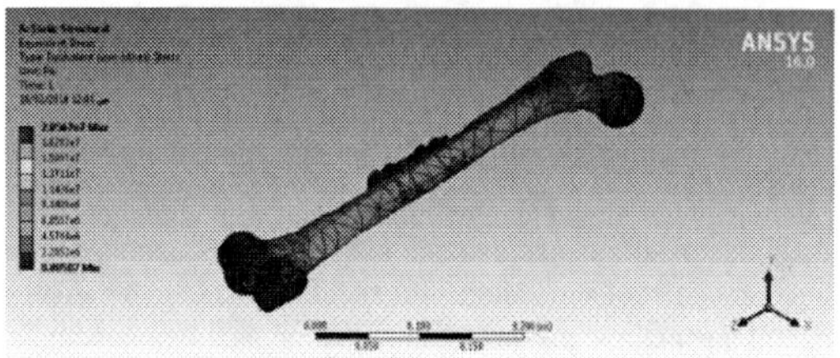

Figure 9 Contour plot for the maximum von Mises stresses of stainless steel type 316L.

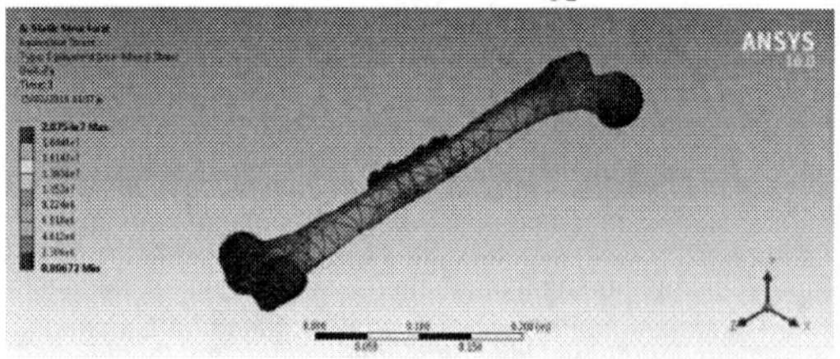

Figure 10 Contour plot for the maximum von Mises stresses of Ti6Al4V titanium alloy.

5 Conclusions

1- The tensile properties of corn natural fiber have been studies using Weibull analysis. The experimental data indicates that the corn natural fiber have a good tensile strength and modulus of elasticity.

2- The mechanical properties of biopolymethylmethacrylate resin are not affected by immersion in PBS at 37 °C., whereas in case of bioepoxy resin larger drops where occurred due to plastization of the material.

3- The alkali treatment by 1% NaOH of corn natural fibers reinforced polymethylmethacrylate green composites give better mechanical properties in PBS conditions than in case of bioepoxy based resin.

4- The experimental results showed that the alkali treatment of corn natural fiber with 1% NaOH increased the values of pH of green composites of both groups in all compositions.

References

1. Clary E.M, and Roe S.C., 1996. In vitro biomechanical and histological assessment of pilot hole diameter for positive-profi le external skeletal fi xation pins in canine tibiae. Vet Surg, 25(6): 453–462.

2. Perren S.M., 2003. Backgrounds of the technology of internal fixators. Ingjury, International journal care injuried, 34(S-B): 1-3.

3. Buddy R.D., Allamn S.H., Frederic J.S., Jack E.L., 1996. Biomaterials: An introduction to materials in medicine. San diego: Academic press.

4. Eglin D., Alini M., 2008. Degradable polymeric materials for osteosynthesis tutorial. Eur Cell Mater, vol. 16, pp. 80-91.

5. Lamis R. Darwish, Mahmoud Farag, Mohamed Tarek El-Wakad, Mohamed Emara, 2013. The use of starch matrix-banana fiber composites biodegradable maxillofacial bone plates. International conference on biology, medical physics, biochemistry and biomedical engineering, pp. 70-76.

6. D. Chandramohan, 2014. Analysis on natural fiber bone plates. European journal of experimental biology, vol. 4(2), pp. 323-332.

7. Javad Malekani, Beat Schmutz, Yuantong Gu, Michael Schuetz, Prasad, 2012. Orthpedic bone plates: Eevolution in structure, implementation technique and biomaterial. GSTF International journal of engineering technology (JET), vol. 1, No. 1, pp. 135-140.

8. Satyanarayana K.G., 2015. Recent developments in 'green' composites based on plants fibers-preparation, structure property studies. Journal of bioprocessing and biotechniques, vol.5, Issue 2.

9. Jasmin P. Jose, Sant Kumar Malhotra, Saba Thomas, Kuruvilla Joseph, Koichi Goda, Meyyarappallil Sadasivan Sreekala, 2014. Advances in polymer composites: macro-and microcomposites-state of the art, new challenges, and opportunities. Polymer composites, vol. 3, 1st ed., Wiley-VCH Verlag GmbH and Co. KGaA.

10. Josmin P. Jose, Sant Kumar Malhotra, Sabu Thomas, Kuruvilla Joseph, Koichi Goda, and Meyyarappallil Sadasivan Sreekala, 2014, "Advances in polymer composites: Macro-and microcomposites-state of the art, new challenges, and opportunities", Polymer composites, Vol. 3, 1st edition, Wiley-VCH Verlag GmbH & Co. KGaA.

11. R. Joffe, J. Andersons and E. Sparnins, 2013. Applicability of Weibull strength distribution for cellulose fibers with highly non-linear behavior, http://www.researchgate.net/publication/259486218_Applicability_of_Weibull_strength_distribution_for_cellulose_fibers_with_highly_non_linear_behavior.

12. E.M. Sheafi and K. E. Tanner, 2014. Effects of test sample shape and surface production method on the fatigue behaviour of PMMA bone cement", Journal of the mechanical behavior of biomedical materials, Vol. 29, pp. 91-102.

13. Nele D., et al., 2010. Assessment of the tensile properties of coir, bamboo and jute fibre. Composites Part A, 41, pp. 588–595.

MATEC Web of Conferences **83**, 09003 (2016)
CSNDD 2016
DOI: 10.1051/matecconf/20168309003

Design and analysis of composite skew plate bending

L. Ikharrazne [1]

[1]EMMA Group, Laboratory of Engineering Science and Applications; ENSAH BP 03, Ajdir Al -Hoceima,Morocco

Abstract. This paper generalizes and extends a numerical formulation within the method to simulate the behaviour of skew composite plate bending. It is of pivotal importance to note that despite the spread use of skew plates, the published literature shows an important lack of scientific works regarding these particular structures. The main goal of the development described in this work is to provide a numerical tool that could be used to obtain approximate solutions for bending problem of skew composite plates subjected to different loading and various support conditions. Both homogeneous and fibre-based materials are investigated. The skew plate behaviour depend on several parameters namely, the skew angle, boundary conditions and constitutive material. Dependence on these parameters is investigated trough numerical results.

1 Introduction

The use of laminated composite in many engineering applications has been expanding rapidly in the past four decades due to their higher strength and stiffness to weight ratios when compared to most metallic materials. In particular, laminated composite skew plates are frequently used in various engineering applications and have a spread economical application in different area of science including aerospace, marine, mechanical, and modern highways.

For example, enormous amounts of money and resources are invested every year in design and construction of new roads. The use of composite skew plates in civil engineering, particularly in skew bridges, can lead to the economical solutions and enables the construction of straight, aesthetic and safe highways needed to carry on increasing speeds of present-days traffic. Despite this practical interest, problems of skew plate bending have not yet been correctly addressed in the published literature. Relatively little is to be found in the engineering literature regarding composite skew plates, partially due to the lack of analytical solutions and to the numerical difficulties encountered with regard to modelling composite plate behaviour specially when dealing with a highly skew angle. Therefore, the design of composite skew plates is at present based on numerical data provided by computational models, and influenced by comparisons with rectangular and square plates which can be solved analytically. A high quality numerical approximation of skew plate behaviour is obtained by the use of a numerical approach like BEM, FEM and FDM.[1-5]

2 Problem formulations

The method offers a solution based on the small deflection theory of plate bending and the classical elasticity. This simplified theoretical model was designed for thin lamina of orthotropic material with different properties in different directions

Consider a linear elastic composite material in which is locally orthotropic and globally orthotropic. This means that in each point of the structure the stress-strain relation is given as:

$$\begin{Bmatrix} \varepsilon_1 \\ \varepsilon_2 \\ \varepsilon_3 \\ \gamma_4 \\ \gamma_5 \\ \gamma_6 \end{Bmatrix} = \begin{bmatrix} \frac{1}{E_1} & \frac{-v_{12}}{E_1} & \frac{-v_{13}}{E_1} & & & \\ \frac{-v_{21}}{E_2} & \frac{1}{E_2} & \frac{-v_{23}}{E_2} & & & \\ \frac{-v_{31}}{E_3} & \frac{-v_{32}}{E_3} & \frac{1}{E_3} & & & \\ & & & \frac{1}{G_{23}} & & \\ & & & & \frac{1}{G_{31}} & \\ & & & & & \frac{1}{G_{12}} \end{bmatrix} \begin{Bmatrix} \sigma_1 \\ \sigma_2 \\ \sigma_3 \\ \tau_4 \\ \tau_5 \\ \tau_6 \end{Bmatrix} \quad (1)$$

The elastic constants involved in the calculations were E_1 and E_2, the Young's moduli in the longitudinal and transverse directions of a unidirectional composite, G_{12}, the longitudinal-transverse shear modulus, and v_{12} the effective Poisson's ratio.

The strength and stiffness of a composite build-up depends on the orientation sequence of the plies. For an orthotropic thin plate under plan stress, the bending moments are expressed in terms of transversal displacement as:

DOI: 10.1051/matecconf/20168309003

$$\begin{Bmatrix} M_x \\ M_y \\ M_{xy} \end{Bmatrix} = -\begin{bmatrix} D_{11} & D_{12} & 0 \\ D_{12} & D_{22} & 0 \\ 0 & 0 & D_{66} \end{bmatrix} \begin{Bmatrix} \frac{\partial^2 \omega}{\partial x^2} \\ \frac{\partial^2 \omega}{\partial y^2} \\ \frac{\partial^2 \omega}{\partial x \partial y} \end{Bmatrix} \quad (2)$$

Where ω is the local deflection and D_{ij} are the bending rigidities.

$$D_{11} = \frac{E_1 h^3}{12(1-\nu_{12}\nu_{21})},\ D_{22} = \frac{E_2 h^3}{12(1-\nu_{12}\nu_{21})}, \quad (3)$$

$$D_{12} = \frac{\nu_{12} E_2 h^3}{12(1-\nu_{12}\nu_{21})} = \frac{\nu_{21} E_1 h^3}{12(1-\nu_{12}\nu_{21})} \text{ and } D_{66} = \frac{G_{12} h^3}{12}$$

The problem requires the solution of the homogeneous differential equation of bending given by:

$$D_{11}\frac{\partial^4 \omega}{\partial x^4} + 2(D_{12} + 2D_{66})\frac{\partial^4 \omega}{\partial x^2 \partial y^2} + D_{22}\frac{\partial^4 \omega}{\partial y^4} = p(x,y) \quad (4)$$

Where p(x,y) is the transversal loading.

3 Numerical implementation

In the following section, we describe an efficient approach for the analysis of skew orthotropic plates. Obviously, it turns out that an adequate meshing is essential for such problems. Figure 1 shows a triangular mesh that is aligned with the skewed geometry of the plate. As it is frequently done, an appropriate coordinate system adapted to the constraints of the problem is chosen. Therefore, triangular coordinate system (x_1,x_2,x_3) is used instead of ordinary cartesian rectangular coordinates as shown in figure 1.

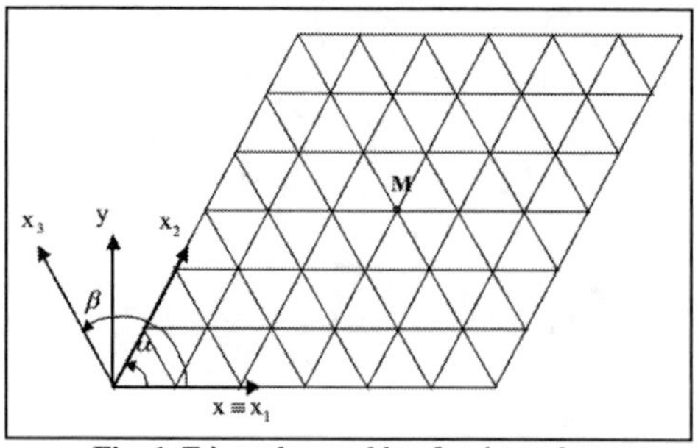

Fig. 1. Triangular meshing for skew plate

Hence, Laplace's harmonic operator is expressed in triangular co-ordinates by the following equation:

$$\nabla^2 = \frac{\partial^2}{\partial x^2} + \frac{\partial^2}{\partial y^2} = (1 + \cot\alpha \cot\beta)\frac{\partial^2}{\partial x_1^2} - \frac{\cos\beta}{\sin\alpha \sin(\beta-\alpha)}\frac{\partial^2}{\partial x_2^2} + \frac{\cos\alpha}{\sin\beta \sin(\beta-\alpha)}\frac{\partial^2}{\partial x_3^2} \quad (5)$$

Using central finite difference for partial derivatives, particular finite difference schemes are obtained for different operators in figure 2 and 3.

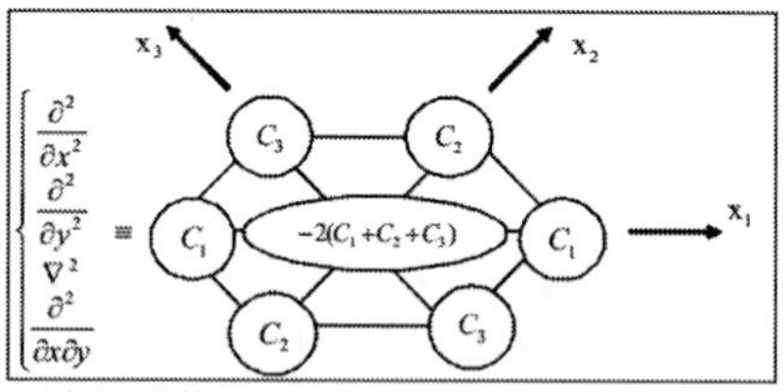

Fig. 2. Finite difference schemes for second order operator differential equations

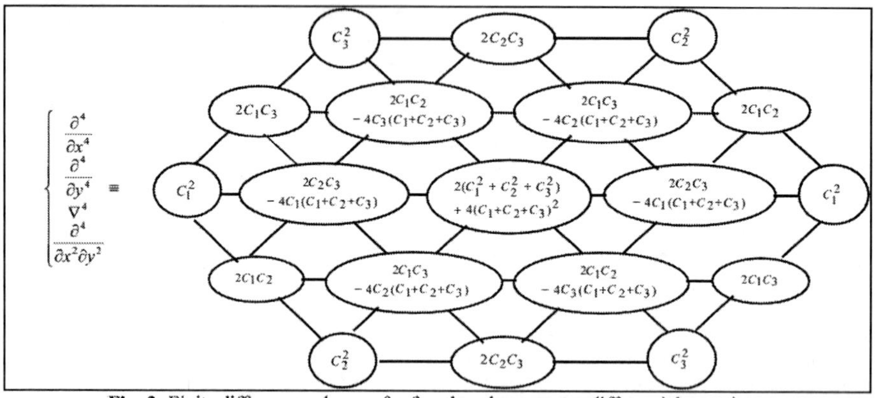

Fig. 3. Finite difference schemes for fourth order operator differential equations

Where coefficients C_1, C_2 and C_3 are expressed by:

Table 1. Expressions of C_1, C_2 and C_3 coefficients

	$\frac{\partial^2}{\partial x^2}, \frac{\partial^4}{\partial x^4}$	$\frac{\partial^2}{\partial y^2}, \frac{\partial^4}{\partial y^4}$	∇^2 et ∇^4	$\frac{\partial^2}{\partial x \partial y}, \frac{\partial^4}{\partial x^2 \partial y^2}$
C_1	$\frac{1}{\Delta X_1^2}$	$\frac{\cot\alpha \cot\beta}{\Delta X_1^2}$	$\frac{1+\cot\alpha \cot\beta}{\Delta X_1^2}$	$\frac{-\sin(\alpha+\beta)}{2\Delta X_1^2 \sin\alpha \ \sin(\beta)}$
C_2	0	$\frac{-\cos\beta}{\Delta X_2^2 \sin\alpha \ \sin(\beta-\alpha)}$	$\frac{-\cos\beta}{\Delta X_2^2 \sin\alpha \ \sin(\beta-\alpha)}$	$\frac{\sin(\beta)}{2\Delta X_2^2 \sin(\alpha)\sin(\beta-\alpha)}$
C_3	0	$\frac{\cos\alpha}{\Delta X_3^2 \sin\beta \ \sin(\beta-\alpha)}$	$\frac{\cos\alpha}{\Delta X_3^2 \sin\beta \ \sin(\beta-\alpha)}$	$\frac{-\sin(\alpha)}{2\Delta X_3^2 \sin(\beta)\sin(\beta-\alpha)}$

DOI: 10.1051/matecconf/20168309003

Table 2. Central deflexions of simply supported isotropic skew plate under uniform loading

α (degree)	Central deflection ωc (cm)								
	Mesh(4x4)		Mesh(6x6)		Mesh(12x12)		Mesh(16x16)		Analytical (Morley)
	Present	FEAP	Present	FEAP	Present	FEAP	Present	FEAP	
90°	0,683	0,691	0,686	0,698	0,688	0,696	0,688	0,694	0,688
85°	0,691	0,681	0,685	0,690	0,682	0,687	0,681	0,686	0,680
80°	0,680	0,653	0,667	0,664	0,658	0,662	0,657	0,661	0,656
60°	0,480	0,403	0,453	0,426	0,438	0,435	0,436	0,436	0,434
50°	0,331	0,252	0,307	0,274	0,295	0,288	0,293	0,290	0,292
40°	0,191	0,129	0,174	0,145	0,165	0,157	0,164	0,160	0,162
30°	0,084	0,050	0,075	0,057	0,070	0,065	0,070	0,067	0,069

4 Numerical examples

4.1 Simply supported isotropic skew plate

As a first example, an isotropic thin skew plate is first investigated since we can compare with existing solutions from literatures. The edges of the plate are all simply supported.

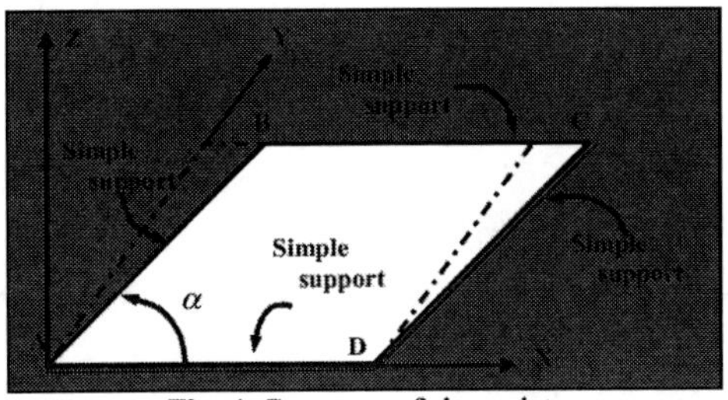

Fig. 4. Geometry of skew plate

The plate has length a in x and y direction and thickness h in z directions. The data assumed for geometric parameters and material properties of the skew plate are: side length a=254 cm, thickness h=2,54 cm; Young's modulus E=210 GPa , Poisson's ratio, ν=0,3. The plate is subjected to uniformly distributed transversal pressure of intensity, q=1,95 kN/m^2.

The maximal deflexion is expected to be at the center of the plate. In table 2, numerical values of central deflexions of investigated skew plates is represented for different meshes and various skew angles (α varying from 30° to 90°). Other numerical analyses are also carried out by using FEAP program version 7.5 [8]. Four node quadrilateral plate elements with tree degrees of freedom on each node are used for calculation. Obtained numerical results are shown in table 2 and compared with an analytical solution given by Morley[6].

A very good agreement is achieved between finite element data and our numerical results. From results given in Table 2, we can observe that the proposed numerical method yields a high level of accuracy using regular meshes

A much more refined meshing of (12x12) and (16x16) grids is required to obtain closer values to Morley's analytical solutions.

4.2 Simply supported orthotropic skew plate

In this section, we are interested to analyse orthotropic skew plate for two different materials E-Glass/Epoxy and Graphite/Epoxy with simply supported boundary conditions.

The table 3 and 4 below shows the data for geometric parameters and the specifications of mechanical properties of E-Glass/Epoxy and Graphite/ Epoxy.

Table 3. Geometric parameters of the plate

Parameter	Value
Side length a	254cm
Thickness h	2,54 cm
Uniform load q	6,89 kN/m^2

Table 4. Mechanical properties of E-Glass/Epoxy WR/5920 and Graphite/Epoxy T300/5208

	E-Glass/Epoxy WR/5920	Graphite/Epoxy T300/5208
E_1 (GPa)	23,72	128
E_2 (GPa)	21,92	13
G_{12} (GPa)	16,54	6,4
ν	0,13	0,3

Using the numerical model proposed in this paper, a computer program package capable of simulating the mechanical behaviour of orthotropic skew plates is developed using Compaq / Digital Visual Fortran. Finite-difference schemes on grids consisting of regular triangles are used for modelling skew plate as described in the previous section.

DOI: 10.1051/matecconf/20168309003

Results on tables 6 and 7 show numerical solutions obtained respectively for E-Glass/Epoxy WR/5920 and Graphite/Epoxy T300/5208 composite plates. The central deflection has been reported for simply supported rhombic plates for different skew angles (α angle) and different grids.

Numerical results given by our model are checked against the only existing analytical solutions of uniformly loaded thin square plate. Under the same conditions of the problem, the analytical solutions are shown in table 5.

Table 5. Analytical solution for central deflexion of thin square plate

Material type	Central deflection ωc (cm)
E-Glass/Epoxy	2,881
Graphite/Epoxy	1,914

Table 6. Central deflexions of simply supported E-Glass/Epoxy composite skew plate

E-Glass/ Epoxy	Central deflection ωc (cm)				
α (degree)	Mesh (4x4)	Mesh (6x6)	Mesh (8x8)	Mesh (12x12)	Mesh (16x16)
90°	2,858	2,871	2,875	2,878	2,879
80°	2,955	2,876	2,838	2,794	2,762
70°	2,874	2,702	2,615	2,505	2,424
60°	2,593	2,351	2,226	2,067	1,955
50°	2,086	1,824	1,688	1,519	1,404
40°	1,364	1,158	1,050	0,919	0,834
30°	0,631	0,530	0,478	0,416	0,376

Table 7. Central deflexions of simply supported Graphite/Epoxy composite skew plate

Graphite/ Epoxy	Central deflection ωc (cm)				
α (degree)	Mesh (4x4)	Mesh (6x6)	Mesh (8x8)	Mesh (12x12)	Mesh (16x16)
90°	0,747	0,751	0,752	0,753	0,753
80°	0,776	0,751	0,742	0,735	0,731
70°	0,732	0,687	0,670	0,652	0,640
60°	0,606	0,544	0,519	0,492	0,475
50°	0,400	0,336	0,309	0,282	0,267
40°	0,189	0,142	0,123	0,105	0,097
30°	0,053	0,036	0,029	0,024	0,023

From these tables, it is clear that the central displacement of square plate (α=90) compare well with analytical solutions given in table 5. Furthermore the numerical values of central deflexions are much closer to the analytical solutions with refined meshes of (12x12) and (16x16).

5 Conclusions

In this brief note we have presented a numerical study on the bending problems of skew composite plates. Typical cases of skew plates are analysed with various skew angles. The validity of the proposed procedure is confirmed by numerical examples. Two quite different examples are selected for the investigation. Hence, central deflexions obtained for the two examples are compared to available analytical solutions and other finite element solutions. A very close agreement is observed in all cases.

References

1. J. N. Reddy, *A review of refined theories of laminated composite plates*, Shock Vib. Dig. **22,** 3-7 (1990)
2. J. N. Reddy, *A simple higher-order theory for laminated composite plates*, J. Appl. Mech., ASME **51,** 745-752 (1984)
3. N. J. Pagano, S. J. *Hatfield, Elastic behavior of multi-layered bi-directional composites*, AIAA J. **10** (7) 931-933 (1972)
4. S. Timoshenko, S. Woinowsky-Kriegner, *Plates and shells*. McGraw-Hill, New York, (1959)
5. O. C. Zienkiewicz,Y. K. *Ceung, The finite element method for analysis of elastic isotropic and orthotropic slabs*. Proc. Inst. Civ. Eng. **28**, 471-488, (1964).
6. C. T. Morley, *Skew plates and structures*, MacMillan Company, New York (1963).
7. S. G. Lekhnitskii, *Anisotropic plates*, Gordon and Breach, Science Publishers (1968)
8. R. L. Taylor, FEAP - A Finite Element Analysis Program. Version 7.5 Dept. of Civil and Environmental Engineering, User Manual, Berkeley, California 94720-1710, January (2003).

Analysis of the effect of an obstacle on the wind energy potential

N. Amahjour[1], and A. Khamlichi[2]

[1]Departement of Physics, Faculty of Sciences, University Abdelmalek Essaadi, Tetouan 93030, Morocco
[2]Departement TITM, National School of Applied Sciences, University Abdelmalek Essaadi, Tetouan 93030, Morocco

Abstract. Assessment of the wind energy potential for a given site requires a primitive evaluation of wind speed for a specified period of time. Wind speed distribution depends on the type of air flow in the atmospheric boundary layer and varies as function of the altitude and the site topography. It is also affected by the presence of obstacles, the curvature radius of obstacle, the roughness of the ground and temperature. In this study, circular motion of the atmosphere that is induced by the existence of an obstacle having the form of a hill is investigated by using a 2D model. The hill profile was approximated by a half-ellipse curve. Computational fluid dynamics (CFD) was performed for the turbulent air flow modelled by the Navier-Stokes equations coupled with the K- ε turbulence model. To solve the equations, COMSOL software that is based on the finite element method was used. The obtained results were compared to those of a flat ground without obstacles. A noticeable modification of the available wind energy resulted for a wind turbine which is placed downwards of a hill like obstacle.

1. Introduction

The increasing concerns with environmental and economical issues are driving the search for more sustainable electrical sources. Wind energy along with solar energy, biomass and wave energy are all possible solutions for environmentally friendly energy production. Wind energy has shown however to be the most prominent renewable energy source for regions where wind speed fulfils ideal characteristics in terms of magnitude and stability. In these last decades, the installation of a great number of wind turbine generators has effectively reduced environmental pollution, fossil fuel consumption, and the costs of overall electricity generation. Although wind is only an intermittent source of energy, it represents a reliable energy resource from a long-term energy policy viewpoint. At a specific wind farm, the available electricity generated by a wind power generation system depends on the mean wind speed, the direction of wind and the standard deviation of wind speed. These quantities vary also as function of the altitude and topographic details of the wind turbine location of installation [1].

Evaluating the amount of the wind energy potential for a given site requires an earlier estimation of wind velocity that is susceptible to be captured by the wind turbine rotor. This needs to be performed during a sufficiently large period of time based on accurate height distribution of wind speeds on that location. Wind velocity distribution is transient in nature and depends on the type of air flow in the atmospheric boundary layer. It varies as function of the altitude and the site topography. It is also affected by the presence of obstacles, the curvature radius of obstacle, the roughness of the ground and temperature.

The air flowing across an obstacle having the form of a solid wall surface will stick to that surface. This is due to adherence forces and viscosity of the fluid which produce the non-slip condition at the wall. As a result a boundary layer is formed near the obstacle where the velocity varies from zero at the wall to a finite value at some sufficient distance from that wall. There are actually many procedures that can be used to study the boundary layer characteristics in terms of skin-friction, dissipation, turbulence and loads that can be generated.

The pioneering work is attributed to Prandtl [2]. He introduced an important assumption that enables to reduce the Navier-Stokes equations. Using the concept of dimensional analysis he has determined the similarity parameters governing the laminar boundary layer [3,4]. Computational Fluid Dynamics (CFD) [5] is a modern approach that is used to solve numerically the turbulent air flow equations around complex obstacles. CFD is based on the Navier-Stokes equations which are coupled to a turbulence closure model such the k-ε model [6]. The main advantage of CFD is that it is a very compelling, non-intrusive, virtual modeling technique with powerful visualization capabilities. Engineers can evaluate the performance of a wide range of system configurations on the computer without the time, expense, and disruption required to make actual changes onsite. This technique has widely been applied to various engineering applications such as automobile, aircraft design and civil engineering, where CFD data was utilized to evaluate various parameters in order to meet the required design criteria.

In this work, simulation of the turbulent flow occurring around an obstacle that is present on a flat ground is performed. A two-dimensional model describing air circulations that are induced by the presence of the obstacle chosen to have the form of an ellipse is developed. Appropriate boundary layer conditions are applied. The equations are solved by using the finite element based COMSOL software package [7].

Like in every finite element calcualtion, the precision of the obtained solution has to be assessed. This depends in the actual analysis on the location of the boundary fixing the domain of the air and also the mesh used. Convergence of the finite element model is studied here asymptotically by modifying the frontier of the domain and the mesh size until a stationnary solution is obtained. The results are compared then with those corresponding to the case of a flat ground without any obstacle. This is done at the aim of catching the spatial heterogeneities of the air flow which is due to obstacles. The results are useful in particular for re-estimating the profiles of wind speed as affected by flow turbulence around complex geometries, which in their turn can be used to accurately predict wind energy potential for a given site location of the wind turbine.

2. Geometry and mathematical formulation

The geometry of the problem consists of a horizontal ground with a hill form obstacle that can be assimilated to an ellipse. The domain of fluid to be modeled is part of the problem and has to be determined through an iterative procedure which assures a stationary solution for a frontier located at sufficiently large distance from the obstacle.

The CFD modeling is implemented by using the COMSOL package software. The air flow near a flat plate with an obstacle is typically turbulent. The partial differential equations that describe the stationary fluid motion are the well known Navier-Stokes equations which write:

$$\rho\nabla.(U)=0 \tag{1}$$

$$\rho(U.\nabla)U=\nabla[-P+\mu(\nabla U+(\nabla U)^{t})]+F \tag{2}$$

where U is the velocity vector, ρ the density, μ the dynamic viscosity, P the pressure, F the density of volumetric forces acting in the fluid and ∇ the gradient operator.

The Reynolds-averaged approach is based on decomposing the velocity as $u_j=\bar{u}_j+u'_j$ where $\bar{u}_j$ is the average velocity vector and u'_j the velocity vector fluctuation. The average continuity and momentum equation for steady state incompressible flow are given by:

$$\rho\nabla.(\bar{U})=0 \tag{3}$$

$$\rho(\bar{U}.\nabla)\bar{U}=\nabla[-P+\mu(\nabla\bar{U}+(\nabla\bar{U})^{t}]-\rho\bar{u}'(\nabla\bar{u}')+F \tag{4}$$

Equation (4) has the same form of the Navier-Stokes equation (2), but now it has an additional term, the turbulent Reynolds stress term, $-\rho\bar{u}'(\nabla\bar{u}')$ representing the influence of the fluctuation on the average flow. In order to close Eqs. (3) and (4), the turbulent Reynolds stresses must be determined. These can be directly modeled through the Reynolds Stress Models (RSM) which involves calculation of the individual Reynolds stresses by using differential transport equations.

Another closure procedure is provided by the K-ε turbulence model.

2.1. The K-ε turbulence model

An extended K-ε turbulence model is used here. It contains the standard two transport equations as proposed by Launder and Spalding and an improved transport equation for the rate of dissipation of the turbulent kinetic energy [8]. This model was found to give more effective response to the energy production rate than does the standard K-ε turbulence model. The model equations write:

$$\rho(\bar{U}.\nabla)K=\nabla\left[\left(\mu+\frac{\mu_t}{\sigma_k}\right)\nabla K\right]+P_k-\rho\varepsilon \tag{5}$$

$$\rho(\bar{U}.\nabla)\varepsilon=\nabla\left[\left(\mu+\frac{\mu_t}{\sigma_\varepsilon}\right)\nabla\varepsilon\right]+C_{\varepsilon1}\frac{\varepsilon}{K}P_K-C_{\varepsilon2}\rho\frac{\varepsilon^2}{K} \tag{6}$$

$$\mu_t=\rho C_\mu\frac{K^2}{\varepsilon} \tag{7}$$

$$P_K=\mu_t\left[\nabla\bar{U}+(\nabla\bar{U})^t\right]\nabla\bar{U} \tag{8}$$

The adaptation of the standard model constants which is applicable to flows with high Reynolds numbers is given in table 1.

Table 1. Values of the K-ε model constants

C_μ	$C_{\varepsilon1}$	$C_{\varepsilon2}$	σ_k	σ_ε
0.09	1.44	1.92	1	1.3

2.2. Boundary layer condition at the ground

For a turbulent shear flow which is steady, the mean-velocity profile in a small region near the surface is described by a relationship called the wall law:

$$\frac{u_t}{u^*} = \frac{1}{\kappa}\ln\left(\frac{z}{z_0}\right) + B \qquad (10)$$

where u^* is the friction velocity, κ is the von Karman constant ($\kappa = 0.42$), B is an empirical constant related to the thickness of the viscous sublayer ($B = 5.2$ in a boundary layer over a smooth flat plane; for rough walls smaller values of B are obtained). The constant z_0 called roughness is the frontier between the logarithmic zone and the inner K-ε zone.

The determiantion of z_0 is considered usually through combining the logarithmic wall function as given in Eq. (10) and the numerical integration approach used. Denoting u_t, u_n as respectively the tangential and normal velocity components, and $\partial u_t / \partial n$ the normal derivative of the tangential velocity, the wall law writes

$$u_n = 0 \qquad (11)$$

$$\frac{u_t}{\sqrt{\frac{\partial u_t}{\partial n}}} = \sqrt{\frac{\eta}{\rho}}\frac{1}{\kappa}ln\left(\frac{z_0}{\sqrt{\frac{\partial u_t}{\partial n}}}\sqrt{\frac{\rho}{\eta}}\right) + 2.4\sqrt{\frac{\eta}{\rho}} \qquad (12)$$

with the following implicit condition to be satisfied on z_0

$$20\sqrt{\frac{\eta}{\rho}} < \frac{z_0}{\sqrt{\frac{\partial u_t}{\partial n}}} < 100\sqrt{\frac{\eta}{\rho}} \qquad (13)$$

2.3. Boundary conditions

Table 2 summarizes all the boundary conditions considered here.

Table 2: Considered boundary conditions

	Parameter	Boundary condition
The wall law	height roughness z_0 [m]	$\frac{u_t}{u_*} = \frac{1}{\kappa}\ln\left(\frac{z}{z_0}\right) + B$
Input section	Logarithmic velocity profile [m/s]	$u_x = u$ $u_y = 0$
Upper edge of the domain	Uniform velocity [m/s]	$u = u_{uniform}$
Output section	Pressure [atm]	$P = P_{atm}$

2.4. Mesh used

For un unbounded domain, the primary difficulty in using CFD at a practical level is the uncertainty of creating a frontier and an initial mesh that is susceptible to produce a sufficiently accurate solution for the actual flow problem. Here, a parametric study has been performed in order to find the adequate domain and mesh size.

Let's consider the problem of a flat ground with an obstacle having the form of a half-ellipse: $125\,m$ of great axis and $50\,m$ of small axis. The air flowing on this ground has density $\rho = 1.225 kg/m^3$ and and dynamic viscosity $\eta = 1.7894 \times 10^{-5} kg/m/s$.

The domain geometry that has yielded convergence is presented in Figure 1. It is limited by a rectangle having height $h = 180\,m$ and width $\ell = 280\,m$. This figure shows the mesh of the first model that corresponds to the case of a flat gournd with an obstacle. The same mesh size was used for the problem of flat ground without obstacle, and in both cases, the finite element used has triangular geometry and linear dinterpolation function.

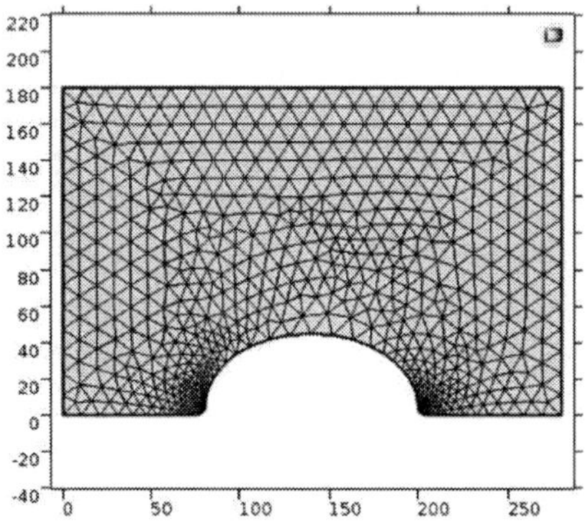

Fig 1. Mesh of the flat ground with obstacle.

3. Results and discussion

Taking an input air flow at the left boundary of the fluid domain that has the asymptotic magnitude of velocity $u_\infty = 18.5\,m/s$, simulations have been performed according to section 2 indications. An iterative procedure based on the condition of obtaining an invariable solution with regards to the frontier location and mesh size has been applied.

Figure 2 gives the obtained isovalues of the air speed magnitude for the case of ground with an elliptic like obstacle.

Figure 3 gives the results in terms of isovalues of air speed magnitude for the problem of flat ground without obstacle.

Figure 4 gives the air speed magnitude as function of the location on the cut-lines that are shown on figures 3 and 4 as dashed lines.

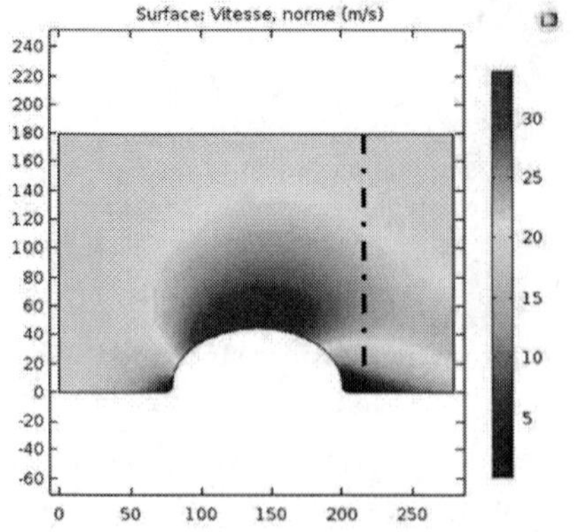

Fig.2. Isovalues of the air speed magnitude for the ground with an obstacle, the dashed line defines a cut-line.

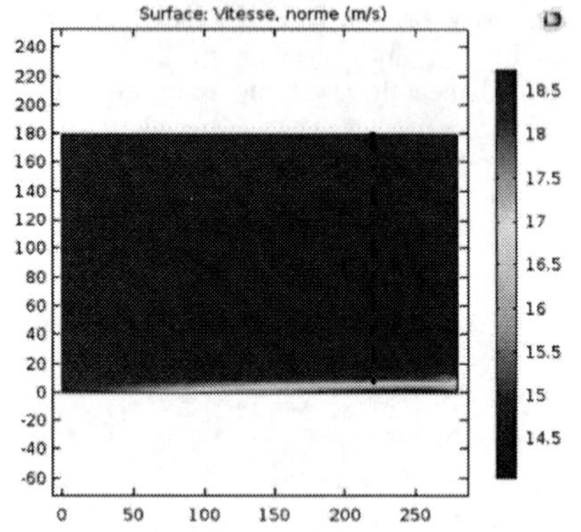

Fig.3. Isovalues of the air speed magnitude for the flat ground without obstacle, the dashed line defines a cut line.

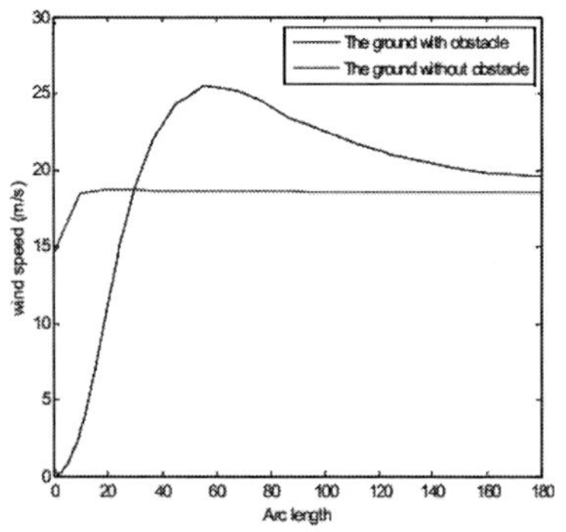

Fig.4. Air speed magnitude as function of the location on a cut-line; comparison between the ground with and without obstacle

From figure 3, one can see that air flow undergoes and acceleration due the obstacle effect. The air velocity jumpt from as the air velocity increases up to 65 m/s .

Figure 5 shows that the air flow velocity is largely affected by the obstacle. It is reduced behind the obstacle and amplified above the obstacle.

The obtained results make it possible also to find the shear stress and turbulence intensity which can be used to make accurate estimation of wind energy as function of the wind turbine location.

4. Conclusions

The local circular motion as induced by the existence of an obstacle having the form of an ellipse has been analyzed under stationary air flow assumption. Computational fluid dynamics based on solution of the turbulent air flow equations that consist of the Navier-Stokes equations coupled to the K-ε turbulence model was performed. The boundary conditions at the ground surface were applied by using an improved wall law. COMSOL software package was used. Accuracy of the model was assessed as function of both the location of the boundary fixing the truncated computational domain and the mesh size. The obtained results for the ground with an obstacle have shown significant discrepancies in comparison with the case of a flat ground without obstacle. The presence of a hill modifies drastically the air flow pattern and influnces hence to a large extent the extractable wind energy. This influence can be determined according to the actual obstacle geometry and location of the wind turbine.

References

1. P. Bahattacharya, *wind energy management* **1**, InTech, (2011). DOI: 10.5772/1530.
2. A.D.Young, *Boundary layers*, BSP Professional Books, Oxford, 180-182 (1989).
3. J.D. Anderson, *Fundamentals of aerodynamics*, 4th Edition McGraw-Hill, (2007).
4. Y.A. Cengel, *Heat Transfer: A Practical Approach*, 2nd Edition McGraw-Hill, (2007).
5. J.H. Ferziger, M. Peric, *Computational Methods for Fluid Dynamics*, 3rd Edition, Springer third, (2002). DOI: 10.1007/978-3-642-56026-2
6. J. Sumner, C.S. Watters, C. Masson. CFD in Wind Energy: The Virtual, Multiscale Wind Tunnel **3,** 989-1013 (2010). DOI: 10.3390/en3050989.
7. *Introduction to COMSOL Multiphysics,* version 5.1COMSOL, Sweden, (2015).
8. B.E. Launder, D.B. Spalding, The numerical computation of turbulent flows, *Computer Methods in Applied Mechanics and Engineering* **3**, 269-289 (1974).

MATEC Web of Conferences **83**, 09005 (2016) DOI: 10.1051/matecconf/20168309005
CSNDD 2016

Solution of inverse localization problem associated to multistatic radar system

M. Boutkhil[1], A. Driouach[1], A. Khamlichi[2]

[1]Department of Physics, Faculty of Sciences, 93030 Tetouan, Morocco
[2]Department TITM, National School of Applied Sciences, 93030 Tetouan, Morocco

Abstract. This work deals with the problem of inverse localization by a target with the aim to retrieve the position of the target, given the intensity and phase of the electromagnetic waves scattered by this object. Assuming the surface cross section to be known as well as the intensity and phase of the scattered waves, the target position was reconstructed through the echo signals scattered of each bistatic. We develop in the same time a multistatic ambiguity function trough bistatic ambiguity function to investigate several fundamental aspects that determine multistatic radar performance. We used a multistatic radar constructed of two bistatic radars, two transmitters and one receiver.

1 Introduction

Research in radar engineering over the last few decades has been mainly focused on improving target detection, parameter estimation accuracy, and reliable target classification in monostatic radar and bistatic radar. Generally, the transmitter and receiver share a common antenna, which is called a monostatic radar system. While a bistatic radar consists of separately located transmitting and receiving sites [1]. Despite recent advancement in radar component technology such as antennas, transmitters, receivers and processors, modern high requirements cannot always be met by traditional radars in many cases which gave birth to multistatic radar system [2].

Multistatic radar systems differ from traditional monostatic systems, or bistatic systems characterized by a separated transmitter and receiver; they consist of multiple transmitters and/or receivers.

A multistatic system consisting of multiple transmitters and one or more receivers can transmit multiple waveforms from collocated or distributed antennas [3], possibly illuminating a larger area than a monostatic system. The power of the transmitting antenna is a limiting factor on the extent of radar's visibility [4]; the use of multiple transmitting antennas to illuminate an area may help overcome the physical limitations of amplifier power and antenna aperture sizes [5].

Multistatic radar has been an area of intense research in recent years due to the inherent flexibility and importance of radar as an all-weather electromagnetic sensor and the additional theoretical advantages of a multistatic radar system. Traditional monostatic radars radiate electromagnetic energy from an antenna, the field propagates through space and is reflected in many directions from objects in the environment [6], and some of the reflected energy is received by the antenna of the radar, and through processing techniques information about the intercepted objects is extracted. Radar systems can be used to detect and track targets, detect moving targets in clutter rich environments, form images of stationary or moving targets or scenes, recognize meteorological conditions, and classify targets.

In this paper we aim to detect and localize two moving targets using multistatic radar with two bistatic radars, two transmitters and one receiver. We will simulate the first case with a polarimetric bistatic radar and secondly with a multistatic radar using two type of polarization, linear polarization and circular polarization.

2 Modeling and method

Radar can be classified by the topology of transmitting and receiving stations (i.e. by their locations). Both monostatic radar and bistatic radar employ a single transmitter and a single receiver. In the monostatic case, the transmitter and receiver are co-located and may share one common antenna, whereas in the bistatic case, the transmitter and receiver are spatially separated [7].

2.1 Bistatic Radar Model

Bistatic radar uses separate antennas at different locations for transmission and reception. Figure 1 shows a typical geometry of a bistatic radar, where a transmitter and a receiver are placed at two sites, separated by a distance L, normally comparable with the target distance, known as the baseline [8].

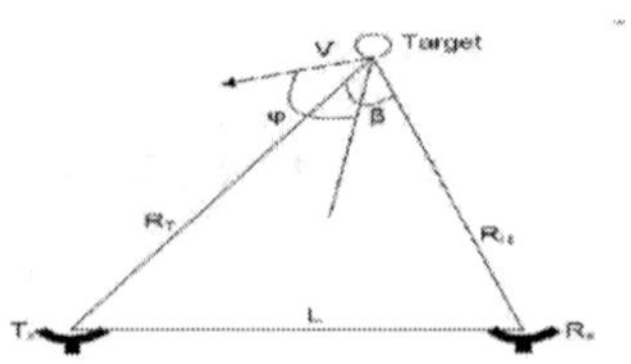

Fig.1 . bistatic radar topology

The angle between the transmitter and receiver with the vertex at the target is known as bistatic angle. In a bistatic radar system, the transmitter, receiver and target form the bistatic triangle, and furthermore the bistatic plane [5]. Unlike monostatic radar, bistatic radar usually measures the range sum, which equals to the total path length of the transmitter to target range R_t and the target to receiver range R_r.

$$R_t + R_r = \frac{1}{2}c\tau \tag{1}$$

where τ is the total delay from transmission to reception, and c is the speed of radio wave propagation.

In a bistatic radar system, when the transmitter and receiver positions are fixed, the bistatic Doppler fb at the receiver site generated by the moving target is calculated by the time rate of change of the total path length of the transmitted signal, normalized by wavelength:

$$f_b = \frac{\frac{dR}{dt}}{\lambda} = \frac{\frac{d(R_t+R_r)}{dt}}{\lambda} \tag{2}$$

where $\frac{dR}{dt}$ is the projection of target velocity onto the transmitter-to-target line-of-sight, given by:

$$\frac{dR_t}{dt} = V\cos(\phi - \beta/2) \tag{3}$$

Then the total Doppler shift is given by:

$$f_b = \frac{2V\cos\phi\cos(\beta/2)}{\lambda} \tag{4}$$

In the bistatic case, the transmitter and receiver are not collocated as in the monostatic case. The transmitter to target range and target to receiver range are R_t and R_r, respectively. The bistatic RCS depends on many factors such as look angle. It is not equal to monostatic RCS, and is often found to be close to the monostatic value measured along the bistatic bisector [2]. Similar to the derivation of the monostatic radar equation, the bistatic radar range equation is given by

$$R_t R_r = \left[\frac{P_t G_t G_r \sigma_b \lambda^2}{(4\pi)^3 k T_s B (SNR) L}\right]^{\frac{1}{2}} \tag{5}$$

Compared with monostatic radar range equation, a major difference here is that the range product replaces the range square.

2.2 Multistatic Radar

The idea of using several spatially separated, cooperative transmitting and receiving stations for effective energy use and better information retrieval has long been in the mind of radar engineers. However, there is no uniform terminology describing radar systems designed with this concept in mind. A couple of terms appear in the literature with slightly different definitions, such as multistatic radar [2] [9] [5], multisite radar [3], radar network [1], distributed radar [7] and netted radar [1]. Multistatic radar is used in this paper.

A typical arrangement of multistatic radar is showed in Figure 2. It is composed of three elements: radar stations, data processing units and the communication links between radar nodes. The basic components of netted radar are the traditional monostatic and bistatic radars.

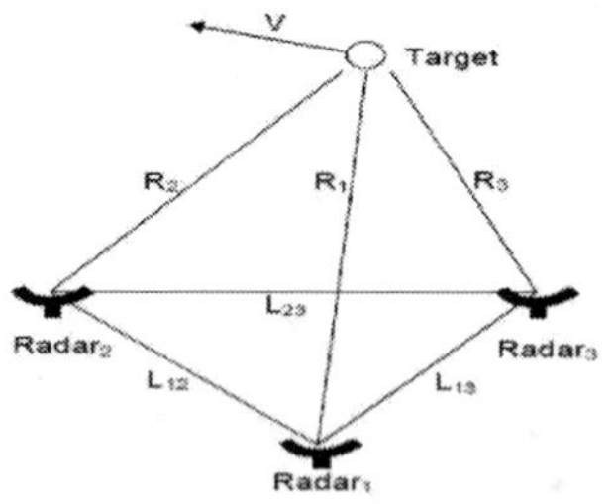

Fig.2 . Multistatic radar topology

Each station can be transmitting, receiving or both transmitting and receiving, giving rise to three operational modes: multiple monostatic operation, multiple bistatic operation and a combination of monostatic and bistatic operation. Most of the data received in radar stations are sent to the central unit for final processing.

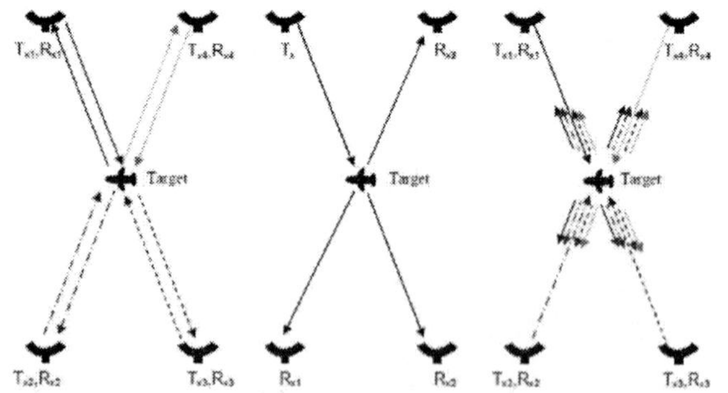

Fig.3 . Multistatic radar operation modes

As seen in figure 2 multistatic radar can be categorized as a multiple monostatic network, a multiple bistatic network, and a mixture of the two. In the multiple monostatic case, each radar can transmit a specific signal and receive the return originated from this unique transmitted signal only [1]. An example of the multiple bistatic case is a radar network comprising one common transmitter and N spatially separated receivers where each transmitter-receiver pair forms bistatic radar. Finally, when each station in the network can transmit and receive signals originated from any stations in the network, the system is fully netted.

2.2 Ambiguity function

The ambiguity function arose from detection and parameter estimation problems concerning a slowly fluctuating point target being observed in an additive white Gaussian noise [1]. It is widely recognized that the ambiguity function is an important tool to qualitatively assess radar performance in terms of resolution and clutter rejection. It should be noted that for a radar system the concept of resolution is different from accuracy. Resolution refers to the radar ability to distinguish different targets which are close to each other in terms of range and Doppler. It is obvious that the further separated in range and Doppler the two targets, the easier they will be distinguished one from the other. By passing through the received signal to a matched filter, the output is defined as ambiguity function, which is given by:

$$|\chi(\tau, v)| = \left|\int_{-\infty}^{\infty} u(t)u^*(t-\tau)\exp(j2\pi vt)\,dt\right| \quad (6)$$

The practical importance of the ambiguity function is that it describes the output signal from a matched filter when the input signal is time delayed by τ and Doppler shifted by v with respect to the nominal signal with which the matched signal is supposed to give the maximum output [4].

The two key parameters for ambiguity function calculation are the time delay τ , and the frequency shift v. From this definition it is seen that the ambiguity function associates two system parameters determining the radar performance, range and velocity, with two signal parameters, delay and Doppler. Therefore, it provides a tool to evaluate how well one can find the range and velocity of targets and how the design of signal can help.

As a useful tool, the ambiguity function is normally represented by graphical plots. Typically it is plotted in three dimensions with respect to the delay-Doppler plane. However, it is more meaningful to plot it on a range-velocity plane, because these two are the useful parameters for measurement, and extremely useful to show the influence of system geometry on the shape of ambiguity function.

3 Results and discussion

We used Matlab 2016 to simulate a multistatic radar system with two polarimetric bistatic radars to estimate the range and speed of two targets. The system operates at 300 MHz, using a linear FM waveform whose maximum the range resolution is 50 meters and the time-bandwidth product is 20.

Figure 4 shows the multistatic radar configuration with two polarimetric bistatic radars with two transmitters and one receiver, and the detected targets with the estimated positions.

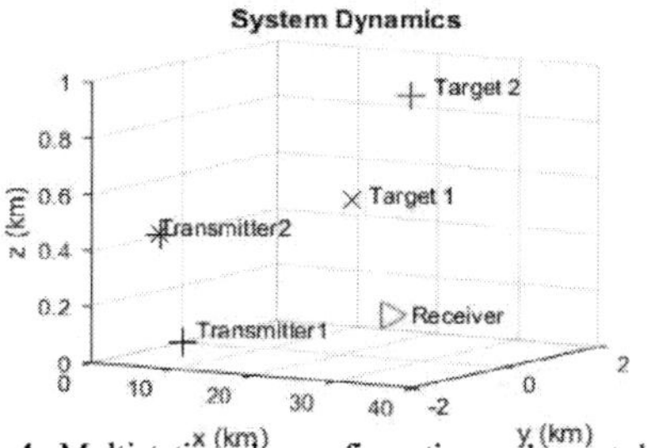

Fig.4 . Multistatic radar configuration and target detected positions

The Range-Doppler map simulation only shows the return from the first target. This is probably no surprise since both the transmit and receive array are vertically polarized and the second target maps the vertically polarized wave to horizontally polarized wave. The received signal from the second target is mostly orthogonal to the receive array's polarization, resulting in significant polarization loss.

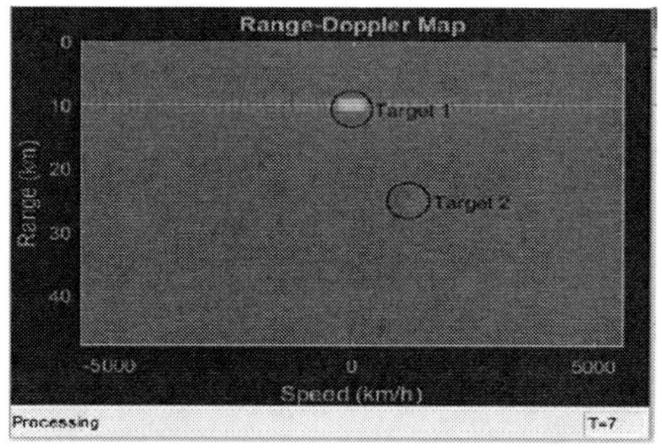

Fig.5. Multistatic radar range Doppler map with linear polarization

Vertical dipole is a very popular choice of transmit antenna in real applications because it is low cost and have an omnidirectional pattern. However, the previous simulation shows that if the same antenna is used in the receiver, there is a risk that the system will miss certain targets. Therefore, a linear polarized antenna is often not the best choice as the receive antenna in such a configuration because no matter how the linear polarization is aligned, there always exists an orthogonal polarization. In case the reflected signal bears a polarization state close to that direction, the polarization loss becomes huge.

One way to solve this issue is to use a circularly polarized antenna at the receive end. A circularly polarized antenna cannot fully match any linear polarization. But on the other hand, the polarization loss between a circular polarized antenna and a linearly polarized signal is 3 dB, regardless what direction the linear polarization is in. Therefore, although it never gives the maximum return, it never misses a target. A frequently used antenna with circular polarization is a crossed dipole antenna.

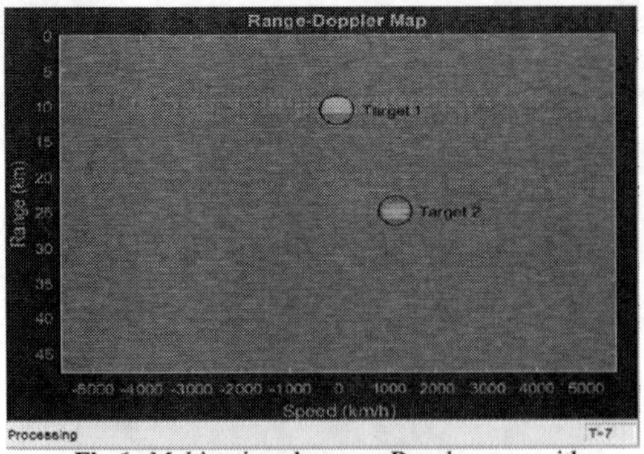

Fig.6 . Multistatic radar range Doppler map with circular polarization

The range-Doppler map now shows both targets at their correct locations.

References

1. Yu Teng, *Fundamental Aspects of Netted Radar Performance*, Electronic & Electrical Engineering UCL , (2010)
2. S. Kingsley S. Quegan. *Understanding Radar Systems*. McGraw-Hill Book Company Europe, Amsterdam, The Netherlands, (1992).
3. T. A. Seliga, F. J. Coyne. Multistatic radar as a means of dealing with the detection of multipath false targets by airport surface detection equipment radars. In Proceedings of *the 2003 IEEE Radar Conference,* 329–336, (2003).
4. H. Rohling, A. Hoss, U. Lubbert, M. Schiementz. *Multistatic radar principles for automotive Radar Net applications*. In German Radar Symposium (2002).
5. H. Steyskal, J. K. Schindler, P. Franchi, R. J. Mailloux. Pattern synthesis for TechSat21 -a distributed spacebased radar system. In IEEE Proceedings, *Aerospace Conference*, volume **2**, 725–732, (2001).
6. A. V. Mrstik. Multistatic-radar binomial detection. IEEE Transactions on Aerospace and Electronic Systems, **14**,103–108, (1978).
7. D. Bruyeren N. A. Goodman. Optimum and decentralized detection for multistatic airborne radar. IEEE Transactions on Aerospace and Electron ic Systems, **43**,806–813, (2007).
8. C. Yin, S. Xu, D. Wang. Location accuracy of multistatic radars (TRn) based on ranging information. In Proceedings, *1996 CIE International Conference of Radar*, 34–38, (1996).
9. W. Yang C. Wei. Target location and speed estimation by multistatic radar system using maximum likelihood approach. In WCCC-ICSP 2000. *5th International Conference on Signal Processing* Proceedings, **3**, 1964–1967, (2000).
10. L. Paradowski. Position estimation from netted radar systems with sensor position uncertainty. In 1996 Proceedings, *CIE International Conference of Radar*, 609–613, (1996).

MATEC Web of Conferences **83**, 09006 (2016) DOI: 10.1051/matecconf/20168309006
CSNDD 2016

Analyzing the effect of large rotations on the seismic response of structures subjected to foundation local uplift

N. El Abbas[1], A. Khamlichi[2], and M. Bezzazi[1]

[1]Faculty of Science and Techniques Tangier, Department of Physics, Box 416, 90 000 Tangier, Morocco
[2]ENSA Tetouan, Department TITM, Box 2222, 93030 Tetouan, Morocco

Abstract. This work deals with seismic analysis of structures by taking into account soil-structure interaction where the structure is modeled by an equivalent flexible beam mounted on a rigid foundation that is supported by a Winkler like soil. The foundation is assumed to undergo local uplift and the rotations are considered to be large. The coupling of the system is represented by a series of springs and damping elements that are distributed over the entire width of the foundation. The non-linear equations of motion of the system were derived by taking into account the equilibrium of the coupled foundation-structure system where the structure was idealized as a single-degree-of-freedom. The seismic response of the structure was calculated under the occurrence of foundation uplift for both large and small rotations. The non-linear differential system of equations was integrated by using the Matlab command ode15s. The maximum response has been determined as function of the intensity of the earthquake, the slenderness of the structure and the damping ratio. It was found that considering local uplift with small rotations of foundation under seismic loading leads to unfavorable structural response in comparison with the case of large rotations.

1 Introduction

The effect of the foundation uplift on the dynamic response of structures has been investigated by many researchers. Housner [1] was the first to study the problem of structures with uplift in detail and to observe some favorable effect of uplift on structural response magnitude. Meek [2] studied the effects of tipping-uplift on the response of a single-degree-of-freedom (SDOF) system and reported that allowing the SDOF system to tip/uplift altered its natural frequency and led to significant reductions in base reactions and in transverse deformations. Further Meek [3] performed analysis of a core stiffened buildings Meek and concluded that in comparison with a fixed-base core-braced structures, tipping greatly reduces the base shear and moment when subjected to seismic excitation.

Considering the flexibility of the structure and the soil to be represented as a Winkler foundation with large rotations leads to considerable difficulties in the governing equations of motion of the coupled soil-structure system. This is why few studies were dedicated to the complete representation of soil-structure interaction by equations of motion under the hypothesis of large rotation of foundation and the occurrence of $P-\Delta$ effect [4]. Instead simplified equations, consisting of only small rotations of the foundation uplift, have been considered [5].

The first objective of the present paper is to perform analysis of the effect on the seismic response of the structure that result from local uplift of foundation by considering both large and small rotations, but within the context of small deformation of the structure. The seismic response will be determined as function of the intensity of the earthquake, the slenderness of the structure and the damping ratio in vertical vibration of the system with its foundation mat bonded to the supporting elements. Then discrepancies that appear on the response when comparing the small base rotation case and the large base rotation case will be assessed.

The considered coupled soil-structure model takes into account the degrees of freedom related to mat lateral displacement, base vertical displacement and base rotation with this last being large. Derivation of the equations is first conducted then integration of the obtained system of ordinary differential equations is performed.

2 Materials and methods

2.1 Modeling of soil-structure interaction

The structure is assumed as a beam like mat which can be further characterized by its first mode of vibration. The structure is like this represented by a one degree of

 DOI: 10.1051/matecconf/20168309006

freedom linear system of mass m, lateral stiffness k and lateral damping c. The mat is supposed to be mounted on a rigid foundation basis that is assumed to react as a rigid rectangular plate of negligible thickness. The foundation mass denoted m_0 is taken to be uniformly distributed; the total moment of inertia is designated by I_0.

The soil-structure interaction takes place at the interface separating the rigid footing and the foundation soil. This interaction can be described by distributed springs and damping elements over the entire width of the foundation, figure 1 gives a schematic representation of the coupled system. The horizontal slippage between the mat and supporting elements is assumed to be negligible. The stiffness per unit length k_w and damping per unit length coefficient c_w of the foundation model are assumed constant and independent of displacement amplitude or excitation frequency. The base excitation is specified by the horizontal and vertical accelerations due seismic excitation. Under the influence of this excitation, the foundation mat may uplift through an angle θ and undergo a vertical movement v defined at its centre of gravity in the unstressed position.

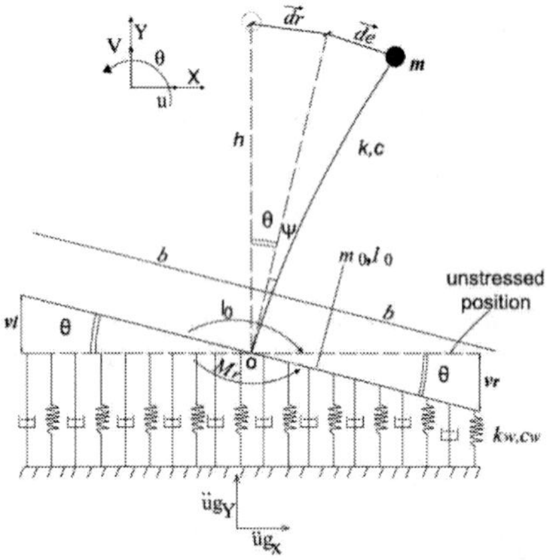

Fig. 1. Flexible structure on Winkler foundation

In figure 1, h designates the height of the structure from the base, M_r the total moment acting on the base mat, $\vec{d}_r$ the rigid horizontal acceleration, $\vec{d}_e$ the elastic horizontal displacement of the mat tip relative to the base, $\ddot{u}_g$ the seismic acceleration, v the vertical displacement of the centre of gravity of the base mat, θ the angle of rotation of the mat base, ψ the angle rotation of the structure, b half width of foundation mat.

Let us introduce the following notations:

$\omega=\sqrt{k/m}$ natural frequency of the rigidly supported structure,

$\omega_v=\sqrt{2k_w b/(m+m_0)}$ vertical vibration frequency of the system with its foundation mat bonded to the supporting elements,

$\alpha=h/b$ slenderness ratio,

$\beta=\omega_v/\omega$ frequency ratio,

$\gamma=m/m_0$ foundation mass to superstructure mass ratio,

$\xi=c/(2m\omega)$ damping ratio of the rigidly supported structure,

$\xi_v=c_w b/[(m+m_0)\omega_v]$ damping ratio in vertical vibration of the system with its foundation mat bonded to the supporting elements.

2.2 Equations of motion

The equations of motion of the entire system are derived by taking into account the equilibrium of the coupled foundation-mat system. The free body diagram of the system with inertial forces is shown in figure 2. The three equilibrium equations are:

- Equilibrium of forces acting on each degree of freedom in the horizontal direction: $\sum F_x=0$
- Equilibrium of forces in the vertical direction: $\sum F_y=0$
- Equilibrium of moments about the center of the foundation of the mat: $\sum M_z=0$

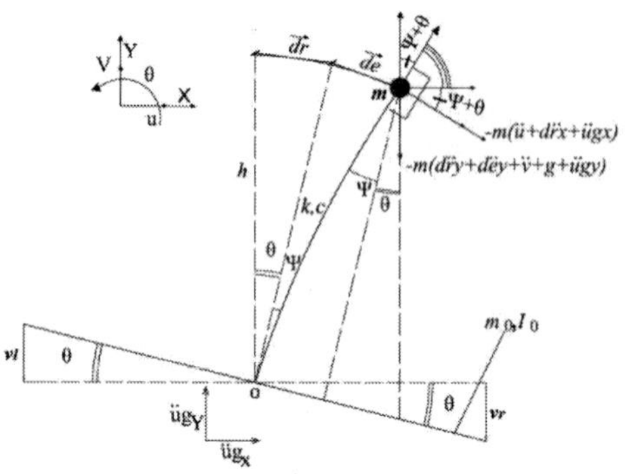

Fig. 2. Free body diagram of the system with uplift showing the considered dependent and independent degrees of freedom

2.2.1 Equations of motion for large rotations

Considering the equilibrium of forces in the lateral direction, the equation of motion in terms of the mat tip lateral displacement writes:

$$m(\ddot{d}_{ex}+\ddot{d}_{rx})\cos(\theta+\psi)+c\dot{d}_{ex}+kd_{ex}=-m\ddot{u}_{gx} \tag{1}$$

with

$$\ddot{d}_{rx}=\frac{h\ddot{\theta}}{2}\left[3\cos\left(\frac{3\theta}{2}\right)-\cos\left(\frac{\theta}{2}\right)\right]+\frac{h\dot{\theta}^2}{4}\left[-9\sin\left(\frac{3\theta}{2}\right)+\sin\left(\frac{\theta}{2}\right)\right] \tag{2}$$

The equilibrium of forces in the vertical direction can be written as

$$(m+m_0)(\ddot{v}+\ddot{d}_{ey}+\ddot{d}_{ry})+F_v=-(m+m_0)g+(m+m_0)\ddot{u}_{gy} \quad (3)$$

with

$$F_v=\int_{-b}^{b}(k_w v_{(\ell,r)}+c_w\dot{v}_{(\ell,r)})dx \quad (4)$$

$$\ddot{d}_{ry}=\frac{h\ddot{\theta}}{2}\left[\sin\left(\frac{\theta}{2}\right)-3\sin\left(\frac{3\theta}{2}\right)\right] +\frac{h\dot{\theta}^2}{4}\left[\cos\left(\frac{\theta}{2}\right)-9\cos\left(\frac{3\theta}{2}\right)\right] \quad (5)$$

$$\ddot{d}_{ey}=-h\cos(\theta)\left(-\psi^2\dot{\theta}+\dot{\psi}+2\dot{\psi}^2+2\psi\ddot{\psi}+\dot{\psi}\dot{\theta}+\psi\ddot{\theta}\right) +h\sin(\theta)\left(4\psi\dot{\psi}\dot{\theta}+\psi\dot{\theta}^2+\psi^2\ddot{\theta}-\ddot{\psi}\right) \quad (6)$$

The vertical displacements at the edges of the foundation mat, see figure3, measured from the initial unstressed positions are:

$$v_i=v\pm x\sin(\theta),\ i=l,r \quad (7)$$

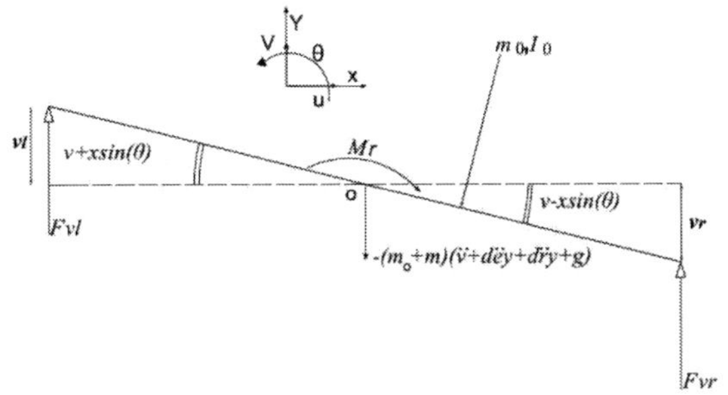

Fig. 3. Free body diagram for the base

In equation (4), F_v is the total vertical force acting on the base mat. This force is obtained as

$$F_v=k_w\int_{-b}^{b}(v\pm x\sin(\theta))dx+c_w\int_{-b}^{b}(\dot{v}\pm x\dot{\theta}\cos(\theta))dx \quad (8)$$

Because the Winkler foundation cannot extend above its initial unstressed position an edge of the foundation mat would uplift at the time instant when [1]:

$$v_i(t)>0,\ i=l,r \quad (9)$$

Calculating the integral in equation (8) yields the following equation:

$$\ddot{v}=-(1-\varepsilon_1^2)\varepsilon_2\xi_v\beta\omega\frac{b}{2}\left[\dot{\theta}\cos(\theta)+\frac{k_w}{c_w}\sin(\theta)\right] -(1+\varepsilon_1)\xi_v\beta\omega\left(\dot{v}+\frac{k_w}{c_w}v\right)-\left(\ddot{d}_{ry}+\ddot{d}_{ey}\right)-g+\ddot{u}_{gy} \quad (10)$$

with

$$\varepsilon_1=\begin{cases}1 & \text{contact at both edges}\\ \dfrac{\varepsilon_2 v}{b\sin(\theta)} & \text{one edge is uplifted}\end{cases} \quad (11)$$

$$\varepsilon_2=\begin{cases}-1 & \text{left edge uplifted}\\ 0 & \text{contact at both edges}\\ 1 & \text{right edge uplifted}\end{cases} \quad (12)$$

Taking the resultant moment about the centre of the foundation of the base mat, the following equation is readily obtained:

$$I_0\ddot{d}_{rx}+hm\left(\ddot{d}_{rx}+\ddot{d}_{ex}\right)\cos(\theta+\psi)+M_r=mh\ddot{u}_{gx} \quad (13)$$

where M_r is the resistant moment which represents the global action of spring and dashpot system acting on the foundation base. It is derived by considering the forces applied on the free body diagram of the base mat as:

$$M_r=k_w\int_{-b}^{b}\left[\pm vx\pm x^2\sin(\theta)\right]dx+ c_w\int_{-b}^{b}\left[\pm x\dot{v}\pm x^2\dot{\theta}\cos(\theta)\right]dx \quad (14)$$

The integral in equation (14) results in the following equation:

$$\ddot{\theta}=\frac{1}{\frac{h}{2}\left[3\cos(\frac{3\theta}{2})-\cos(\frac{\theta}{2})\right]}\Bigg\{-\frac{h\dot{\theta}^2}{4}\left[-9\sin\left(\frac{3\theta}{2}\right)+\sin\left(\frac{\theta}{2}\right)\right] +\frac{3h}{\gamma b^2}\left(2\xi\omega\dot{u}+\omega^2u\right)-(1-\varepsilon_1^2)\frac{3\varepsilon_2c_w}{2m_0}\left(\dot{v}+\frac{k_w}{c_w}v\right) -(1+\varepsilon^3{}_1)\frac{c_wb}{m_0}\left[\dot{\theta}\cos(\theta)+\frac{k_w}{c_w}\sin(\theta)\right] \quad (15)$$

The equations of motion of the system in case of large rotation are formed by equations (1), (10) and (15).

2.2.2 Equations of motion for small rotations

The equations of motion of the system under hypothesis of a small rotation of the foundation are obtained by using the same approach used in the case of large rotation and by letting the following approximations:

$$\cos(\psi+\theta)\approx 1 \quad (16)$$

$$\sin(\psi+\theta)\approx \psi+\theta \quad (17)$$

The three final equations of motion are then:

$$m(d_{ex}+h\ddot{\theta})+c\dot{d}_{ex}+kd_{ex}=-m\ddot{u}_{gx} \quad (18)$$

DOI: 10.1051/matecconf/20168309006

$$\ddot{v}=-(1-\varepsilon_1^2)\varepsilon_2\xi_v\beta\omega\frac{b}{2}\left(\dot{\theta}+\frac{k_w}{c_w}\theta\right)-$$
$$(1+\varepsilon_1)\xi_v\beta\omega\left(\dot{v}+\frac{k_w}{c_w}v\right)-(\ddot{d}_{rx}+\ddot{d}_{ex})-g+\ddot{u}_{gy} \quad (19)$$

$$\ddot{\theta}=-\dot{\theta}^2\left(-9\frac{3\theta}{2}+\frac{\theta}{2}\right)+\frac{12}{\gamma b^2}\left(2\xi\omega\dot{u}+\omega^2 u\right)$$
$$-(1-\varepsilon_1^2)\frac{6\varepsilon_2 c_w}{hm_0}\left(\dot{v}+\frac{k_w}{c_w}v\right)-(1+\varepsilon^3{}_1)\frac{4c_w b}{hm_0}\left(\dot{\theta}+\frac{k_w}{c_w}\right) \quad (20)$$

with

$$\varepsilon_1=\begin{cases}1 & \text{contact at both edges}\\ \dfrac{\varepsilon_2 v}{b\theta} & \text{one edge is uplifted}\end{cases} \quad (21)$$

and ε_2 having the same definition as in equation (12).

Finally, the equations of motion of the system in the both cases are:

- For large rotation they are formed by equations (1), (10) and (15).
- For small rotation they are formed by equations (18), (19) and (20).

These systems of ordinary differential equations are highly nonlinear. Their numerical integration can be achieved iteratively as the form of this system is not a priori known because of the conditions corresponding to equations (11), (12) et (21).

Integrating the three-non-linear ordinary system of differential equations by using the Matlab command ode15s enables to calculate the response of the structure and to perform parametric studies.

3 Results

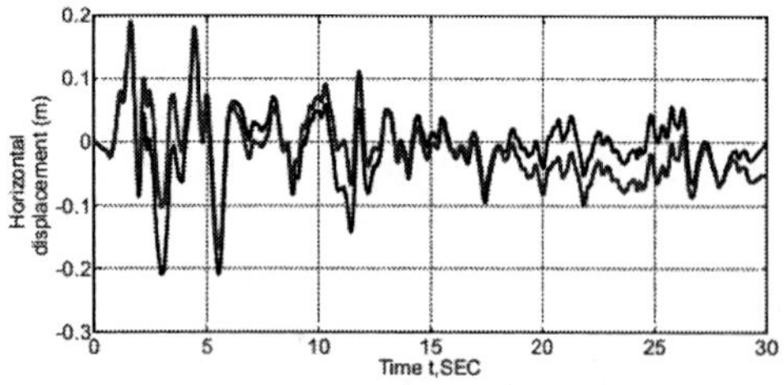

(a)

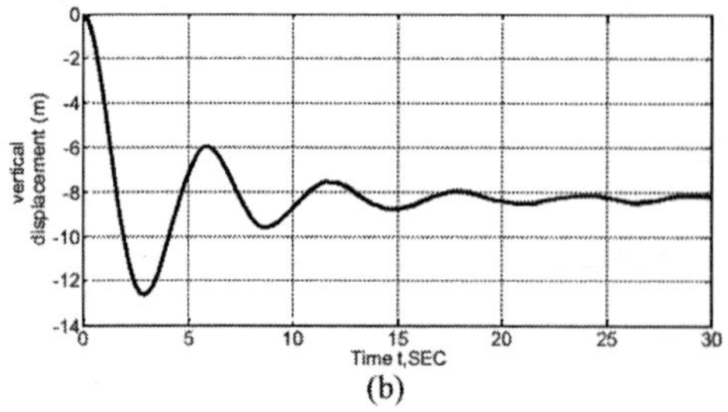

(b)

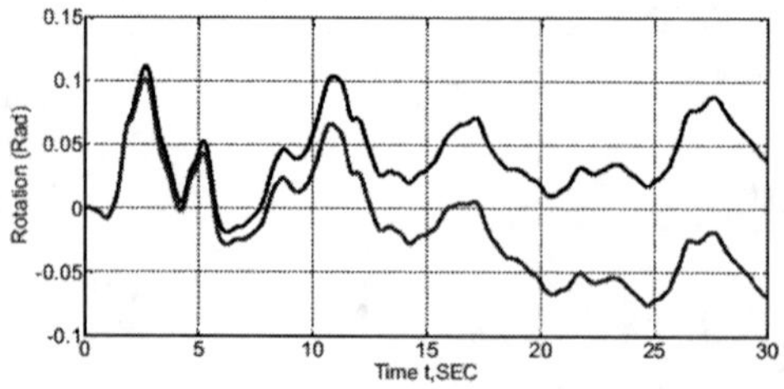

(c)

Fig. 4. Response of the structure under El Centro ground motion: (a) horizontal displacement; (b) vertical displacement; (c) base rotation; blue color is for large rotation of the base and black corresponds to the small rotation of the base

The seismic responses of the considered system are shown in figure 4 for the two hypothesis small and large rotation of foundation. The results present in terms of the lateral displacement of the structure, the foundation rotation and vertical movement to its center of gravity.

3 Conclusions

The effect of base uplift on the maximum response of a flexible structure which was taken to set up on a Winkler like foundation has been determined as function of the slenderness of the structure and the damping ratio in vertical vibration of the system with its foundation mat bonded to the supporting elements.

The obtained results lead to some discrepancies between the two cases: large and small rotations. Since the numerical cost is almost the same for the two hypotheses, the general case of large rotations can be considered in order for instance to integrate the $P-\Delta$ effect.

References

1. G. W. Housner, J. Bulletin of the Seismological Society of America, **53**, 403-417 (1963)
2. J.W. Meek, J. Struct. Div., **101**, 1297-1311 (1975)
3. J. W. Meek , J. Earthquake Engineering & Structural Dynamics, **6**, 437-454 (1978)
4. G. Oliveto, I. Cali, A. Greco, J. Earthquake Engineering & Structural Dynamics, **32**, 369-393 (2003)
5. M. Apostolou, N. Gerolymos, J. Bulletin of Earthquake Engineering, **8**, 309-326 (2009)

MATEC Web of Conferences **83**, 09007 (2016) DOI: 10.1051/matecconf/20168309007
CSNDD 2016

An adaptive control for a variable speed wind turbine using RBF neural network

E. El Mjabber[1], A. El Hajjaji[1], and A. Khamlichi[2]

[1]Faculty of Sciences, Department of Physics, 93030 Tetouan, Morocco
[2] National School of Applied Sciences, Department TITM, 93030 Tetouan, Morocco

Abstract. In this work, a controller based on Radial Basis Functions (RBF) for network adaptation is considered. The adaptive Neural Network (NN) control approximates the nonlinear dynamics of the wind turbine based on input/output measurement and ensures smooth tracking of optimal tip speed-ratio at different wind speeds. The wind turbine system and this controller were modeled and a program to integrate the obtained coupled equations was developed under Matlab/Simulink software package. Then, performance of the controller was studied numerically. The proposed controller was found to effectively improve the control performance against large uncertainty of the wind turbine system. comparison with nonlinear dynamic State feedback control with Kalman filter controller was performed, and the obtained results have demonstrated the relevance of this RBFNN based controller.

1 Introduction

Wind power generation has experienced these last decades a rapid development at both the hardware and software levels. The main challenge has been to get better performance of wind power conversion and to guarantee high quality standard of the produced electrical energy. Control of variable wind speed turbines has always been performed by considering different zones of wind speeds. In the following, focus is on the below power rated regime or partial load regime for which the generator power does not reach the nominal value. In this zone, wind speed remains lower than that of the rated power speed. For a given wind speed, there exists then an optimal rotor speed or equivalently an optimal generator speed that maximizes the power coefficient and hence the energy extracted from wind. The blade pitch angle is kept constant in this zone and equal to its optimal value. For wind speeds exceeding the rated wind speed, control strategy of wind turbine has a different objective which consists in maintaining constant the rated power output and consequently the rotor speed. This is performed by pitch control which limits the extracted mechanical power from wind. It is carried out by changing the angle of the blades around their longitudinal axis so as to limit the mechanical power at just a level providing a constant output power equal to its nominal value [1].

In order to maximize power extraction in the below rated power regime by tracking the optimal tip speed ratio, a generator torque based controller is generally used. As the wind turbine system suffers from nonlinearities and uncertainties, it is unfeasible to control this system within the framework of deterministic approaches. Deriving control for these systems can be better performed within the framework of modern approaches which are less sensitive to the mathematical model used for the system.

Various methods have been proposed for this purpose. A variable speed wind turbin e nonlinear control of the system was introduced by Boukhezzar and Siguerdidjane [2]. This controller is based on a nonlinear methodology and uses a special wind speed predictor. Although efficiency of this controller was found to be higher than those previously proposed, it reached only 74%. This required also a rather high generator torque and induced high fluctuations.

In [3] a proportional integral (PI) based torque control procedure is used and the optimal gains were tuned by particle swarm optimization and fuzzy logic theory.

Artificial neural network (ANN) based control techniques were found to be capable in many applications and even when the model of the system is poorly known. One type of ANN based methods is the radial basis function neural network (RBFNN) which uses radial basis function as activation functions. An online training neural network-based algorithm for wind speed estimation and adaptive control of wind turbine system for maximum power extraction was presented in [4]. The authors have used a PI controller and designed a radial basis function (RBF) neural network for approximating the mechanical torque and tuning the gains. The neural network identifier enabled to obtain the wind speed from the actual estimated optimal mechanical torque. This information was supplied to a block-back stepping controller in order to regulate the rotor by tracking its optimal speed value. Another control strategy which is based on RBFNN was introduced in [5]. The authors have proposed an algorithm which adjusts the generator torque output to its optimal value for any random wind speed input. In [6], an approach based on neural network was proposed. The controller which is based on the cloud RBF neural network and approximate dynamic programming was designed to track the maximum wind power speed without any need for wind speed detection.

The objective of this work is to design a controller which makes use of the general concept of RBFNN in order to track the maximum power in the partial loading regime of variable wind speed turbine, while achieving better efficiency and stability of the generator torque.

2 Modelling and control

2.1 Modelling of the wind turbine system

Wind energy is converted in wind turbines to kinetic and then to electric power. The extracted power from wind is limited by the actual system aerodynamics. It can be calculated as:

$$P_m = \frac{1}{2}\rho\pi R^2 C_p\left(\frac{\omega R}{v},\beta\right)v^3 \tag{1}$$

where ρ is the air density, R the rotor radius, C_p the power coefficient, ω the rotor angular speed, v is the wind speed and β is the pitch angle.

Thus, any change in the rotor speed or the wind speed induces change in tip-speed ratio leading to power coefficient variation. Without control, the generated power will permanently fluctuate and the system efficiency will be very bad. The aerodynamic torque coefficient is related to the power coefficient as follows

$$T_m = \frac{P_m}{\omega} \tag{2}$$

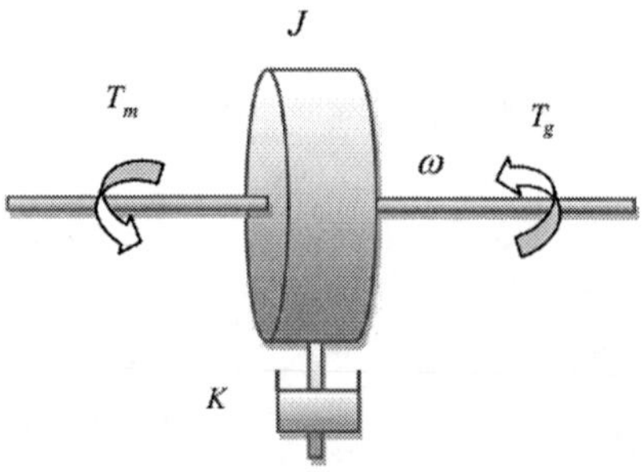

Fig. 1. One degree of freedom model of the wind turbine

The power coefficient depends on the actual wind turbine and C_p is given as function of $\omega R/v$ and the pitch angle β . An explicit expression of this coefficient can be found in [2].

Considering a one degree of freedom approximation of the wind turbine motion as presented in figure 1, the mechanical equation writes [2]

$$\dot{\omega} = \frac{1}{J}\left(T_m - T_g - K\omega\right) \tag{3}$$

where J is the turbine total inertia, T_g the generator torque and K the turbine total external damping.

2.2 Adaptive RBF controller design

In many classical control systems, some information of the system under study is required in the form of group of algebraic and differential equations. These equations express the relation between the inputs and outputs of the system. However, these mathematical models are not easy to develop especially for complex system. They may contain parameters which are difficult to measure, or may change during the system operation. Classical control theory also suffers from some limitations due to the assumptions made in designing the control systems such as linearity and time-invariance. These problems can be overcome by using artificial neural network (ANN) based control techniques.

Here, the RBFNN based method is used. The RBFNN have typically three layers: an input layer, a hidden layer with nonlinear RBF activation functions and a linear output layer [7]. The output of the hidden layer ψ_j is a Gaussian function having the form

$$\psi_j(X) = \exp\left(-\frac{\left\|X - C_j\right\|^2}{2\sigma_j^2}\right) \tag{4}$$

where X, C_j and σ_j are the input vector, centre and width of j_{th} hidden layer respectively.
The output layer writes as follows

$$f(X) = W^T\psi(X) \tag{5}$$

The training process of the RBFNN is done by the following algorithm [4]
Step 0: Initialize all weights at random.
Step1: Calculate each element of the output vector f by equation (5)
Step 2: Calculate the error ε_i of each neuron in the output layer by the following equation:

$$\varepsilon_i = f_i - \hat{f}_i,\ \text{i=1,2,…,m}$$

where $\hat{f}$ is the desired output vector of the ith neuron in the output layer
Step 3: Update the weights using the following equation:

$$W_{ij}(n+1) = W_{ij}(n) + \alpha\varepsilon_i(n)\psi_j, \quad (6)$$

i=1,2,...,m, j=1,2, ... ,k

Step 4: calculate the total error ε_T according to:

$$\varepsilon_T = \sum_{n=1}^{k}\sum_{j=1}^{k}\left(y_j(n) - \hat{y}_j(n)\right)^2 \quad (7)$$

Step 5: Return to step 2 and repeat the calculations until ε_T is less than the desired error.

We focus in this work on performing torque control for optimal power extraction at the below rated wind speed, region 2. The controller should optimize the energy capture from the wind by tracking the optimal rotor speed, with in addition reducing to the least the fluctuations affecting the generator torque. In this region, the blade pitch angle β_{opt} and the tip speed ratio λ_{opt} are set to their optimal values. The rotor speed must then be adjusted continuously according to the actual optimal rotor speed ω_{opt} which is given by

$$\omega_{opt} = \frac{\lambda_{opt}}{R}v \quad (8)$$

This adjustment is performed by using the generator torque T_g.

The wind turbine dynamical Eq. (3) can be written under the following standard affine form

$$\dot{\omega} = f(\omega) + g\,u + d \quad (9)$$

with

$$f(\omega) = \frac{\rho\pi R^2}{2J}\frac{C_p(\beta,\lambda)v^3}{\omega} - \frac{K}{J}\omega \quad (10)$$

the plant nonlinear function and where u is the control input, d an outer disturbance satisfying $|d| \le D$ and $g = -\frac{1}{J}$.

Using the RBFNN, we train a nonlinear approximation of the system function f so that

$$f(\omega) = W^T\psi + v \quad (11)$$

and v is an approximation error of the neural network with $|v| \le v_{max}$.

A first-order dynamic response is selected for the rotor speed tracking error:

$$\dot{e} + \zeta e = 0 \quad (12)$$

where ζ is a constant gain, the error is fixed by the actual difference between rotor speed ω and optimal rotor speed ω_{opt}:

By using Eq. 9, Eq. 12 and Eq. 13, the control law is considered under the following form:

$$u = \hat{T}_g = \frac{1}{g}\left[-\hat{f}(e) + \zeta e + \dot{\omega}_{opt}\right] \quad (14)$$

The unknown nonlinear function $\hat{f}(e)$ can be represented through the RBFNN as:

$$\hat{f}(e) = \hat{W}^T\psi(e) \quad (15)$$

To ensure the stability of the controller defined by equation (12), the Lyapunov method is suggested.

Proof.

Considering the following stability function:

$$V = \frac{1}{2}e^2 \quad (16)$$

The derivative of the above function writes

$$\dot{V} = e\dot{e} \quad (17)$$

By using Eq. (12), one arrives at:

$$\dot{e} = \zeta e \quad (18)$$

$$\dot{V} = -\zeta e^2 \quad (19)$$

Then, the system is stable if $\zeta > 0$, as $\dot{V} < 0$.

3 Results and discussion

For the simulation of the RBFNN based controller, the Controls Advanced Research Turbine (CART) is considered. This wind turbine has a nominal power rating of 600 kW and a hub height of 36 m. It is a 43 m diameter, two-bladed, teetered hub machine. This wind turbine was considered in [2] to study performance of the Nonlinear Dynamic State Feed-Back Control with Kalman estimation (NDSFCK).

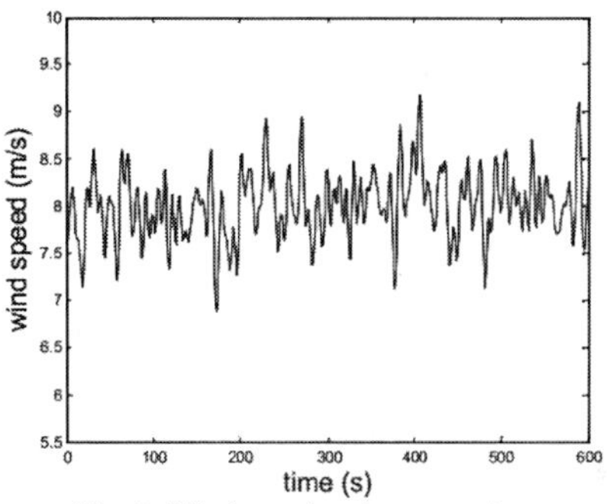

Fig. 2. Wind speed at the system input

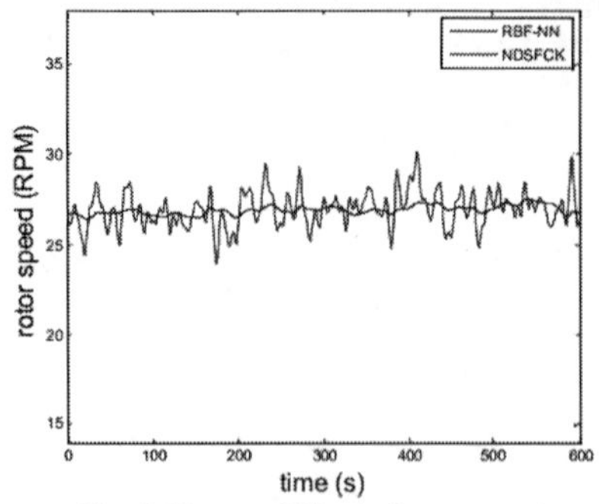

Fig. 3. Time variations of rotor speed

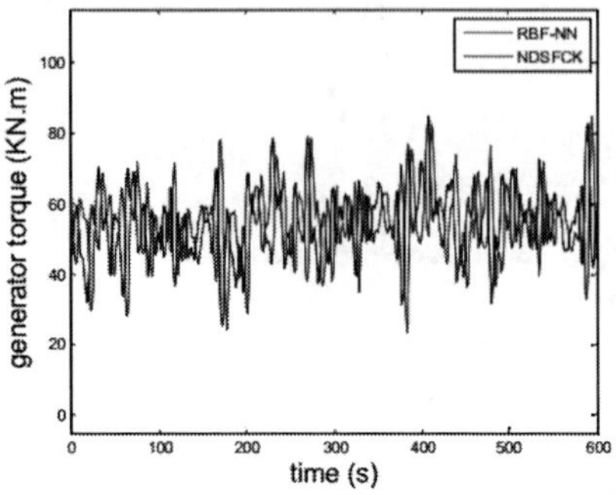

Fig.4 . Time variation of the generator torque

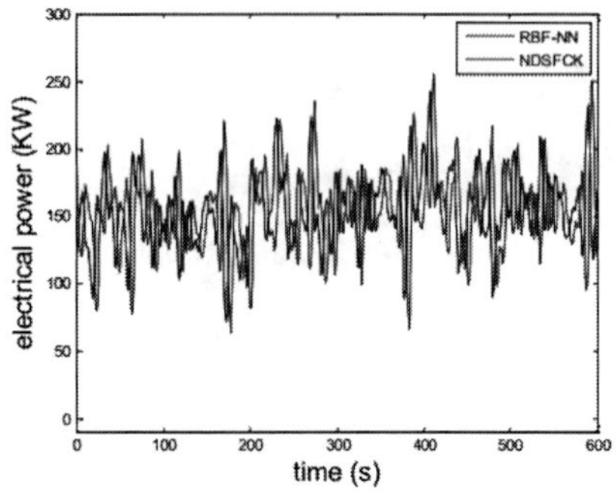

Fig.5 . Time variation of the electrical power

Table 1. Performance comparison between RBFNN based controller and that based on the NDSFCK [2]

Control strategy	RBFNN	NDSFCK
Std(Tg)(kN.m)	9.4	9.8
Max(Tg)(kN.m)	84	85
Efficiency (%)	76	74

Wind data used for the simulation is given in figure 2. The mean wind speed is $8m/s$.

Figure 3 gives rotor speed as function of time. Figures 4 and 5 give time variations of the generator torque and electrical power. Table 1 gives a comparison between the RBFNN and the NDSFCK controllers. Figure 3 shows that the obtained rotor speed is very stable with fluctuations that are considerably reduced in comparison with the rotor speed obtained through the NDSFCK. As shown in figure 4 the generator torque is very stable for the RBFNN based controller. Its mean value is lower than that of the NDSFCK. Figure 5 shows that better tracking of the optimal rotor speed is performed by the RBFNN controller, so more efficient electrical power quality is obtained. From Table 1, one can note that a gain of 2% is achieved by the RBFNN controller in comparison with the NDSFCK controller, with a reduce standard deviation of the generator torque.

4 Conclusions

A RBFNN based controller was proposed for variable wind speed turbine control by acting on the rotor speed. This was considered in the partial load zone. The controller gave good results, and performed better than the NDSFCK controller by increasing efficiency of energy, while it required lower generator torque with lower fluctuations.

References

1. D. Kumar, K. Chatterjee, A review of conventional and advanced MPPT algorithms for wind energy systems, Renewable and Sustainable Energy, **55,** 957-670 (2016).
2. B. Boukhezzar, H. Siguerdidjane, Nonlinear control of a variable-speed wind turbine using a two-mass model, IEEE Transactions on Energy Conversion, **26**,149-162 (2011).
3. M. Sheikhan, R. Shahnazi, A.N. Yousefi, An optimal fuzzy PI controller to capture the maximum power for variable-speed wind turbines, J. Neural Comput & Applic, **23**, 1359-1368 (2013).
4. R. Poutlangari, R. Shahnazi, M. Sheikhan, RBF neural network based PI pitch controller for a class of 5-MW wind turbines using particle swarm optimization algorithm, Expert systems with applications, **51,** 641-648 (2012).
5. F. Jaramillo-Lopez, G. Kenne, F. Lamnabhi-Lagarrigue, A novel online training neural network-based algorithm for wind speed estimation and adaptive control of PMSG wind turbine system for maximum power extraction, Renewable Energy, **86,** 38-48 (2016).
6. Q.W. Zhong, J.J Wen, L.R. Zhao, C.H. Wu, Maximum wind power tracking based on cloud RBF neural network Renewable Energy, **86**, 466-472 (2016).
7. J. Liu, *Radial basis function neural network control for mechanical systems*, (Springer, London, 2013).

MATEC Web of Conferences **83**, 09008 (2016) DOI: 10.1051/matecconf/20168309008
CSNDD 2016

Effect of atmospheric stability on the estimation of wind power potential

M. M. Ichenial [1], A. El Hajjaji [1], and A. Khamlichi [2]

[1]Department of Physics, Faculty of Sciences at Tetouan, 93030, Morocco
[2]Department TITM, National School of Applied Sciences, Tetouan 93030, Morocco

Abstract. Analyzing the impact of atmospheric stability on vertical wind profile enables to assess the available wind resource in order to determine optimal locations for wind turbines. Different methods have been proposed to study the degree of stability of the atmosphere. These include the Richardson number, the profile method and the Monin-Obukhov length. Focus is here on the stability issue and its influence on the wind shear coefficient which plays an important role in processing experimental data with regards to the atmospheric surface-layer turbulence. Data collected from the Cabauw meteorological mast were analyzed in order to determine the characteristics of the wind resource at that location and to estimate the wind resource for power generation as affected by stability.

1 Introduction

Wind energy has evolved tremendously during the last few decades. Most of wind projects are developed based on the mean wind speed at the hub height of the turbine without any consideration of atmospheric stability effect. However, significant dependence of wind resource on this stability issue has been recognized [1,2]. Stability is linked to the influence of the thermal effects. Unstable stratification occurs when there is a lot of surface heating. The warm air rises and the result is a thick boundary layer with heavy turbulence in the air. If the adiabatic cooling effect causes the rising air to become colder than its surroundings, its vertical motion will be suppressed. This is called a stable stratification. It occurs at cold nights when the ground surface is cold. In this situation, the movement of the air is dominated by friction with the earth surface and the increase in mean wind speed with height is large. In a neutral atmosphere, adiabatic cooling of the air as it rises is such that it remains in thermal equilibrium with its surroundings. This is often the case with strong winds. For wind energy utilization, neutral stability is the most desirable situation [3].
The mean wind speed constitutes only a rough measure of the wind resource and does not provide a high accurate estimate of energy production. The wind resource must also be characterized by the variations in speed and direction, as well as air density, in time and space [4]. Sufficient temporal and spatial dimensions of wind resource assessment are required for suitable wind farm design. Focus is here on the stability issue as it affects the wind shear coefficient for a particular site where stability is found to be significant. Three stability criteria are applied and the modified wind shear coefficients are obtained and used to evaluate the estimated energy production for that site.

2 Stability of atmospheric boundary layer

The atmospheric boundary layer is the lowest part of the atmosphere and its characteristics are directly influenced by contact with earth's surface. Here, physical quantities such as velocity, temperature, and relative humidity can change rapidly in space and time. An important parameter in the characterization of the wind resource is the variation of horizontal wind speed with height above the ground. One would expect the horizontal wind speed to be zero at the earth's surface and to increase with height in the atmospheric boundary layer. This variation of wind speed with elevation is called the vertical profile of the wind speed or vertical wind shear. The variation of 'instantaneous' profiles is related via the similarity theory of boundary layers. On the other hand, the changes in long- term averages as a function of height are related to the statistics of occurrence of various influencing factors, such as atmospheric stability and surface roughness.
A particularly important characteristic of the atmosphere is its stability – the tendency to resist vertical motion or to suppress existing turbulence. The stability of the atmospheric boundary layer is a determining factor for the wind speed gradient or wind shear that are experienced in the first few hundred meters above the ground. Numerous classification systems have been developed to define the atmospheric stability. In this study we use the flowing methods:

DOI: 10.1051/matecconf/20168309008

- Monin- Obukhov similarity theory [5];
- Richardson Numbers [6-9];
- Profile Methods [10,11].

These methods are applied in the following to a site which is located near the Cabauw village in the center of Netherlands for which wind data is available.
This location was chosen because its topography is very suitable for boundary layer measurements. The surroundings are flat within a radius of at least 20 km.

3 Wind data from the Cabauw site area

The Cabauw tower is located at the geodesic coordinates 51°58'16"N and 04°55'36"E. The foot of the tower is at 0.5m below mean sea level. The distance to the sea is about 45 km. The weather in the area is characterized by a moderate maritime climate, with a prevailing westerly circulation. Meteorological data for this site are from Rotterdam Airport: KNMI meteorological station. These data are related to Greenwich Mean Time. As identified in 1987, the meteorological roughness height of the site was fixed to the value0.15cm.

Measurements are available as 10 min averages over several years. Here data of the years 2011, 2012 and 2013 are used. Wind velocity and direction are measured at 10, 20, 40, 80, 140 and 200m altitude. The considered data refer here to wind velocity at the usual observation height of 10m and at 80m hub height. Wind shear will be presented for this height range as well as the range 40-140m which corresponds to a common rotor of height between 40 and 120 m.

During the three years (2011-2013) period, annual mean values of wind speed, turbulence intensity, wind power and total wind energy were as follows: 7.2 m/s, 14.3%, 110.25 W/m^2, 72.91 kW/m^2 respectively.

3.1 Analysis of stability

The measurements from the Cabauw tower are pre-processed before analyzing stability according to the three methods mentioned above. The results are filtered with a stationary filter that excludes the values measured when the mean conditions of air are changed rapidly and for which the airflow cannot be considered as a stationary field [12].

The pie chart of the stability parameter is displayed in figure1. One can see from this figure that except the Profile Method 1, the other methods used to estimate the atmospheric stability give closer results.
A continuous display of the stability distribution will enable to have a better idea about the atmospheric stability of the chosen site. A comparison is conducted in the following at multiple levels [13]:

- Atmospheric stability distribution per hour of the day;
- Atmospheric stability distribution per month;
- Atmospheric stability distribution per wind direction.

Wind data acquired during the period 2011-1013 for Cabauw was processed form the stability viewpoint.

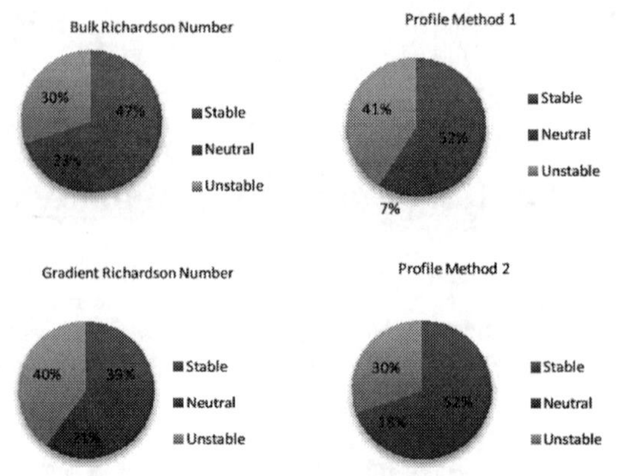

Fig. 1. Comparison of atmospheric stability distribution divided into three classes.

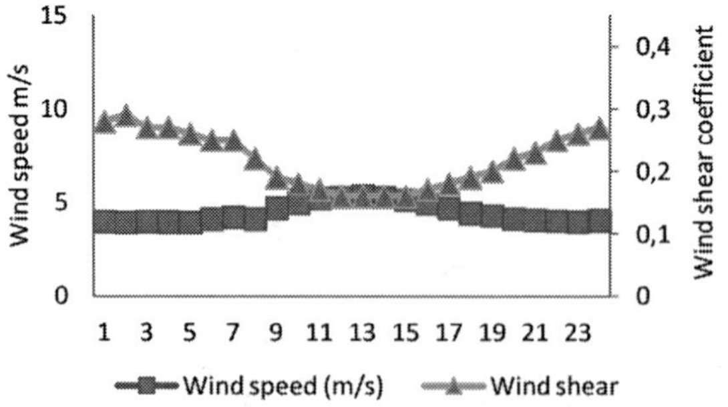

Fig. 2. Variation per hour of the day by stability class of 10-min annual mean wind speed and wind shear observed.

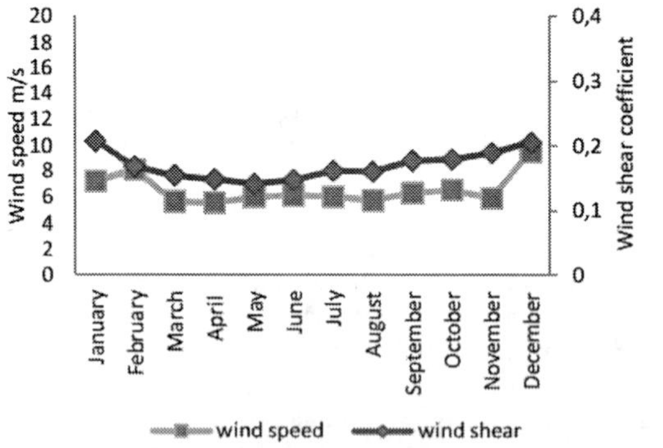

Fig. 3. Monthly variation by stability class of 10-min annual mean wind speed and wind shear coefficient observed

Figure 2 gives variation per hour of the day by stability class of 10-min annual mean wind speed and WSC observed.
Figure 3 gives monthly variation by stability class of 10-min annual mean wind speed and wind shear coefficient observed.

DOI: 10.1051/matecconf/20168309008

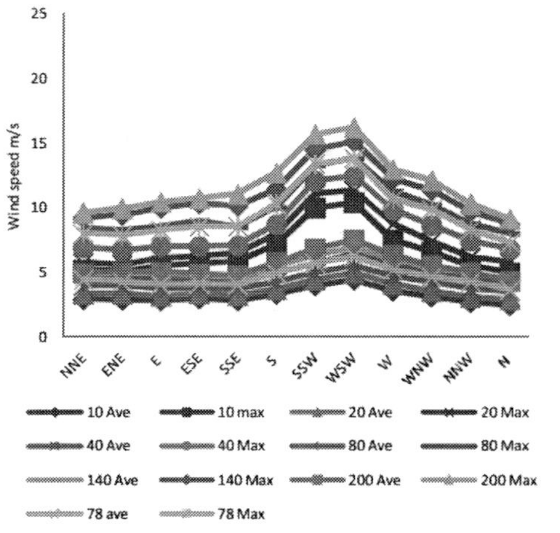

Fig. 4. Variation per wind direction by stability class of 10-min annual mean wind speed and wind shear coefficient observed.

Figure 4 gives variation per wind direction by stability class of 10-min annual mean wind speed and wind shear coefficient observed.

Figure 2 shows that according to the Bulk Richardson Number, the atmospheric stability is mainly very stable with almost no unstable and very unstable condition during the night. Comparing this with estimation of the Profile Method 2, the very stable conditions occur less frequently and near neutral and stable conditions occur more frequently. The Profile Method 1 describes the atmosphere with almost only extreme conditions and therefore compares badly with the Bulk Richardson Number method and Profile Method 2.

Figure 3 shows that according to Bulk Richardson Number and Profile Method 2 small fluctuations between each month can be observed. With in the months of March, April, May and June a peak of around 30% of very unstable conditions is observed. The Profile Method 1 and Gradient Richardson Number give a much higher estimation of the very unstable conditions, with the peak occurring in the same months. This indicates seasonal influences on the atmosphere stability, which are visible in the distribution of the unstable and very unstable conditions.

Figure 4 shows that the Profile Method 1 gives a higher frequency of occurrence for very unstable conditions and lower frequency of occurrence for unstable and neutral conditions compared to the Bulk Richardson Number, Gradient Richardson Number and Profile Method 2.

3.2 Effect on power production

The effect of atmospheric stability as determined for the wind data from the Cabauw site has been analyzed for the reference wind turbine: Gamesa G80-2.0MW.

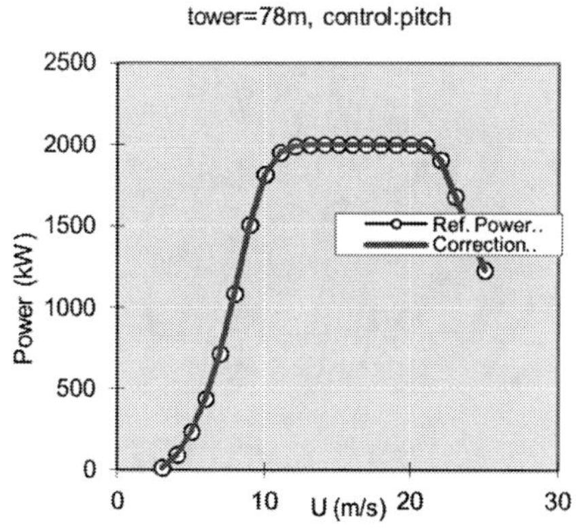

Fig. 5. Power as function of the wind speed for the wind turbine, Gamesa G80-2.0MW.

Figure 5 gives the Gamesa G80-2.0MW power characteristic.

Processed results from the Cabauw area have shown that the shear exponent in the lower atmosphere boundary layer (<200m) in day time varies in the interval 0.1 to 0.2. This corresponds to a wind velocity ratio U_{80}/U_{10} which varies between 1.2 to 1.5.

At night the situation is quite different as the shear exponent has a much wider range with values up to 0.8, but more usually between 0.2 and 0.6 showing that the site indeed was suitable to study the effect of atmospheric stability on wind potential.

Figure 6 gives the estimated energy as function of wind direction.

Figure 7 gives the estimated wind turbine energy production per month.

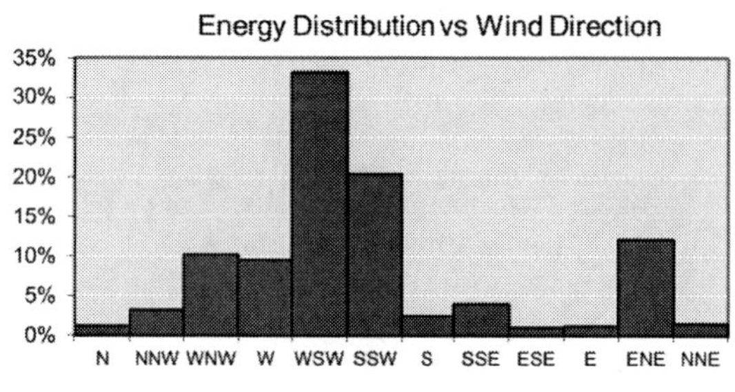

Fig. 6. Energy distribution as function of wind direction for the Gamesa G80-2.0MW in Cabauw site

Taking into account atmospheric stability criteria, figure 7 shows that wind turbine energy production is largely affected by the wind shear coefficient which results in form the state of stability. The wind shear coefficient determines the vertical wind speed distribution over the rotor plane. The predicted wind energy production is like this different from that one associated to a constant overage wind shear coefficient.

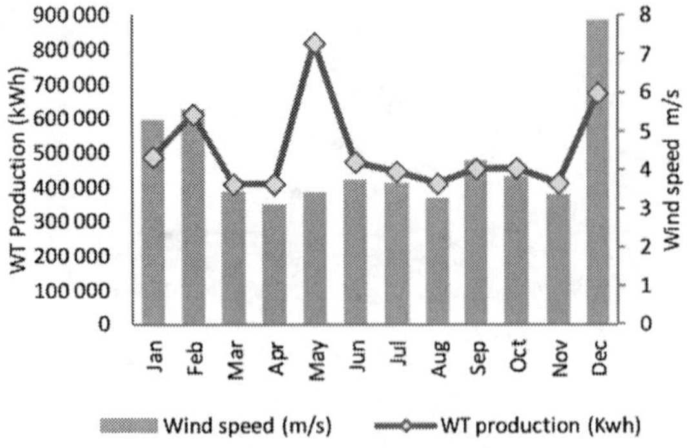

Fig. 7. Variation of the wind turbine energy production per month for the Gamesa G80-2.0MW in Cabauw site

4 Conclusions

Atmospheric stability is a key issue for estimating the wind resource that is available for a given site. The stability state modifies the wind shear coefficient and energy as it can be captured by a standard wind turbine.

In this work, various stability criteria have been applied to measurement data for a particular site. The obtained results in terms of energy production have shown that stability can yield significant variations in comparison with the constant profile method based on the average wind shear coefficient.

References

1. A. Sathe, J. Mann, T. Barlas, W.A.A.M. Bierbooms, G.J.W. Bussel, Influence of atmospheric stability on wind turbine loads, Wind Energy, **16**, 1013-1032 (2013)
2. B. Lange, S. Larsen, J. Højstrup, R. Barthelmie Importance of thermal effects and sea surface roughness for offshore wind resource assessment, Journal of Wind Engineering and Industrial Aerodynamics, **92**, 959-88 (2004)
3. J. Rohatgi, An Analysis of the Influence of Atmospheric Stability on Vertical Wind Profiles Its Influence on Wind Energy and Wind Turbines, Wind Eng. 319–332 (1996)
4. H. Erich, *Wind Turbines: Fundamentals, Technologies, Application, Economics* 2nd edition, (Springer, Germany, 2006)
5. T. Foken, 50 Years of the Monin -Obukhov Smilarity Theory, Boundary-Layer Meteorology, **119**, 431-44 (2006)
6. A. Sathe, W. Bierbooms, *Influence of Different Wind Profiles due to Varying Atmospheric Stability on the Fatigue Life of Wind Turbines,* (The Science of Making Torque from Wind, IOP Publishing, 2007)
7. B. Lange, S. Larsen, J. Hojstrup, R. Barthelmie, The influence of thermal effects on the Wind Speed Profile of the Coastal Marine Boundary Layer. Boundary-Layer Meteorology, **112**, 587–617, (2004)
8. M. Mohan, T.A. Siddiqui, Analysis of various schemes for the estimation of atmospheric stability classification, Atmospheric Environment, **32,** 3375-3781 (1998)
9. A. Grachev, C. Fairall, Dependence of the Monin-Obukhov Stability Parameter on the Bulk Richardson Number Over the Ocean, Journal of Applied Meteorology, **36**, 406-415 (1996)
10. A. A. M. Holtslag. Estimates of diabatic wind speed profiles from near-surface weather observations, Boundary-Layer Meteorology, **29**, 225–250 (1984)
11. M. Motta, R. J. Barthelmie, P. Volund, The influence of non-logarithmic wind speed profiles on potential power output at Danish offshore sites, Wind Energy, **8**, 219–236 (2005)
12. D.A. Spera, *Wind Turbine Technology: Fundamental Concepts of Wind Turbine Engineering*, second ed., (ASME Press, New York, 2009)
13. J.J.W. Hutschemaekerrs, comparison of classification systems to define atmospheric stability and their impact on wind turbine design, (Delft, 2014)

DOI: 10.1051/matecconf/20168309009

Reinforcement of RC structure by carbon fibers

B. Kissi[1], Y. Riyad[2], I. Mrani[2], B. EL Ghoulbzouri[3], M.A. Parron[4], and Y. El khatib[1]

[1]University Hassan II, ENSAM Casablanca, Morocco
[2]University Chouaib Doukkali, FS El-jadida, Morocco
[3]National School of Applied Sciences of Al Houceima, Morocco
[4]University of Cadiz, Ramon Pujolavenue, Algeciras 11202, Spain

Abstract. In recent years, rehabilitation has been the subject of extensive research due to the increased spending on building maintenance work and restoration of built works. In all cases, it is essential to carry out methods of reinforcement or maintenance of structural elements, following an inspection analysis and methodology of a correct diagnosis. This research focuses on the calculation of the necessary reinforcement sections of carbon fiber for structural elements with reinforced concrete in order to improve their load bearing capacity and rigidity. The different results obtained reveal a considerable gain in resistance and deformation capacity of reinforced sections without significant increase in the weight of the rehabilitated elements.

1 Introduction

The disorders occurring in structures are often due to the degradation of the materials used, or the change in the load of the structure (excess load).The deterioration of materials (including reinforced concrete) over time, or the structural modification of buildings spur the development of efficient technologies of reinforcement.

The overriding criteria in the evaluation of these technologies are the sustainability of the technology solution and the lowering of the implementation constraints [1].

The method which has emerged in recent years is the reinforcement by composite materials [2]. This method uses the bonding of composite plates (Textile unidirectional or bidirectional reinforcement, strips, carbon fibers...) or in situ direct lamination of composites (textile fabrics, glass or carbon) associated to thermosetting polymer matrices (epoxy resins, polyester ...).

This technique of reinforcement is an innovative alternative to the traditional solution using bonding metal plates. Its use is relatively recent. It dates back to about a decade in Civil Engineering [3].It began to be marketed in the 1980s in the United States, Japan and Europe. The development of this technology in the USA and Europe is due to the need of reinforcing the aged structures or the ones which their functionality has changed.

The reinforcement of beams can occur in two ways [4]; the first way is the resumption of flexural strains; two reinforcing techniques are used to take up the bending efforts: the bonding plates or bonding strips.

Gluing plates on strained concrete areas is the oldest methods, it suffers from a big disadvantage: its lack of adaptability. It presents a difficulty of implementation, in fact the adhesion between the plate and the reinforced element must coincide perfectly.

The bonding of strips technique appeared in 1980. It is similar to the other method, the only difference is that the bonded element is adjustable to the reinforced structure.

The second way is the resumption of shear strains; it is done using reinforcements arranged transversely to the axis of the beam to strengthen the transverse reinforcement.

Again, the reinforcements may be in the form of strips or plates but the predominant method is the bonding strips. FRP (fiber reinforced polymer) consists of fibers arranged in one direction. Care should be taken to place the reinforcements in the right direction to take up the forces.

There are three different arrangements of reinforcements for taking up the shear efforts of a beam [5]: The bonding of reinforcement on the four faces of the beam (a), the bonding of reinforcement on the three sides of the beam as "U" (b), the bonding of reinforcement on both sides of the beam web (c).

DOI: 10.1051/matecconf/20168309009

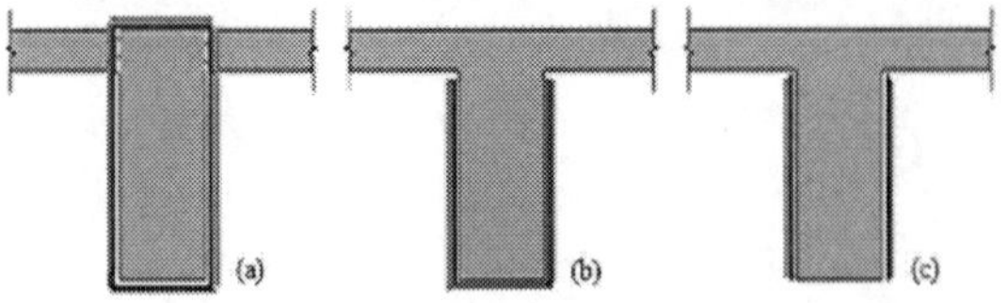

Fig. 1. *Disposition of reinforcement strips on the beam (FRP shear reinforcement)*

For the reinforcement of columns, the first studies began after the Loma Prieta earthquake in San Francisco bay, California at 1989.The first publications on this subject come from (CHAI and al 1991. Priestley and al.1992). This type of reinforcement is obtained by confining the reinforced concrete members with CFRP strips. When the reinforcement is applied around a column; its compressive strength and ductility are improved. The improvement of compressive strength of concrete appears only after the appearance of cracks and the beginning of the expansion. In this case, the CFRP wrap allow to the column to take up an additional compressive forces.

When the reinforcement is applied around a column, its compressive strength and ductility are improved. The improvement of compressive strength of concrete appears only after the appearance of cracks and the beginning of the expansion. In this case the FRP wrap allow to the column to take up an additional compressive forces.

When an axial load is applied to the column, the concrete expands laterally, in particular due to the Poisson effect, creating tensile stresses in the reinforcement, which opposes the increase in section. The reinforcing wrap confines then the section of the column. The result is an increase of the bearing capacity and ductility of structure.

The functioning of the reinforcement of this type of elements is different from the reinforcement of horizontal elements such as slabs or beams. In fact, for slabs and beams, reinforcements must increase the properties already provided by the frames. The properties of steel and concrete are not changed. In the case of reinforcement of the columns, the reinforcements make change the apparent characteristics of the concrete.

To increase the compression resistance for the elements, FRP strips are used in the same direction of the frames. There are three methods for the application of strips [4]: The reinforcement of a full post (a), the partial reinforcement with continuous spiral winding (b), and the partial reinforcement by rolling rings (c)

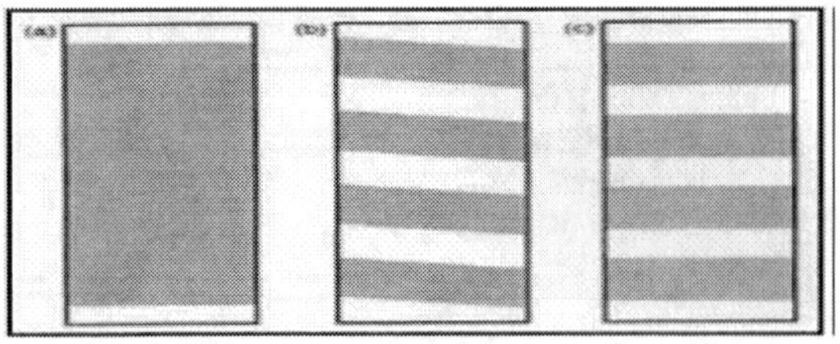

Fig. 2. Different type of reinforcements disposition on a column [4]

2 Equation of the problem

2.1 Vertical elements

The method we will follow to design the necessary reinforcement is as follows: [6]

- Determination of the nominal and factored capacities of the existing column. :

$$\phi P_{n(max)} = \phi \times 0.85 \times \left(0.85 \times f'_c \times \left(A_g - A_{st}\right) + f_y \times A_{st}\right) \quad (1)$$

where: f'_c : compressive strength of the concrete, A_g : column section, A_{st} : steel section, f_y : yield strength of steel, Φ : strength resistance factor

- Determine the amount of strengthening required. :

Formula (2) gives the new loading of the column according to the live load (G) and the dead load (Q):

$$P_u = 1{,}4 \times G + 1{,}7 \times Q \quad (2)$$

To determine the new load to be carried by the column we use the formula (3):

$$P_n^{requis} = \frac{P_u}{\phi} \quad (3)$$

We determine the required confined concrete compressive strength f'_{cc}:

$$f'_{cc} = \frac{\left(\frac{P_n^{requis}}{0.85}\right) - f_y \times A_{st}}{\phi \times \psi_f \times \left(A_g - A_{st}\right)} \quad (4)$$

The required confining pressure f_l is determined from the quadratic equation:

$$f_l^2 \times \frac{4}{(f'_c)^2} + f_l \frac{4\times\left(\frac{f'_{cc}}{f'_c}+1.25\right)-40}{f'_c} + \left(\frac{f'_{cc}}{f'_c} + 1.25\right)^2 = 0 \quad (5)$$

Finally, we calculate the necessary number of reinforcement layers:

$$n = \frac{f_l \times D}{2 \times t_f E_f \varepsilon_{fe}} \quad (6)$$

For a rectangular column, D is given by Lam and Teng's model [8]:

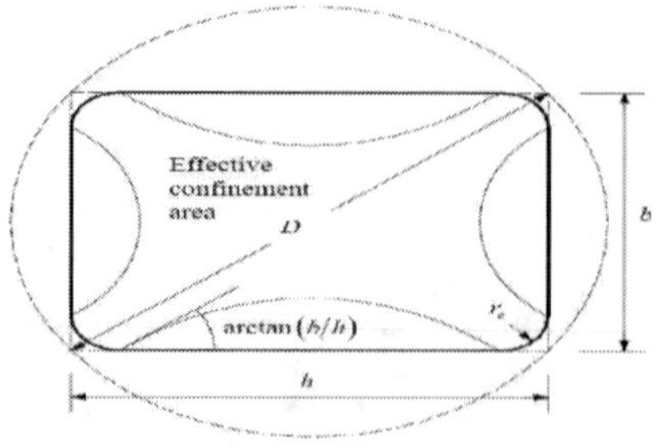

Fig. 3. Lam and Teng's model for FRP -confined concrete in rectangular columns [7]

2.2 Horizontal elements

By knowing the efforts applied to the studied element (beam, slab), we can calculate the strains in concrete and frames and finally determine the required reinforcement section [6]:

Based on the balance equation, we determine the necessary section of reinforcement using the equation (18):

$$A_f = \frac{\frac{M_d}{\phi} - A_s \times f_{yd} \times (d - 0.4 \times x)}{\sigma_f \times \psi_f \times (h - 0.4 \times x)} \quad (7)$$

where: Ψ_f: strength reduction factor, Φ: resistance factor, σ_f: Stress in the reinforcements, x: position of the neutral axis, A_s: steel section, M_d: flexural Moment, $f_y d$: yield strength of steel

3 Result and discussion

The aim of the study is to design a reinforcement of a beam in the first floor of a structure (this structure is a villa with an area of 386 m²). The table below summarizes the data of this beam:

Table 1. Beams data, load details, and material properties

Length (m)	3.78	Live Load(t/m)	1	Concrete compressive strength (MPa)	500
Width (cm)	25	Dead Load (t/m)	0	Yield strength(MPa)	25
Height (cm)	25	Load at ULS(t/m)	1.35	Steel section (cm²)	2.50

The table 2 summarizes the results for a load increase ranging from 10% to 70%:

Table 2. Table showing the various parameters calculated based on load increase

Load increase	*10%*	*20%*	*30%*	*40%*	*50%*	*60%*	*70%*
Load at ULS (t/m)	*1.48*	*1.62*	*1.75*	*1.89*	*2.02*	*2.16*	*2.29*
Flexural moment (t.m)	*2.65*	*2.89*	*3.13*	*3.38*	*3.62*	*3.86*	*4.10*
Reinforcement section (cm²)	*0.18*	*0.40*	*0.70*	*1.09*	*1.61*	*2.72*	*4.09*

Based on the table above, we established a curve representing the variation of the load based on the reinforcement section, as you can see in the figure below:

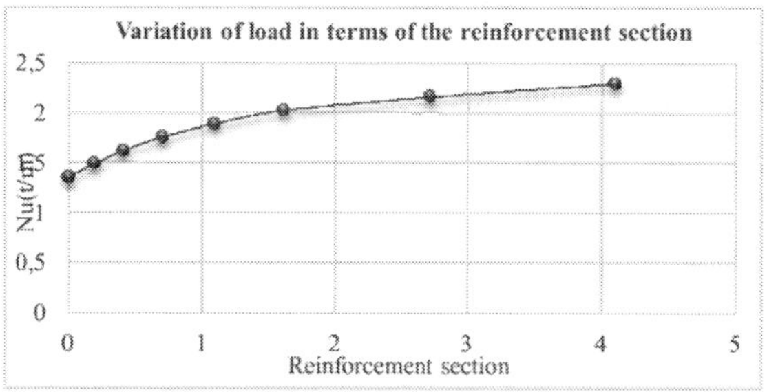

Fig. 4. Curve of the variation loads in terms of the reinforcement section

From the graph above, we can see that by using the same section of reinforced concrete and increasing the reinforcement section (section of CFRP), the beam could support a load that exceeds 70% of the load on which it was dimensioned .This proves that the reinforcement by carbon fiber is the most suitable solution to strength the structures having changed functionality or the aging structures. (A good ration weight/resistance, its resistance to fatigue, corrosion or even fire resistance).

4 Conclusion

The results obtained in this study are very encouraging; reinforcement by Carbon Fiber-Reinforced Polymer (CFRP) shows a part of its efficiency and leads to optimally reinforced structure. Depending on the type of stress, we now know the areas requiring reinforcement and the disposition of this latter.

However, the project we had the chance to conduct requires far more in-depth analysis.

In an additional study, it would be interesting to develop other methods and calculation models of reinforcement sections so that they can adapt to different types of sections.

References

1. H. Belhannachi, *Rehabilitation et renforcement des poteaux en beton arme*. PhD, Mentouri-Constantine University, (2009) - p: 14&15.
2. J. Miranda Vizuete, *techniques et systemes de renfort des structures en beton*, (2000) - p: 1.
3. A. Kassoul, *rehabilitations des constructions en beton arme*, Chapitre III - p: 17.
4. J. G. Teng, J. F. Chen, S. T. Smith, L. Lam; FRP: *Strengthened RC Structures*, Wiley, (2002).
5. A. Cecília, R.P. Ferreira, J. Cunha, *Analysis of the distribution of reinforcement with carbon fibers in concrete beams*, (2013) - Ciência & Engenharia (Science & Engineering Journal) , ISSN 1983-4071 22 (2): p. 47 – 56.
6. L.C. Bank, *Composites for construction – Structural Design with FRP Materials*– John Wiley & Sons, (2006) - p: 324-333.
7. T. Jiang, *FRP-Confined reinforced concrete columns: Analysis, behavior and design*. PhD, the Hong Kong Polytechnic University, July (2008) - p: 39.

On-line vibration monitoring of bearing faults in induction machine using Cyclic Spectral Analysis

F. Tafinine

Laboratoire LTII, Faculté de Technologie, Université A.Mira de Béjaia , 06000 Bejaia, Algérie

Abstract. A robust signal processing method for developing an on-line bearing fault detection based on Cyclic Spectral Analysis is presented in this paper. The exploitation of cyclostationarity has recently been proved extremely fruitful to conceive high accuracy diagnostics tools. Cyclostationarity encompasses a subclass of nonstationary signals which exhibit some cyclical behaviour; it ideally fits the property of many rotating and reciprocating machine vibrations, due to the inherent periodic modulations that these sustain during operation. The traditional spectral analysis is not dedicated for non-stationary signals and for an *on-line* diagnosis. The proposed technique is tested with different bearing conditions, the results show that this technique is especially useful for non-stationary features extraction; it gives a good basis for an integrated induction motor condition monitor.

1 Introduction

Asynchronous machines are far more widely used in the industrial world for their simplicity, their robustness and their cost. The importance of detection and diagnosis of faults in machines have expanded considerably due to the increased complexity of plant equipment and high costs associated with failure and shutdown. Faults detection and isolation problems can be solved by different techniques: model based, signal based or knowledge based methods. Faults can be due to electrical or mechanical failures. The most common failure in an asynchronous machine is bearing failure [1].

Vibration analysis techniques were used for the monitoring and diagnosis of huge rotating machines. Different methods were used. Time domain statistical method such as energy monitoring (energy increases as the machine's condition deteriorate), frequency domain method such as spectrum (characteristic frequencies associated with the failure appear in the spectrum), parametric models (the parameters of the model change with the machine's condition). Vibrations can be measured by attaching displacement or velocity transducers, and thus requires direct access to the machine. In recent years much work has been carried out on the investigation of possible fault operations of electrical machines. The usual methods for detection of faults are based on vibration analysis [3-6,10-11] or on Motor current signature analysis (MCSA) [1- 2, 12-13].

The bearings condition monitoring has received considerable attention for many years [3, 9, 14]. Many methods are available for detecting faults in bearing systems. The majority of these methods assume that faults in the bearings produce impulses [10, 11].

The Fourier analysis may be the most commonly used approach for bearing fault diagnosis. However, it is not appropriate because: first, vibration signal is generally *non-stationary* [5], and second, this approach cannot be applied for an on-line diagnosis, because the operator must inspect the total vibration signal spectrum in order to detect the frequency of defects.

In this paper, we will present the detection of the outer race defect using *Cyclic Spectral Analysis*. The results obtained confirm the effectiveness of this method in the detection of the frequency of motor bearing damage.

A non-invasive technique is now proposed. It is very practical and can be implemented at low cost. This paper is organized as follows: Section 2 presents experimental setup and the bearing faults characteristic frequencies. Section 3 reviews the theoretical background of the proposed diagnostic procedure. First we introduce the principles of the spectral domain analysis using Fourier transform. We emphasize on how the relevant information can be extracted from the spectral density using the FFT algorithm. Second, we describe the Cyclic Spectral Analysis and how it can be used for diagnosis by extracting the relevant frequencies characterizing the motor operating mode. Finally, a comparative approach of the proposed diagnostic procedure and several usual vibratory diagnostic methods are presented in section 4. And the conclusions are given in section 5.

2 Experimental setup

We have used a motor of 380 V; 50 Hz; 1.1 kW; number of pair poles (P=2). The experiments are realized with several levels of load and speed. Tests are conducted for a healthy bearing then with an outer race fault. The signals are measured using multiple vibration sensors installed on the induction machine. Note that the all measurements are taken at stationary regime.

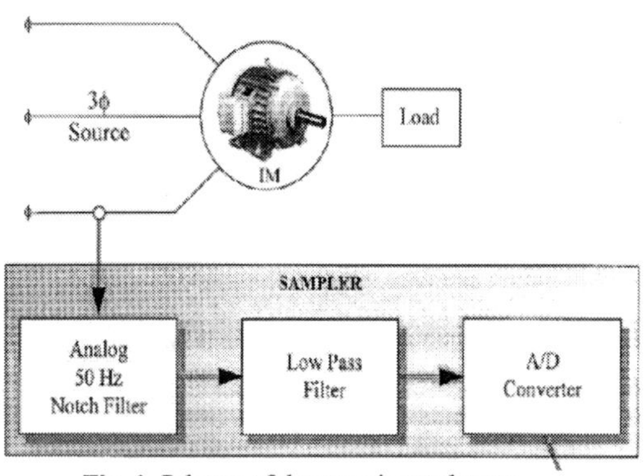

Fig. 1. Scheme of the experimental setup

2.1 Bearing failure

The vibrations frequencies depend: on the rotation speed " f_r ", on the geometrical parameters of the bearing (figure 2) and on the type of defects [2, 9]. The vibration frequencies of bearing failure are given in the equations bellow [7, 8]:

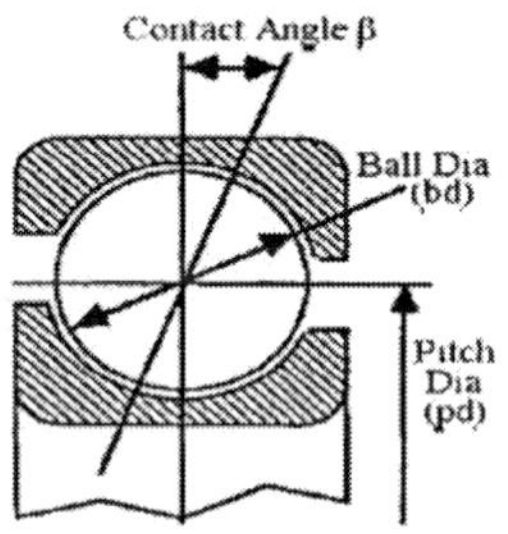

Fig. 2. Geometrical parameters of the bearings.

Outer race bearing fault frequency F_O:

$$Fo=\frac{N}{2}\,f_r\left[1-\frac{BD}{PD}\cos\beta\right] \tag{1}$$

Inner race bearing fault frequency F_i:

$$Fi=\frac{N}{2}\,f_r\left[1+\frac{BD}{PD}\cos\beta\right] \tag{2}$$

Cage defect frequency F_c:

$$Fc=\frac{1}{2}\,f_r\left[1-\frac{BD}{PD}\cos\beta\right] \tag{3}$$

Ball defect frequency F_b:

$$Fb=\frac{PD}{2BD}\,f_r\left[1-\left(\frac{BD}{PD}\cos\beta\right)^2\right] \tag{4}$$

where

« BD » is the ball diameter,

« PD » is the pitch diameter,

« β » is the contact angle of the balls on the race

« N » is the number of balls.

2.2 Frequency characteristics of failure

The Benchmark is equipped with an RS bearing type 6028-2RS. The defect frequency at the speed of 3004 rpm is 4.58 Hz (see table 1).

Table 1. Frequency characteristics of the bearing type 6028-2RS.

	Frequency (Hz)			
Speed (rpm)	f_c	f_b	f_o	f_i
3004	0.28	5.43	4.58	5.81

3 Applications and results

3.1. Classical spectral analysis

The classical method converts the sampled signal to the frequency domain using a FFT (Fast Fourier Transform) algorithm. The generated spectrum includes only the magnitude information about each frequency component. Signal noise that is present in the calculated spectrum is reduced by averaging a predetermined number of generated spectra. The power spectral density is illustrated in the figure 3.

The result shows a periodic modulation of 4.6 Hz (sideband spacing = 4.6 Hz) in the defect machine spectrum. This modulation is due to the outer race defect.

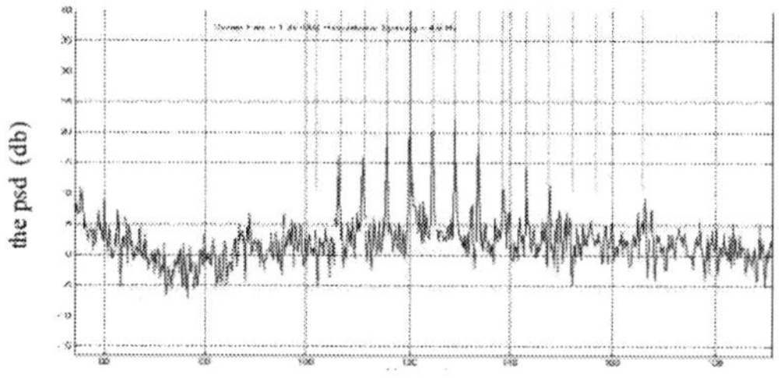

Fig. 3. PSD of the vibration signal with outer race defect.

3.2 Cyclical spectral analysis

The *cyclic power spectrum*. $S_x(f;\alpha)$ of signal $x(t)$, is defined as [17], [18]:

$$S_x(f;\alpha)=\lim_{L\to\infty}\frac{1}{L}\mathrm{E}\left\{X_l\left(f+\frac{\alpha}{2}\right)X_L\left(f-\frac{\alpha}{2}\right)^*\right\} \tag{5}$$

where $X_L(f)$ is the Fourier transform of $x(t)$ limited in the interval L.

The definition of the *cyclic coherence* is [15], [16]:

$$|\gamma_x(f;\alpha)|^2 = \frac{|S_x(f;\alpha)|^2}{S_x\left(f+\frac{\alpha}{2}\right)S_x\left(f-\frac{\alpha}{2}\right)} \qquad (6)$$

This expression presents a correlation coefficient between two components at frequencies $f+\alpha/2$ and $f-\alpha/2$. It takes values between 0 and 1, where 0 expresses nonlinear correlation, and 1 in the case of a linear relationship between the two components [17].

The diagnostics information can be revealed in the α*-frequency* domain, the magnitude of the cyclic coherence can be used as a measure of the fault severity [19].

Cyclic coherence of the damaged bearings (figure 4) illustrates significant cyclostationary. A fundamental spectral line at α = 4.5 *Hz* without sidebands is, according to table 1, the characterization of an outer-race fault. In the cyclic coherence, some bands located within [0; 3] *Hz* might be due to structural resonances that are excited by the impacts of the rolling elements on the faults (ie. F_c, f_r). These observations agree with the frequency band in the power spectral densities in figure 3.

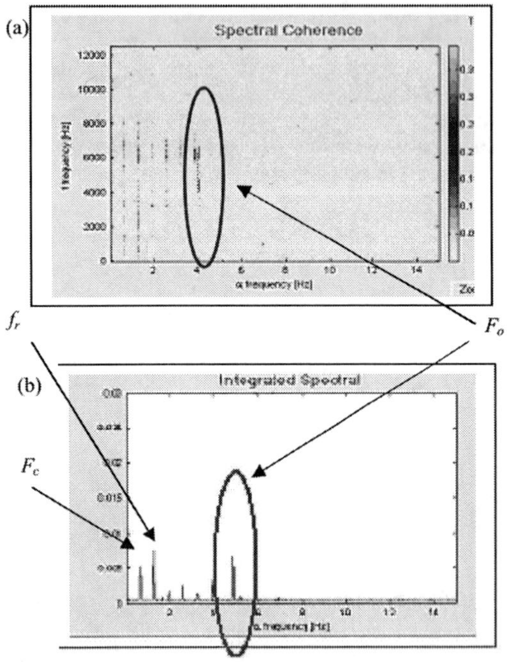

Fig. 4 Estimated cyclic spectral coherence (a) and integrated spectral coherence (b) of the vibration signal with an outer-race fault.

4 Comparison of the two analysis tools

Vibrations produced by faulty rolling-element bearing are essentially random cyclostationary [18]. In order to detect the frequency of defects it is habitual to demodulate the signal in different frequency bands in the total frequency axis. This method is the classical technique in the spectral analysis (PSD). The *Cyclic Spectral Analysis* can replace this empirical process by a precise methodology to quickly find the frequency of the faults.

The new diagnosis procedure overcomes the disadvantages related to motor parameters dependency. In contrast to the usual vibrations signal spectral analysis, the proposed diagnosis method allows also the ability to extract *quickly* the characteristic frequencies that distinguish the motor operating modes.

The *cyclic coherence* is actually the cyclic spectral quantity that offers the best fault detection tool, it reveals the discrete spectral signature of the fault. Its magnitude serves as a relative measure of the fault severity. These observations have been supported by experimental results.

5 Conclusions

This paper has investigated a method for detecting bearing faults using the cyclical spectral analysis. Machine fault features extraction without prior information of the fault type is important for diagnosis and maintenance planning. The *Cyclic Spectral Analysis* technique is qualified as a parsimonious method for exploring the information in the entire spectrum of the vibrations signal in the *cyclic frequency* domain. The resulting procedure is fined suitable for an on-line diagnosis, without manual research of the frequency band containing the fault.

Acknowledgment

The author wishes to express his gratefulness to the members of the Roberval Laboratory (UTC France) for their support in the experimental setup.

References

1. R.R. Schoen, T. Habetler, F. Karman, R. Bartheld, IEEE Trans. Ind. App. **31**, 1274 (1995)
2. M.E.H. Benbouzid, IEEE Trans. Ind. Electron. **47**, 984 (2000)
3. R. Boustany, J. Antoni, in Proc. of the 16th IFAC World Congress, Prague, Czech Republic, (2005)
4. F. Zhipeng, C. Fulei, Shock and Vib. 1155 (2015)
5. Y. Yang, D. Yu, J. Cheng, Measurement, **40** 943 (2007)
6. R. Boustany, J. Antoni, Mech. Syst. Sig. Process. **19**, 1245 (2005)

7. B. Raison, G. Rostaing, O. Butscher, C.S. Maroni, in Proc. *28th* Annual Conference of the Industrial Electronics Society, Sevilla, Spain, 2002.
8. W.T. Thomson, in Proc. of the IEEE Symposium on Diagnostics for Electrical Machines Power Electronics and Drives (SDEMPED), Gijon, Spain, 1999.
9. T.A. Harris, *Rolling bearing analysis*, (4th Edition); John Wiley & Sons, Inc, New York, USA, 2001.
10. N.G. Nikolaou, I.A. Antoniadis, Mech. Syst. Sig. Process. **16,** 667 (2002)
11. T.C. Lim, R. Singh, J. Sound vib. **139**, 179 (1990)
12. M. Blodt, *Condition Monitoring of Mechanical Faults in Variable Speed Induction Motor Drives*, Ph.D. Thesis, INP Toulouse, France, (2006)
13. M. Blodt, M. Chabert, J. Regnier, J. Faucher, IEEE Trans. Ind. App. **42**, 1454 (2006)
14. R. Isermann, Control Eng. Pract. **5**, 639 (1997)
15. J. Antoni, F. Bonnardot, A. Raad, M. El Badaoui, Mech. Syst. Sig. Process. **18**, 1285 (2004)
16. R.B. Randall, J. Antoni, S. Chobsaard, Mech. Syst. Sig. Process. **15**, 945 (2001)
17. A. Raad, J. Antoni, M. Sidahmed, Mech. Syst. Sig. Process. **22**, 574 (2008)
18. A.C. McCormick, A.K. Nandi, J. Sound vib. **12**, 225 (1998)
19. C. Capdessus, M. Sidahmed, J.L. Lacoume, Mech. Syst. Sig. Process. **14**, 371 (2000)

AUTHOR INDEX

AUTHOR INDEX